Compensating Ecological Damage
Comparative and Economic Observations

Jing Liu

Compensating Ecological Damage
Comparative and Economic Observations

intersentia

METRO

Ius Commune Europaeum

Jing Liu
PhD Thesis

Compensating Ecological Damage
Comparative and Economic Observations

Intersentia Ltd
Trinity House | Cambridge Business Park | Cowley Road
Cambridge | CB4 0WZ | United Kingdom
Tel.: +44 1223 393 753 | Email: mail@intersentia.co.uk

ISBN 978-1-78068-174-0
D/2013/7849/68
NUR 823

Cambridge – Antwerp – Portland
www.intersentia.com | www.intersentia.co.uk

Cover photo © Phototreat – iStockphoto

Layout by Marina Jodogne.

British Library Cataloguing in Publication Data. A catalogue record for this book is available from the British Library.

ACKNOWLEDGEMENTS

In 2009, when I moved from the other end of this continent to Maastricht and started my doctoral research, four years seemed an endless luxury to dive deep into one research topic and also too long period away from my family and friends. At the moment of finalizing the dissertation, however, I start to realize that doctoral research provides only a modest start in grasping a subject and every extra hour with my colleagues and friends here is appreciated. One of the joys of completion is to reflect on the journey and remember all the people who have helped and supported me along this long but fulfilling road.

This thesis is the result of PhD research at Maastricht University and is supported by the China Scholarship Council (CSC). I am sincerely grateful to the Faculty of Law of Maastricht University and the CSC for giving me the wonderful opportunity to carry out this research. I am particularly aware of the debt I owe to my supervisor Professor Michael Faure, who is both a great guide in knowledge and a caring guide in life for me. His inspiration, unrivalled understanding of law and economics and environmental law, as well as his sincere commitment to scholarship greatly enlightened me. He is also a close friend who made me feel at home since the beginning of my stay in Maastricht. As the first Chinese PhD student in Faculty of Law, coming to Europe for the first time, I found that the beginning was not that easy. Michael and his family – considerate Wang Hui (we call 'teacher mother' in Chinese) and lovely son Tony – help me enormously with their enthusiasm and hospitality. The wonderful time we had together to celebrate many Chinese and European holidays: Spring Festival, Mid-autumn day and Christmas will always be a precious and beautiful memory in my mind.

My deep gratitude and appreciation go to the members of my assessment committee for the enormous time and patience they spend to read the manuscript and for making valuable comments: Professor Marjan Peeters, Professor Chris Backes and Professor Marleen van Rijswick. Their insightful suggestions helped me greatly to improve my thesis. Additional thanks to Marjan for several inspiring discussions, guidance on my understanding of European environmental law and feedbacks in Ius Commune Workshops. I also would like to thank wholeheartedly other members who agree to sit in my committee and to participate in my defense:

Professor Nicolas de Sadeleer, Dr. Niels Philipsen, Professor Lucas Bergkamp, Professor Qin Tianbao and Professor Francisco Djedjro Melèdje. It is indeed a great honour for me. Additional thanks to Professor Qin Tianbao, who has been my teacher since I was a bachelor student, guided my journey to Maastricht and keeps supporting me all the way. Special thanks also to Dr. Niels Philipsen, who helps me to understand the fantastic area of law and economics.

There are also many other people in and around the faculty who support me warmly and to whom I owe my sincere gratitude. I am really grateful for the helpful METRO team of Yleen Simonis, Chantal Kuypers, Elke Hundhausen, Marina Jodogne and Marjo Mullers. Additional thanks to Marina, who was extremely helpful making the layout of the book and helping me catch up with the schedule. Many thanks also go to Kevine Kindji, Emma Dermawati, Laura Tilindyte, Mehdi Piri Damagh, Sachin Badkas, Tobias Heldt, Sarah Schoenmaekers, Thomas Biermeyer, Wenqing, Taotao, Huizhen, Yuan, Li Xuesong, Xiaowei, Guang and Jingjing, Viorelia Gasca and Katharina Eisele for the great pleasure they brought me and reminding me that I am not alone even so far away from home.

I would like to sincerely thank the officers and professors I interviewed for their wiliness to share their time and knowledge. I would particularly like to thank Professor Ludo Veuchelen, Professor Benjamin Van Rooij, Danny Van Welkenhuyzen, Harko Kremers, Professor Wang Canfa, Professor Wang Jin, Professor Li Zhiping, Li Zeyan, Kathinka Furst, Chen Ling, Lu Qingbin, Yang Yaxiao for making my interviews in Europe and China possible.

It is also a great pleasure for me to have many other friends in and around Maastricht, and also back in China, accompany me during the journey. Yang Liu, Wang Long, Han Xing, Lin Shenping, Mao Kai, Shi Xiaona, Chen Xiahong, Zhao Jinjin, Hou Jun, Zhou Chen, Guo Lei all brought me beautiful memories in the past four years.

Special thanks to Professor Wang Shuyi, my supervisor during my master studies. I am indebted for his unconditional support both when I was in China and when I am here. My gratitude also goes to Dr. Du Zhihua and Liu Zhubin, who guided my way to research in the fantastic area of law and kept a close eye on me all through the years.

Finally, I must express my gratitude to my father Liu Chaoyin and my mother Li Zhongying. They give me a home where I can always go back to and they made me become who I am today. Without their love, everlasting support and endless care, this research would simply not have been accomplished.

TABLE OF CONTENTS

ACKNOWLEDGEMENTS V

LIST OF ABBREVIATIONS XVII

LIST OF FIGURES AND TABLES XXI

CHAPTER 1: INTRODUCTION 1

1. Background 1
2. Research Questions and Problem Definitions 4
3. Methodology 6
4. Structure 10

Part I Compensation Models for Ecological Damage: Theoretical Framework 13

CHAPTER 2: THE DEFINITIONS AND CHARACTERISTICS OF ECOLOGICAL DAMAGE 17

1. Introduction 17
2. The Choice of a Concept – Environmental Damage, Ecological Damage and Natural Resources Damage 18

2.1. Relevant Concepts in Legal Documents and Literature 18
2.1.1. Concepts in Legislation and Treaties 18
2.1.1.1. Environmental Damage 18
2.1.1.2. Pollution 20

2.1.1.3. Damage to the Environment, Impairment of the Environment and other Related Concepts 21
2.1.1.4. Natural Resources Damage 23
2.1.2. Concepts in Literature 23
2.1.2.1. Environmental Damage and Pollution Damage 23
2.1.2.2. Natural Resources Damage 24
2.1.2.3. Ecological Damage 24
2.2. The Choice of the Concept 25
2.2.1. Environment, Natural Resources and Ecology 26
2.2.2. The Choice and Definition of Concepts 28

3. The Characteristics of Ecological Damage 29
3.1. The Concept of Damage in Tort Law 29
3.1.1. Damage and Relevant Concepts 29
3.1.2. Definition of Damage 30
3.1.3. Pecuniary and Non-Pecuniary Losses 31
3.2. Ecological Damage: Does it Fit into the Traditional Tort Regimes? 34
3.2.1. Damage to Owned Natural Resources 34
3.2.2. Damage to Un-owned Natural Resources 36
3.2.2.1. Cost of Restoration 36
3.2.2.2. No Restoration 37
3.2.3. Other Characteristics of Ecological Damage 38

4. Definition of Compensation 40

5. Conclusion 41

CHAPTER 3: LEGAL RULES CONCERNING THE PREVENTION OF ECOLOGICAL DAMAGE: ECONOMIC OBSERVATIONS 45

1. Introduction 45
1.1. Regulation and Liability 47

2. Liability Rules for Ecological Damage 50
2.1. The Criteria in Designing Optimal Liability Rules 50
2.2. Standing 51
2.2.1. Legal Analysis 52
2.2.1.1. Standing in the U.S. 52
2.2.1.2. Standing in the EU 53
2.2.2. Tested in Economics 61
2.2.2.1. Public Authority 62
2.2.2.2. Environmental NGOs 63
2.2.2.3. Individuals 64
2.3. Liable Parties 65

2.3.1. Economic Analysis ... 65
2.3.2. Applied to Ecological Damage ... 66
2.3.2.1. Definition of Liable Party ... 66
2.3.2.2. Limitation of Victims Claiming against other Parties and Recourses ... 68
2.4. Liability Standards ... 69
2.4.1. Economic Analysis ... 69
2.4.1.1. Primary Costs ... 69
2.4.1.2. Tertiary Costs ... 72
2.4.2. Liability Standards for Ecological Damage ... 72
2.5. Causation and Multiple Tortfeasors ... 75
2.5.1. Economic Analysis ... 75
2.5.1.1. Uncertain Causation ... 75
2.5.1.2. Multiple Tortfeasors ... 77
2.5.2. Applied to Ecological Damage ... 78
2.5.2.1. Uncertain Causation ... 78
2.5.2.2. Multiple Tortfeasors ... 79
2.6. Damages ... 80
2.6.1. Economic Analysis ... 80
2.6.2. Applied to Ecological Damage ... 82

3. The Evaluation of Liability Rules ... 84
3.1. How Do Liability Rules Contribute to Prevention and Compensation? ... 85
3.1.1. Compensation ... 85
3.1.2. Prevention ... 86
3.2. Remedies – Compensation Mechanisms ... 87
3.2.1. Compensation ... 87
3.2.2. Prevention ... 88
3.2.3. Risk Spreading and Costs ... 89

4. Conclusion ... 90

CHAPTER 4: ANALYSIS OF SPECIFIC COMPENSATION MECHANISMS ... 95

1. Introduction ... 95
1.1. A Brief Overview of Compensation Mechanisms for Ecological Damage ... 95
1.2. Criteria for Compensation Mechanisms: Compensation, Prevention and Costs ... 98

2. Liability Insurance ... 103
2.1. The Basic Rationale of Liability Insurance ... 103
2.1.1. Risk Aversion and the Law of Large Numbers ... 103
2.1.2. Insurability ... 105
2.1.3. Problems of Insurance ... 106
2.1.3.1. Moral Hazard ... 106
2.1.3.2. Adverse Selection ... 107

2.2. Liability Insurance for Ecological Damage ... 108
2.2.1. Insurability of Liability for Ecological Damage ... 108
2.2.1.1. Characteristics of Ecological Damage Influencing Insurability ... 108
2.2.1.2. Liability Rules Influencing Insurability ... 109
2.2.2. Moral Hazard and Adverse Selection ... 112
2.3. The Evaluation of Liability Insurance ... 114
2.3.1. Primary Costs (Prevention) ... 114
2.3.2. Secondary Costs (Compensation) ... 115
2.3.3. Tertiary Costs ... 118
2.3.4. Interaction between the Different Costs Reduction Goals ... 118
2.4. Compulsory Insurance ... 120

3. First-party and Direct Insurance ... 122
3.1. Introduction: First-party Insurance and Direct Insurance ... 122
3.2. The Feasibility of First-party Insurance and Direct Insurance in the Environmental Context ... 124
3.2.1. First-party Insurance ... 124
3.2.2. Direct Insurance ... 126
3.2.2.1. Ecological Damage ... 126
3.2.2.2. Traditional Damage ... 127
3.3. The Evaluation of First-party Insurance and Direct Insurance ... 128
3.3.1. First-party Insurance ... 128
3.3.1.1. Primary Costs (Prevention) ... 128
3.3.1.2. Secondary Costs (Compensation) ... 128
3.3.2. Direct Insurance ... 129
3.3.2.1. For Ecological Damage ... 129
3.3.2.2. For Traditional Damage ... 131

4. Risk-sharing Agreements ... 131
4.1. Introduction to Risk-sharing Agreements ... 131
4.2. Specific Institutional Arrangements of a Risk-sharing Agreement ... 133
4.2.1. Examples of Risk-sharing Agreements ... 133
4.2.2. Institutional Arrangements ... 136
4.3. Evaluation of Risk-sharing Agreements ... 137
4.3.1. Primary Costs ... 137
4.3.2. Secondary Costs ... 140
4.3.3. Tertiary Costs ... 141

5. Environmental Funds ... 142
5.1. Introduction: the Various Types of Environmental Funds ... 142
5.1.1. Funds with or without Risk-spreading Functions and Ways to Finance the Funds ... 144
5.1.2. Funds Established for the Benefits of Victims and Injurers ... 145
5.1.3. Relationship to Liability Rules and Insurance ... 146
5.2. Evaluation of the Various Funds ... 149

5.2.1. Environmental Funds to Cover Liability 152
5.2.1.1. Environmental Funds as One Type of Financial Guarantees 152
5.2.1.2. Funds Offering an Upper Layer of Compensation 152
5.2.2. Complementary Funds 153
5.2.3. Environmental Funds Irrespective of Liability 153
5.3. Conclusion 154

6. Security Mechanisms Provided by the Liable Party or by a Third-party Guarantor 155
6.1. The Various Security Mechanisms 155
6.1.1. Self-insurance and Captives 155
6.1.2. Mortgages and Liens 157
6.1.3. Deposits, Trust Funds and Escrow Agreements 157
6.1.4. Corporate Guarantees, Bank Guarantees and Surety Bonds 157
6.2. Analysis: From the Perspective of Compensation and Prevention 158
6.2.1. Self-insurance and Captives 159
6.2.2. Mortgages and Liens 160
6.2.3. Deposits, Trust Funds and Escrow Agreements 161
6.2.4. Corporate Guarantees, Bank Guarantees and Surety Bonds 161

7. The Use of the Capital Market to Provide Coverage 162

8. Conclusion: Comparison between Individual Instruments and Policy Recommendations 165
8.1. Liability Insurance 166
8.2. First-party Insurance and Direct Insurance 167
8.3. Risk-sharing Agreement 167
8.4. Environmental Funds 168
8.5. Guarantees by the Polluters 169

CHAPTER 5: COMPENSATION MODELS FOR ECOLOGICAL DAMAGE 173

1. Introduction 173

2. Proposed Models to Prevent and Compensate for Ecological Damage 174
2.1. Compensation, Liability Rules, and Insolvency 174
2.1.1. Recalling Characteristics of Ecological Damage 174
2.1.2. Mandatory Financial Security 175
2.2. Models of Compensation 177
2.2.1. Model 1: Mandatory Financial Security 177
2.2.2. Model II: Voluntary Financial Security 182
2.2.3. Model III: Optimal Financial Cap 184

3. Towards a Multilayered Approach to Compensate for Ecological Damage 185

3.1. Factors Influencing the Choice of Instruments 186
3.2. When the Liable Party is Identifiable 187
3.3. When the Liable Party is not Identifiable 189

Part II Empirical Part 191

CHAPTER 6: COMPENSATION SYSTEMS UNDER THE INTERNATIONAL REGIME 195

1. International Compensation System for Oil Pollution 195
1.1. Liability Rules under the International Regimes on Compensation for Oil Pollution 195
1.1.1. Liability Rules under the CLC Regime 196
1.1.1.1. The Adoption of the 1969 CLC and the 1971 Fund Convention 196
1.1.1.2. The 1984 and 1992 Protocols 197
1.1.2. Liability Rules under the Bunker Convention 199
1.2. Compensation Instruments for Oil Pollution Damage 200
1.2.1. Protection and Indemnity Policies 201
1.2.2. Compensation Funds 202
1.2.2.1. The 1971 Fund 202
1.2.2.2. The 1992 Fund and the winding up of the 1971 Fund 203
1.2.2.3. The 2003 Supplementary Fund 204
1.3. Critical Analysis 205

2. The International Compensation System for Nuclear Damage 207
2.1. International Regimes for Compensation of Nuclear Damage 209
2.1.1. Origins of International Regimes 209
2.1.2. First Generation of Nuclear Liability Conventions 210
2.1.2.1. Strict Liability 211
2.1.2.2. Channeling of Liability 212
2.1.2.3. Limited Liability 212
2.1.2.4. Financial Security 213
2.1.2.5. Exclusive Jurisdiction 213
2.1.2.6. Public Funding 214
2.1.3. The Second Generation of Nuclear Liability Conventions 214
2.1.3.1. General Issues 214
2.1.3.2. The Coverage of Environmental Damage 217
2.2. Evaluation from a Law and Economics Perspective 221
2.2.1. Liability 221
2.2.2. Financial Security 223
2.2.2.1. Public Funds 223
2.2.2.2. Insurance 224
2.2.2.3. Pooling 226

CHAPTER 7: COMPENSATION SYSTEMS IN THE U.S. 231

1. Compensation System for Nuclear Damage in the U.S. 231

1.1. Liability for Nuclear Damage 232
1.2. The Requirement of Financial Protection 236
1.2.1. Financial Requirement for Third-party Liability 236
1.2.2. Financial Requirement for Property Damage 238
1.2.3. Compensation under the Convention on Supplementary Compensation 238
1.2.4. Financial Requirement for Decommissioning Nuclear Facilities and Financing for Nuclear Waste Disposal 239
1.3. The Instruments to Compensate for Nuclear Damage 240
1.3.1. Nuclear Liability Insurance 240
1.3.1.1. Basic Content 240
1.3.1.2. Risk Differentiation 243
1.3.2. The Retrospective Premiums Scheme 244
1.3.3. The Mutual Pool for Property Damage 246
1.4. Critical Analysis 247
1.4.1. Combined Use of Regulation, Liability and Compensation Instruments 248
1.4.2. Is there a Subsidy for the Nuclear Sector? 249
1.4.3. Compulsory Financial Requirements 252
1.4.4. Multiple Layers of Compensation Instruments 253
1.4.5. Use of Insurance and Risk-sharing Pools 254
1.4.6. Private Interest Approach 255

2. The Compensation System for Natural Resources Damage in the U.S. 257
2.1. Introduction 257
2.2. Compensation System for Oil Pollution 258
2.2.1. Liability for Oil Pollution 259
2.2.1.1. Liability under the OPA 259
2.2.1.2. State Statutes 263
2.2.2. Financial Responsibility and Trust Fund 263
2.2.2.1. Financial Responsibility: General 264
2.2.2.2. P&I Clubs and Insurance 264
2.2.2.3. Trust Fund 266
2.3. Compensation System for Damage Caused by Hazardous Substances 266
2.3.1. Liability under the CERCLA 267
2.3.1.1. Response Action and Natural Resources Damage 267
2.3.1.2. Responsible Parties and the Scope of Liability 269
2.3.1.3. Stringent Liability and the Relief of Liability 270
2.3.2. Financial Responsibility and the Trust Funds under the CERCLA 272
2.3.2.1. Financial Responsibility: General 272
2.3.2.2. Insurance 273
2.3.2.3. Trust Funds 274
2.4. Empirical Data about Natural Resources Damage in the U.S. 275
2.5. Economic Analysis 276
2.5.1. Liability Rules 277

2.5.1.1. Liability Rules under the OPA 277
2.5.1.2. Liability Rules under the CERCLA 279
2.5.2. Financial Responsibility 282
2.5.2.1. Financial Responsibility under the OPA 282
2.5.2.2. Financial Responsibility under the CERCLA 284
2.5.3. Trust Funds 285
2.5.3.1. Trust Fund under the OPA 285
2.5.3.2. Trust Fund under the CERCLA 285

CHAPTER 8: COMPENSATION SYSTEMS IN THE EU AND SOME EU MEMBER STATES 287

1. The Compensation System under the ELD 287

2. Compensation Mechanisms Used in Member States 293
2.1. Liability Insurance in Germany 294
2.2. Direct Insurance in the Netherlands 296
2.3. Risk-sharing Agreements in Germany 299
2.3.1. Nuclear Liability 299
2.3.2. Pooling System 301
2.4. Environmental Funds in the Netherlands 303

3. Critical Analysis 304

CHAPTER 9: CONCLUSION OF PART II 309

1. A Comparison of the Systems 309

2. The Compensation Models 316

3. The Choice between Different Compensation Instruments 317

4. The Divergence of Practice from Theoretical Models and Indicators 318
4.1. Inefficiency of Liability Rules 318
4.2. Broad Use of Environmental Funds 318
4.3. Implementation of Financial Requirement 318

Part III Towards an Efficient Compensation SYstem in China 321

CHAPTER 10: COMPENSATION SYSTEM FOR ECOLOGICAL DAMAGE IN CHINA AND POLICY RECOMMENDATIONS 323

1. Introduction 323
1.1. Theory: Environmental Liability in Past and Present 324

1.1.1. Basis of Liability ... 324
1.1.2. Scope of Liability: Is Natural Resources Damage Covered? ... 326
1.1.3. Standing ... 329
1.1.4. Causation, Multiple Tortfeasors and Burden of Proof ... 332
1.1.5. Quantification ... 333
1.2. Practice ... 335
1.2.1. Who Acts after an Accident? ... 335
1.2.2. Barriers to Access to Justice ... 336
1.2.3. NGOs ... 339
1.2.4. Remedies ... 341
1.3. Summary ... 343

2. Environmental Insurance ... 344
2.1. Theory ... 344
2.1.1. Statutory Background ... 345
2.1.2. Theoretical Insurance Options ... 346
2.2. Practice ... 347
2.2.1. Insolvency Risk - Compulsory Insurance? ... 347
2.2.2. Environmental Insurance in Practice ... 348
2.2.3. Difficulties and Limits ... 352
2.3. Summary ... 353

3. Compensation for Vessel-induced Marine Oil Pollution ... 354
3.1. Theory ... 354
3.1.1. Scope of Compensable Damage and Quantification of Damage ... 354
3.1.2. Standing ... 358
3.1.3. Mandatory Financial Security ... 359
3.1.4. Compensation Funds ... 360
3.2. Practice ... 361
3.2.1. Filing Claims ... 361
3.2.2. Compensation via P&I Clubs ... 362
3.2.3. Remedies ... 364
3.2.4. Cases ... 365
3.2.5. Challenges ... 367

4. Economic Analysis of the Chinese Compensation System for Natural Resources Damage ... 367
4.1. Liability Rules ... 367
4.2. Insurance ... 369
4.3. Marine Oil Pollution ... 370

5. Recommendations to Improve the Compensation System ... 371
5.1. Legal Framework ... 371
5.2. Practical Issues ... 372

CHAPTER 11: CONCLUSION AND POLICY RECOMMENDATIONS 375

1. Summary 377

2. Proposed Models to Prevent and Compensate for Ecological Damage .. 378

3. Indicators to Choose Compensation Instruments and a Multilayered Compensation Approach 382

4. The Compensation System in China and Policy Recommendations 386

5. Limits and Future Research 388

SUMMARY: COMPENSATING ECOLOGICAL DAMAGE: COMPARATIVE AND ECONOMIC OBSERVATIONS 391

SAMENVATTING: COMPENSATIE VOOR ECOLOGISCHE SCHADE: RECHTSVERGELIJKENDE EN ECONOMISCHE BESCHOUWINGEN 395

POLICY DOCUMENTS, GUIDELINES, REPORTS 399

LEGISLATION 405

BIBLIOGRAPHY 413

CURRICULUM VITAE 449

LIST OF ABBREVIATIONS

AEC	Atomic Energy Commission
ANI	American Nuclear Insurers
APPA	Air Pollution Prevention Act (China)
ART	Alternative Risk Transfer
AtG	Atomgesetz
AVB	Aansprakelijkheidsverzekering Bedrijven (Liability Insurance Policy)
Bofas	Bodemsaneringsfonds voor tankstations
BP	British Petroleum
CERCLA	Comprehensive Environmental Response, Compensation, and Liability Act (US)
CGAA	Coast Guard Authorization Act
CGMTA	Coast Guard and Maritime Transportation Act
CLAPV	Centre for Legal Assistance to Pollution Victims
CLC	International Convention on Civil Liability for Oil Pollution Damage
CMC	Chinese Commercial Maritime Code
COFR	Certificate of Financial Responsibility (US)
CPL	Civil Procedural Law (China)
CRISTAL	Contract Regarding an Interim Supplement to Tanker Liability for Oil Pollution
CRS	Congressional Research Service
CRTD	Civil Liability for Damage Caused during Carriage of Dangerous Goods by Road, Rail and Inland Navigation Vessels
CSC	Convention on Supplementary Compensation for Nuclear Damage
CSMAA	China Shipowners Mutual Assurance Association
CWA	Clean Water Act (US)
DOE	Department of Energy (US)
DOI	Department of Interior (US)
ECtHR	European Court of Human Rights

EDA	Environmental Damage Prevention and Remediation Act (Germany)
EGA	Environmental Guarantee Account
EII	Environmental Impairment Insurance
EIL	Environmental Liability Insurance Policies
ELD	Environmental Liability Directive
ELINI	European Liability Insurance for the Nuclear Industry
EMANI	European Mutual Association for Nuclear Insurance
eNGO	Environmental Non-Governmental Organization
ENO	Extraordinary Nuclear Occurrence
EPA	Environmental Protection Act (US)
ERF	Engineering Rating Factor
ERICAM	Environmental Risk Internalisation through Capital Markets
EU	European Union
GAO	General Accounting Office (US)
GDP	Gross Domestic Product
GDV	German Insurance Association
GONGO	Government-Organized Non-Governmental Organization
GPCL	General Principle Civil Law (China)
GPL	General Public Liability Policies
GTPL	General Third Party Liability Policies
HNS	International Convention on Liability and Compensation in Connection with Carriage of Hazardous and Noxious Substances by Sea
IAEA	International Atomic Energy Agency
Ibs	Interimwet Bodemsanering
IMO	International Maritime Organization
INPO	Institute of Nuclear Power Operations
INTERTANKO	International Association of Independent Tanker Owners
IOPCF	International Oil Pollution Compensation Fund
LRTAP	Long-Range Transboundary Air Pollution
Lugano Convention	1993 Lugano Convention on Civil Liability for Damage Resulting from Activities Dangerous to the Environment
MAERP	Mutual Atomic Energy Reinsurance Pool
MAS	Milieu-aansprakelijkheidsverzekering Samenwerkingsverband (Environmental Liability Insurance)
MEP	Ministry of Environmental Protection
MEPA	Maritime Environmental Protection Act (China)
MOE	Minister of the Environment
MS	Member States (of the European Union)
MSA	Maritime Safety Agency
MSV	Milieuschadeverzekering (Dutch Environmental Damage Insurance)
NEA	OECD Nuclear Energy Agency
NEIL	Nuclear Electri Insurance Limited

NELIA	Nuclear Energy Liability Insurance Association
NGO	Non-Governmental Organization
NOAA	National Oceanic and Atmospheric Administration
NPL	National Priority List
NRC	Nuclear Regulatory Commission
NRD	Natural Resource Damage
OCIL	Oil Casualty Insurance Ltd.
OECD	Organization for Economic Co-operation and Development
OJ	Official Journal (of the European Union)
OPA	Oil Pollution Act (US)
OSLTF	Oil Spill Liability Trust Fund
PAA	Price-Anderson Act (US)
P&I Clubs	Protection and Indemnity Clubs
PLA	Public Liability Action
RCRA	Resource Conservation and Recovery Act (US)
RRG	Risk Retention Group
SFP	Secondary Financial Protection
SMCRA	Surface Coal Mining and Reclamation Act
SME	Small or Medium Enterprises
SOA	State Oceanic Agency
SUBAT	Stichting Uitvoering Bodemsanering Amovering Tankstations ((Foundation for the Remediation of Petrol Stations) (The Netherlands)
SWPPA	Solid Waste Pollution Prevention Act (China)
TEU	The Treat on European Union
TFEU	The Treaty on the Functioning of the European Union
TLL	Tort Liability Law (China)
TOVALOP	Tanker Owners Voluntary Agreement concerning Liability for Oil Pollution
UHV	Environmental Impairment Liability Insurance (Germany)
UK	United Kingdom
USV	Allgemeine Versicherungsbedingungen für die Umweltschadensversicherung (Environmental Damage Insurance) (Germany)
U.S.	United States
VAN	Verenigde Assurantiebedrijven Nederland
WANO	World Association of Nuclear Operators
Wbb	Wet Bodembescherming
WPPA	Water Pollution Prevention Act (China)
WTP	Willingness to Pay

LIST OF FIGURES AND TABLES

List of Figures

Figure 1: Model I: Mandatory Financial Security..........178
Figure 2: Model II: Voluntary Financial Security..........182
Figure 3: Model III: Optional Financial Cap..........184

List of Tables

Table 1: Comparison among Various Compensation Instruments..........170
Table 2: Liability Caps under the OPA..........262
Table 3: Liability Caps under the CERCLA..........271
Table 4: Comparison of different Compensation Systems..........311
Table 5: The Types of Vessels and Required Financial Security in China..........360

Chapter 1

INTRODUCTION

1. Background

The environment provides human beings with a respectable and prosperous life. Both are, however, threatened by rapid industrialization and urbanization. Traditional tort law provides only a limited remedy to the damage to the environmental components subject to property rights. The components of the environment which are not privately owned have only been regarded as compensable in some legal systems recently. For a long time the environment has been thought to be of little value, apart from its property interests. Regulation has been employed to prevent damage to the environment. However, regulation only achieves limited success. The environment, as a public good, is vulnerable to the tragedy of the commons. Numerous significant accidents across the world have alarmed people. Apart from personal injury and property damage, the manmade damage to the environment itself can also be astonishing. For example, the 1989 Exxon Valdez accident led to 10.8 million gallons of crude oil being spilled into Prince William Sound. The oil eventually covered 11 million square miles of ocean, leading to 2.1 billion dollars to be paid for the cleanup.[1] This accident was once the largest accidental marine oil pollution case, however, it was succeeded by the Deepwater Horizon accident in 2010, which caused approximately five million barrels of oil leakage.[2] One can also find many examples of environmental disasters the other side of the Atlantic, such as the mine accident Aznacollar, in Spain in 1998,[3] the Erika oil spill on the French coast in 2002[4] and the accident at the MAL

1 Exxon Valdez Oil Spill Trustee Council's website: <www.evostc.-state.ak.us/facts/qanda.cfm>, last accessed on 8 May 2013.

2 National Commission on the BP Deepwater Horizon Oil Spill and Offshore Drilling, The Amount and Fate of the Oil, Working Paper No. 3, <www.oilspillcommission.gov/-sites/default/files/documents/Updated%20Amount%20and%20Fate%20of%20the%20Oil%20Working%20Paper.pdf>, last accessed on 8 May 2013.

3 Brans 2001, p. 179.

4 For details about the Erika case, see Foley & Nolan 2008.

alumina factory near Kolontar in Hungary in 2010.[5] In a world facing threats from increasingly costly environmental disasters, how to prevent and compensate for such damage is an important question to answer.

Thus far, more attention has been paid to traditional damage, including bodily injuries, property damage and economic losses caused by environmental accidents. This research, however, focuses on damage to the environment itself. Before discussing this research topic, the terminology needs to be clarified. Several terms have been employed by various legislation and academic works. For example, at the legislative level, three well-known terms can be noted: 'environmental damage' is used in Directive 2004/35/EC of the European Parliament and of the Council on Environmental Liability with regard to the Prevention and Remedying of Environmental Damage (ELD),[6] 'natural resources damage' is mainly used in the Comprehensive Environmental Response, Compensation, and Liability Act (CERCLA),[7] and 'impairment of the environment' can be found in the 1993 Lugano Convention on Civil Liability for Damage Resulting from Activities Dangerous to the Environment (Lugano Convention).[8] In addition to these examples, 'ecological damage' is also a commonly used term in academic literature.[9] The overlaps and nuances between those concepts are examined in more detail in Chapter 1. Let me briefly explain why this research prefers to use the notion 'ecological damage': 'Natural resource damage' is a concept based on U.S. federal acts, which distinguish it from response costs/removal costs.[10] The latter concepts, however, are also relevant when we discuss the damage to the environment, thus they are also included in this research. 'Environmental damage' is usually defined broadly to incorporate traditional damage caused by environmental accidents. To avoid possible misunderstandings, this research chooses the term 'ecological damage'. This refers to the physical impairment of nature or, more specifically, damage to soil, water, air, climate and landscape, the flora and fauna living in them, and the interactions between them.[11] It concerns damage to both the public parts of the environment (public natural resources) and the parts subject to property rights (private natural resources).

Specific legislation has been developed for ecological damage since several decades ago. For example, a comprehensive legislative framework has been established for the compensation of natural resources damage in the U.S. through the Clean Water Act (CWA), the CERCLA and the Oil Pollution Act (OPA).[12] EU

5 Redsludge Tragedy, < http://redsludge.bm.hu/>, last accessed on 8 May 2013.

6 Directive 2004/35/EC of the European Parliament and of the Council on Environmental Liability with regard to the Prevention and Remedying of Environmental Damage, *OJ* 2004 L143/59.

7 Comprehensive Environmental Response, Compensation, and Liability Act of 1980 (CERCLA, also called Super fund law), 42 U.S.C. §§9601-9674.

8 Art. 2(7) of the Lugano Convention, 32 International Legal Materials, 1228 (1993).

9 Brans gives an overview of these concepts. Brans 2001, p. 12-22.

10 Brans 2001, p. 67.

11 Lahnstein 2003, p. 307.

12 Clean Water Act of 1977 (CWA), 33 U.S.C. §§1251-1387, and Oil Pollution Act (OPA) of 1990, 33 U.S.C. §§2701-2761.

started to harmonize its legislation on liability for environmental damage through the ELD in the recent decade. There are some similarities between these systems, such as the combined use of public and private law. In other words, on the one hand, there is a regulatory requirement to respond to and restore the damaged environment (either by liable parties or by public authorities); on the other hand, the polluters are held liable for such damage. Hence, the liability for ecological damage is not pure civil liability anymore.[13] Besides, in addition to liability, some compensation instruments, such as insurance, guarantees, environmental funds, are either mandated or encouraged to be used to cover the potential damage.[14] All those characteristics make the regime of ecological damage compensation a very comprehensive and intricate system. Although experience has been accumulated to deal with this question, the existing system still faces challenges from time to time. The Deep Horizon accident triggered debates on the sufficiency of the OPA system to compensate for catastrophic oil spills, especially for offshore facilities.[15] The development of the CERCLA was more cumbersome, leading to fierce criticism on its liability system and chilling effects on the redevelopment of brown fields.[16] The ELD was only passed in 2004 and was transposed into all Member States' domestic law in 2010, three years later than the required date.[17] This system is comparatively new. It allows Member States to extend its scope by allowing a broader explanation of 'environmental damage' or allowing strict liability applying to more activities than listed under Annex III. However, the transposition of ELD shows that only a limited number of Member States have extended the scope.[18] It is reported that the awareness of the ELD's actual influence among many stakeholders is low and there have been few ELD cases so far.[19] Hence, it is doubtful to what extent the ELD will have sufficient preventive effect.

In response to the increasing awareness of the necessity to deal with ecological damage and the development of legislation, more literature has addressed, on the one hand, the liability issue and, on the other hand, the financial security systems. Scholars in the EU did extensive research on the legislative history of the ELD and its transposition and implementation in the Member States.[20] While their

13 Winter, Jans, Macrory & Krämer 2008, p. 165.

14 For example, under the OPA and CERCLA compensation systems, there are mandatory requirements to provide financial security and financial security under the ELD is voluntary. Addressed in Chapters 7 and 8.

15 Addressed in Chapter 7, Section 2.

16 McMorrow 2004, p. 1094-1095.

17 See De Smedt & Faure 2010.

18 CEA, Navigating the Environmental Liability Directive. A Practical Guide for Insurance Underwriters and Claims Handlers, April 2009, p. 10, available at: <www.-insuranceeurope.eu/uploads/Modules/Publications/1240585425_eld-best-practice-guide-update.pdf>, last accessed on 8 May 2013. For the transposition of the ELD into domestic law, see also: Weissenbacher 2005; Coroner 2006; Slabbinck, Descamps & Bocken 2006; Jendroska & Bar 2007; Winter, Jans, Macrory & Krämer 2008; Alberton 2007; Fogleman 2009.

19 Interview with Mr. Kremers, Verenidge Assurantiebedrijven Nederland (VAN), 8th October, 2012, Maastricht, Interview transcript on file with the author.

20 See, for example, Alberton 2007; Coroner 2006; Jendroska & Bar 2007; Rehbinder 2007 and Slabbinck, Descamps & Bocken 2006.

counterparts in the U.S. focused on those of CERCLA, OPA and the assessment regulations under them.[21] In addition, various compensation mechanisms began to gain increasing attention.[22] Financial security under the ELD and its implementation were analyzed.[23] Some general discussions about specific mechanisms also took place.[24] Those studies hence focused on either liability or specific compensation instruments. They either explained or evaluated the regime in a given jurisdiction.

Those valuable studies provide the basis for this research. However, they either focus on a specific jurisdiction or a specific type of compensation mechanism, such as insurance or environmental funds. From a theoretical point of view, less attention has been paid to the question how to choose between different instruments, such as regulation, liability rules and compensation mechanisms, or how they can interact to combat ecological damage. This research tries to provide an integrated study by answering the question how different instruments mentioned above can be mobilized and can interact to, on the one hand, prevent ecological damage more efficiently and, on the other hand, to restore the damaged environment better. Three theoretical models, with different interaction between regulation, liability rules and compensation mechanisms (mainly based on whether compulsory financial security is required), are proposed in this research to try to realize the aims of prevention and restoration. Some criteria are also provided in choosing among different compensation mechanisms.

In addition to the integrated theoretical research, this thesis also tries to add to the literature by studying how ecological damage is compensated in China from both a theoretical and a practical point of view. Existing research pays more attention to the compensation system in the U.S. and the EU. However, the corresponding research in the developing world is still weak. China, as a major emerging economy, has experienced high speed economic development in recent decades; however, China is suffering serious environmental problems as well. In such a society which still struggles to find a balance between economic development and environmental protection, how damage to the environment itself is tackled is an increasingly interesting topic. This research, based on both literature review and field studies, investigates the area of the compensation systems for ecological damage in China and to give some policy recommendations, considering the theoretical models and China's specific problems.

2. Research Questions and Problem Definitions

As far as ecological damage is concerned, the primary question is how to prevent such damage. Regulation has long been in place to tackle various environmental

21 See, for example, Swanson, 2001; Patrick Thomas III 1992; Letourneau & Welmaker 2000; Schmid 2008 and Smith 1993.

22 Many authors have paid attention to the insurance and environmental liability, such as Zhu 2007 and Faure & Grimeaud 2003, p. 123-192.

23 See, for example, Bocken 2006; Bocken 2009 and Peeters & Woerd 2006.

24 Such as the discussion about environmental funds which is not confined into practice in the EU. See Hawke & Hargreaves 2003.

issues, including ecological damage. However, excessive damage still happens from time to time, challenging the effectiveness of human reactions. Of course, preventing ecological damage does not mean eradicating it totally regardless of the costs that may arise. Such an ambition is neither feasible nor desirable. From an economic point of view, instead of eradicating ecological damage, the aim is to achieve better prevention at lowest costs. In addition to regulation, economics literature also regards liability rules as potential tools to prevent damage. In other words, liability rules should be designed in such a way as to create deterrent incentives for the potential damage creators. When liability, especially strict liability, is used (which is quite often the case for ecological damage) a question immediately emerging is how to deal with the insolvency risk (which again is a quite realistic risk for ecological damage). To alleviate the insolvency risk and guarantee the deterrent effect of liability rules, some instruments such as insurance and guarantees come into play.

Regulation and liability rules mentioned above focus on the prevention of ecological damage. However, they neither mean to nor are able to prevent all damage. Ecological damage may still materialize from time to time. How to restore the damaged environment or, more broadly, how to compensate for such damage hence creates another challenge. Here, the compensation instruments (liability insurance, first-party and direct insurance, risk sharing agreements, environmental compensation funds, guarantees and capital market) again can play an important role. This research tries to consider both sides of the question: prevention and compensation. In other words, the research question is: *how can better prevention and compensation of ecological damage be achieved by mobilizing different instruments: regulation, liability rules and compensation mechanisms?* Of course, these various instruments have their own advantages and weaknesses and reality is full of complexities. This research does not mean to be so ambitious to provide a panacea for all problems brought by ecological damage or create a single model fitting all situations. Instead, this research identifies several scenarios which may happen in practice and, accordingly, designs three models with different interactions among those instruments. It also recognizes several factors influencing the choice between compensation mechanisms in different situations. To answer the questions mentioned above, some sub-questions need to be answered first: *what is ecological damage, and how is it different from traditional damage? Which legal instrument is more suitable to prevent and compensate for ecological damage: regulation or liability rules? Are they satisfactory to realize these aims; if not, are other compensation mechanisms the solutions? How should liability rules and individual compensation mechanisms be designed to achieve these goals of prevention and compensation? How do those instruments interact?*

To establish whether and how the theoretical models work in practice, this thesis explores some compensation systems in practice. Such systems include: the international compensation systems for oil pollution and nuclear damage, the compensation systems for nuclear damage and natural resources damage in the U.S. and the ELD system in the EU as well as a few examples of compensation mechanisms used in EU Member States. Why those systems are chosen and how the comparative study is carried out is explained in the following section dealing with the methodology of this thesis.

In addition to the theoretical findings and the western practice, this research also examines the status quo in China, including both legislation and practice. Increasingly, environmental costs have accompanied the spectacular economic growth in China in recent years. The Chinese Environmental Economics Accounting Research Report estimated that the costs of environmental degradation was 1,274,57 billion RMB in 2008, representing 3.9 per cent of the gross domestic product (GDP).[25] An earlier World Bank report showed that the environmental costs might reach 8 per cent of total GDP in China.[26] In such a developing economy with both alarming environmental damage and strong development demands, how can a tradeoff be achieved between these two goals poses an important question. Such a tradeoff is also reflected in the compensation system for ecological damage. At the legislative level, general environmental liability rules are based on both civil laws and environmental statues. Several important liability principles are reiterated in the Tort Liability Law promulgated in 2009. However, as for ecological damage specifically, the law is unclear. Apart from the case of marine oil pollution, there is neither a general obligation to restore the damaged environment for polluters nor explicit liability for such damage. The unclear legislation and lack of incentives for both companies and public authorities to restore the environment make practice even more obscure. Too often one sees the damaged environment unrestored or only restored by the government without direct recourse from the actual polluters. The only area where legal action seems to work better is marine oil pollution, where compensation for ecological damage is not only in place to some extent on paper but also in practice. However, one needs not despair, many changes are under way, in terms of both legislation and practice. Against this background, to understand the existing compensation system for ecological damage in China and to improve the system, considering both the theoretical models and the specific situation in China, necessary tasks are to be undertaken.

3. Methodology

Both economic analysis and comparative studies are used in this thesis. Economic analysis can be used both to design the theoretical compensation models for ecological damage and evaluate the practice in different jurisdictions. Different schools of law and economics exist and this research mainly applies the neo-classical economic analysis of tort law and insurance. The neoclassical economic analysis comprises the first generation of law and economics and is used broadly in different areas of law, including property law, contract law, tort law and so on. This school is criticized and complemented by some other schools, such as Public Choice theory, Institutional law and Economics, behavioral law and economics and so on. However, neoclassical economics analysis is the only approach providing a clear normative analysis of tort and insurance. The other schools, such as the behavioral

25 The Chinese Environmental Economics Accounting Research Report (2008) has been completed, <www.caep.org.cn/ReadNews.asp?NewsID=2761> (Chinese), last accessed on 8 May 2013.

26 World Bank 1997, p. 23.

law and economics, although providing a valuable supplement to traditional law and economics, fail to provide a systematic framework for the analysis of tort and insurance and have limited normative implications.[27] Hence this research mainly employs neoclassical law and economics in its analysis. Especially Calabresi's well-known social costs of accidents theory is employed to design liability rules for ecological damage and to compare the potential of different compensation mechanisms to reduce social costs. According to Calabresi, accidents can lead to three categories of social costs: the primary costs (precaution costs plus the actual damage), secondary costs (concerning loss spreading) and tertiary costs (administrative costs).[28] The compensation system needs to be designed in a way to minimize the total social costs.

In addition, public choice theory is also used in this research to explain some of the divergence of the practice from theoretical models. The empirical part of this research shows that practice does not always comply with theoretical models, especially when liability rules are designed. Those divergences can be explained to some extent by public choice theory. According to public choice theory, although they claim to serve the public interest, the legislators, generally being politicians, are unavoidably influenced by rent-seeking behavior and lobbying of different interest groups. As for the compensation regime, potential accident victims, risk creators, lawyers, commercial and professional interests, as well as public officials, may all influence the legislative process.[29]

Comparative study is employed in the empirical part of this research to show how different systems deal with the prevention and compensation of ecological damage. As mentioned earlier, some international, U.S. and EU systems are discussed in this research. The U.S. compensation system for natural resources damage and the ELD system in the EU are chosen because liability for ecological damage is explicitly established in those systems. The other systems under discussion do not specifically deal with ecological damage. They are also covered here because, on the one hand, the instruments for traditional damage and ecological damage are not always separable (for example the international regime for oil pollution covers both preventive/restoration costs and personal injury and property damage); on the other hand, given the limited experience in the field of ecological damage, some instruments used for general liability provide helpful insight as well (such as the capacity of the nuclear compensation system to cover catastrophic damage). The ELD scheme is still young, which means that in many Member States, the specific system for ecological damage is still under development. How they deal with general environmental liability therefore sheds light on the possibilities of the future compensation systems for ecological damage. For the examination of practice in the Member States, a patchwork approach is followed: only some examples of the compensation mechanisms are examined. When discussing those systems, this thesis explores both liability rules and

27 Faure 2010.

28 Calabresi 1970, p. 26-31.

29 Ogus 2007, p. 38-40.

compensation mechanisms (financial security). The comparison here does not intend to provide a tentative checklist of all the similarities and differences between those systems. It is employed to show the different application of the theoretical models in practice and to understand the divergences of practice from theory. With 'empirical', I focus on how the legal framework for ecological damage is established and which types of compensation instruments are in place in different jurisdictions. To the extent possible, I also provide a discussion on the effect the specific systems. However, a full evaluation of the practical effect of the compensation systems is not conducted due to the limited data and scope here. For the discussion of the American experience, I relied mostly on desk research since existing literature already provides good analysis of the practical issues. For the research on the EU system, I complemented the desk research with interviews, especially interviews with commercial and mutual insurers.

To better understand the compensation mechanisms used in EU Member States, I have conducted research interviews with commercial insurers, mutual insurance companies and The Center for Nuclear Energy with respect to the nuclear insurance in Europe and direct insurance in the Netherlands. However, these interviews are very limited and were conducted only as a complement to the existing literature. In other words, only when there is a gap or only old information in literature, interviews were used. For example, literature has shown that direct insurance started to develop in late 1990s. However the literature was published only a few years after the adoption of this type of insurance. To understand how this product developed over the years and its effect, I conducted an interview with a representative from *Verenidge Assurantiebedrijven Nederland*, the insurance association which designed direct insurance and maintained this insurance pool in the 1990s and still continues to support this product provided by its affiliated insurance companies. More interviews were conducted regarding the compensation system in China. This is because, on the one hand, less literature exists with respect to ecological damage in China and, on the other hand, enforcement of environmental law is often regarded as problematic in China. This hence requires a field study to understand the actual compensation system in practice. The interviewees in China included environmental law professors, relevant public authorities (Ministry of Environmental Protection; the Guangzhou Oceanic Agency), One Maritime Court which has a comparatively long history in dealing with marine pollution litigation, some scientists with expertise in environmental health and soil restoration and insurers.[30] However, one has still to admit that those interviews are still limited. On the one hand, those interviews were only conducted in Beijing and Guangzhou. This is because the interviews with national agencies and insurers in Beijing help to give an overview of the national environmental policies and insurance market. Guangzhou is the capital city of the Guangdong Province, one of the best representatives with both high-speed economic growth and increasing environmental losses. In spite of these reasons, the interviews conducted in only those two cities are far from enough to reveal the holistic picture

30 The details about the interviewees are addressed in Chapter 10.

of the compensation system in China. Besides, these interviews are mainly qualitative surveys focusing on specific areas of the compensation system but not a collection of quantitative data. The interviews were of a semi-structured nature. That means that the respondents received a questionnaire in advance and that questionnaire guided the interview. Afterwards the interviews were summarized in a transcript which is on file. This transcript constituted the basis for using the interviews in this thesis. Hence, these interviews alone can neither provide answers about the real environmental quality in China nor can they explain the actual improvements brought by regulation, liability rules or insurance. Nevertheless, these interviews provide useful insight into the practice of the restoration of environmental damage in environmental litigation, into existing problems and into the development of environmental insurance. In addition, the limited number of interviews means that the possible biases of interviewees cannot be totally eradicated. But to the extent possible, the author tries to minimize the impact of biases. For example, to answer the question how environmental disputes are resolved and to what extent the damaged environment (such as soil) is restored, the author tries to obtain information from interviewees from different backgrounds: government officials in charge of environmental protection; professors in environmental law; NGOs with experience of environmental claims and courts. To understand the environmental insurance market in China, I also interviewed (re)insurers; law professors and officials. Furthermore, I also collected information on available insurance products and their practice from the website of insurance companies providing them and media reports. The information from different aspects on the same questions can help to provide a comparative comprehensive and balanced picture of the Chinese compensation system.

When discussing the compensation system in China, one has to notice that it is still a system in its early development. At the legislative level, many new statutes and regulations are developing, such as the liability rules for soil pollution, ecological damage assessment rules and so on.[31] At the practical level, ecological damage started to attract more attention; some compensation mechanisms, such as environmental insurance and oil pollution funds, are from a more recent time.[32] Since the Chinese compensation system is still in full evolution, it is not easy to provide a full picture of the status quo in China or to predict the outcome of such evolution. However, it is still important to sketch the main characteristics of the existing system, to point out the problems so far and to provide insights to the direction of future developments.

Another limitation of this research concerns the choice of the jurisdictions in Europe in the empirical part. This research discusses briefly the ELD system in the EU and some examples of compensation mechanisms used in its Member States, such as environmental liability and the nuclear mutual pool in Germany, direct insurance and an environmental fund in the Netherlands. This research does not intend to give a comprehensive picture of how ecological damage is compensated in

31 *Ibid.*
32 *Ibid.*

the EU Member States. A cherry-picking approach is used to shed some light on the compensation system in the EU. This is because the empirical research does not intend to provide a full description of the compensation system in the EU but to show how practice corresponds with the theoretical models. Instead of discussing the whole picture of compensation in a Member State (including liability rules and all its available compensation mechanisms), in this thesis only one example is chosen for each type of compensation instruments discussed in the theoretical part. The theoretical part discusses liability insurance; direct insurance; risk-sharing pools, compensation funds and the capital market. The capital market also started to develop, but mainly in the area of natural disasters rather than man-made disasters. Very limited experience exists for the capital market tools to cover environmental liability. Hence in discussing the practice in the EU Member States, this thesis discusses one example for each of the other four mechanisms. The German Environmental Liability Act of 1990 requires compulsory insurance from some operators. However, this provision remains largely unimplemented. This example shows compulsory environmental liability is not always desirable or feasible. Direct insurance in the Netherlands started to develop in the late 1990s, as a replacement for the traditional liability insurance. This product has developed for more than a decade and experience has been accumulated to support the analysis of its pros and cons. For the risk-sharing pool, it is used more frequently in the area of marine oil pollution and nuclear damage. The Protection & Indemnity Club is an example of the former and is discussed in the part of international systems. Risk-sharing pools for nuclear liability in Germany provide an interesting example of the latter, which makes the compensation for a catastrophe possible. As for the compensation fund, the Dutch example is again interesting because it makes the decontamination of historical pollution possible and avoids serious preventive distortion. In addition, the choice of jurisdictions also reflects the language limits of the author. Since the author's language skill is limited to English and Chinese, research mainly focused on the areas where English literature could provide sufficient and fruitful discussions.

4. Structure

The structure of this research follows directly from the research problems defined above. This research was undertaken from both a theoretical and an empirical perspective; hence the thesis is divided into three parts. After the introduction (Chapter 1), Part I will examine the characteristics of ecological damage and liability rules, which will be used as a basis to guide the design of a compensation scheme in theory. Part II will test the assumptions proposed in Part I through the practice in the international regime, the U.S. and the EU. After the theoretical and empirical study, the thesis will address the situation in China, which constitutes Part III: there the question will be asked what is the existing situation for ecological damage compensation in China and how can this compensation system be improved?

The theoretical framework will be developed in Chapters 2 to 5 of Part I. Chapter 2 will focus on a definition of the research subject: ecological damage, its definition and characteristics and especially those making it different from traditional damage. The different concepts used to refer to the damage to the environment in both legislation and academic literature will be analyzed.

Chapter 3 will provide a short discussion of the possible legal rules concerning the prevention of ecological damage—both regulation and liability rules. Economic analysis shows that a combined use of regulation and liability rules is desirable. Then the thesis unravels the elements of liability rules for ecological damage: standing, responsible parties, strict liability and negligence, causation, compensation scope and damage assessment. The limitations of regulation and liability rules in both prevention and compensation are analyzed, making compensation mechanisms necessary both regulation and liability rules.

In the first section of Chapter 4, individual compensation instruments, including the most popular one - liability insurance and the alternatives - first-party and direct insurance, risk sharing agreements, environmental funds, guarantees and capital markets are examined respectively. Economic analysis shows their different potential to reduce the three categories of social costs. A short comparison of these instruments is also given.

Chapter 5 summarizes and concludes Part I. It tries to integrate the instruments discussed above: regulation, liability rules and compensation instruments. It identifies different scenarios and proposes three compensation models accordingly, mainly based on whether compulsory financial security is required or not. Based on a comparison among the compensation mechanisms in Chapter 3, this research identifies some factors influencing the choice of mechanisms. As for the use of those compensation mechanisms, this research does not intend to choose an optimal one in each situation but to establish a multilayered compensation system, considering the balance between the factors identified.

After the theoretical analysis in Part I, an empirical study follows in Part II to test the theoretical models and factors in practice. Chapter 6 discusses the international compensation systems for oil pollution and nuclear damage. The U.S. compensation systems for natural resources damage and nuclear damage are sketched in Chapter 7, followed by the EU systems in Chapter 8. Chapter 9 summarizes Part II and discusses the coherence and divergences between theoretical models and practices.

Based on the theoretical analysis and empirical examination, the Part III, Chapter 10, looks at another part of the world, China. The Chinese compensation system for ecological damage is examined, both at the legislative and at the practical level. After a description of the existing situation, theoretical models are used to evaluate the Chinese compensation system and a few policy recommendations are provided.

After the theoretical discussions and empirical study of western countries and China, some concluding remarks are made and policy recommendations are provided in Chapter 11.

PART I

COMPENSATION MODELS FOR ECOLOGICAL DAMAGE: THEORETICAL FRAMEWORK

COMPENSATION MODELS FOR ECOLOGICAL DAMAGE: THEORETICAL FRAMEWORK

To answer the research question of how to better prevent and compensate for ecological damage, the thesis will look at both theoretical as well as empirical research. Part I focuses on the theoretical analysis. This research first defines the research subject, ecological damage, and discusses some of its characteristics which make the traditional tort system an insufficient solution. Further, this part examines the possible legal instruments to deal with ecological damage respectively and, based on the individual research, some compensation models with the interaction of these instruments are proposed.

Damage to the environment has long been regarded as not compensable due to the lack of market value of the environment itself. It only began to gain importance when the irreversibility and aggravation of the damage became significant and the non-economic value of the environment was recognized by society. But both the legislators and scholars still disagree on the term to define this damage; and the meaning of the possible terms are not univocal. Chapter 2 chooses 'ecological damage' to express the subject of this research and tries to define it. Then the traditional concept of 'damage' is explored to see if ecological damage can be fit into the traditional tort regimes. Firstly, this chapter gives a brief overview of the concepts used in legislation and literature to refer to the damage to the environment. Among these expressions, three terms are broadly used: environmental damage, natural resources damage and ecological damage. The comparison between those heading words, environment, natural resources and ecology, shows more similarities than distinctions. Thus, the choice between these concepts depends on the custom and unity of usage. Environmental damage is often broadly defined to encompass the damage caused via the environment. Natural resources damage is mainly used in the U.S., and it is distinguished from removal costs/response costs which are the immediate response to environmental accidents and are also relevant in this research. Ecological damage, which is a popular usage in Europe, is chosen in this research, and it is more environment-oriented and specific. In the second part of Chapter 2, 'damage' in traditional tort law is briefly analyzed where the individual interest is involved and the measure of loss is determined by the diminution in economic value. Thus the term 'ecological damage' cannot be fit into the traditional tort regimes: for damage to owned natural resources, the loss of non-economic value of resources is often neglected in assessment; for damage to un-owned natural damage, lack of an injured individual interest is a major obstacle to compensation. In view of the incompatibility of ecological damage with traditional damage, specific instruments should be developed to prevent and compensate for ecological damage.

Both regulation and liability can be used to prevent ecological damage. The choice between the two instruments is discussed in Chapter 3. Considering the factors of information asymmetry, insolvency risk, the threat of a liability suit and administrative costs in the context of ecological damage, one can find neither regulation nor liability rules dominating the other. A combined use of both instruments is desirable. Different regulatory instruments can be employed to prevent ecological damage and a detailed discussion of them is outside the limited scope of this research. The focus of Chapter 3 is liability rules. This chapter applies Calabresi's social costs theory to guide the design of liability rules: standing, responsible parties, liability standards, causation and assessment. Though this part is theory oriented, some examples of how these liability components are adopted in different jurisdictions are briefly mentioned. These examples do not intend to give a tentative description of the empirical world but to simply show the possible options in designing liability rules. After elaborating the concrete liability components, this chapter analyzes the capacity of regulation and liability rules to prevent and compensate for ecological damage. It is found that the combination of regulation and liability rules can still not achieve the goals, which makes the introduction of compensation instruments desirable.

In Chapter 4, specific compensation mechanisms are analyzed in detail. This section discusses six types of mechanisms: liability insurance, first-party and direct insurance, risk sharing agreements, environmental funds, guarantees and capital markets. They are either commonly used in practice, or proposed in the literature as one potential solution. Calabresi's social costs theory is also employed here to show the capacity of these mechanisms to reduce three types of social costs. Limited examples are also provided here to explain the function of the individual compensation mechanisms. After the individual discussions, a short comparison of those mechanisms is given.

Chapter 5 tries to propose some compensation models with the combined use of regulation, liability rules and compensation instruments according to different scenarios that may arise. In these models, different compensation mechanisms can be used. Based on the comparison in Chapter 4, some factors influencing the choice of compensation instruments are indentified. These factors do not aim at selecting one single optimal instrument in a certain context but help to establish a multilayered compensation system with a suitable mix of mechanisms.

Chapter 2

THE DEFINITIONS AND CHARACTERISTICS OF ECOLOGICAL DAMAGE

1. Introduction

The last century has not only seen the prosperity of traditional environmental tort law but also the development of rules about damage to the environment. Since then, many notions have been developed to refer to this recently recognized damage in both legislation and literature. Hot debates over its definition, scope and evaluation arose and are still ongoing. The definition is of fundamental importance to determine the instruments that are useful for its prevention and compensation, including the type and scope of the liability, and relevant regulatory instruments.

This chapter gives a brief overview of the concepts used in legislation and treaties to denote the damage to the environment. Early treaties discussed damage to the environment under the general heading of 'damage' without further definition; later various terms such as 'environmental damage' or 'natural resources damage' were introduced into both international treaties and domestic legislation. To distinguish damage to public environmental components from those privately appropriated, sub-concepts such as 'impairment of the environment' or 'damage to the environment itself' were also developed. Similar concepts and the differentiation between damage to public and privately-owned natural resources can also be found in literature. One goal of this chapter is to define damage to the environment. To choose between the different notions, this chapter analyzes the heading word of the notions: environment; natural resources and ecology. Based on the comparison and custom in usage, the term 'ecological damage' was chosen, which comprises not only damage to public natural recourses, but also damage to owned natural resources.

Another objective of this chapter is to examine whether the notion of 'ecological damage' can fit into traditional tort law. To answer this question, the first step is to understand the concept 'damage' in traditional tort law, its scope and categories. On this basis, the features of two types of ecological damage – damage to public natural resources and damage to owned natural resources, are analyzed. The analysis shows that there are many obstacles to compensating ecological damage within the traditional tort law regime: unidentifiable injurers and victims, a long time lapsing between the activities and the appearance of damage, proof of causation, and difficulties in quantifying the damage. Thus the introduction of 'ecological damage' is more than a political slogan; specific rules are necessary to prevent and compensate this damage.

The characteristics of ecological damage makes specific rules necessary: instead of monetary compensation, the most important remedy for ecological damage is to clean up the pollution and to restore the damaged environment to its original status. The analysis below shows that the restoration-based approach can provide a more complete assessment of the values of the damaged environment and hence promote the internalization of the full costs. Only when restoration is not possible or disproportionately expensive, monetary evaluation shall be used. In addition to liability rules, some other instruments can also be used to promote the restoration/compensation of ecological damage, such as insurance, risk-sharing pools and compensation funds. Therefore, in this thesis, the term 'compensation' has a broader implication than its traditional meaning (monetary compensation) and this chapter also gives a definition of this term.

2. The Choice of a Concept – Environmental Damage, Ecological Damage and Natural Resources Damage

2.1. *Relevant Concepts in Legal Documents and Literature*

2.1.1. Concepts in Legislation and Treaties

As mentioned in the introduction, the subject of this research is the loss or damage to the environment, rather than personal injury, property damage and economic loss caused via the environment. This damage may be caused both to the privately appropriated environment and to the public environment. Various concepts have been developed to refer to damage to the environment in international treaties and in domestic legislation.

2.1.1.1. Environmental Damage

Specific terms are also developed in legislation, of which, 'environmental damage' is widely used both at the regional and at the national level. One of the examples is the ELD. Under the ELD, a harmonized liability system is established for 'environmental damage' but not for traditional damage such as personal injuries, property damage and economic losses. Three categories of damage are covered under the heading of 'environmental damage':

> (a) damage to protected species and natural habitats, which is any damage that has significant adverse effects on reaching or maintaining the favourable conservation status of such habitats or species;
>
> (b) water damage, which is any damage that significantly adversely affects the ecological, chemical and/or quantitative status and/or ecological potential, as defined

> in Directive 2000/60/EC, of the waters concerned, with the exception of adverse effects where Article 4(7) of that Directive applies;
>
> (c) land damage, which is any land contamination that creates a significant risk of human health being adversely affected as a result of the direct or indirect introduction, in, on or under land, of substances, preparations, organisms or micro-organisms;[1]

According to the definition, although specifically focusing on damage to the environment rather than on traditional damage caused by the same environment-related activities, the ELD is restrictive in application. Only biodiversity damage, water damage and land damage are classified as 'environmental damage'. Those categories are subject to both geographical and threshold limitation. Biodiversity damage should be the damage to any species or habitat protected under Council Directive 79/409/EEC of 2 April 1979 on the Conservation of Wild Birds,[2] Council Directive 92/43/EEC on the Conservation of Natural Habitats and of Wild Fauna and Flora (the Habitat Directive)[3] or under any national law. The term 'biodiversity' is narrower than that in the Convention on Biological Diversity,[4] which encompasses not only flora and fauna but also the interrelationships and variability between living organisms.[5] Water damage is also limited by the categories defined in Directive 2000/60/EC of the European Parliament and of the Council establishing a framework for the Community action in the field of water policy (Water Framework Directive).[6] In addition to the geographical limitation, the biodiversity and water damage should also be 'significantly adverse'. As regards land damage, the threshold is determined by its risk for human health instead of the adverse changes to the land itself.

In the Act on Compensation for Environmental Damage of Finland, the scope of environmental damage is much broader. It refers to a loss caused by activities carried out in a certain area and resulting from:

1. pollution of the water, air or soil;
2. noise, vibration, radiation, light, heat or smell; or
3. other similar nuisance.[7]

1 Art. 2(1) of the ELD.

2 Council Directive 79/409/EEC of 2 April 1979 on the Conservation of Wild Birds, *OJ* 1979 L103/1. It has been amended since then and this Directive, together with its amendments were codied as Directive 2009/147/EC of the European Parliament and the Council of 30 November 2009 on the Conservation of Wild Birds, *OJ* 2010 L20/7 (Wild Birds Directive).

3 Council Directive 92/43/EEC of the Council of the European Communities of 21 May 1992 on the Conservation of Natural Habitats and of Wild Fauna and Flora, *OJ* 1992 L206/7.

4 Art. 2 of the CBD, 1760 UNTS 79; 31 ILM 818.

5 Hattan 2002, p. 4.

6 Directive 2000/60/EC of the European Parliament and of the Council establishing a framework for the Community action in the field of water policy, *OJ* 2000 L327/1.

7 Section 1, the Act on Compensation for Environmental Damage (737/1994) translation version is available at: <www.finlex.fi/fi/laki/kaannokset/1994/en19940737.pdf>, last accessed on 8 May 2013.

The act differentiates environmental damage from bodily injury and material loss, and confines its application to the former.[8]

2.1.1.2. Pollution

The term 'pollution' and 'pollution damage' are concepts used frequently by international treaties. For example, the International Convention on Civil Liability for Oil Pollution Damage 1969 (the CLC 1969) defined 'pollution damage' as:

> 'loss or damage caused outside the ship carrying oil by contamination resulting from the escape or discharge of oil from the ship, wherever such escape or discharge may occur, and includes the cost of preventive measures and further loss of damage caused by preventive measures.'[9]

The convention gives no further guidance to interpret the scope of the words 'loss or damage'. Although personal injury or property damage are not explicitly mentioned in the CLC 1969, they seemed to be covered by the vague definition of pollution damage.[10] However, it is unclear whether pure economic losses would fall within this definition. Whether environmental damage *per se* is included is a matter left to the interpretation of national courts according to the transpositive legislation.[11] As envisaged in the 1969 conference, the liability regime in the CLC 1969 alone did not suffice to provide full compensation and an international oil pollution compensation fund (IOPC fund) should also be developed to contribute to the compensation.[12] Thus the International Convention on the Establishment of an International Fund for Compensation for Oil Pollution Damage was concluded in 1971 (Fund Convention 1971).[13] The Practice of the IOPC fund has considered losses or damage suffered by persons who depend directly on earnings from coastal or sea-related activities as recoverable economic losses.[14] This practice was later accepted by the Protocol of 1992 to Amend the International Convention on Civil Liability for Oil Pollution Damage 1969 (1992 CLC)[15] and the Protocol of 1992 to Amend the International Convention on the Establishment of an International Fund for Compensation for Oil Pollution Damage (1992 Fund Convention).[16] As incorporated into Article I(6) of the 1992 CLC, pollution damage refers to:

8 Section 5, the Act on Compensation for Environmental Damage.

9 Art. I(6) of the CLC 1969, 973 UNTS 3, RMC I. 7.30, II. 7.30.

10 Sandvik & Suikkari 1997, p. 62 and Mason 2003, p. 3.

11 Wilkinson 1993, p. 83.

12 Wang 2008, p. 75.

13 International Convention on the Establishment of an International Fund for Compensation for Oil Pollution Damage, 1110 unts 57, Cmnd 5061.

14 Wetterstein 1997, p. 37.

15 The Protocol of 1992 to Amend the International Convention on Civil Liability for Oil Pollution Damage 1969 (1992 CLC) , Misc 36 (1994), Cm 2657, RMC I, 7.51, II. 7.51.

16 The Protocol of 1992 to Amend the International Convention on the Establishment of an International Fund for Compensation for Oil Pollution Damage, RMC I.7.111, II,1,7,111, Misc 37 (1994), Cm 2658.

> (a) loss or damage caused outside the ship carrying oil by contamination resulting from the escape or discharge of oil from the ship, wherever such escape or discharge may occur provided that compensation for impairment of the environment other than loss of profit from such impairment shall be limited to costs of reasonable measures of reinstatement actually undertaken or to be undertaken;
>
> (b) the costs of preventive measures and further loss or damage caused by preventive measures.

Scholars deduced from the formulation that loss of profit from impairment of the environment is recoverable, which encompasses not only consequential loss claims but also claims for pure economic loss.[17] According to the definition, *impairment of the environment* is also compensable and is assessed via reasonable measures of reinstatement which is actually undertaken or to be undertaken. This formulation still aroused criticism over the difficulties in determining 'reasonable' measures and the reluctance to compensate spills producing 'irreparable environmental damage'.[18] Similar prescriptions can also be found in the 1979 Convention on Long-Range Transboundary Air Pollution (LRTAP)[19] and in the 1977 London Convention on Civil Liability for Damage from Pollution by Oil resulting from Exploration for and Exploitation of Seabed Mineral Resources (1977 London Convention).[20] In such definitions, the damage is confined to a particular environment (such as air) or resulted from specific sources (such as oil). Not only damage to the environment, but also traditional damage such as personal injury and property damage are included in the definition, either explicitly (in the LRTAP) or implicitly (in the 1977 London Convention).

2.1.1.3. *Damage to the Environment, Impairment of the Environment and other Related Concepts*

In addition to the notions of 'environmental damage' and 'pollution damage', some of the relevant concepts are also used by international and national legal

17 Consequential loss claims refers to loss of earnings by owners or users of property contaminated by oil; pure economic loss means loss of earnings suffered by parties whose property has not been damaged, such as coastal hoteliers, fishery concerns. See Mason 2003, p. 3 and Wetterstein 1997, p. 38.

18 Sandvik & Suikkari 1997, p. 65; Maffei 1991, p. 390.

19 LRTAP defines the term 'air pollution' as: 'the introduction by man, directly or indirectly, of substances or energy into the air resulting in deleterious effects of such a nature as to endanger human health, harm living resources and ecosystems, material property and impair or interfere with amenities and other legitimate uses of the environment.' Art. 1(a), Convention on Long-Range Transboundary Air Pollution, 18 International Legal Materials 1442 (1979).

20 In this convention, 'pollution damage' means: 'loss or damage outside the installation caused by contamination resulting from the escape or discharge of oil from the installation and includes the cost of preventive measures and further loss or damage outside the installation caused by preventive measures.' Art. 1.6 of the 1977 London Convention, 14 International Legal Materials 1450(1977).

documents. 'Damage to the environment' is used in the rules concerning genetically modified organisms in the English Environmental Protection Act 1990. According to the act, 'Damage to the environment' is 'caused by the presence in the environment of genetically modified organisms which have (or of a single such organism which has) escaped or been released from a person's control and are (or is) capable of causing harm to the living organisms supported by the environment.'[21] Hence the concept 'damage to the environment' only refers to harm to the living organisms supported by the environment but not to traditional damage suffered by individuals.

The 1993 Lugano Convention on Civil Liability for Damage Resulting from Activities Dangerous to the Environment (Lugano Convention) makes a distinction between the 'impairment of the environment' and 'damage to property', 'personal injury' and 'loss of life'.[22] Since damage to property is excluded from impairment of the environment, it can be deduced that the latter is confined to damage to the environmental components that are not appropriated. The assessment of 'impairment of the environment' is determined by reinstatement measures actually undertaken or to be undertaken. The same differentiation can be found in the 1992 CLC and the amended proposal for an EC Council Directive on Civil Liability for Damage Caused by Waste.[23]

In the Convention on Civil Liability for Damage Caused during Carriage of Dangerous Goods by Road, Rail and Inland Navigation Vessels (CRTD), the term 'loss or damage by contamination to the environment' is used. It limits 'compensation for impairment of the environment other than for loss of profit' to 'costs of reasonable measures of reinstatement actually undertaken or to be undertaken'.[24] This makes it a similar concept to that under the Lugano Convention and the 1992 CLC. In some conventions, there is even no specific heading for such damage. However, they may include some components which are relevant for damage to the environment. For example, in the Basel Protocol on Liability and Compensation for Damage Resulting from Transboundary Movements of Hazardous Waste and Their Disposal, no specific name is given to such damage but it allows the compensation for 'costs of measures of reinstatement of the impaired environment'.[25]

21 Art. 107(3) of English Environmental Protection Act 1990.

22 Art. 2(7) of Lugano Convention.

23 Art. 2.1, Proposal for a Directive on Civil Liability for Damage Caused by Waste, *OJ* C251/3 (1989); as amended COM(91) 219 final *OJ* 1991 C192/6 (23 July 1991).

24 Art. 1.10 of the Convention on Civil Liability for Damage Caused during Carriage of Dangerous Goods by Road, Rail and Inland Navigation Vessels, UN Doc. ECE/TRANS/79.

25 Art. 2(2) of Basel Protocol on Liability and Compensation for Damage Resulting from Transboundary Movements of Hazardous Waste and their Disposal, UN Doc. UNEP/CHW.1/WG/1/9/2.

2.1.1.4. *Natural Resources Damage*

The notion of natural resources damage is mainly used in U.S. legislation. A typical example is the OPA, which prescribes six kinds of compensable damage: damage to natural resources, real or personal property, subsistence use, revenues, profits and earning capacity and public services.[26] In U.S. legislation, nature resources means land, fish, wildlife, biota, air, water, ground water, drinking water supplies, and other such resources 'belonging to, managed by, held in trust by, appertaining to, or otherwise controlled' by the United States, any state or local government or Indian tribe.[27] The scope of natural resources is defined by their relationship to the public authority but not by the ownership. Thus not only the public natural resources, but also the privately owned ones are encompassed under the heading of 'natural resources damage'.[28] Legislation gives tentative guidance on the evaluation of natural resource damage, and the damages contain three parts:

a. the cost of restoring, rehabilitating, replacing, or acquiring the equivalent of, the damaged natural resources;
b. the diminution in value of those natural resources pending restoration; plus
c. the reasonable cost of assessing those damages.[29]

2.1.2. Concepts in Literature

In addition to the legislation, scholars also developed concepts to define this new type of damage, of which, 'environmental damage', 'ecological damage', and 'natural resources damage' are again the most frequently used.

2.1.2.1. *Environmental Damage and Pollution Damage*

'Environmental damage' can be defined in at least two different ways. In the broader sense, it refers to the damage caused via the environment, including not only damage to public natural resources but also damage to the owned parts of the environment and even consequential losses, such as pure economic loss, cleanup costs and personal injury.[30] This concept is not very specific – without differentiation damage to the environment from other traditional environment-related damage. In the narrower sense, damage to persons or property is excluded. The term 'environmental damage' means 'damage done to the components of the environment, as well as the loss of diminution of environmental values caused by

26 33 U.S.C. §2702(b)(2).
27 15 C.F.R. §990.30.
28 Privately-owned natural resources may be relevant when there is a 'substantial degree of government regulation, management, or other form of control over the property'. Pure private natural resources are not incorporated. See Boyd 2003, p. 265.
29 33 U.S.C. §2706(d)(1).
30 Brans 2001, p. 13.

the deterioration or destruction of such components.'[31] To avoid ambiguity, several sub-concepts are developed, such as damage to the environmental itself, damage to the environment *per se*, pure environmental damage and impairment of the environment. They are used in contrast to damage to the privately-owned environmental components. Thus, though some privately-owned environmental components have significant ecological value, they are still excluded from those sub-concepts.

The term 'pollution damage' is often used interchangeably with the term environmental damage in the literature.[32] An examination of the history of legislation concerning pollution reveals the shift in defining pollution damage: a first generation of definitions focuses on activities influencing the usefulness of the resource to man, while a second generation shifts its emphasis to resources' ecological effects.[33] As with 'environmental damage' in the broader sense, the concept of pollution damage includes both damage to the environment and traditional damage (personal injury and property damage) suffered by individuals.

2.1.2.2. *Natural Resources Damage*

'Natural resources damage' is a notion used mostly by American scholars and is based on the U.S. legislation (especially the CERCLA and the OPA), which gives elaborate rules on its scope and assessment. It refers to damage to natural resources that are 'managed by, held in trust, appertaining to or otherwise controlled' by federal and state trustees.[34] In the U.S., some natural resources, such as land and minerals, can be purely private property, while many other important resources, including wildlife, birds, and fish, as well as most rivers and coastal waters are not 'owned' by anyone.[35] However, they may both be of enormous ecological value to the general public and hence they are contained in the notion of 'natural resources damage'. Not only is the injury or destruction of natural resources included, the damage suffered by the public at large due to the injury or destruction of the resources also comprises an important part.[36]

2.1.2.3. *Ecological Damage*

The term 'ecological damage' is mainly used in Europe. In Europe, three approaches are used to define this term.[37] The first approach is based on the legal status of the

31 Barboza 1997, p. 76.

32 Larsson 1999, p. 158.

33 The first generation of definitions often focuses on damage caused by specific sources; contemporary definitions tend to be more general, addressing the whole environment, and shift their focus from human uses to ecological effects. Larsson 1999, p. 172, 175.

34 42 U.S.C. §9601(16); 33 U.S.C. §2701(20).

35 Brighton & Askman 1997, p. 177.

36 The damage suffered by the public means the loss of natural resources service, such as its ecological, esthetic functions, but not the pure economic losses from the damage to the natural resources. See Brans 2001, p. 21.

37 Brans 2001, p. 17-19.

environmental components — only damage to the natural resources not subject to property rights is included. The privately owned natural resources, which may have equally important ecological value, are excluded. According to the second approach, ecological damage is limited to the damage to the natural resources lacking market value. For example, when discussing the 'Patmos' case, Maffei defined 'ecological damage' as 'a kind of damage devoid of an economic value but based upon a legal interest of the State in protecting the quality of the public domain *per se*'.[38] Under this approach, many of un-owned natural resources and some owned natural resources, having an ecological value that is not reflected in the market value of the property, are included. However, the criterion is not very clear. Moreover, lack of market value is an important factor influencing assessment of the damage, but it is of secondary importance in determining whether this type of damage fits in the tort law system.[39] Under the third approach, the term 'ecological damage' refers to the damage caused to the environment, regardless of the existence of property rights. An example is that given by Professor M.R. Will who identifies some characteristics in the definition of ecological damage: 'a detrimental physical change of the environment, that does not — or not only — affect personal property, health or economic interests, but injures the public interest in the conservation and use of the environment.'[40] A similar opinion can also be seen in Bocken's article. Although he does not give a direct definition, when discussing the compensation of ecological damage in Belgium, he includes both 'damage to individual property which also presents a public interest in view of its ecological value' and 'damage to the ecosystem as such or to parts of the environment which have not been individually appropriated by man'.[41] The coverage of this approach is more complete, and the environment is regarded as a unity. Besides, the term 'pure ecological damage' is often used to point out the damage to the environmental components that have not been appropriated.[42]

2.2. *The Choice of the Concept*

As can be seen from the legal documents and academic research, a variety of concepts are used to denote the damage done to the environment, of which 'environmental damage', 'natural resources damage' and 'ecological damage' are the most frequently used. 'Pollution' and 'pollution damage' are terms often used interchangeably with 'environmental damage', and they are often connected with damage caused by certain sources or activities.[43] 'Damage to the environment per

38 Maffei 1991, p. 381.
39 Brans 2001, p. 18-19.
40 Marticke 1992, p. 28.
41 Bocken 1997, p. 150-152.
42 Bocken 1997, p 157.
43 Such as the definition of pollution (pollution damage) in conventions concerning the protection of the marine environment. See Art. 1(1) of the Convention for the Prevention of Marine Pollution from Land-Based Sources 1974, 1546 UNTS 119; 13 ILM 352 (1974); UKTS →

se', 'impairment of the environment' and 'pure ecological damage' are the sub-concepts of 'environmental damage' or 'ecological damage', carrying the same meaning: damage to the environment components not subject to private property rights. Taking into consideration the similar meaning of 'pollution damage' and 'environmental damage', and the limited scope of the sub-concepts, the choices of concepts in this research are made from 'environmental damage', 'natural resources damage' and 'ecological damage'. The first step towards this task is to compare the leadwords of these terms – environment, natural resources and ecology.

2.2.1. Environment, Natural Resources and Ecology

In a general sense, 'environment' is defined as: 'the surroundings or conditions in which a person, animal or plant lives or operates'. 'The environment' is defined as 'the natural world, as a whole or in a particular geographical area, especially as affected by human activity'.[44] In a legal sense, the term 'the environment' is usually defined by dividing it into different categories. Salter identifies three types: the natural environment (protection of environmental media), the manmade environment (including the cultural heritage), and the human environment (including regulations on food content, products, safety issues, leisure and economic health).[45] Legislation varies in the scope of 'the environment'. The term used in the New Zealand Environment Act of 1986 contains all these three categories:

1. ecosystems and their constituent parts;
2. all natural and physical resources;
3. the social, economic, aesthetic and cultural conditions which affect the environment or which are affected by changes to the environment.[46]

While in the English Environment Protection Act 1990, 'the environment' consists of the natural environment and manmade environmental parts.[47]

Brans discusses three types of definitions for 'the environment', according to the breadth of the scope: the narrow category is limited to natural resources; the broader one contains 'characteristic aspects of landscape and property forming part of the cultural heritage'; and an even broader one covers 'public services and uses

1978, No. 64; for the convention concerning air and atmosphere; see, for example, Art. 1(a) of the 1979 Geneva Convention on Long-range Transboundary Air Pollution, 1302 UNTS 217.

44 Soanes & Stevenson 2003, p. 581.

45 Cited in Larsson 1999, p. 156.

46 Art. 2, New Zealand Environment Act of 1986, No. 127, available at: <www.-legislation.govt.nz/act/public/1986/0127/latest/viewpdf.aspx>, last accessed on 8 May 2013.

47 According to Section 1(2) of English Environmental Protection Act, 'environment' is defined as: 'of all, or any, of the [media] the air, water and land; and the medium of air includes the air within buildings and the air within other natural or man-made structures above or below ground.'

stemming from the existence of natural resources'.[48] However, few legal instruments have extended so broadly as to include the last two categories.[49]

The term 'natural resources' means 'materials or substances occurring in nature which can be exploited for economic gain'.[50] Compared to 'the environment', which emphasizes the system equilibrium and quality of the nature, the term 'natural resources' focuses on the economic utility and sustaining the quantity of resources. It often contains specific kinds of resources, such as water, air, soil, flora and fauna. In the U.S., legislation defines the term 'natural resources' broadly to include 'land, fish, wildlife, biota, air, water, ground water, drinking water supplies, and other such resources belonging to, managed by, held in trust by, appertaining to, or otherwise controlled by the United States, any State or local government, or any foreign government.'[51] In addition to the specific elements, public uses or services of these elements, which are often neglected, are also contained in the concept of 'natural resources'.[52] The New Zealand Environment Act of 1986 uses the term 'natural and physical resources', which 'includes water, air, soil, minerals, hydrocarbons, and energy, all forms of flora and fauna (whether native to New Zealand or introduced) and any building, structure, machine, device, or other facility made by people.'[53] In this concept, public use and service are not included.

The ecology is defined as 'the branch of biology that deals with the relations of organisms to one another and to their physical surrounding'.[54] A highly relevant concept is the ecosystem, which is 'the meeting ground on which species interact, the integrated networks of biotic and abiotic elements through which materials and information flow, and that supports our continued existence on the planet'.[55]The Biological Diversity Convention defines ecosystem as 'a dynamic complex of plant, animal and micro-organism communities and their non-living environment interacting as a functional unit.' In the New Zealand Environment Act, ecosystem means: 'any system of interacting terrestrial or aquatic organisms within their natural and physical environment.'[56] Rather than emphasizing the specific

48 Brans 2001, p. 10.

49 Art. 2(7) of Lugano Convention; Art. 1(2) of the 1992 UN/ECE Convention on the Protection and Use of Transboundary Watercourses and International Lakes, 31 International Legal Materials 1312 fall into the second category; the U.S. domestic laws such as CERCLA and OPA fall into the third category.

50 Soanes & Stevenson 2003, p. 1172.

51 42 U.S.C. §9601(16).

52 'Service' provided by natural resources damage is explicitly contained when the legislation defines its goal. For example, Art. 1 of NOAA rule mentions the goal of OPA like this: 'The goal of the Oil Pollution Act of 1990 (OPA), 33 U.S.C. 2701 *et seq.*, is to make the environment and public whole for injuries to natural resources and services resulting from an incident involving a discharge or substantial threat of a discharge of oil (incident).' 15 C.F.R. §990.10. Thus not only the injury to the specific resource itself but also the loss of its service should be compensated.

53 Art. 2, New Zealand Environment Act of 1986.

54 Soanes & Stevenson 2003, p. 552.

55 Levin & Pacala 2003, p. 63.

56 Art. 2 of New Zealand Environment Act of 1986.

components, the notion 'ecosystem' focuses on the unity of the nature and the interaction of the components.

The terms 'environmental', 'ecological' and 'natural resources' have only slight distinctions. 'Environment' focuses on the quality of nature; 'ecology' emphasizes the value of the ecosystem in function; while 'natural resources' attaches more importance to the quantity of specific elements and their economic value. These three terms are overlapping and may even be the same in their extensions and forms of carrier. On the one hand, the extension of the environment depends on the center being surrounded. Thus the category of 'natural environment' contains the natural factors and their interactions, which has the similar extension to 'ecosystem'. On the other hand, if the term 'natural resources' is broadly explained to encompass public use and service, as is the case in the U.S. legislation, its scope is also analogous to that of the two other terms. Not only the economic value of a specific resource, but also the ecological, aesthetic and cultural value, embodied via the public use and service are included in the notion of 'natural resources'. Hence the choice of concepts in damage depends on custom and the consistency in the usage.

2.2.2. The Choice and Definition of Concepts

'Environmental damage' is a widely used concept. As discussed earlier, this term is usually used to incorporate all types of damage caused by an environmental tort, including both damage to the environment and traditional damage suffered by individuals. With such a broad scope, a fundamental distinction between environment-related types of damage, which can be covered in traditional tort law and damage to the environmental itself, which involves collective interests rather than individuals, is neglected.[57] 'Environmental damage' can also be defined with a narrower coverage, as is the case in some legal documents, such as the ELD and the Act on Compensation for Environmental Damage of Finland. However, in view of the wide use of its broader definition, such a term may be misleading sometimes.

'Natural resources damage' is mainly used in the U.S. However, in the U.S., a distinction is made between the concepts 'natural resources damage' and 'respond costs/removal costs'.[58] In case of hazardous waste/oil release or threat of release, the first action taken is an emergency measure to clean up and to prevent further damage (called respond costs and removal costs under the CERCLA/OPA). Only when the emergency measures failed to restitute the damaged environment, further restorations are needed and the natural resources damage assessment starts. Hence, both the concept of 'national resources damage' and 'respond costs/removals costs' are relevant when discussing damage to the environment. The use of the term 'natural resources damage' in this research hence may suggest excluding 'respond costs/removal costs'.

As far as the term 'ecological damage' is concerned, it might also be obscure since it can be used under three approaches. It is chosen in this research not because

57 Brans 2001, p. 12.
58 *Ibid.*, p. 67.

it is necessarily superior to the other terms, since none of them are clear in their scope without delimitation. The choice is mainly based on the fact that it is more widely used in Europe. In this research, the broadest approach is used to define it: both damage to the public natural resources and those being appropriated are included. However, because not all privately-owned natural resources are of interest to the general public, only those natural resources that have an ecological value that exceeds the personal interests of the owner should be included here.[59]

3. The Characteristics of Ecological Damage

3.1. *The Concept of Damage in Tort Law*

A fundamental question arises before analyzing the compensation system: is ecological damage different from traditional damage, so that a separate regime is necessary? It is of particular importance to ensure that the introduction of the concept 'ecological damage' is legally meaningful rather than only a political slogan. The first step towards answering this question is to clarify the meaning of the term 'damage'.

3.1.1. Damage and Relevant Concepts

There is no easy way to define 'damage'. In the Oxford Dictionary of English, 'damage' is defined as: 'physical harm that impairs the value, usefulness, of normal function of something.'[60] But as legal terminology, the word may carry many different meanings. There are some relevant notions which need clarification here: damages, loss, harm and injury. As discussed in *Swansea Corpn v Harpur*, 'the word "damages" and "damage" have more than one meaning, and great care has to be exercised in examining the context in which they severally appear.'[61] 'European codifications have rarely attempted to define it in detail, either generally or particularly.'[62] Nevertheless, a fundamental difference exists between damage and damages: damage means 'the interference with a legally recognized interest, whereas damages represent the sum awarded for the violation of such an interest'.[63]

The term 'damage' and the relevant terms, such as 'harm', 'loss' and 'injury' are not used consistently by scholars and judges. To define 'the environment' and 'environmental damage', Larsson discusses the separation between damage and compensable damage. From her point of view, damage to the environment is 'defined in the instruments of environmental law, and includes all adverse effects on man, his artifacts and the environment'; whereas compensable damage is 'defined in schemes of restitution and liability, and embraces only economic losses,

59 *Ibid.*, p. 19.

60 Soanes & Stevenson 2003, p. 436.

61 *Swansea Corpn v Harpur* [1912] 3 KB 493 at 505, CA, per Fletcher Moulton LJ.

62 Von Bar 2000, p. 6.

63 Deakin, Johnston & Markesinis 2003, p. 782.

or other harm expressed in economic terms.'[64] In her article, loss and harm are used in a non-legal context (economic terms); damage is a legal concept in environmental law rather than a liability rule. Thus any harm subject to regulation or other environmental instruments is considered to be damage; even it is not compensable under liability rules. According to Christian Von Bar, all European legal systems distinguish harm/loss and damage for the principle of prohibition of excessive liability. When analyzing compensation in respect of the sentimental value of things, he mentioned 'those supporting protecting the sentimental value of a thing by delict law regarded pain suffered upon the loss of a pet as damage: those against, only as loss.'[65] According to him, damage is a concept concerning liability, equal to 'compensable damage' as defined by Larsson. To clarify the role of harm in environmental law, Albert C. Lin deemed harm as a widely accepted justification for ordering damages and injunctive relief or the exercise of government regulation.[66] In his article, the term 'harm' was also used as a legal concept.

As far as the relationship between 'damage' and 'injury' is concerned, Scottish judges thought: '"injury" is limited to actionable wrong, while "damage", in contrast with injury, means loss or harm occurring in fact, whether actionable as an injury or not.'[67] A Canadian case defines the difference between these words differently: '"damage" has, I think, a more restricted meaning than "injury" as the latter word may mean a wrong which "damage" never does. The word "damage" includes "injury" when the latter word is used to denote physical harm to persons.'[68]

From the above it follows that some basic distinctions between damages, damage, loss/harm and injury exist. Damages are the sum awarded for violating the legally protected interest. Loss and harm are factual concepts, referring to the status of being harmed and not necessarily leading to compensation. 'Injury' can be either a legal or a factual concept, depending on the explanation. In spite of these discrepancies in defining 'damage' and relevant terms, there is some consistency. The concepts should be understood in a certain context: non-legal or legal context, and if in the latter context, in which domain of law. To clarify these concepts, in this thesis, 'harm' and 'loss' will be used as natural terms and the term 'damage' will be used as a concept in liability rules. In general, both the loss in contractual liability and tortuous liability are discussed under the heading of 'damage'. However, as far as the subject of this research, ecological damage, is concerned, only damage caused by tort is relevant here.

3.1.2. Definition of Damage

With the clarification of the nature of damage, it is still not an easy task to define 'damage'. No statutory definition of the term 'damage' is to be found in the

64 Larsson 1999, p. 158-159.
65 Von Bar 2000, p. 6-7.
66 Lin 2006, p. 901-902.
67 *Crofter Hand Woven Harris Tweed Co Ltd v Veitch* [1942] AC 435 at 442, HL, per Lord Simon LC.
68 *Provincial Secretary-Treasurer v York* (1957) 16DLR (2d) 198 at 201-205, NBCA, per Bridges.

European legal systems with the exception of Austria,[69] where 'damage is called every detriment which was inflicted on someone's property, rights or person.'[70] Scholars also try to define 'damage'. According to U. Magnus, it should be defined broadly to suit all situations: damage is 'any negative modification in the injured party's legally protected sphere.'[71] Christian Von Bar bases damage on the principle of prohibition of excessive liability: law cannot impose damages on each and every act of carelessness; 'where and when damages should be awarded should be determined by the relationship between the violation of a right and the actual loss.'[72] Thus several factors are relevant to determine the existence of damage: whether loss is prerequisite and the scope of violated rights or interest. Generally speaking, loss is necessary for an action for damages, irrespective of whether damages are awarded only for violations of absolute rights and objects of legal protection or for all loss arising from conduct deemed 'fautif' by the courts.[73] However, the English law deserves more attention here, which comprises two kinds of wrongs: wrongs which are actionable *per se* and those for which proof of loss is an essential ingredient. For those actionable *per se*, any infringement is automatically wrongful whether or not any loss is suffered; and in response to that, nominal damages are awarded to vindicate the claimant's interest. For the latter category, including torts of negligence, nuisance, deceit and the economic torts, no cause of action arises unless loss is suffered by the claimant.[74]

There is no unity towards the scope of rights and interest, of which the violation can be regarded as compensable. As an example in European legal systems, the Austrian legislation limits damage to harm caused to property, rights, or persons, but the scope is still not clear in this expression. Little dispute arises over the compensability of individual's protected rights and interests but how about those of a collective nature? Can loss of sensitive value be regarded as damage to property? Difficulties also arise regarding infringements of personality rights and protected interests. To answer this question, and to understand to which rights and interests the harm can be recognized as damage, the categories of damage – pecuniary and non-pecuniary loss should be examined.

3.1.3. Pecuniary and Non-Pecuniary Losses

Pecuniary and non-pecuniary losses are two main types of damage, but the distinction is not always clear. In general, damages can be awarded if the injured party's patrimony is diminished. Non-pecuniary damage can be defined as losses 'that are suffered by damaging goods or interests which have in themselves no economic price or value on a financial market'[75] or 'which are no damage to a

69 Magnus 2001, p. 191.

70 Austrian General Civil Code, Art. 1293, cited in Magnus 2001, p. 10.

71 Magnus 2001, p. 191.

72 Von Bar 2000, p. 6-9.

73 Von Bar 2000, p. 9.

74 Tettenborn, Wilby & Bennett 2003, p 25.

75 Lindenbergh & Van Kippersluis 2009, p. 215.

person's assets or wealth or income and which are therefore incapable of being quantified in objective financial manner by reference to a market.'[76] The difference between the two kinds of damage lies in whether they can be translated into monetary terms directly.

As for the scope and assessment of the two types of damage, differences exist between common law systems and civil law systems. In common law systems, non-pecuniary loss can be divided into three subheadings: pain and suffering, loss of amenities and the loss of expectation of life. Whereas in civil law system, the non-pecuniary loss includes 'the sufferance, harm or injury caused to one's reputation and honor, disgraces and physical deformities, suppression of pleasures and satisfactions which life can provide to those who are mentally and physically healthy, and frustrations of all kinds.'[77] To further understand the scope of damage, an analysis will be made according to a few legal systems.

Article 6:95 of the Dutch Civil Code distinguishes between patrimonial damage and non-patrimonial damage, the latter being recoverable only to the extent that the law imposes an obligation to pay damages.[78] Pecuniary damage comprises all financial and material losses incurred, such as damage to goods and persons, loss of profit, impairment of earning capacity, the loss of value of real property due to a nuisance, the reasonable cost of preventing or mitigating damage and assessing costs.[79] Article 6: 106 BW allows the awards of non-pecuniary loss in the following situations: liable party causing damage intentionally physical personal injury, damage to the reputation of a living person or to the memory of a deceased person, harm to the person in any other way. Outside the ambit of tort law and contract law, non-pecuniary loss can also be awarded in case of unlawful detention, unlawful seizure of a drivers license by the police, invasion of privacy by unlawful registration of personal data and breach of a collective bargaining agreement.[80] Unlike the patrimonial damage, which is usually assessed through the diminution of value, non-patrimonial damage 'is assessed in accordance with the principles of equity.'[81]

In Germany, the main distinguishing feature between pecuniary and non-pecuniary damage is whether the lost resources have some economic value and whether they can be measured in terms of money or in a subjective way.[82] According to the German Civil Code, compensation for non-pecuniary loss is allowed without any restriction as far as restitution in kind is concerned, but it is

76 Rogers 2001, p. 246.

77 Benedek 1998, p. 612-614.

78 Art. 6:95, Dutch Civil Code (Burgerlijk Wetboek), 'Damage to be repaired pursuant to a legal obligation to pay damages consists of loss to property, rights and interests and any other prejudice, to the extent that the law confers a right to damages therefore.' A translation version can be seen: Warendorf, Thomas & Curry-Sumner 2009, p. 660.

79 Brans 2001, p. 256.

80 Wissink & Van Boom 2001, p. 155.

81 Magnus 2001, p. 148.

82 *Ibid.*, p. 94.

demanded only as provided by law as far as compensation in money is concerned.[83] The statute allows non-pecuniary damage in three cases only: for pain and suffering from bodily injuries in tort law, for ruined holidays under a package tour contract and for sex discrimination under an employment contract. Monetary compensation for grave infringements of the right of personality was further added by court.[84] Pecuniary damage is calculated by the difference between the present state of the victim's sphere and that which would have presently existed had the damage not occurred. To assess the non-pecuniary damage, the idea of 'fair compensation' or 'satisfaction' is introduced, which suggests a desire for a certain degree of uniformity of awards.[85]

The classification in England deserves more attention here. Pecuniary damage covers those losses which are concerned with a person's wealth, whether money, intangible property and tangible property.[86] The term 'non-pecuniary damage' is the remaining category, which makes its coverage much wider than that in other countries mentioned above. It covers, for example, 'pain and suffering' and 'loss of amenity' in personal injury cases; grief and bereavement, distress and humiliation, inconvenience, disappointment and loss of reputation.[87]

As can be seen from the above analysis, a violation of individuals' rights and interests, when accompanied with loss, can cause both pecuniary and non-pecuniary damage. As far as the collective interests are concerned, the violation of these rights and interests is not considered to be damage in any of these countries. Research about tort law in Austria, Belgium, England, France, Germany, Greece, Italy, the Netherlands, South Africa and the U.S. shows that compensation of collective damage is only a feature of French law, where procedural methods are provided to claim damages in the interest of others or in the common interest.[88]

It is less controversial to identify the scope of violated rights and interests for pecuniary damage, while the nations differ about the scope for non-pecuniary damage. The broadest coverage is that in England, which involves in personal injuries (pain and suffering), loss of amenity, bereavement, disappointment, distress, humiliation and loss of enjoyment, inconvenience and damage to reputation.[89] However, this only concerns damage arising from injuries to the body, health and personality of the injured party, which has limited significance in incorporating the harm to environmental interests as non-pecuniary damage.

One issue which deserves more attention here is that the borderline between pecuniary and non-pecuniary damage is not always clear. There are some losses on the borderline between those two types, such as pure loss of use, pure loss of

83 Arts. 249, 253, The German Civil Code (BGB), translation copy available at: <www.gesetze-im-internet.de/englisch_bgb/>, last accessed on 8 May 2013; see Magnus & Fedtke 2001, p. 109.

84 Magnus 2001, p. 94.

85 Markesinis & Unberath 2002, p. 916.

86 Magnus 2001, p. 59.

87 Tettenborn, Wilby & Bennett 2003, p. 80.

88 Magnus 2001, p. 202.

89 Tettenborn, Wilby & Bennett 2003, p. 80.

earning capacity, pure loss of holiday or free time, damage to things which remain unrepaired and frustrated expenses. Loss of use will be relevant to ecological damage. The loss of use may lead to pecuniary damage such as loss of profit for the time of repair, and non-pecuniary damage such as the amenities of the use. Solutions vary widely concerning compensation for pure loss of use. U.S. law provides for compensation irrespective of whether a substitute had been hired or profits were lost or not. As mentioned above, to indentify the loss of service (use) the injured resources could have provided is an important method to measure damage. Under English law, there is some reluctance to apply this nowadays in the case of private property. German law has adopted a split solution: lost use of essentially needed things such as a house, flat, car or ship is indemnified, whereas lost use of other things is not.[90]

3.2. *Ecological Damage: Does it Fit into the Traditional Tort Regimes?*

As has been discussed above, ecological damage contains damage to un-owned natural resources and damage to owned natural resources, which have an ecological value that exceeds the personal interests of the owner. Natural resources can provide many different ecological and human services. Both the public and individuals may suffer from the damage. Thus different kinds of rights and interests may be influenced by damage to the environment: damage to owned natural resources results in property loss of the owners for the diminution in market value of natural resources. The public suffers from a reduction of the service provided by both owned and un-owned natural resources. Human health and life may be injured. Those who depend on the damaged natural resources may suffer pure economic loss.

'Damage' in a legal context is a normative notion and not so much a factual one.[91] Thus it is not necessarily possible to consider the loss mentioned above as damage. Whether a loss can constitute damage is a legal question to be answered on the basis of legal norms. This research will analyze damage to owned natural resources and damage to un-owned natural resource respectively to see to what extent they fit into the concept of 'damage' in traditional tort law.

3.2.1. Damage to Owned Natural Resources

If an environmental accident hits the owned environmental components (natural resources), various kinds of loss may arise: personal injury and damage to property for the owners; pure economic losses suffered by those who depend on earnings from the environment, such as fishermen or hoteliers; the public is also impacted by the loss of ecological value.

90 Magnus 2001, p. 198.
91 Brans 2001, p. 253.

Cases of personal injury do not raise unusual problems under traditional tort law. Personal injury and loss of life can be measured by money, which means that it can be classified as pecuniary damage.

Legal practice varies considerably from country to country on the question of compensating pure economic losses; but traditionally there has been a reluctance to award compensation for it.[92] The trend of strict limitation has been relaxed. An example is the International Oil Pollution Compensation Fund established under the 1971 Fund Convention, which allows compensation for recoverable economic losses 'suffered by persons who depend directly on earnings from coastal or sea-related activities.'[93] The revised 1992 CLC and Fund Convention further clarified that loss of profit from impairment of the environment is also compensable.[94]

Although caused via the environment, and compensable, personal injury and pure economic losses are not damage to the environment, and thus cannot be defined as ecological damage. The owners' property loss is relevant and pecuniary: individual interests are intervened with and loss is assessed through a diminution of market value.

As discussed above, damage to owned natural resources is recoverable in traditional tort law; however, there are still some obstacles to full compensation and restoration for ecological damage.

Firstly, the owner is not obliged to sue for damage he suffered and to use the damages to restore the damaged environment in civil law. For the natural resources having ecological value that exceeds the interests of the private owner, it is the public at large that will suffer from the lost public service of natural resources.

Another key issue involves the measurement of damage. Both restitution in kind and compensation for lost value are available in tort law. The choice between them is primarily concerned with assessing damage. The European legal systems 'seek a solution by evaluating the interests concerned and incorporating all the facts of the case, instead of incorporating with inflexible doctrinal premises.'[95] 'English common law bases its decision on reasonableness'; continental European law addresses the problem at the starting point of fairness: restitution in kind is excluded 'if specific redress would be excessively onerous for the debtor'[96] or 'if restitution in kind is possible only through disproportionate outlay',[97] the top limits of restoration costs may be a new replacement,[98] percentages of the market value of the injured thing[99] or lost value.[100] Thus the availability of restoration of injured natural resources is subject to the limitation of its cost. Since the nature resources

92 Wetterstein 1997, p. 33.

93 *Ibid.*, p. 37-38.

94 Art. I, 6(a) of the 1992 International Convention on Civil Liability for Oil Pollution Damage, Misc 36 (1994), Cm 2657, RMC I. 7.51, ii. 7. 51. These conventions will be further discussed below in Chapter 6, Section 1.

95 Von Bar 2000, p. 151.

96 Art. 2058 II of Italian Civil Code, cited in Von Bar 2000, p. 153-154.

97 Art. 251(2) of BGB.

98 For example in Greece, see Von Bar 2000, p. 153-154.

99 German, Austrian, see Von Bar 2000, p. 154-155.

100 *Ibid.*, p. 156.

covered by ecological damage have ecological value in excess to its market value, the meaningful remedy for it is to restore the environment to the state it would be in had damage not happened, rather than monetary compensation for its market value.

Moreover, the problems in indentifying injurers and proving causation also prevent full compensation of ecological damage. For damage to the owned natural resources, the owners of injured resources, who suffer property loss, are the victims. As far as the injurers are concerned, problems may arise: causation is often not easy to establish, in view of a lack in scientific certainty between certain activities and losses. Thus the injurers who are responsible for certain damage are not easy to indentify. In traditional tort law, there are high standards about the level of probability that is necessary to establish causation. German law requires a high level close to certainty; in the Netherlands, courts have broad discretion on the level of probability to establish causation; in England, Ireland and Scotland, the claimant should prove that, 'the defendant's activity was, at least, a material contributor to the damage.'[101] While for ecological damage, whether some activities, such as emission of certain chemicals can lead to specific harm, say death of plants, reduction in fish production in certain area is often not scientifically certain. Even in the cases that causation is established, difficulties still appear when multiple actors are involved. Who and to what extent his activities contribute to the damage is not easy to identify.

3.2.2. Damage to Un-owned Natural Resources

As far as damage to un-owned natural resources is concerned, problems arise in integrating it into traditional classification of damage, since no individual interests are interfered with. Collective harm is seldom recognized as compensable under the European legal system. Thus it can neither be covered under the head of 'pecuniary damage' nor 'non-pecuniary damage'. To make this damage recoverable, it should be individualized. Restoration is such a method to transform the damage to un-owned natural resources into a pecuniary loss that is suffered by the person who took the measures.

3.2.2.1. Cost of Restoration

To ensure full compensation of ecological damage, it would be helpful to examine which values natural resources have and which are recoverable. American scholars have paid much attention to the values of natural resources. According to them, there are at least three kinds of values: use value, existence value, and intrinsic value. Use value is the value derived from people's actual use of the environment, which can be evaluated through market values, such as for fishing or hunting. Existence value is the value natural resources have to humans beyond use value, comprising three parts: option value, vicarious value and inter-temporal value.

[101] Hinteregger 2008, p. 609.

Intrinsic value is the value of natural resources independent of humans for their status as natural creatures or objects.[102] But scholars disagree on which values are compensable: some American scholars deem them all as recoverable;[103] while according to some European scholars, the compensable scope is limited. For example, existence value is a doubtable category, for everything may be assigned with existence value, and the value can be negative.[104] In spite of the differences, there is some consistency: in addition to the use value, some non-use values which cannot be transformed into money directly are also compensable. Restoration is such a method which makes compensation of both use value and non-use value possible: 'the environment is restored as far as possible so that fishing, berry-picking, swimming, etc. — the exercise of public rights — are possible again.'[105]

Although the individualization of collective harm seems constructive from a pragmatic point of view, some authors have therefore questioned the legal basis of this solution. Taking restoration measures does not give standing to public authorities or others to claim compensation for the cost incurred in taking restoration measures.[106] As far as the un-owned natural resources are concerned, no subject can be identified as the claimant based on ownership. Whether and to what extent, public authorities and NGOs can be authorized as claimants depends on the specific legislation and cases law in respective legal systems. In the U.S., public authorities are given standing on the basis of trusteeship and *parens patriae*. In the Netherlands, development of case law has granted NGOs and government standing to sue under some limitations.[107] However, the authorizations are based on the duties of government or other interests rather than property rights, which requires the legal development outside of the domain of traditional tort law.

3.2.2.2. *No Restoration*

Although granting standing to public authorities or NGOs can transform the loss into pecuniary damage, they are not always obligated to restore the injured natural resources. They may lack in incentives to take restoration measures, or willingness to recover such costs.

Moreover, there is irreparable damage to natural resources. 'Reasonability' or costs of the restoration measures also limit their availability: the reasonability criterion is adopted in many international treaties, such as in the 1992 protocol to the CLC (Article 2.3) and the CRTD (Article 1.10). In the U.S., restoration should not be disproportionally expensive. Thus if restoration is not possible or disproportionally expensive, the compensability of ecological damage arises again. Under this condition, the injured natural resources are equally valuable and no individual interests are directly intervened. However, such losses cannot be

102 Cross 1989, p. 280-296.
103 Cross 1989, p. 280.
104 Bergkamp 2001, p. 336.
105 Wetterstein 1997, p. 48.
106 Brans 2001, p. 260.
107 Brans 2001, p. 261.

classified as either pecuniary damage or non-pecuniary damage in traditional tort law. Hence it makes the losses unrecoverable.

3.2.3. Other Characteristics of Ecological Damage[108]

The above analysis shows that traditional tort law is neither sufficient in compensation for privately-owned natural resources nor public natural resources. One of the characteristics mentioned above which prevents full compensation through traditional tort law is that for ecological damage, the primary remedy is restoration but not monetary compensation. This section addresses some other characteristics ecological damage has which also influence the design of the compensation system.

Firstly, the emphasis on restoration leads to more government intervention. When ecological damage or imminent threat of damage occurs, measures should be taken to prevent further damage, not simply to clean up and restore the area. Liable parties are usually the first to know that damage or the threat of damage exists. They can thus take immediate preventive and cleanup measures. However, liable parties do not necessarily have the expertise to restore the natural resource effectively, or may tend to externalize costs by taking insufficient restoration measures. Public environmental protection authorities or the trustee of specific natural resources may have better information and expertise concerning certain natural resources and proper restoration. Therefore, it is a common practice for public authorities to intervene in environmental restoration.[109] A mix of administrative and civil elements may influence the operation of compensation instruments, and requires cooperation between public authorities and the providers of compensation instruments.

Secondly, ecological damage can be caused by different activities, including both industrial pollution and non-pollution activities. Under both CERCLA and OPA in the U.S., a release or substantial threat of release triggers response action and liability.[110] Under the ELD regime both contamination and non-contamination events can trigger restoration requirements for damage caused to protected species, natural habitats, and water.[111] This distinction illustrates the divergent policies concerning liability for cleanup and damage. The U.S. system only regulates damage caused by pollution but neglects to regulate non-pollution. The ELD established the same remedial criteria for damage caused by pollution and non-pollution events.[112] It is worth noting that the majority of existing compensation

108 This section is based on the article Faure & Liu 2012, p. 269-275.

109 See Boyd 2003, p. 266-267.

110 See 42 U.S.C §9604, 9607; 33 U.S.C. §2702.

111 The ELD defines damage to protected species, natural habitats, and water according to the extent and scope of resources damage. The ELD indentifies some activities (activities listed in Annex III) as triggers of environmental damage. These activities include non-pollution events, such as 'water abstraction and impoundment of water subject to prior authorization in pursuance of Directive 2000/60/EC'. Section 6 of Annex III , the ELD.

112 *Ibid.*, Annex III.

mechanisms covers only pollution events and non-pollution events are still largely unregulated in this area.[113]

Another issue worth noting is that an environmentally harmful activity usually does not create natural resources damage alone. Natural resources damage is often accompanied by personal injuries and property damage. In these situations, the injurers face liability not only for ecological damage but also for relevant traditional damage. For example, CERCLA and OPA holds firms liable for natural resources damage, as well as for all response costs, personal injury and property damage, and lost government revenue.[114] Identified firms are required to provide financial responsibility for all these liabilities, instead of exclusively for ecological damage.[115] Many compensation instruments are hence designed to compensate for damages comprehensively beyond ecological damage. When a specific instrument is analyzed, attention needs to be paid to its coverage: whether it covers only natural resources damage or also traditional damage.

A fourth characteristic involves the stages where ecological damage is created. An environmentally harmful activity can create environmental risks during its construction, operation and even in the post-closure phrase. Regulation is the primary instrument used before the operation stage. Before the start of a new project, the operator may need to apply for a license. Considering the potential environmental impacts of the project, an environmental impact assessment may be required. Such regulation mainly aims to prevent the creation of pollution, as the project developer is often required to account for potential environmental impacts of their activity, and if necessary take preventive actions.[116] Even with such regulatory requirements, project construction may still create ecological damage. Some regulations even go so far as to ask for compensation for such potential damage at the beginning of the project. One example is the Habitats Directive. It requires Member States to take compensatory measures to ensure the environmental protection and overall coherence when a project is justified for overriding public interest but negatively impacts the environment.[117] The Commission published a non-binding guidance document on this issue.[118] According to this document, 'compensatory measures' need to be independent of

113 Bio Intelligence Serv.: Study on the Implementation Effectiveness of the Environmental Liability Directive (ELD) and Related Financial Security Issues 47 (2009), available at: <http://ec.europa.eu/environment/enveco/others/pdf/implementation_efficiency.pdf>, last accessed on 8 May 2013 [hereinafter Bio Intelligence Serv report 2009]. There are several explanations for this, such as insurers' lack of experience concerning environmental insurance policies and the lack of public awareness that the ELD covers non-pollution events.

114 See Boyd 2003, p. 259.

115 *Ibid.*, p. 281.

116 See Council Directive 85/337/EEC, on the Assessment of the Effects of Certain Public and Private Projects on the Environment, preamble & Art. 5, *OJ* 1985 L175/40.

117 *Ibid.*, p. 11.

118 Commission Guidance Document on Art. 6(4) of the 'Habitats Directive' 92/43/EEC: Clarification of the Concepts of Alternative Solutions, Imperative Reasons of Overriding Public Interest, Compensatory Measures, Overall Coherence, Opinion of the Commission, (January 2007), available at: <http://ec.europa.eu/environment/nature/natura2000/-management/docs/art6/guidance_art6_4_ en.pdf>, last accessed on 8 May 2013.

the project,[119] and the measures consist of restoration or enhancement in existing sites, habitat recreation, and proposing a new site under the Habitats and Birds Directive.[120] This explanation shows that the term 'compensatory measures' has similar implications as remedial actions under the ELD.[121] After the construction of a facility or initiation of a project, ecological damage may be created during operation. This is the usual stage that liability rules and compensation instruments are discussed. Some industrial facilities may continue to pose environmental risks even during and after the closure of such facilities. For example, after the useful life of a nuclear power plant expires, the remaining radioactive substances and the facility itself continues to pose substantial risks for human beings and for the environment. Proper closure and post-closure monitoring are important requirements for certain industrial activities. Additionally, a facility may still create ecological damage after closure, and compensation instruments may be used to internalization costs. A detailed discussion of the interaction between regulation, liability rules and compensation instruments in all the three stages of activities is beyond the scope of this article. Since most practice (in terms of liability rules and established compensation instruments) focus on the damage caused during operation, and complexities of the two other scenarios, this research focuses on compensation models for ecological damage created during the operation of a plant.

4. Definition of Compensation

Different remedies can be awarded in case of an injury through a tort, including damages in terms of money and restoration in kind. Damages are defined as 'a money payment to compensate the victim, that is to say, to restore him, so far as money can, to the position he would have been in if the wrong complained of had not been committed' under the Principles of European Tort Law.[122] The Principles of European Tort Law also allow for restoration in kind as one remedy, if 'it is possible not too burdensome to the other party'.[123] Although both damages and restoration can be used as remedies to damage under tort, the former is the most common remedy. When the term 'compensation' is used, one usually thinks of monetary compensation. However, as far the ecological damage is concerned, 'compensation' should be explained more broadly.

The above discussion on the characteristics of ecological damage shows that the environment has different types of values and a restoration-based approach can provide a better assessment of its various values. In case of ecological damage, instead of awarding monetary compensation, the foremost issue is to clean up the pollution and to restore the damaged environment to its initial status. Only when restoration is not possible or disproportionately expensive, other monetary evaluation methods are used. This approach is followed in both the U.S. and the

119 *Ibid.*, p. 10.

120 *Ibid.*, p. 14.

121 Art. 2(11) of the ELD.

122 Magnus, Rogers & Moréteau 2005, p. 149-150.

123 Magnus, Rogers & Moréteau 2005, p. 159.

ELD to evaluate and remediate ecological damage. In the U.S., the CERCLA and the OPA establish the basic liability framework for ecological damage. For example, in case of release or imminent threat of release of oil, removals shall be taken to minimize or mitigate damage, including containment and removal of oil or a hazardous substance from water and shorelines.[124] If there is still damage left after removals, measures should be taken to evaluate natural resources damage, which, according to the discussion above, include restoration costs, interim losses and assessment costs.[125] The ELD covers three types of ecological damage: damage to protected species and natural habitats, water damage and land damage. In case of (threat) of damage, the operators shall take preventive measures and remedial measures.[126] Preventive measures means 'any measures taken in response to an event, act or omission that has created an imminent threat of environmental damage, with a view to preventing or minimizing that damage' and remedial measures means 'including mitigating or interim measures to restore, rehabilitate or replace damaged natural resources and/or impaired services, or to provide an equivalent alternative to those resources or services as foreseen in Annex II'. Hence, the ELD is not a pure civil liability regime but also focuses on the restoration of the damaged environment.

Although for ecological damage restoration is the focus, when restoration is not possible, damage cannot be recovered to the initial status or there is loss pending the restoration, monetary compensation is still needed. Hence the term 'restoration' is too narrow to cover the full remedies for ecological damage. Moreover, this thesis focuses on two functions of the legal regime: the *ex ante* prevention and *ex post* compensation for ecological damage. Hence the term 'compensation' is used in this thesis. It is worth noting, however, that this term should be explained broadly, including firstly to restore the environment, and when restoration is not possible or feasible, the monetary compensation can be used.

5. Conclusion

Damage caused via the environment has attracted much attention for some time; and with the value of the environment itself increasingly recognized by the public, the research and rules about damage to the environment began to develop. In national legislation and international treaties, different concepts are used for the new category: environmental damage, pollution damage, natural resources damage and damage to the environment. They are defined differently in international treaties and domestic legislation. 'Environmental damage' can refer to damage to the environment rather than property loss and personal injury, or damage to limited kinds of environmental components; 'pollution damage' is often defined broadly, containing both damage to the environment and property loss and personal injury; the term 'natural resources' is mainly used in the U.S. To differentiate damage to

124 33 U.S.C. §2701(30).

125 *Supra*, Section 2.1.1.4.

126 Arts. 5, 6 of the ELD.

public natural resources from owned natural resources, the concept 'impairment of the environment' was developed. In addition to the notions mentioned above, the term 'ecological damage' is also used in literature.

One task of this chapter is to choose between these concepts and delimitate the research subject. Although the differences in the legal status of natural resources may lead to distinctions in legal basis of *locus standi* and assessment, they are overweighed by the commonalities: both kinds of natural resources may have equal ecological values and deserve equal protection. Thus the chosen concept should contain both damage to public natural resources and damage to owned natural resources which have ecological value in excess of their market value to the owners. So the choice should be made between 'environment damage', 'natural resources damage' and 'ecological damage'. The literal comparison of the heading words: 'environment', 'natural resources' and 'ecology' shows that they are similar in complication and scope. Given the similarity, the custom of usage should be taken into account to determine the choice. Since environmental damage is often defined broadly and also contains other damage caused via the environment; natural resources damage is mainly used in the U.S. and is distinguished from 'removal costs/response costs', in this research, the term 'ecological damage' is chosen. It contains both the damage to un-owned components of the environment and to owned parts which have ecological value in excess of the personal value to the owners.

After the delimitation, this chapter tried to find out whether ecological damage can fit into traditional tort law. In traditional tort law, the relationship between legally protected rights and interests and actual loss determines whether a loss is qualified as damage. Both pecuniary and non-pecuniary damage requires that rights or interests are individually violated, which presents the major difficulty for harm to un-owned natural resources to be recognized as compensable damage. Restoration helps to individualize the intervened collective interests, and thus the damage can be classified as pecuniary damage. However, some damage is so serious that it is irreparable; costs of the measures also influence its availability. If this is the case, the damage is also not compensable. Moreover, the difficulties in indentifying the victims also prevent the compensation of damage to public natural resources. The public natural resources are owned by no one, so the claimant cannot be based on ownership, which is the usual case in property damage. Different legal systems provide various solutions: the public trust principle is used in the U.S. to authorize the public authority and Indian tribes to claim for the damage; in some legal systems, the environmental association can also be regarded as victims. For example, in the Netherlands, a public association can bring an action with the aim to protect collective interests, if it promotes the interests according to its articles of association.[127] Despite those practices, the debate and difficulties in indentifying victims of ecological damage are still ongoing, and the solutions may go beyond the boundaries of traditional tort law.

[127] Brans 2001, p. 44.

Even for the damage to owned natural resources, where claims may be based on ownership, problems will also arise when the owners are not willing to claim compensation. Assessment through diminution of market value also leaves much significant ecological value uncompensated. The proof of causation and identification of injurers still prove to be obstacles for compensation. The level of probability to establish causation in traditional tort law is usually quite high for the cases of ecological damage. Even if the causal link between one kind of activity and damage is established, the involvement of multiple actors still results in a problem to indentify whose activity and to what extent a certain activity is responsible for specific damage.

Moreover, some other characteristics also mean that traditional tort law is not a suitable solution for ecological damage, such as the involvement of government during the prevention and remediation of damage. However, this does not mean that all instruments discussed in this research should be specific for ecological damage. In reality, some liability principles (such as a total limit/strict liability) may apply to both ecological damage and traditional damage; although some special arrangement needs to be made for ecological damage, such as the links between administrative requirements, standing issues and assessment rules. Compensation mechanisms, such as insurance, may also cover both types of damage. Both pollution and non-pollution activities can lead to ecological damage and damage can be caused in different stages. In this research, focus is put on the damage caused by pollution during the operation stage.

This research discusses the compensation system for ecological damage. However, it should be borne in mind that the term 'compensation' should be defined broadly to include both restoration of the environment and if the restoration is not possible or feasible, monetary compensation.

Chapter 3

LEGAL RULES CONCERNING THE PREVENTION OF ECOLOGICAL DAMAGE: ECONOMIC OBSERVATIONS

1. Introduction

In addition to liability rules regulatory instruments have also been employed to prevent ecological damage, including both the traditional instruments such as environmental standards, licenses and the newly developed economic incentive instruments. The protection of traditional rights such as property rights and personal rights in environmental law is realized through the protection of the environment itself. However, no matter how exquisitely the prevention measures are designed, ecological damage still happens from time to time. Compensation of ecological damage is of great importance both from the perspective of justice and efficiency.

From the perspective of justice, activities leading to ecological damage cause loss to either individuals or the public. Property rights of the owners are interfered with when damage is caused to owned natural resources; as for damage to un-owned natural resources, the public use and enjoyment of resources are harmed. As Coleman said, 'the duty of wrongdoers in corrective justice is to repair the wrongful losses for which they are responsible.'[1] Thus the wrongdoers who caused ecological damage should pay for the loss. Less disputable than other concepts such as distributive justice and retributive justice, the concept of corrective justice has been labeled the core basis of liability according to traditional legal research.[2] One special feature of ecological damage is that in addition to compensating the victims, compensation rules also help to remedy the injured environment. It is because, on the one hand, the public interests are better recovered through the restoration of the environment; on the other hand, the environment has its own intrinsic values which deserve protection and restoration.

1 Coleman 1992, p. 324.
2 Bergkamp 2001, p. 112.

Rather than focusing on justice, the law and economic perspective attaches more attention to efficiency and to the deterrent function of tort law.[3] From this perspective, compensation is an instrument rather than a goal: it should be designed to promote efficiency or, in other words, to give preventive incentives to potential tortfeasors. To examine how compensation can improve efficiency, the theory of market failure needs to be discussed briefly here. According to modern welfare economics, a competitive market solution will correspond to a Pareto-efficient solution 'which maximizes, subject only to tastes, technology and initial endowments, that particular welfare function.'[4] In such an ideal competitive market, a legal intervention is not necessary. However, in the real world, markets fail to perform efficiently under certain conditions involving externalities, public goods, information failure and monopoly, as well as other malfunctions of competition.[5]

Many components of the environment are public goods. Two characteristics are often involved: non-exclusiveness, which means that excluding other's consumption is either impossible or prohibitively expensive, and non-rivalry, that is a good may be enjoyed by some without diminution of the amount available for others. Thus the problem of free riders can arise and transaction costs for bargaining can be high enough to prevent achieving efficiency. Therefore legal intervention would be helpful to promote efficiency in the market for public goods. This can precisely be the case with public natural resources.

A negative externality is the effect an injurer inflicts on a third party, with the costs not or only partially reflected in the price of the productive activity.[6] Pollution can be perceived as a typical externality: the negative influence on victims and the environment, the costs of which is not considered by the producers in their daily activities. Externalities can lead to market distortions, making the allocation of resources move away from Pareto-efficiency. Transforming an externality into a private good is a method of internalizing it. But intervention is not always necessary: some interactions that satisfy the definition of externality are efficiently handled in markets, while others are not.[7] The Coase Theorem provides an insightful summary of this:

> 'When transaction costs are zero, an efficient use of resources results from private bargaining, regardless of the legal assignment of property rights.
>
> When transaction costs are high enough to prevent bargaining, the efficient use of resources will depend on how property rights are assigned.'[8]

Ecological damage can be deemed to be of a reciprocal nature: polluters' activities injure the victim's rights; while at the same time, the presence of neighbors also limits the polluters' freedom. Thus whether the polluters have the right to pollute or

3 Faure & Grimeaud 2003, p. 19.
4 Bator 1958, p. 351.
5 Randall 1983, p. 131-132.
6 Zweifel & Tyran 1994, p. 44.
7 Randall 1983, p. 138.
8 Cooter & Ulen 2012, p. 85.

victims can get compensation for damage determines the allocation of property rights between them. Ecological damage often involves public natural resources, which means that there are no individual owners; moreover, the causation and burden of proof are serious obstacles to identifying the responsible parties. These factors make bargaining between the tortfeasors and victims involving ecological damage rather expensive. Thus assignment of property rights is essential to achieve efficiency. To set a proper threshold for compensation is such a way to allocate property rights.

Although both justice and efficiency require ecological damage to be compensated, their requirements for the scope of compensation are different. From the perspective of corrective justice, the wrongdoers should always give full compensation to the victim for the damage he caused. Compensation itself is the goal. As far as efficiency is concerned, full compensation does not always lead to efficiency. For example, in the negligent liability rules, the standard of due care depends on the marginal costs of taking more care and the marginal benefits of the reduction in damage. Thus the standard of care which is optimal according to the efficiency criterion may be in contrast to the requirement of corrective justice. For the latter, the wrongdoer needs to pay for the damage he caused regardless whether he is at fault or not.

The above analysis shows the importance of compensation rules for ecological damage. On the one hand, compensation promotes the achievement of justice; on the other hand, it also gives incentives to potential polluters to behave carefully to prevent the damage. To realize the goals of prevention and compensation of ecological damage, different instruments are developed, including regulation and liability rules.

1.1. *Regulation and Liability*

Law and economics literature presents intensive research on the choice between regulation and liability rules as instruments of environmental policy. From the perspective of law and economics, scholars try to find the instrument which can control pollution with the least social costs. The total social costs contain both the costs of taking precaution and the external damage costs.[9]

In choosing an optimal instrument to internalize externalities, law and economics scholars differentiate between so-called unilateral accident situations and bilateral accident situations.[10] A unilateral accident situation is one where only one party (usually identified as the injurer or tortfeasor) can influence the accident risk. In a bilateral accident situation, not only the injurer but also the other party (usually referred to as the victim) can *ex ante* take efficient measures to reduce either the probability of the accident or also the amount of the damage. As will be indicated below, the optimal instrument to internalize externalities depends on whether the accident situation is unilateral or bilateral. The choice of the instrument relies on the

9 Burrow 1999, p. 228.

10 See particularly Shavell 1980, p. 1-25.

comparison of either the social costs themselves or the dead weight costs every instrument may cause.[11]

Scholars compared regulation with liability and tried to find out which was preferable in particular circumstances. They established various models to analyze and compare the instruments and thus came to divergent conclusions. White and Wittman studied four different programs: a fixed price incentive for pollution reduction (effluent tax), a system of quantity control, and two types of liability rules (negligence and strict liability) in a model with imperfect information. On this basis, they drew the conclusion that a strict liability rule was preferable to both quantity and price controls, and negligence liability was either as good as or better than regulation. On the contrary, Innes came to another conclusion: he based his assumption on the model of unilateral accidents and argued that *ex-ante* regulation was more efficient than imposing *ex-post* liability for harm.[12] There are still other scholars who prefer to combine the two kinds of instruments. Burrow argues: 'Any attempt to establish the general theoretical superiority, in efficiency terms, of either of the instruments over the other is doomed to failure.'[13] This is because the superiority of the instruments is context specific. Thus, rather than contending that one instrument dominates another, some determinants of the relative desirability were given by Shavell.[14] According to him, four criteria determine the desirability of liability and regulation: the difference in knowledge about risky activities between private parties and a regulation authority, the possibility of private parties being incapable of paying for the full magnitude of harm, the threat of suit for harm and the administrative costs incurred by private parities and by the public.

These four determinants can also apply to the choice of the instrument for ecological damage. With regard to the first determinant, neither of the two instruments can dominate the other. Ecological damage can be caused by accidents and gradual pollution, dangerous activities and daily beneficial behaviors. It also involves highly technological problems. Thus for some dangerous activities, or big enterprises, the operators can posses much better information about risks and how to reduce them, such as the transnational chemical companies, and ships which transfer oil and dangerous substances. Some small operators may be lacking relevant information and technology. Compared to them, the government is superior in collecting information. Turning to the capability of compensation, the costs of ecological damage could be huge compared to the capital of the injurers, especially the small responsible operators. As for the third criterion, owing to the difficulties in determining *locus standi*, causation and the burden of proof, the threat of suit harm is often not high enough to give the efficient incentives. However, the liability rules are superior to regulation when administrative costs are considered, since in the former case, costs only arise when the suit is litigated; while monitoring

11 An example of the comparison of social costs is Burrow 1999, an example of the comparison of dead weight costs is White & Wittman 1983, p. 413-425.

12 Innes 2004, p. 29-48.

13 Burrow 1999, p. 227.

14 Shavell 1984a, p. 357-374.

and enforcement costs are always there no matter the private parties abide with it or not.

Those four determinants are designed from the public interest perspective. The private interest perspective also helps to compare regulation with liability rules. Although the authorities who take either of the instruments contend to serve the public interest to prevent or compensate ecological damage, neither can get rid of the influence of various lobbying groups. Generally speaking, regulation is more vulnerable to lobbying groups. Their effort may influence the setting of environmental standards and the enforcement of regulation instruments: environmental organizations ask for stringent standards and strict enforcement; industry will lobby for loose rules; and even lawyers can benefit from complicated regulation.[15] On the contrary, liability rules are applied by judges, who are more independent. Moreover, it only applies when there is litigation. Thus, from the private interest perspective, in an independent judicial system, it is less possible for liability rules to deviate from the public interest—giving the efficient incentives to prevent ecological damage. However, this may not be true when the judicial system is not independent from the administrative system and the former tends to align its interest with the latter.

According to the above analysis, as far as ecological damage is concerned, there is no optimal choice between regulation and liability rules.

In view of the deficiencies of both regulation and liability, neither leads private parities to exercise the socially desirable levels of care. Instead of competing with each other, 'it is often socially advantageous for the two means of controlling risk to be jointly employed-for parties to be required to satisfy a regulatory standard and also to face possible liability.'[16] In such a model, regulation can complement liability rules to give potential tortfeasors enough incentives to prevent damage; moreover, the best knowledge about risky activities can be used to determine the optimal level of care. Thus, to establish a system to prevent ecological damage, regulation and liability rules should be combined to give efficient incentives for prevention.

However, as mentioned above, both prevention and compensation are important goals of legal rules concerning ecological damage. Most traditional regulation instruments aim at prevention rather than compensation. Instead, liability rules play a role in both prevention and compensation. On the one hand, as discussed above, regulation alone cannot provide optimal incentives to prevent ecological damage, and liability rules offer a beneficial complement. On the other hand, liability rules help to compensate the victims and contribute to realize one indispensable value of law: corrective justice; remedying the damaged environment promotes the recovery of both public interests as well as the intrinsic value of the environment itself.

15 Lawyers may lobby for the level of legal complexity that maximizes their total income. For lawyers, their choice of legal complexity depends on the tradeoff between the increase in income from legal fees when the law becomes more complex versus the drop in income due to the decreasing amount of litigations. See White 1992, p. 381-395.

16 Shavell 1984b, p. 271.

Regulation of ecological damage has long been established as one part of environmental regulation and cannot be easily differentiated from regulation of broadly defined environmental damage. On the contrary, as discussed in the last chapter, substantial obstacles in traditional tort law fail to set up effective liability rules for ecological damage. In view of the significance of liability rules and inefficient traditional solutions, this chapter focuses on the liability rules for ecological damage rather than on regulation instruments.

2. Liability Rules for Ecological Damage

2.1. *The Criteria in Designing Optimal Liability Rules*

As mentioned in the introduction, liability rules are important from the perspective of both corrective justice and efficiency. This section tries to determine an optimal liability regime for ecological damage. To fulfill this objective, the first step is to outline the criteria in designing liability rules.

Compensation is a desirable consequence of liability; however, it is not always accepted as the goal of tort law.[17] The inherent structure of liability rules makes it impossible to provide compensation to every victim: the requirement of causation narrows down the scope of liable parties; only under strict liability, victims can always get compensation (if no defense applies); while under negligence rules, compensation is not awarded if no fault by the actors can be found. In addition, some other arrangements such as insurance and funds may be cheaper than liability in offering victims compensation. From the perspective of law and economics, the primary goal of liability rules is prevention but not compensation.[18] Compensation, although appreciated and also essential, needs to be realized through other institutions and will be addressed in the following chapters.

Law and economics literature has discussed intensively the deterrent function of tort law, starting with Calabresi's well-known book The Costs of Accidents.[19] An optimal liability regime should minimize the total social costs. Calabresi distinguished three different kinds of social costs: primary, secondary and tertiary costs. Primary costs are the costs of avoiding damage plus losses that finally occur. To minimize primary costs is a balance between expected damage and avoidance costs. The optimal status will be achieved when the marginal costs of accident avoidance is equal to the marginal benefits of decreasing the damage. Secondary costs are the costs of loss spreading, which requires risk spreading or the primary costs to be shifted to the less risk-averse party. Tertiary costs refer to the costs of administrating the legal system.[20] Although he admits that a hard distinction between those three types of costs would be arbitrary; the division is very practical in analyzing the functions of different costs reduction methods.[21] Hence, such a

17 Bergkamp 2001, p. 70-72.

18 Faure & Grimeaud 2003, p. 19.

19 Calabresi 1970.

20 Calabresi 1970, p. 340.

21 *Ibid.*, p. 29-31.

division is followed in discussing liability rules for ecological damage in this research. Taking into account the comprehensive requirements to realize efficiency, first-best solutions cannot always be reached.[22] For example, no efficient rules exist to create both incentives for an optimal care level and for adopting an optimal activity level in bilateral accidents.[23] It is even more complicated when considering the balance of the three types of costs: liability rules are not an efficient institution if distortive incentive effects are less costly that the saving costs of the judicial system and insurance coverage.[24] Thus, rather than establishing an optimal and certain liability regime, this section tries to find some criteria in liability design.

Both economic analysis and legal analysis are employed to design liability rules in this section. The legal issues related to liability rules are comprehensive and broad. In this section, only some main areas are discussed: standing, liable parties, liability standards, causation and the scope of compensation. To develop these aspects respectively, legal analysis is helpful to provide possible alternatives by discussing concrete legislation. Liability rules are complicated issues, which need to be considered not only alongside economic criteria, but also in accordance with legal rationales and orders in specific jurisdictions. The design of liability rules in this section follows a two-step procedure (with the exception of Section 2.2): firstly, an economic perspective is followed to decide under which conditions a possible alternative is more efficient in the minimization of total social costs; then the general rationale is applied under the context of ecological damage. Legal analysis is used either in offering possible alternatives or in testing the economic rationale in liability for ecological damage.

2.2. *Standing*

Locus standi is the 'right of an individual or a group … to have a court enter upon an adjudication of the issue brought before that court.'[25] It is the admissibility condition to file a case and the requirement to keep out the frivolous and inefficient procedures. There is a trade-off in the design of *locus standi*: on the one hand, the standing should not be too restrictive to become a substantial obstacle in pursuing judicial protection; on the other hand, the standing should not be enlarged so that it would lead to an inefficient use of the judicial system and over-deterrence for potential injurers.

In many legal systems, substantial barriers exist in recovering damages for injury to, destruction of, or loss of un-owned natural resources. In these systems, interference of individual interests is a condition of standing. However, for damage to public natural resources, it is the public at large rather than specific individuals that suffer the loss. The restrictive conditions for standing need to be relaxed to allow responding claims. This enlargement depends highly on the institutional design of specific countries. The legitimacy is the prerequisite for the design of

22 Schäfer 2000, p. 573.
23 Shavell 1987, p. 312.
24 Schäfer 2000, p. 576.
25 Stein 1979, p. 3.

alternatives for standing. Therefore instead of performing a theoretical economic analysis and then applying it to ecological damage, this section works the other way around by first exploring a few solutions and proposals in practice, and then testing those alternatives through economic analysis.

2.2.1. Legal Analysis

To examine the legal alternatives of standing for ecological damage, the practices in the U.S. and in the EU are explored briefly here. The difficulties in providing a legal solution for public natural resources have been overcome in the U.S. by federal statutory law. Even before the enacting of federal legislation, common law has offered state trustees limited *locus standi* for natural resources damage. However, compared to the federal law, the standing is more limited under common law.

2.2.1.1. Standing in the US

In the U.S., common law has provided standing for public authorities to bring a claim to the courts for public natural resources damages. A state may use the so-called public trust doctrine to file a litigation to recover damages for injured natural resources.[26] Under the public trust doctrine, a state can claim for the damage to natural resources on which it has trust duties for the interests of its citizen. It was only used to protect fish in navigable waters and land in the early stage and later expanded to non-navigable waters and streams as well as parks, land, wet lands and wild life.[27]

In addition, the doctrine of *parens patriae* also allows states to bring a suit to protect natural resources. This doctrine is originally used to endow the governments the power to act as guardian for persons legally unable to protect their own interests and was then expanded to allow the claim for injuries to quasi-sovereign interest.[28] The quasi-sovereign interests incorporate the health, and welfare of its citizens, that state's environment and natural resources, and the state's general economy.[29] Unlike the public trust doctrine which solves the problems of both standing and cause of action, the *parens patriae* doctrine does not provide cause of action for compensation for natural resources damage. Public nuisance is a usually revoked cause of action in lawsuit brought under *parens patriae* doctrine.[30]

In spite of the common law doctrines which provide solutions to standing for public natural resources, there are still some limitations to provide comprehensive protection for various resources. Firstly, in many states, a major limitation of the public trust doctrine is that it cannot be applied to pollution to groundwater and to

26 Kanner 2005, p. 88-99.
27 Lazarus 1986, p. 649-650.
28 Ratliff 1999, p. 1850-1851.
29 Brans 2001, p, 55.
30 Brans 2001, p. 56.

the aquifer.[31] Another restriction is that it does not involve the trusteeship of a federal authority.[32]

Compared to the common law, American federal legislation provides a broader scope of standing in case of damage to natural resources. In CERCLA and OPA, federal, state trustees and Indian tribes 'shall act on behalf of the pubic... as trustee of natural resources' to recover natural resources damage.[33] The involved natural resources contains: 'land, fish, wildlife, biota, air, water, ground water, drinking water supplies, and other resources belonging to, managed by, held in trust by, appertaining to, or otherwise controlled' by trustees.[34] Trusteeship thus provides enough interest links between public authorities and tribes with injured natural resources to overcome the difficulties in standing. Moreover, trusteeship also helps to attach specific authority to certain resources instead of a general authorization. Many federal public authorities have been authorized to act as the trustees of specific types of natural resources damage, such as Department of Agriculture, Department of commerce, Department of Defense, Department of Energy and Department of the Interior.[35] The clarification of rights in the federal legislation makes the claim litigation much easier.

It is interesting to see that, not only public authorities, but also individuals have a role to play. Citizen suits provide an avenue to individuals to pursue a claim for injury to public natural resources. However, they cannot file the claims for damages directly. The only avenue for them is to file a suit to compel a trustee to seek natural resources damages.[36]

2.2.1.2. *Standing in the EU*

In EU, there are also several approaches which might help to solve standing issues for ecological damage. The first approach to be mentioned is the standing under the ELD, which allows restoration and compensation for three types of environmental damage (ecological damage). The EU is also a party to the Convention on Access to Information, Public Participation in Decision-Making and Access to Justice in Environmental Matters (the Aarhus Convention),[37] which establishes a high standard of access to justice in environmental matters. A third approach is via the European Court of Human Rights (ECtHR), where environmental cases can be brought if they also concern the human rights of the plaintiffs.

The effort to establish a comprehensive compensation framework for environmental damage started long before the adoption of the ELD. The White

31 Schmid 2008, p. 488.
32 Kanner 2005, p. 129.
33 42 U.S.C. §9607(f), 33 U.S.C. §2702(b).
34 42 U.S.C. §9601(16), 33 U.S.C. §2701(21).
35 Lee 2002.
36 Kanner & Ziegler 2006, p. 134.
37 The Aarhus Convention was adopted on 25 June 1998 and entered into force on 30 October 2001. The Convention on Access to Information, Public Participation in Decision-Making and Access to Justice in Environmental Matters, 2161 UNTS 447.

Paper on Environmental Liability in the EU (White Paper) was adopted in 2000,[38] which aimed to adopt a liability regime not only for the damage to the environment, but also for traditional damage. For the damage to environment, the White Paper adopts a two-tier approach to grant standing. Since the protection of the environment is a public interest, the Member States (public authorities) have the primary duty to ensure the restoration of damage and claim for the costs. Public interest groups can file a claim if the Member State does not act at all or does not act properly. In other words, public interest groups can make a claim against the polluters if the state fails to do so. Moreover, in urgent cases, they can ask the court for an injunction or costs of prevention actions they have taken directly from the polluters without to request action by a public authority first.[39] Instead of establishing a civil liability regime, the Environmental Liability Directive creates a system of a more public law nature. It does not include liability for traditional damage any more but focuses on damage to the environment. Under the ELD, the operators have an obligation to take preventive action or remedial action in case of (an imminent threat of) damage and shall bear the costs of such actions.[40] The public authorities may also take actions themselves and if so, shall claim the costs from the liable operators.[41] Public interest groups can also play a role but to a more limited extent. According to Article 12, eligible 'natural or legal persons' can submit observations to the competent authorities relating to instances of (the threat of) environmental damage or request the competent authority to take action under the ELD.[42] This article limits 'natural or legal persons' to be parties:

> '(a) affected or likely to be affected by environmental damage or
> (b) having a sufficient interest in environmental decision making relating to the damage or, alternatively,
> (c) alleging the impairment of a right, where administrative procedural law of a Member State requires this as a precondition.'[43]

It is also mentioned specifically that environmental NGOs can be the parties mentioned above.

> 'The interest of any non-governmental organization promoting environmental protection and meeting any requirements under national law shall be deemed sufficient for the purpose of subparagraph (b). Such organizations shall also be deemed to have rights capable of being impaired for the purpose of subparagraph (c).'[44]

Hence eligible environmental NGOs can also submit observations or request the competent authority to take action under the ELD. The public authorities' decision,

[38] EC, White Paper on Environmental Liability, COM(2000) 66 final, 9 February 2000.
[39] EC, White Paper on Environmental Liability, COM(2000), 66 final, Section 4.7.
[40] Arts. 5, 6 of the ELD.
[41] Art. 8 of the ELD.
[42] Art. 12(1) of the ELD.
[43] *Ibid.*
[44] *Ibid.*

act or failure to act is subject to review: the persons mentioned above 'shall have access to a court or other independent and impartial public body competent to review the procedural and substantive legality of the decisions, acts or failure to act of the competent authority under this Directive.'[45] This mirrors the rule under the Aarhus Convention.[46] This broad procedural right of environmental NGOs, however, can be narrowed by the following paragraph:

> 'This Directive Shall be without prejudice to any provisions of national law which regulate access to justice and those which require that administrative review procedures be exhausted prior to recourse to judicial proceedings.'[47]

Therefore the procedural right of NGOs to sue public authorities in the court depends to a large extent on the domestic law of the Member States.[48] Moreover, the ELD does not authorize the NGOs to take action directly against the polluters.

Another approach to solve the standing issue in environmental cases is via the Aarhus Convention. The Aarhus Convention establishes three pillars on procedural rights concerning environmental matters: access to information, public participation and access to justice. According to the Aarhus Convention, the public can request environmental information from the government.[49] The public means 'one or more natural or legal persons, and, in accordance with national legislation or practice, their associations, organizations or groups.'[50] Hence, to be able to ask environmental information, the public does not need to be affected, to have a sufficient interest at stake or to allege the impairment of a right. Furthermore, Member States shall ensure the public concerned (the public affected or likely to be affected by, or having an interest in, the environmental decision making) the right to participate in decisions on environmental related activities.[51]

For access to justice, it allows two types of claims. The first type allows 'members of the public concerned (a) having a sufficient interest or, alternatively, (b) maintaining impairment of a right, where the administrative procedural law of a party requires this as a precondition have access to a review procedure ... to challenge the substantive and procedural legality of any decision, act or omission subject to' relevant provisions under the Aarhus convention.[52] It also prescribes the standing of NGOs explicitly: 'any non-governmental organization meeting the requirements referred to in article 2, paragraph 5 (non-governmental organizations promoting environmental protection and meeting any requirements under national law)' can be deemed as the eligible party.[53] Under the second type, 'where they

45 Art. 13(1) of the ELD.
46 Art. 9(2) of the Aarhus Convention.
47 Art. 13(2) of the ELD.
48 For the discussion about NGOs' standing to sue under the ELD, see Winter, Jans, Macrory & Krämer 2008, p. 174-175.
49 Art. 4 of the Aarhus Convention.
50 Art. 2(4) of the Aarhus Convention.
51 Art. 6 of the Aarhus Convention.
52 Art. 9(2) of the Aarhus Convention.
53 *Ibid.*

meet the criteria, if any, laid down in its national law, members of the public have access to administrative or judicial procedures to challenge acts and omissions by private persons and public authorities which contravene provisions of its national law relating to the environment.'[54] Therefore if private parties violate national environmental law and the members of the public satisfy the requirements of domestic law, the Aarhus Convention allows the public to take action against the private parties.[55] To ensure that the domestic requirement does not seriously impede the right of access to justice, the Convention further requires that the procedure 'shall provide adequate and effective remedies, including injunctive relief as appropriate, and be fair, equitable, timely and not prohibitively expensive.'[56] Therefore, the standing of NGOs is broader under the Aarhus Convention than under the ELD. The NGOs can only take action against public authorities, but not the polluters themselves under the ELD. However, both types of actions can be taken under the Aarhus Convention.

The EU signed the Aarhus Convention in 1998 and formally adopted it in 2005 via the Council Decision 2005/370/EC.[57] After the signing of the Aarhus Convention, the EU started to align its legal framework with the requirements of the Aarhus Convention. The first two pillars of the Aarhus Convention are transposed into EU law through two directives: 2003/4/EC (Access to Information Directive)[58] and 2003/35/EC (Public Participation Directive).[59] The preamble to Directive 2003/35 explicitly refers to the access to justice provision of the Aarhus Convention. It guarantees access to justice in two areas: cases related to Environmental Impact Assessment as regulated by Council Directive 85/337[60] and cases related to integrated pollution prevention and control, as regulated by Directive 96/61.[61] There have already been several cases adjudicated in the ECJ according to the 2003/35 Directive, concerning the requirements for NGOs to bring

54 Art. 9(3) of the Aarhus Convention.

55 This provision creates a challenge for many members of the Aarhus Convention. Difficulties exist regarding its interpretation. See Pánovics 2010, p. 142-143.

56 Art. 9(4) of the Aarhus Convention.

57 Council Decision 2005/370/EC, of 17 February 2005 on the Conclusion, on Behalf of the European Community, of the Convention on Access to Information, Public Participation in Decision-making and Access to Justice in Environmental Matters, *OJ* 2005 L124.

58 Directive 2003/4/EC of the Parliament and of the Council of 28 January 2003 on Public Access to Environental Information and Repealing Council Directive 90/313/EEC, *OJ* 2003 L41/26.

59 Directive 2003/35/EC of the European Parliament and of the Council of 26 May 2003 Providing for Public Particpation in Respect of the Drawing up of Certain Plans and Programes Relating to the Environment and Amending with regard to Public Participation and Access to Justice, *OJ* 2003 L156/17.

60 *OJ* L175 of 5 July 1985, amended by Directive 97/11/EC, *OJ* 1997 L73 and now revised and superseded by Directive 2011/92/EU of the European Parliament and of the Council of 13 December 2011 on the Assessment of the Effects of Certain Public and Private Projects on the Environment *OJ* 2012 L26/1.

61 *OJ* L257 of 10 October 1996, now Directive 2008/1/EC of the European Parliament and of the Council of 15 January 2008 concerning Intergrated Pollution Prevention and Control, *OJ* 2008 L24/8.

a case to court. For example, the case *Djurgården-Lilla Värtans Miljöskyddsförening v Stockholms kommun genom dess marknämnd* (the Djurgårdem case)[62] dealt with a preliminary ruling from a Swedish court on the admissibility of NGOs to appeal environmental decisions in the court. In this case, a local Swedish NGO appealed a decision concerning the construction of a tunnel but it was held to be inadmissible by the local court. This was because it failed to fulfil the conditions of the Swedish Environment Act, which required the NGO to have to have at least 2,000 members. The ECJ held that the public concerned as defined under the Directive 2003/35: 'must be able to have access to a review procedure to challenge the decision by which a body attached to a court of law of a Member State has given a ruling on a request for development consent, regardless of the role they might have played in the examination of that request by taking part in the procedure before that body and by expressing their views.' It further held that the Directive 'precludes a provision of national law which reserves the right to bring an appeal against a decision on projects which fall within the scope of that directive solely to environmental protection associations which have at least 2,000 members.' Hence, this case influenced national procedural law on the scope of NGO's standing. The case *Bund für Umwelt und Naturschutz Deutschland, Landesverband Nordrhein-Westfalen eV v Bezirksregierung Arnsberg* provides another similar example.[63] However, those cases concern only the challenges against decisions of public authorities involving environmental impact assessments. To what extent the NGOs can have access to justice against private parties for their activities influencing the environment is untouched.

The EU tried to adopt a package of legislation to implement the access to justice provisions of the Aarhus Convention, including the Proposal for a Regulation on the application of the provisions of the Convention on Access to information, Public Participation in Decision-making and Access to Justice in Environmental Matters to EC institutions and bodies,[64] the Proposal for a Council Decision on the conclusion, on behalf of the European Community, of the Convention on access to information, public participation in decision making and access to justice regarding environmental matters,[65] and the Proposal for a Directive on access to justice in environmental matters.[66] The Proposal of a Regulation tried to implement the Aarhus Convention with respect to EU institutions and bodies. The

62 Case C-263/08 *Djurgården-Lilla Värtans* [2009] ECR I-9967 of 15 October 2009.

63 Case C-115/09, *Bund für Umwelt und Naturschutz Deutschland, Landesverband Nordrhein-Westfalen eV v Bezirksregierung Arnsberg* [2011] ECR 1-03673.

64 Comission Proposal for a Regulation on the application of the Provisions of the Convention on Access to information, Public Participation in Decision-making and Access to Justice in Environmental Matters to EC institutions and bodies, COM(2003) 622 final.

65 The Proposal for a Council Decision on the Conclusion, on Behalf of the European Community, of the Convention on Access to Information, Public Participation in Decision Making and Access to Justice regarding Eenvironmental Matters, COM(2003) 625 final.

66 Commission Proposal for a Directive on Access to Justice in Environmental Matters, COM(2003) 624 final, 24 October 2003.

Regulation was adopted in 2006[67] and has lead to hot discussions about its sufficiency and effectiveness among scholars.[68] The Proposal of a Directive tried to implement the Convention in Member States. This research focuses on liability for ecological damage. Therefore the standing to take action against polluters who cause ecological damage is more relevant here and this section focuses on the Proposal of a Directive, but not on the Regulation. The strong opposition from Member States results in the proposal being, politically speaking, considered 'dead'[69] and this directive does not seem likely to be adopted in any time soon.[70] However, it has some interesting provisions for the standing issue of environmental NGOs.

The Proposal of a Directive allows the members of the public to have access to environmental proceedings against private persons.[71] Therefore, members of the public can take action directly against the private persons. This proposal defines the eligible 'members of the public' in the same ways as the Aarhus Convention: they need to either 'have a sufficient interest or maintain the impairment of a right', where required by law.[72] The Proposal of a Directive distinguishes NGOs from the 'members of the public'. More precisely, it introduces the concept of a 'qualified entity', which means 'any association, organisation or group, which has the objective to protect the environment and is recognized according to the procedure laid down in Article 9.'[73] Environmental NGOs hence fall into the category of 'qualified entities'. Even if they do not satisfy the requirements for 'members of the public' mentioned above, 'Member States shall ensure that qualified entities recognised in accordance with Article 9 have access to environmental proceedings, including relief ... if the matter of review in respect of which an action is brought is covered specifically by the statutory activities of the qualified entity and the review falls within the specific geographical area of activities of the entity.'[74] Article 8 and 9 establish substantive criteria and procedural requirements for an entity to be recognized as a 'qualified entity' to have access to justice. These restrictive provisions caused concerns after the publication of the Proposal of the Directive. In its first reading, the European Parliament proposed to broaden the definition of 'qualified entity'.[75] The European Economic and Social Committee also expressed its comments on the concept of qualified entity and tried to extend it to legal persons

67 Council Regulation No. 1367/2006 on the application of the provisions of the Convention on Access to information, Public Participation in Decision-making and Access to Justice in Environmental Matters to Community institutions and bodies, *OJ* L264/13.

68 Crossen & Niessen 2007; Poncelet 2012, p. 303-307.

69 Jans & Vedder 2012, p. 236.

70 Pánovics 2010, p. 135; Krämer 2009, p. 17.

71 Commission Proposal for a Directive on Access to Justice in Environmental Matters, Art. 3.

72 Commission Proposal for a Directive on Access to Justice in Environmental Matters, Art. 4(1).

73 Commission Proposal for a Directive on Access to Justice in Environmental Matters, Art. 2(1)(c).

74 Commission Proposal for a Directive on Access to Justice in Environmental Matters, Art. 5.

75 European Parliament Legislative Resolution on the Proposal for a European Parliament and Council Directive on Access to Justice in Environmental Matters, *OJ* 2004 C103 E/626.

having environmental protection as one of their objectives.[76] In addition to the restrictive definition of qualified entities (NGOs), another important issue is which type of cases NGOs can bring in the court, especially against private parties. Can they challenge the legitimacy of the private parties' activities and seek for injunction or can they claim for monetary compensation if the private parties cause environmental damage? To answer these questions, Article 3 deserves more attention here:

> 'Member States shall ensure that members of the public, where they meet the criteria laid down in national law, have access to environmental proceedings in order to challenge acts and omissions by private persons which are in breach of environmental law.'[77]

This provision uses the expression 'access to environmental proceedings' and the term 'environmental proceedings' is defined broadly to contain 'the administrative or judicial review proceedings in environmental matters other than proceedings in criminal matters.'[78] A potential limit exists with the aim of the proceedings: challenging 'acts and omissions by private persons which are in breach of environmental law.'[79] However, ecological damage does not always involve a breach of environmental law. If ecological damage is caused by a lawful action of private persons, whether members of the public can claim for the damage in the court is less clear. Moreover, if the members of the public want to bring environmental cases in the court, they should also 'meet the criteria laid down in national law.'[80] Therefore the issue of access to justice against private persons still lies to a large extent in the hands of Member States. This is because the possibility for private persons to challenge acts by other private persons in courts is mainly the competence of Member States. According to the Subsidiary principle, this proposal does not want to prejudge the detailed provisions to be laid down by Member States.[81]

Another approach to bring an environmental case to the court in Europe is through the ECtHR. The ECtHR is established under the European Convention of Human Rights. Both procedural and substantive rights are important for a case to be accepted by the ECtHR. For the procedural right, Article 34 of the convention allows persons, NGOs or group of individuals to bring a case to the court. The applicant should be 'the victim of a violation by one of the High Contracting Parties

76 Opinion of the European Economic and Social Committee on the Proposal for a Directive of the European Parliament and of the Council on Access to Justice in Environmental Matters, *OJ* 2004 C117/55.

77 Commission Proposal for a Directive on Access to Justice in Environmental Matters, Art. 3.

78 Commission Proposal for a Directive on Access to Justice in Environmental Matters, Art. 2(1)(f).

79 Commission Proposal for a Directive on Access to Justice in Environmental Matters, Art. 3.

80 *Ibid.*

81 Commission Proposal for a Directive on Access to Justice in Environmental Matters, Explanantory Memorandum, point 6.

of the rights set forth in the Convention or the Protocols hereto.'[82] In other words, parties whose right under the Convention are not affected cannot apply to the court. The European Commission on Human Rights has pointed out that:

> 'the applicant cannot complain as a representative for people in general, because the Convention does not permit such an actio popularis. The Commission is only required to examine the applicant's complaints that he himself was the victim of a violation.'[83]

This approach is still followed today. Though it is possible for NGOs to file a case in certain circumstances, their standing is very limited. For example, in *Asselbourg and Greenpeace v Luxembourg,* when deciding the standing of Greenpeace, the ECtHR held that an NGO cannot itself claim to be the victim of a violation, by only act as a representative of its members or employees.[84]

With regard to the substantive rights the applicant can resort to, several articles are relevant, such as Article 2 (right to life) and Article 3 (prohibition of torture). The most frequently invoked provision in environmental cases is Article 8, concerning the protection of private and family life. There have already been several cases brought to the ECtHR in which the judges declared that the pollution violates the human right to respect private and family life and the home.[85] However, these cases focused more on individual interests rather than on collective interests, usually concerning noise, industrial pollution causing danger to people's health and passive smoking. However, as for the damage to the environment itself, the role of the ECtHR is more limited. One interesting case is *Kyrtatos v Greece,*[86] where the plaintiffs complained that urban development in Tinos Island led to the destruction of their physical environment (such as its scenic beauty and changing from a natural habitat for wildlife to a tourist place) and hence affected their private life. However, the Court rejected their claims based on Article 8 because it found that the plaintiffs had not been directly affected. The court held that 'Neither Article 8 nor any of the other Articles of the Convention are specifically designed to provide general protection of the environment as such; to that effect, other international instruments and domestic legislation are more pertinent in dealing with this particular aspect.' The influence on the environment, including the damage to the birds and other protected species was not of such a nature as to directly affect their private life

82 Convention for the Protection of Human Rights and Fundamental Freedoms, Art. 34.

83 *X Association v Sweden* (1982) 28 DR 204, 206.

84 *Asselbourg and 78 Others and Greenpeace Luxemburg v Luxemburg* (App no 29121/95) ECtHR 29 June 1999.

85 For example, in the case of *López Ostra v Spain*, the serious pollution caused by fumes from a waste-treatment plant affected the plaintiff's well being and caused health problems. The European Court of Human Rights found it to violate the right to respect private and family life and the home (Art. 8 of the 1950 European Convention for the Protection of Human Rights and Fundamental Freedoms, 87 UNTS 103). *López Ostra v Spain* ECHR 9 December 1994, Ser, Application No. 16798/90; see also of *Öneryıldız v Turkey*, ECHR 18 June 2002, Application No. 48939/99; *Giacomelli v Italy*, ECHR 2 November 2006, Application No. 59909/00.

86 *Kyrtatos v Greece* (2005) 40 EHRR 16.

under Article 8. However, it is not always clear what parts of the environment are of sufficient interest to individuals so that damage to them could amount to an interference of private life. For example, in this case, the Court also held that if there were a destruction of a forest area in the vicinity of the applicants' house, the applicant's life might be found more directly affected.[87] Moreover, there was one dissenting opinion in the case. Judge Zagrebelsky argued that a degradation of the environment could amount to a violation of a specific right under the Convention. The damage to the swamp next to the applicants' property led to a corresponding deterioration in the quality of their life.[88]

To be eligible in the ECtHR, one needs to be a victim and his right under the Convention should be violated. The ECtHR can include some environmental elements into existing human rights, but does not provide protection to the general environment. It is argued that the ECtHR is not the right forum for environmental litigation, especially public interest litigation, given their insufficient expertise to litigate complex environmental cases and the heavy case load.[89] For ecological damage, it is the environment itself that suffered the losses, which does not necessarily influence individuals significantly and directly. Therefore, the ECtHR is not a promising vehicle to bring a claim for ecological damage.

2.2.2. Tested in Economics

Traditionally, the parties in litigation should have a direct interest at stake. The standing requirements can act as a gatekeeper for filing cases.[90] This can be justified from an economics perspective: not to overburden the courts or create over deterrence of operators. The limited standing requirements, however, do not always hold true, for example, in common law public nuisance cases or environmental cases.[91] The environment is a public good, involves many stakeholders. Hence the verdict concerning environmental protection also takes on the nature of a public good and would be undersupplied by a rational victim.[92] In this case, a broadened scope of standing can help to overcome the undersupply problem and increase efficiency.

The above analysis shows that instead of sticking to the traditional standing requirement, there are at least three approaches available to grant standing for natural resources. The public authorities, environmental NGOs and individuals are all possible plaintiffs in ecological damage cases. The choice of standing needs to allow claims for significant ecological damage and, at the same time, to avoid numerous frivolous and meaningless cases. To strike such a balance, on the one hand, the subject needs to have enough incentives to file such a claim and, on the

87 *Ibid.*, paragraph 53.
88 *Kyrtatos v Greece*, partly dissenting opinion of judge Zagrebelsky.
89 Schall 2008, p. 444-449.
90 Van Aaken 2005.
91 Stein 1979.
92 Van Aaken 2005.

other hand, the created costs, including information costs, administrative costs and enforcement costs should be kept reasonable.

An ideal claimant should be someone who has both the willingness and capacities to bring a lawsuit. Whether a specific subject can represent the impaired environment or whether an obligation to protect and restore the environment exists plays an important role in determining the willingness to bring a lawsuit. A reasonable design should make the information requirement and initiating costs not so burdensome to weaken the capacities of the initiating parties to engage in such a procedure.

Moreover, the costs arising with different initiating parties also need to be considered. To file a litigation places a burden on the aspect of information: the detection and assessment of damage, proving the causation between liable parties' activities and the damage (at least *prima facie*), and finding the fault of injurers under negligence rules. Administrative costs are another part of the costs to file such a case: an organization may be set up to serve the purpose of environmental protection and filing a claim; individuals may gather to take a class action. Furthermore, public authorities, NGOs and individuals play different roles in enforcing and monitoring the compliance of the judgment.

2.2.2.1. *Public Authority*

Public authorities are in a good position to claim for ecological damage in view of their capacity in information collection and damage assessment: the detection of ecological damage needs effective environmental monitoring; damage assessment is a time-consuming and costly procedure. Small NGOs and individuals may not have expertise and financial resources to do so.

Public authorities have the duty to monitor environmental status and establish databases for various environmental elements. They also have the advantage to collect intellectual and capital resources to proceed with damage assessment. However, the advantages in capacity do not always hold true, especially in developing countries, where the lack of capacity in environmental agencies is often reported.[93] Moreover, public authorities may not always have incentives to do the assessment and bring a claim. Local authorities may be reluctant to adopt strict measures because they are 'often particularly vulnerable to lobbying by industrial interest groups as a consequence of the controlling role the latter often play in the socio-economic interests of local communities.'[94] Furthermore, the public authority is also prone to being captured by lobby groups. If an obligation is placed on the public authority to protect the environment and engage in the restoration when injured, the authority may be better stimulated to file a case. However, this interferes with another problem: the clear division of duty between different public authorities. A general provision to grant standing to public authorities without

93 For example, insufficient capacity of environmental protection agencies, especially in western areas is often reported. See Schwartz 2003; Van Rooij 2003.

94 Faure, Goodwin & Weber 2010, p. 120.

clearly defined duties may lead to the shuffling of responsibility among different authorities. Explicit duty, on the contrary, promotes the incentives to bring litigations for ecological damage. From this perspective, the institutional arrangement in the U.S. is desirable: many different public authorities are designated as trustees for natural resources, and a claim for the damage can be brought by the trustees, such as the Department of Defense, the Department of Justice, the Department of the Interior, the Department of Agriculture, the Department of Commerce and so on.[95] The duty is explicitly divided among them: for example, the Department of Commerce is the trustee for the natural resources 'that are found in, under, or using waters navigable by deep draft vessels, tidally influenced waters, or waters of the contiguous zone, the exclusive economic zone, and the outer continental shelf', the trusteeship of Department of the Interior include: 'migratory birds; anadromous fish; endangered species and marine mammals; federally owned minerals; and certain federally managed water resources'.[96] As far as the costs are concerned, since public authorities are responsible for daily environment monitoring and management, and in many jurisdictions, it is the public authority that takes or guides the damage assessment and restoration measures, the information costs for them to bring litigation are relatively lower. When both the parties are to file a claim and execute (guide) the restoration, it is easier to keep continuity and reduce the enforcement costs.

2.2.2.2. *Environmental NGOs*

For the environmental NGOs, they may have a strong incentive to bring a lawsuit for ecological damage since environmental protection is usually the aim in their articles of association. As public interest groups, they are also less susceptible to influence from lobbying activities. However, legal arrangements are necessary to alleviate them from the heavy burden of litigation costs. As mentioned above, the information requirement in the case of a claim for ecological damage is rather high, including technology, experts and evaluation methods. Thus not all NGOs are able to engage in such a task. For example, without daily environmental monitoring and data collection, it would be difficult for the NGOs to determine the baseline of the environment before the damage happens. Accurate assessment of the damage is also a highly technical and burdensome job. In addition, even when the damages are successfully awarded, not all NGOs are eligible to use those financial resources in restoration of the damaged environment. For the NGOs that are well organized, equipped with necessary experts and experienced, it would be beneficial to authorize them the standing to sue. Otherwise, the public authority may be more suitable for such a task.

However, there is less controversy as regards the compensation for their cleanup costs and restoration costs. If the cleanup and restoration measures are taken, the losses are quantified in monetary value, and NGOs have strong

95 40 C.F.R. §300.600(b) and Exec. Order No. 12580, 3 C.F.R. 193, ADMIN. MAT. at 45031.

96 40 C.F.R. §300.600(b) (1),(2).

incentives to claim for the costs they paid. Denying them the access to compensation will discourage them from engaging in prompt and effective responses to ecological damage. Thus, when the liable parties or public authorities fail to respond or to respond timely to the damage and fail to take appropriate cleaning or restoring measures, it is advisable to allow the environmental NGOs to interfere and granting them the right to claim for the costs will promote the incentives for them engaging in cleaning and restoration.

2.2.2.3. *Individuals*

Considering the burden of damage assessment and litigation costs, it is even less promising for individuals to claim for ecological damage. Engaging in the process of damage evaluation and litigation requires the knowledge of environmental sciences, economics and law; it is inadvisable to expect individuals to be experts on all those aspects. Furthermore, since their rights are not directly violated and the interests of specific individuals are usually just slightly influenced, they may also lack incentives to bring cases on behalf of the public.

Although some NGOs and individuals are inferior to public authorities in bringing such a lawsuit as far as the capacities are concerned, there are still some practices and proposals which give them secondary standing. For example, the citizen litigation provisions in the U.S. make is possible for citizens to compel the public authorities to claim for the natural resources damage if the latter failed to do so.[97] As discussed earlier, the public interest groups play different role in addressing environmental damage under the White Paper and the ELD. If Environmental NGOs and individuals have secondary status to claim against injurers, as proposed under the White Paper, it can create additional incentives for injurers to prevent pollution. If requests can in fact be formulated against the public authorities by NGOs or individuals, the public authorities may have more incentives to behave diligently and it can also create indirect preventive incentives for injurers. The concrete design depends on the situtations under the specific jurisdictions, including the attitudes towards public interest litigation, standing of NGOs and the actual role played by NGOs in protecting the environment.

All in all, the choice of standing to sue for ecological damage is a tradeoff between the incentives to litigate and the costs of information, administrative and enforcement. If a clear and definite duty is imposed on the public authority for the protection of specific environment, and they can adopt effective monitoring and regulation, the public authority can be the eligible party to initiate a claim for ecological damage. They also have the advantages of lower administrative and enforcement costs. However, compared to NGOs, it is easier for the public authority to be captured. The NGOs have the willingness to protect the environment and bring a lawsuit in case of damage, and they are more indifferent to lobby groups, but the information costs and litigation costs are burdensome for them. Thus, only in the community where NGOs are more developed and active, they are more

97 Kanner & Ziegler 2006, p. 134.

suitable for engaging in such a sophisticated job. However, it is still beneficial if NGOs and individuals can be offered secondary standing to file a claim if the public authorities fail to do so or to sue public authorities to compel them to make claim in such a situation. Furthermore, if the NGOs take actual cleaning and restoration measures, they also need to be granted standing for the arising costs.

2.3. *Liable Parties*

Indentifying the liable parties is another essential issue in the establishment of liability for ecological damage. Corrective justice requires that 'one who has voluntarily harmed another' 'must pay for any loss he caused.'[98] The liability should be imposed on the right person to realize the value of corrective justice. In addition, the polluter pays principle, a fundamental principle of environmental law, has shifted from the public sphere to civil liability.[99] The integration of this principle also raises the question of indentifying the 'polluter'.

2.3.1. Economic Analysis

The law and economics approach aims at the minimization of the total social costs by giving the parties involved in an accident optimal incentives. The first step towards this objective is to decide to whom the incentives should be given. To indentify the liable parties is a challenging issue in ecological damage cases. Booming science and technology in modern times has created numerous industrial activities dangerous to the environment; however, the causal link between many activities and damage is often unclear. This uncertainty is aggravated by the universal existence of multiple tortfeasors. To ascertain causation is a necessary condition to indentify the liable parties. This question is presented below in the subsection 'causation and multiple tortfeasors'.

In many international environmental agreements, a channeling of liability is adopted. This means that the law indentifies the liable parties and imposes liability exclusively on that specific party, rather than on others, even if the loss was actually caused by them. For example, in the Paris Convention on Third Party Liability in the Field of Nuclear Energy (hereinafter Paris Convention),[100] the operator of a nuclear installation is liable for the damage to persons and property damage where it can be proven that the damage was caused by a nuclear accident. The right to redress against the third party only exists when the third party intentionally caused the harm or a contractual arrangement allows the right of redress.[101] A similar provision is also adopted in the CLC, where liability is channeled to 'the owner of a ship at the time of an incident, or, where the incident consists of a series of occurrences, at the time of the first such occurrence.'[102] Although the CLC does not

98 Gordley 1997, p. 157.
99 De Sadeleer 2006, p. 91.
100 Paris Convention on Third Party Liability in the Field of Nuclear Energy, 956 UNTS 251.
101 Art. 3(6) of the Paris Convention.
102 Art. III, 1 of the 1992 CLC.

preclude the owner's right of recourse against third parties,[103] the definition of 'the owner' is rather limited so that some other parties who can also contribute to pollution are excluded.[104]

Since channeling of liability denies the possibility to claim for damages from some parties who actually caused the damage, it is problematic from the deterrence perspective. The actual injurer, exempted from liability to victims and recourses, is therefore lacking incentives to prevent damage. There are also arguments in favor of channeling, contending that within in the context of the Coase theorem it makes no difference on which parties the liability is imposed on, since the costs can be transferred though contractual arrangement. However, high transaction costs may in practice prohibit the re-allocation of liability and the transitive process may be interrupted as a result of insolvency of one of the parties involved.[105] Another argument to defend channeling is that it can save the victims costs in investigation and indentifying the liable tortfeasors. However, this argument again is not so convincing: the potentiality of channeling in saving investigating costs is limited; and even when there is such an advantage, it is doubtful whether the strength is so significant that it outweighs the benefits in providing victims with more guarantees. Without channeling, the victims can also claim against other parties who contribute to the damage.[106]

2.3.2. Applied to Ecological Damage

To indentify the liable party for ecological damage, it is advisable to avoid the provisions concerning channeling of liability. Two aspects should be considered when judging whether the rules constitute channeling: firstly, how broad the liable party is defined; and secondly, whether victims can claim the damages from other parties and recourses against them are possible. To illustrate these problems, the CERCLA and the ELD are chosen as examples.[107]

2.3.2.1. *Definition of Liable Party*

Liable Party under CERCLA

There are four kinds of potential responsible parties in CERCLA: (1) current owners or operators of a vessel or a facility where hazardous substances were disposed; (2) previous owners or operators of such a vessel or facility; (3) transporters of hazardous substances; (4) generators of the substances.[108] The liability provision

103 *Ibid.*

104 The 1992 CLC even contains a list of the parities that are excluded from being considered the liable party under Art. III, 4.

105 Faure & Wang 2006a, p. 187.

106 Faure & Wang 2006a, p. 188.

107 The compensation systems under the CERCLA and the ELD are discussed in detail in the empirical part, see Chapter 7, Section 2.3 and Chapter 8, Section 1. This section just uses them as examples to show the possible design of liable parties for ecological damage.

108 42 U.S.C. §9607(a).

contains two important concepts: owner and operator. The U.S. Supreme Court clarified the definition of operator under CERCLA:

> 'Under CERCLA, an "operator" is someone who directs the workings of, manages, or conducts the affairs of a facility; more specifically, operator must manage, direct, or conduct operations specifically related to pollution, that is, operations having to do with the leakage or disposal of hazardous waste, or decisions about compliance with environmental regulations.'[109]

In this clarification, 'operator' is defined though its control over a facility and its scope is broad. A parent company operating the facility is directly liable for the damage under the title of 'operator', rather than indirectly liable under the veil piercing doctrines.[110] In other words, there is a clear distinction between direct operator liability and indirect liability under veil-piercing doctrines – the dividing line is whether the parent controls the facility or the subsidiary.

Furthermore the 'operator', the 'owner' of a facility is also defined as the liable party under CERCLA. Although the notion 'owner' is not clarified in CERCLA, courts have accepted a broad definition of it. A variety of 'owners' are held liable by federal judges, such as an owner 'who does not know of and did not contribute to the release of hazardous waste', a passive owner who leases property to the tenants causing the damage, Corporate officers, majority shareholders, and foreclosing creditors.[111] This broad definition of owners, combined with strict, joint and several liability is even criticized for creating inefficient incentives to the actors.[112]

Liable Party under the ELD

Unlike CERCLA, the ELD constricts the liable party to the 'operator', which is defined as:

> 'Any natural or legal, private or public person who operates or controls the occupational activity or, where this is provided for in national legislation, to whom decisive economic power over the technical functioning of such an activity has been delegated, including the holder of a permit or authorisation for such an activity or the person registering or notifying such an activity.'[113]

According to the definition, a party constitutes the 'operator' either for its control over the activity or for its decisive economic power over the technical functioning of

109 Head note 14 at 118 S.Ct. 1878.

110 Betlem 2006, p. 162.

111 Peabody 1994, p. 429.

112 Gergen argued that, 'owners' are defined broadly in CERCLA without differentiation between landowner directly or indirectly involved with the contamination. He contended that the strict, joint and several liability under CERCLA have resulted in relatively few private cleanups and an inadequate level of reporting of hazardous waste contamination by landowners. Based on this situation, he proposed to differentiate the two kinds of owners and impose on them strict liability and negligence-based standard respectively. See Gergen 1994, p. 635.

113 Art. 2(16) of the ELD.

such an activity. The concept is not limited to the permit holder and includes a similar form of direct operator liability as CERCLA.[114] Furthermore, the ELD leaves room for Member States to extend the scope of responsible parties.[115]

According to the above analysis, the liable party in both CERCLA and the ELD, although delimited by some notions, has a broad scope. The definition tries to indentify the liable party though the linkage of managing, directing, conducting operations or control, thus to ensure the party is the one who actually caused the damage.

2.3.2.2. *Limitation of Victims Claiming against other Parties and Recourses*

Neither in CERCLA nor in the ELD is there such a provision as in the Paris Convention that excludes claims against other parties. A third party defense is contained in both legislations to exempt the actors from liability (at least to some extent) if it is a third party who caused the damage. In CERCLA, the potential responsible party is not liable when the damage is caused by 'an act or omission of a third party', thus claims can be brought by victims against the third party.[116]

The provision in the ELD seems to be more ambiguous. It stipulates that 'an operator shall not be required to bear the cost of preventive or remedial actions' if the damage 'was caused by a third party and occurred despite the fact that appropriate safety measures were in place'.[117] With this formulation, an ambiguity is indentified in that the third party defense may only apply when the public authorities have taken preventive/remedial measures and resource the costs from operators.[118] However, if the operators have taken the measures themselves, it is not clear whether they can get reimbursement from the third parties. Betlem even interprets the provision narrowly as follows: even if the measures are taken by public authorities, the operator is only authorized to redress the costs from the third party but not to deny the liability altogether.[119] This rule neither gives appropriate incentives to the operator nor the contributing third party.

In conclusion, the liability for ecological damage should not be channeled. To avoid channeling of liability, on the one hand, the liable party needs to be broadly defined; on the other hand, liability rules should not preclude the victim's claim against another contributing party and recourses from him by the one who actually pays for damages.

114 Betlem 2006, p. 187.

115 Art. 16(1) stipulates that 'The Directive shall not prevent Member States from maintaining or adopting more stringent provisions in relation to ... the identification of additional responsible parties.'

116 42 U.S.C. §9607(b)(3).

117 Art. 8(3) of the ELD.

118 Betlem 2006, p. 187.

119 Betlem 2006, p. 151-152. According to him, Art. 8(3) of the ELD discusses the burden of the costs but does not deal with liability: 'In such cases Member States shall take the appropriate measures to enable the operator to recover the costs incurred.' Hence the operators are still liable parties but can recover the costs from the third parties who cause the damage.

2.4. *Liability Standards*

After addressing the problem of who is responsible for ecological damage, the next problem is under which conditions he should be responsible. Much discussion has taken place concerning the choice between negligence and strict liability both at a general theoretical level and concerning environmental damage. Under negligence, the actor is only responsible when 'fault' can be established, or in the other words, he does not exercise 'due care'. Whereas the term 'strict liability' is generally defined as 'liability imposed on a defendant in the absence of a breach of a duty to exercise reasonable care.'[120] Thus as long as the causation can be established, the actor is liable regardless whether he is at fault or not.

2.4.1. Economic Analysis

As mentioned above, from a law and economics perspective, optimal liability rules are those that can minimize total social costs. The goal of this part is therefore to analyze how strict liability and negligence can affect the primary costs and administrative costs respectively. The secondary costs (risk spreading and risk shifting) have only a limited purpose in the context of ecological damage. Many environmental elements are *res commune* or *res nullius*, or owned communally. Thus damage to publically owned resources is automatically maximally spread. Hence when discussing the social costs brought by different liability standards, this research focuses on the primary costs and tertiary costs.

2.4.1.1. Primary Costs

As far as primary costs are concerned, the literature distinguishes two situations—the unilateral cases, in which only tortfeasors contribute to the loss and bilateral cases in which preventing damage is a function of both tortfeasors' and victims' caretaking. The care level and activity level are both available approaches to minimize the primary costs. In this section, the analysis is focused on those two approaches. Under each approach, discussions are further distinguished into unilateral cases and bilateral cases.

Care Level

To determine the optimal care level, a balance should be achieved between accident avoidance costs and the diminution of actual damages. More care taken by the injurers will reduce the costs of damage, but increase the costs of accident avoidance. The optimum is achieved when the marginal costs of care taking equal the marginal benefits in accident reduction.[121]

Shavell established the model in unilateral cases. According to him, both strict liability and negligence can give incentives to the tortfeasor to take optimal care.

120 Klass 2004, p. 907.
121 Shavell 1987, p. 7.

Under the negligent rule, the actors will follow the optimal care level as long as the due care level established in the legal system equals the optimal level. More care will result in increased costs of care taking while less care will lead to compensation, which both make the primary costs high. As far as strict liability is concerned, the injurer has to pay for the damage regardless which level of care he took, which means that he will always bear the entire primary costs. Therefore, the most efficient way for him is to follow the optimal care level. However, to arrive at this conclusion, a few assumptions are made: the court is able to set the efficient due care level and assess the magnitude of damage accurately. The results change when the assumptions are relaxed. Firstly, consider the court's insufficient capacity in deciding the due care level. Under a negligence rule, the tortfeasor will take less care if the due care level is set below the optimal level. While under strict liability, the tortfeasor will always follow the optimal care level, regardless of the court's corresponding capacity since it is the tortfeasor himself that determines the care level then. Hence the information costs for the courts lead to a preference for strict liability. When the second assumption, that the legal sanction equals the harm actually caused, is relaxed, the conclusion reverses: under the negligence rules, 'equality between harm and sanction is not essential as long as the sanction is sufficiently large so that the private costs of the injurer are minimized by conforming to the legal standard.'[122] Under strict liability, insufficient compensation will lead to a lower care level of the tortfeasor.[123]

Under a bilateral situation, efficient liability rules should give incentives to both the injurer and the victim. Strict liability only gives incentives to injurers to take care but not the victims, since they can always get compensation from the injurers. Combining strict liability with a contributory negligence defense can make the victims take the optimal care to prevent damage as well. Moreover, simple negligence, negligence with the defense of contributory negligence and comparative negligence rule can all lead to socially optimal care.[124] Thus strict liability with a contributory negligence defense and negligence may be equally efficient. However, it is noteworthy that such a conclusion is drawn under the same assumptions as those in unilateral cases. The court's ability in setting the due care level for the victim is essential under both strict liability with a contributory negligence defense and under negligence. The only difference is that the former is indifferent to the court's setting of the due care level for tortfeasors. If the equality between the legal sanction and the damage is not achieved, the potential injurer will take insufficient care under strict liability.[125]

In both unilateral and bilateral cases, the optimal deterrent function will only be realized when the tortfeasor has enough wealth to pay for the compensation. If an insolvency problem arises, the tortfeasor in neither case will have enough incentives to take efficient care. Under strict liability, the injurer will only take care to prevent an accident with a magnitude equal to his total wealth. While under

122 Schäfer & Müller-Langer 2009, p. 8.
123 Schäfer & Müller-Langer 2009, p. 11.
124 Schäfer & Müller-Langer 2009, p. 17-19.
125 Schäfer & Müller-Langer 2009, p. 21.

negligence, it is rational for the injurer to bear the costs of care taking up to his total wealth.[126] Therefore, if the insolvency problem is accompanied by that fact that no financial guarantee is available for the tortfeasor to cover all loss, a negligence rule is preferable.

Activity Level

In addition to care level, liability rules can also minimize primary costs of accidents through their influence on the activity level. The activity level reflects the number of times that a particular injurer engages in a risky situation that could lead to an accident. In a traffic liability situation an activity level would refer to the number of kilometers driven; in a product liability situation the activity level would correspond to the number of products manufactured. Law and economic literature argues that this constitutes an argument in favor of strict liability rather than negligence in unilateral cases. Under the negligence rules, the tortfeasor tends to adopt too high an activity level since, as long as he complies with the due care level, he is free from the compensation liability. As for the strict liability, the tortfeasor is always liable for the damage he caused, whatever care level he chose. He therefore can only reduce the damages though reducing the activity level, given a specific care level.[127] Results change quite drastically in bilateral cases. In this situation, both parties involved in an accident can contribute to the loss. Thus both the tortfeasor and victim should be given incentives to adopt an efficient activity level. Unfortunately, as Shavell argued, neither strict liability with a defense of contributory negligence nor negligence can lead to optimal incentives for both parties.[128] Without a first-best choice, the second-best one depends on whose activity has bigger influence on the accident risk. Strict liability is more favorable when the tortfeasor is the major influencing factor, whereas negligence is appropriate when the victim can contribute more to the accident.

There is another issue which should be mentioned when discussing the influence liability rules impose on activity level. Gilead pointed out that many activities do not only have negative externalities but also have positive spin-off effect.[129] In determining the optimal activity level, both types of externality an activity has should be considered and therefore an increase in activity level is not always socially undesirable. For example, the manufacture of drugs, a highly risky industry as it is, also creates huge social benefits. Such consideration leads to a more cautious attitude towards strict liability in unilateral cases. Its application is often restricted through the delimitation of 'dangerous activity'. For instance, in the ELD, although the term 'dangerous activity' is not explicitly mentioned, the Annex III gives a list of activities of which the operators will be held strictly liable for the damage they cause. In the Restatement of Tort Law in the U.S., strict liability is applied to abnormally dangerous activities, which is limited to activities creating 'a

126 Faure & Grimeaud 2003, p. 35.

127 Shavell 1980, p. 2-3.

128 Shavell 1980, p. 7.

129 Gilead 1997, p. 589.

foreseeable and highly significant risk of physical harm even when reasonable care is exercised by all actors'; and 'not one of common usage'.[130]

2.4.1.2. *Tertiary Costs*

The third economic criterion requires a comparison between the administrative costs of strict liability and negligence. Under strict liability, the judge only has to decide the magnitude of the loss to award the corresponding compensation. Under a negligence rule, the judge needs to find the due care level equal to the optimal level. To define the due care level, the information about avoidance costs and expected damage is essential, which creates more legal uncertainty than under strict liability. In this sense, it is more expensive to make a judgment under negligence rules. Furthermore, under strict liability, it is less costly for a plaintiff to claim for the damages since he does not need to prove the 'fault' of the defendant. However, this may also lead to more cases in view of the relief in the procedural burden of plaintiffs, which will hence lead to increased costs of administrating the legal system under strict liability.[131]

From the analysis above, neither strict liability nor negligence dominates each other in all cases. It should be borne in mind that all the results found above are based on specific assumptions. Suppose a simplified model: judges can decide both the magnitude of the loss and set the due care level efficiently; no insolvency problem arises; and liability can lead to changes in activity level. If the activity causing damage is dangerous so that the activity level should be controlled, strict liability is preferable in unilateral cases. The choice in bilateral cases relies on which party contributes more to the damage. If the activity is not socially undesirable and therefore there is no need to reduce the activity level, a negligence rule would be favorable, where no excessive incentives are given to potential tortfeasors to reduce his activity level. However, the results may change dramatically when the assumptions are relaxed. Inaccurate information for judges about due care levels favors strict liability; the negligence rule is less sensitive to the accurate magnitude of legal sanction. Another advantage of the negligence rule is that it leads to less distortive incentives in the case of the tortfeasor's insolvency. In addition, when deciding the liability rules, the problems of risk aversion and administrative costs should also be taken into consideration.

2.4.2. Liability Standards for Ecological Damage

This part applies the theoretical analysis to ecological damage and tries to find which liability rules are more efficient in this case.

To address the problem of choosing liability standards, the scope and method of defining ecological damage should be reviewed and explored briefly here. As defined in the second chapter, ecological damage refers to damage to both public

130 Restatement (Third) of Torts: Liability for Physical and Emotional Harm, § 20.
131 Schäfer & Müller-Langer 2009, p. 24.

natural resources and appropriated natural resources of which the ecological value exceeds the personal interests of the owner. Traditional damage caused via the environment, such as personal injury and property damage are excluded. Therefore under the heading of ecological damage, it is the public at large rather than individuals that are the victims. The standing of the public authority and NGOs is not based on their identity as victims but their duties or aims in environmental protection. The effect this characteristic has on liability standards is explored later in this part. Another issue which should be mentioned is the tendency to establish liability rules for ecological damage according to the activities causing it. For instance, in the U.S. natural resources legislation, damage is delimitated through making the 'owner or operator of any vessel from which oil or a hazardous substance is discharged'[132] or 'each responsible party for a vessel or a facility from which oil is discharged, or which poses the substantial threat of a discharge of oil, into or upon the navigable waters or adjoining shorelines or the exclusive economic zone'[133] responsible for the incurred actual costs. In this legislation, owning or operating a facility where oil or hazardous substances may be released is regarded as a dangerous activity. Thus, the industries activity level should be controlled. In the ELD, a differentiated approach has been established in setting up the liability standards for environmental damage. For the activities listed in Annex III, which are regarded as dangerous, strict liability is applied. In addition, it also holds the operators of general professional activities responsible for the damage to protected species and habitats.[134] In this case, damage caused by both dangerous activities and other activities can be covered as ecological damage. This part tries to find proper liability rules for ecological damage in accordance with the classification of activities.

First, consider ecological damage caused by dangerous activities, where incentives should be given to tortfeasors to set efficient activity levels. Ecological damage can in most cases be regarded as the consequence of unilateral accidents. Only the injurer can influence the risk. Although in ecological damage cases, assessment of the damage constitutes a challenge for the judges, it is even more difficult for the judges, who are often laymen concerning technical problems, to decide the optimal care level. Compared to the judges, injurers may possess more information concerning the methods and costs to reduce the damage, and thus are in a better position to find out the efficient care level. In view of the information asymmetry, strict liability outperforms negligence in providing an efficient care level as long as the insolvency problem does not arise. When it comes to the activity level, the probability that injurers of ecological damage are detected and held liable is still low. Suppose insurance is available to internalize damages in injurers' production costs so that liability can impose an influence on the activity level; then again this favors strict liability for its incentives for injurers to adopt an efficient activity level.

132 33 U.S.C.A. §1321(f)(1).
133 33 U.S.C. §2702(a).
134 Art. 3(1)(b) of the ELD.

There are some cases where other parties in addition to the injurers can also influence the environmental risk. Nevertheless, as discussed above, victims of ecological damage — the public at large are not the other contributing parties. Other than the traditional victims, some third parties, such as the public authorities in charge of environmental affairs, can take preventive measures and affect the accident risk. Thus ecological damage can be deemed to be a bilateral case on the condition that one considers third parties as victims.[135] Even in this case, the injurers' influence on the accident still prevails over the influence of other parties. The optimal liability rules in this situation are the ones which can give incentives to injurers to have an efficient activity level. Therefore, once again it constitutes a situation where strict liability (with defenses) is favorable.

Secondly, if loss is caused by non-dangerous activities, dramatic changes can be found. Environmental protection law has proposed high standards for industries. There are many normal operations which can be the potential cause of ecological damage but they are not abnormally dangerous or ultra hazardous, thus not socially undesirable. The argument in favor of strict liability due to its influence on injurers' activity levels should be balanced.

The above analysis is focused on how to minimize the primary costs. The influence of liability rules on administrative costs in case of ecological damage should also be borne in mind when making the choice.

Strict liability is more favorable than negligence in view of the fact that injurers possess better information than judges to find the optimal care level in ecological damage cases. In this sense the application of strict liability can save administrative costs. The worry of increasing costs resulting from too many lawsuits under strict liability could be relieved by the difficulties in establishing liability for ecological damage.

Another refinement that needs to be added here is that the possibility of injurers' insolvency is dismissed in the above discussion. As far as ecological damage is concerned, the magnitude of damage could be tremendous, such as in the cases of Exxon Valdez in the U.S. and the Tisza pollution in Romania.[136] The Deepwater Horizon accident lead to the leakage of millions barrels of oil and to the establishment of a $20 billion compensation fund.[137] However, the full scale of the damage may still not have manifested. There are many small enterprises which can also cause huge damage. The two situations combine to create a high risk of insolvency. When insolvency arises, negligence rules are preferable to strict liability to provide efficient care levels. The situation is even worsening by the tendency of big enterprises to split into smaller ones to escape the liability when strict liability is

135 Faure & Grimeaud 2003, p. 31.

136 The Exxon Valdez vessel spilled million gallons of U.S. crude oil into the sea in 1989, leading to a 2.1 billion dollar clean-up cost. Tisza pollution caused by a cyanide spill following another damburst in Baia Mare, Romania in 2000 also brought tremendous damage.

137 BP, BP Establishes $20 Billion Claims Fund for Deepwater Horizon Spill and Outlines Dividend Decisions, available at: <www.bp.com/genericarticle.do?categoryId-=2012968&contentId=7062966>, last accessed on 8 May 2013.

imposed.[138] Thus to solve the under-deterrence problem caused by insolvency, financial guarantees are very important. If enough guarantees are available, the advantage of negligence over strict liability in light of insolvency problems disappears.

Taking into account this analysis and leaving aside the problem of injurers' insolvency, if ecological damage is caused by dangerous activities, strict liability is favorable; as for the case of other activities, negligence is preferred. Moreover, a financial guarantee is of great importance, considering the possibility of insolvency.

2.5. *Causation and Multiple Tortfeasors*

Clarifying the meaning of causation is a problem perplexing tort scholars. The attempt leads to debates in two doctrines: cause-in-fact and proximate cause. The former doctrine determines when an act is part of a causal chain of the injury, and it is often decided by the 'but-for' test. The latter doctrine limits causation by the proximity of an act to the harm. Expanding liability to ecological damage creates challenges for the establishment of causation, which is mainly embodied in two aspects: firstly, it is often uncertain whether or to what extent the ecological damage is caused by a specific factor; and secondly, it may be that several actors create the risks simultaneously, and the damage is not easy to be traced to or shared between them. This section analyses the problems of uncertain causation and multiple tortfeasors from the law and economics perspective, and then applies them to liability for ecological damage.

2.5.1. Economic Analysis

2.5.1.1. Uncertain Causation

Uncertain causation is a usual obstacle in establishing liability. For example, a pollutant can lead to a certain disease in humans. However, taking into account the background risk, it is often uncertain whether a specific victim suffered an injury because of this pollutant. The law and economics literature proposes two approaches to establish liability under uncertain causation: an all-or-nothing criterion or a proportional liability criterion. Under an all-or-nothing criterion, there is either no liability at all or, if the liability is imposed, the compensation equals all the losses. The threshold probability rule is the most common all-or-nothing criterion, implying that the injurer bears full liability when his probability to cause the damage exceeds the threshold and no liability at all when the reverse is the case. Under the proportional liability criterion, the liability will always be imposed

138 An empirical research about the influence of strict liability in Toxic Waste Management in the U.S. shows that in some cases, spills may be more numerous in states that have adopted strict liability. One contributing factor is that large firms tend to spit into smaller ones to reduce their liability. See Alberini & Austin 1999, p. 20-48.

whenever there is a possibility for the actor to cause the damage; but the imposed compensation is reduced proportionally to account for the uncertainty.[139]

The all-or-nothing approach dominates traditional tort law. But beginning in the 1980s, there is a tendency to apply proportional liability in American courts. Law and economics scholars have done intensive research on the choice between the two criteria. Kaye tried to show the superiority of the 'preponderance of the evidence' standard, which incorporates a threshold probability of 50 per cent, over other threshold probability rules and over the proportional liability criterion.[140] He made this argument on the basis that the aim of ascertaining causation is to minimize the *ex post* costs of erroneous liability decisions, ignoring the *ex ante* incentive effects. Another argument in favor of threshold probability rules involves the ability of a high threshold probability rule to produce more evidence to the court.[141] This argument is also made from an *ex post* perspective, considering the incentives of the plaintiff to develop evidence. In addition, Shavell mentioned another advantage of threshold liability rules in saving administrative costs.[142]

Literature which prefers proportional liability rules to threshold probability rules evaluates the effects of the two approaches from an *ex ante* perspective. Rather than adopting the rules minimizing *ex post* error-costs, Shavell regards the creation of optimal deterrence as the objective of the liability regime, and compares the two approaches by their influence on care level and activity level of the actors.[143] Under the threshold probability rules, if the actor's probability of causation is below the threshold, he is then not liable at all and lacks the incentives to take optimal care. Similarly, if the actor has a probability of causation above the threshold, he is then facing full liability and over-deterrence will arise. However, this distortion in incentives is less likely to happen under proportional liability rules. The actor then faces liability equal to the expected loss relative to his activity, which leads to socially optimal levels of care. Moreover, proportional liability is also preferable with respect to activity levels. The threshold probability rules result in over-deterrence from engaging in the activity under strict liability when the probability of causation is above the threshold. Under-deterrence arises under both strict liability and negligence if the probability of causation stays below the threshold.[144]

In view of the above analysis, the discrepancy in preference of threshold probability rules and proportional liability rules depends mainly on whether the *ex post* or *ex ante* perspective is adopted. The *ex ante* approach aims at providing optimal deterrence to minimize total social costs. The legal errors, criteria of the *ex post* approach, do not 'involve a welfare cost *per se*, and their minimization could perhaps be taken as a measure of fairness but not as a proxy for optimal deterrence.'[145] Bearing in mind the prevention as the objective of liability rules, the

139 Ben-Shahar 2009, p. 91.
140 Kaye 1982, p. 487-516.
141 Levmore 1990, p. 691-726.
142 Shavell 1987, p. 117.
143 Shavell 1985, p. 587-609; Shavell 1987, p. 115-118, 123-126.
144 Ben-Shahar 2009, p. 95.
145 Ben-Shahar 2009, p. 93-94.

ex ante approach is more desirable here. Shavell's research has shown that proportional liability is superior to threshold probability rules in reducing the primary accidents costs. But this advantage should be balanced with its weakness with regard to administrative costs. Proportional rules lead to a higher number of litigations since the victims who are less likely to be injured by the defendants can also get proportional compensation. In addition, it is probably too ideal to assume that courts can find out the precise probability of the defendant being the cause of the injury. The choice of these two approaches, therefore, depends on the trade-off between their advantages in primary costs and administrative costs.

When the strength of proportional rules can not offset its weakness in administrative costs, the threshold liability rules should apply. Consider three standards of proof commonly used in legal practice: proof beyond reasonable doubt, preponderance of the evidence, and reversal of the burden of proof. These rules set up different thresholds and place different burdens on plaintiffs (defendants). As the analysis above shows, considering the actor's care level, under threshold probability rules, the actor whose probability of causation is above the threshold will be over-deterred; when the probability is below, under-deterred. As far as the activity level is concerned, there are distortions in providing incentives with only one exception (if the probability of causation is above the threshold, no distortion will arise under negligence). Thus regardless what threshold is set, there will still be distorted deterrent effects, since the actor's probability of causation is either above or below the threshold. Therefore, no optimal solution exists under the threshold approach. If proportional liability is precluded due to its high administrative costs, the choice of standards of proof under the threshold probability approach needs to consider other factors and must be made on a case-by-case basis.

2.5.1.2. *Multiple Tortfeasors*

When the damage results from the actions of multiple parties, the legal system needs to decide how to allocate liability between them. Two approaches are widely adopted: joint and several liability as well as non-joint liability. Under the first approach, a plaintiff can choose to file a litigation against some or all of the injurers; he can also decide to claim for all his losses from one injurer or to recover a portion from each. Under non-joint liability, the plaintiff can recover from one defendant only the share of the damages attributable to that defendant. From the deterrence perspective, literature has developed three models to compare joint and several liability with non-joint liability.

Firstly, assume the injurer is fully solvent, negligence rules apply and the court can establish the due care level equal to the optimal level. Joint and several liability will produce the socially optimal incentives. If one of the multiple parties takes due care, it is rational for the other party to be cautious too, or he will face the full damage, including the part contributed by the non-negligent party. It is also irrational for both parties to be negligent, which means that they as a whole deviate from the efficient care level, and the primary accident costs will increase. Thus at least one party will experience higher private costs than those he causes with no negligence. The analysis is different under a non-joint liability rule. Considering the damage function is convex (for example, the additional damage caused by one unit of pollution increases with increasing amounts of pollution), there is a tendency for the injurers to choose to be negligent.[146] So the joint and several liability is more preferable in this model.

If strict liability is applied where the injurers are fully solvent, under-deterrence will always arise whatever liability rules are adopted. Assume one party takes optimal care, since the damage function is convex, and liability is distributed among each party in proportion to his share of the damage (say the amount of emission), the increase in another party's liability is less than the increase in social loss. The remainder of the increased damage is shifted to the one who takes optimal care. Thus it is rational for both parties to take less care than the optimal level.

The third model involves limited solvency. Kornhauser and Revesz's research shows that neither rule dominates the other in this situation.[147]

2.5.2. Applied to Ecological Damage

2.5.2.1. Uncertain Causation

The economic analysis above shows the advantage of proportional liability in offering the optimal deterrence to injurers. However, it has the drawback of high administrative costs. As far as ecological damage is concerned, the uncertainty of causation is high, and it is difficult for judges to ascertain the precise proportion of an activity contributing to certain harms. This difficulty constitutes an important hindrance to apply the proportional liability rules to ecological damage. Moreover, the conclusion that proportional liability can provide optimal deterrence is based on the assumption that there are lots of similar victims. Each specific victim, of which the damage is either caused by the injurer's activity or not, is either overcompensated (if his damage is not caused by the activity) or under compensated (if his damage is caused by the activity and needs to be fully compensated) under proportional liability. But as the Law of Large Numbers works, if lots of victims' claim together, this makes the injurer responsible for all the social costs he made and leads to optimal deterrence. According to the Law of Large Numbers, 'in repeated, independent trail, the actual incidence of that particular outcome will converge more closely with the probability of that outcome as the number of trials increases.'[148] Applied to the tort cases, with the increasing number of claims, the average of the probability that the damage is caused by a certain act approaches the actual probability. However, it does not hold for ecological damage. Lots of the damage caused by pollution goes undetected; the problems of standing and assessment are also obstacles to ecological damage litigation. Thus the

146 Kornhauser & Revesz 1989, p. 831-884.

147 Kornhauser & Revesz 2009, p. 122-124.

148 Wagner 2009, p. 382.

probability that ecological damage cases are brought to courts is low, and the number of the claims for ecological damage is still limited. Without a sufficient number of homogenous claims, the actual probability that the damage is caused by certain activities may deviate from the theoretical outcome, which makes the proportionate rules less effective.

When it comes to the choice of threshold probability rules, the above analysis reveals that none of the threshold rules can lead to optimal deterrence. As long as the probability of an activity in causing the harm is not equal to the threshold chosen, either over-deterrence or under-deterrence will arise. Scholars proposed some other approaches to determine the causation with regard to toxic pollution. Farber prefers a 'most likely victims' approach to proportional liability rules. He designed a compensation system that 'plaintiffs whose injuries were least likely to have been caused by the defendant receive nothing, while those with the highest causation probabilities get full compensation.'[149] However, this design is also only valid when large numbers of victims exist, which is not the case for ecological damage. Brinker provides another alternative: the substantial-factor approach. This approach enables a plaintiff to recover all his losses as soon as 'he or she has established that a defendant's negligent activities were a substantial factor in causing the plaintiff's injury.'[150] A substantial-factor approach does not mandate threshold levels. Instead, it allows the plaintiff to present whatever evidence there is available. Although this design results in uncertainty, it provides a hint that lots of factors should be combined to determine the causation. As far as ecological damage is concerned, even if the threshold can be properly set, it is not easy for the injurer to decide whether or to what extent his behavior can cause certain damage. Such a procedure makes the potential harm more predictable for the injurer, which is important for effective deterrence. In the U.S. natural resources damage legislation, two types of detailed assessment procedures are established. No explicit threshold is formulated in the legislation; it gives the assessment made in accordance with the assessment regulation a rebuttable presumption.[151] In the assessment regulations, the provisions concerning injury definition and pathway determination,[152] on the one hand, help to establish causation and, on the other hand, makes the possible harms predictable to injurers.

2.5.2.2. *Multiple Tortfeasors*

In the case of ecological damage, the existence of multiple tortfeasors is quite a common phenomenon. As is shown in the economic analysis, there is no dominant preference for either joint and several liability or non-joint liability. The comparison turns on the levels of solvency of the defendants.

Ecological damage sometimes turns out to be huge; at the same time, it is also quite probably that small entrepreneurs contribute to significant damage. These

149 Farber 1986, p. 1221.

150 Brinker 1998, p. 1326.

151 42 U.S.C. §9607(f)(2)(C).

152 43 C.F.R. §11.62-11.63.

situations create a great risk of insolvency. In case of a limited solvency problem, there is no dominant choice between joint and several liability and non-joint liability.

Insolvency risks can be reduced through financial guarantee mechanisms. If the insolvency issue is ignored, joint and several liability is preferable under negligence; and under strict liability, again, neither rule is an efficient choice. As presented above, both strict liability and negligence are available in liability for ecological damage. Thus no first-best choice exists here.

2.6. *Damages*

Awarding efficient damages is another essential issue for liability rules concerning ecological damage. To indentify the injury and quantify it into monetary damages are difficult issues in establishing the liability. From the perspective of law and economics, the objective of damage awards is not full compensation for the victims, as deduced according to common wisdom, but to provide optimal incentives to minimize total social costs. Many factors together shape the optimal damage award, incorporating: the nature of the injury, the type of risk, the liability rule, existing imperfections, etcetera.[153]

This section first discusses the optimal damages theoretically, and then tries to find the efficient damages in liability for ecological damage.

2.6.1. Economic Analysis

Losses can be generally divided into two classifications: the damage to replaceable goods, for which the equivalent commodities are available on the market, and harm to irreplaceable goods, such as death and serious physical injury. For the replaceable goods, the victims can be fully compensated by payment of the market price of the good. On the contrary, the owners of an irreplaceable good cannot perceive any equivalent goods available on the market. Therefore, the approaches to assess the damage and methods to determine the optimal damages are divergent under these two conditions. This section makes the economic analysis respectively according to the types of injury.

For the injury to replaceable goods, the loss can be evaluated through the reduction in market price. Under this situation, it is not such a problem to find out what constitutes full compensation. The question left is whether full compensation is necessary to provide optimal deterrence. Assume both the injurer and the victim are risk neutral. As discussed in the section on strict liability, full compensation is essential to provide the injurer efficient incentives in setting the care level under strict liability. If the damage is underestimated, the costs will not be fully internalized, and the injurer will only take care to prevent the damage up to the damages awarded. However, negligence is not so sensitive to the scope of damages if due care level is efficiently established. A lesser award may also suffice since the

153 Arlen 2000, p. 682.

injurer's liability experiences an increase from zero to expected liability if he decides to reduce his care level from due care to less. But full compensation would be necessary if 'the court requires a causal relationship between the negligence and the losses, so that the injurer only has to compensate the losses that were caused by the negligent character of his behavior.'[154] Under this but-for rule, the polluters have to pay for the care costs and the increase in the damage when he reduces his care-taking from the due care level. The damage that will nevertheless happen even if the optimal care level is adopted is not covered under those liability rules. Thus care costs should be balanced with the damages, and full compensation (increase in the damage when care level changes) is necessary to create optimal incentives.

Next we can take into account the risk attitude of parties in an accident. If insurance is available, the risk averse party, either the injurer or the victim, can spread the risk by buying insurance, which is usually a cheaper risk-spreading instrument than liability. Hence the secondary costs can be minimized. When insurance is not available, the minimization of secondary costs requires placing the burden on the less risk averse party. If victims are risk averse, full compensation under strict liability ensures the victim's optimal risk spreading, where risk-spreading will not be optimal for victims under negligence. If, on the contrary, the injurer is more risk averse, the injurer should be liable for the amount less than the victim's losses, since the risk itself is deemed as a cost by the injurer under strict liability. Under the negligence rules, full compensation also makes injurer's risk spreading optimal, since he can avoid the liability by taking due care.

For the damage to irreplaceable goods, such as death and serious physical injury, both the determination of full compensation and deciding the optimal damages are problematic. To decide the damage of death and serious physical injury, there are, theoretically, two approaches: an *ex ante* approach and an *ex post* approach. The *ex post* approach is a method to decide the actual losses to the victim after the damage. It is also the widely-used approach in legal practice. Usually, the victims of physical injury are compensated for the pecuniary losses caused by the injury plus an award for the 'pain and suffering'. In an *ex ante* damage awards system, the damages award *ex post* are determined by the *ex ante* value of life.[155] This approach implies such a logical inference: the marginal benefit from a unit of wealth is different when the damage happens. For example, the victim of a fatal accident may not derive any utility from *ex post* compensation; a victim that suffered serious physical injury may not enjoy the utility from the same wealth as much as that when he is healthy. From the deterrence point of view, the *ex ante* approach is preferred, for then, the award is equal to the cost to society of the risk imposed rather than victims' actual losses.

Law and economics scholars have worked on finding out the optimal damage from the perspective of deterrence. Several factors have been mentioned in the determination of optimal damage. Firstly, *ex ante* evaluation of the risk depends both on the magnitude of the risk but also on the background risk affecting the

154 Visscher 2009, p. 155. See also Kahan 1989, p. 434-437.

155 Friedman 1982, p. 83.

evaluators. Secondly, the evaluation also depends on the potential victim's willingness to pay to avoid the risk or the compensation they can accept for bear the risk. Scholars also discussed the influence of accuracy in the assessment of awards. Kaplow and Shavell argue that if injurers know the harm they will cause, the accurate assessment of damages will make them internalize the total social costs they imposed. However, 'accuracy in the assessment of harm cannot influence the behavior of injurers ... to the degree that they lack knowledge of the level of harm they might cause when they make their decisions.'[156] In lack of the right information on the harm they might cause, the polluters only have incentives to reduce the damage to the extent they expected the accident to happen, no matter whether the *ex post* evaluation is accurate or not.

Scholars even propose a system of scheduled awards under which a victim is compensated with a pre-determined amount, rather than an award evaluated individually afterwards. Danzon suggests a compensation system based on the average cost of a certain type of risk.[157] This advice is revised by a proposal which decides the award by magnitude and nature of the risk imposed rather than just harm.[158] This advice is based on the opinion that optimal damages should be based on the *ex ante* social costs, not on victim's actual losses. In such a system, not every victim is fairly compensated; however, a large number of cases together can provide the injurer with optimal deterrence to internalize the entire social costs they produce.

2.6.2. Applied to Ecological Damage

Ecological damage is the damage suffered by the public, who bear the loss not only of economic value of the environment, but also the esthetic, cultural and ecological values. Traditionally, ecological damage is only compensable when the environmental elements involved are appropriated, and only for the reduced market price. However, in view of its ecological and other values, the environmental elements are not traditional replaceable good, which can be fully compensated by buying the equivalent goods in the market. At the same time it is also arbitrary to classify ecological damage into traditional irreplaceable good. Unlike death and serious physical injury, for which the victims cannot be restored to the status before damage, the injured environment can often at least be partly recovered. Thus the compensation for ecological damage can be divided into two kinds: to the extent the environment can be restored, it is analogous to replaceable goods; for the part that cannot be fully restored and the loss during restoration, it can be compensated as irreplaceable goods.

Ecological damage cannot be fully compensated by the price of a market equivalent, for the market price of an environmental element often excludes its ecological, esthetic and other values. However, by taking restoration measures, or

156 Kaplow & Shavell 1996, p. 192.

157 Danzon 1984, p. 527-530.

158 Arlen 1993, p. 1106.

when restoration is not possible, replacing the injured environment with one offering the same ecological functions and service, the injured values of the environment can also be compensated. By assessing the costs of these measures, this part of the damage can easily be quantified into monetary awards.

Having the advantage of compensating ecological damage comprehensively, the restoration approach is followed both in America and in the EU. In the U.S. legislation, restoration costs have a broad scope, including 'the cost of restoring, rehabilitating, replacing, or acquiring the equivalent of, the damaged natural resources.'[159] Firstly, actions should be taken to 'return an injured resource to its baseline condition, as measured in terms of the injured resource's physical, chemical or biological properties or the services it previously provided.'[160] For the resources that cannot be restored, its ecological functions and services can also be compensated by replacement and acquisition of the equivalent, which means 'the substitution for an injured resource with a resource that provides the same or substantially similar services.'[161] In the ELD, the restoration-based approach is also followed – environmental damage is compensated by asking the liable operators to bear the costs of prevention and remediation measures. To guide the remediation of damage to water or protected species or natural habitats, the ELD differentiates three types of remediation measures: primary remediation, complementary remediation and compensatory remediation. The first category means 'the remedial measure which returns the damaged natural resources and/or impaired services to, or towards, baseline condition.'[162]

Even when the restoration measures are adopted, sometimes the damaged natural resources and impaired services cannot be recovered to their baseline condition (the condition before damage happens). Moreover, restoration takes time; the services of natural recourses are still lost before they are fully restored. These losses, which are more like irreplaceable goods, cannot be compensated through direct quantification. Thus the approach of willingness to pay and willingness to accept which were proposed to compensate death and serious physical damage could be a useful method here. This method has been adopted in the U.S. In the DOI regulation, the loss of natural resources service pending the restoration is called compensable value. In determining the valuation methodology of compensable value, the regulation stipulates: 'The authorized official may choose among the valuation methodologies listed in this section to estimate willingness to pay (WTP) or may choose other methodologies provided that the methodology can satisfy the acceptance criterion in paragraph (c)(3) of this section.'[163] Thus willingness to pay has been adopted as an important evaluating method for compensable value, which can be regarded as damage to irreplaceable goods.

The economic analysis part has shown that the injurer's knowledge of the harm is essential for providing optimal deterrence. For the ecological damage, due

159 33 U.S.C. §2706(d)(1)
160 43 C.F.R. §11.14(ll).
161 43 C.F.R. §11.14(ii).
162 Annex III, 1 of the ELD.
163 43 C.F.R. §11.83(c)(2).

to the technical complications and uncertain causation, it is not always easy for injurers to predict the harm his activity may cause. Thus a design to ensure the knowledge about potential harm available to injurers would be very important. This design can be achieved by elaborate procedural rules of assessment. In the U.S., the DOI regulation and NOAA regulation provide detailed assessment procedures, which act as guidance to both trustees and liable parties. The rules about baselines, and exposed pathway give guidance about identifying causation and make the potential harm of the injurer's activity more predictable for them.

In addition, the scheduled awards are also embodied in the U.S. natural resources assessment. For example, in the DOI regulation, two types of procedures are established: type A assessment and type B assessment. Type A assessment is a simplified assessment by using the standard procedures and requires minimal field observation to determine damages. The applications of type A assessments are limited by a series of criteria to ensure that the damages covered are of similar kinds and thus can be determined by simplified models.[164] For the other damages, which do not·enjoy so many similarities, type B assessments should be adopted, and more field observations are incorporated into the assessments.

3. The Evaluation of Liability Rules

The increasing ecological damage has brought much loss to the public, which sometimes turns out to be catastrophic and irreversible. In response to the mushrooming damage, legislators and policy makers have made efforts to, on the one hand, prevent ecological damage *ex ante*, and on the other hand, compensate the victims and restore the impaired environment. As mentioned in the introduction to this chapter, prevention and compensation both serve as the goals of legal rules concerning ecological damage.

Regulation is a broadly used and traditional instrument to prevent ecological damage, the scope of which is said to be limited only by the imagination of regulators.[165] Operating largely *ex ante*, regulation instruments can, with various manifestations, such as direct ordering and controlling, market-based mechanisms, administrative guidance and coordination, provide potential injurers with incentives to adopt proper care levels and activity levels to reduce damage. However, this advantage of regulation should be balanced with the high information requirement and substantial administrative costs it may cause.[166] Furthermore, regulation can be more probable to 'be subject to political pressure and to capture by the regulatees and to collusion under different forms.'[167] Thus regulation alone cannot guarantee optimal incentives to prevent ecological damage. What is more, regulation instruments are largely prospective, and do not aim at

164 43 C.F.R. §11.34.

165 Viscusi 1989, p. 70.

166 Shavell drew four criteria in choosing regulation or liability, including information asymmetry, the possibility of insolvency, the threat of suit for harm and the administrative costs. See Shavell 1984a, p. 357-374.

167 Boyer & Porrini 2002, p. 258.

compensation. Therefore, regulation alone is far from an ideal instrument to realize the aims to prevent and compensate for ecological damage.

Compared to regulation, liability rules are a relatively new instrument in response to ecological damage. Liability rules concerning ecological damage were discussed from a theoretical perspective in the last section. This section tries to examine how liability rules can contribute to preventing and compensating for ecological damage, and if prevention and compensation cannot be successfully achieved, which remedies can be designed to cure the insufficiencies.

3.1. *How Do Liability Rules Contribute to Prevention and Compensation?*

3.1.1. Compensation

'The tort system is one of the oldest systems used to manage economic and social activities by allocating the burden of a loss to an injurer.'[168] Liability can be used to reinstate victims at the expense of injurers' financial assets. From this perspective, liability does have a function in compensation for ecological damage. However, the aforementioned arguments have shown compensation is not such a goal of liability rules for ecological damage.

Under liability rules, the obligation to restore and compensate for the impaired environment exists only when some conditions are satisfied to establish liability. As a necessary element to establish liability, the causal link between potential injurers and harm is difficult to prove: scientific uncertainty may exist when judging whether an activity results in certain harm; the existing background risk is not easy to be differentiated from damage actually caused by a certain activity; and there may be many similar activities relevant to the damage, making it difficult to identify the real injurers. Thus, if the causal link cannot be established or the injurers cannot be identified, ecological damage will be left uncompensated and the impaired environment not restored.

The traditional tort system is fault based, under which the injurer is only liable if he failed to satisfy the due care level. If the injurer adopts the due care level in his activities, and the damage still results from his activities, he can escape from liability and the victims cannot get compensation. Even when it comes to strict liability, which has been introduced and widely applied to ecological damage,[169] there is not a sufficient guarantee for compensation. The injurers are liable irrespective of the care level they adopt; however, possible defense, such as force majeure and regulatory compliance, can still restrict the victim's protection.

Even if all the aforementioned requirements are satisfied so that the liability can be established, the victims may only be compensated theoretically. The realization of awarded damages depends on the financial assets of the liable parties. After World War II, the development of some dangerous industries, such as oil

168 Fiore 2009, p. 406.

169 For example, 42 U.S.C. §9607(a), 33 U.S.C. §2702(a), Art. 3(1)(a) of the ELD.

transportation, nuclear devices, and treatment of hazardous substances, has led to many accidents or accumulated damage with enormous losses. These immense damages put a high requirement on the financial status of the liable parties, whose solvency is, unfortunately, far from infinite. Therefore, even when damages are awarded in judgments, the limited solvency of liable parties, unaccompanied by financial guarantee mechanisms, can still fail to guarantee appropriate compensation for ecological damage.

3.1.2. Prevention

The design of liability rules for ecological damage in the last section bears in mind the goals to provide optimal incentives to prevent damage. However, as far as ecological damage is concerned, there are many hindrances to offering optimal incentives.

Information asymmetry is a problem facing legislators and judges in determining optimal incentives. As discussed in last section, setting the due care level equal to the optimal level is essential in giving liable parties efficient incentives under the negligence rule. This has placed a heavy burden on judges who need a comprehensive understanding about risk-reducing alternatives and their costs, the benefits of these alternatives in reducing the accidents losses et cetera to find out the efficient care level. Not as sensitive to the care level set by the judges as the negligent rule, strict liability requires accurate information about the magnitude of losses. To evaluate the ecological damage, which often involves losses independent of the market value, is also a difficult job for the judges. Without this information, liability rules can hardly provide efficient deterrent effects.

In addition to leaving ecological damage uncompensated, the problem of insolvency can also lead to under-deterrence. Under strict liability, the injurer only has incentives to avoid an accident with a magnitude equal to his total wealth.[170] If the losses are larger than his total wealth, the problem of under-deterrence will arise. Compared to that under strict liability, insolvency is less of a problem under negligence. As long as the costs of taking care are less than the injurer's wealth, he will still have incentives to behave carefully. If the injurer's financial assets are not sufficient to cover the costs of care taking, the under-deterrence problem remains.

What is more, the deterrent function of liability rules also relies on the probability of being sued. If all damage caused by a certain actor can be detected and claimed for, the injurer will internalize all the costs he causes and have optimal preventive incentives. However, as far as ecological damage is concerned, much damage is left undetected or is only found out after a long time. Usually, it is the public at large, instead of individuals, that suffers the losses. Hence there are no strong incentives to bring such as lawsuit. In this case, the probability to be sued is low for ecological damage, which can also lead to under-deterrence.

As discussed above, although they are beneficial complements to regulation and provide preventive incentives, and they also have a compensatory function,

170 Faure & Grimeaud 2003, p. 35.

liability rules themselves are neither satisfactory in the prevention nor the compensation of ecological damage.

3.2. *Remedies – Compensation Mechanisms*

The discussion above reveals the insufficiency of liability rules in both prevention and compensation. Thus other compensation mechanisms need to be developed as remedies to ensure that, on the one hand, the costs of damage are internalized so that optimal deterrence can be achieved and, on the other hand, victims have actual access to compensation and the injured environment can be restored.

Various compensation mechanisms have been established or are required/encouraged to be developed in different jurisdictions. In the U.S., compensation mechanisms are introduced under the heading of 'financial responsibility'. For example, in the CERCLA, a mandatory obligation to ensure financial responsibility is imposed on the owner or the operator of each vessel over three hundred gross tons. This responsibility can be realized by any one or any combination of insurance, guarantee, surety bond, or qualification as a self-insurer.[171] Similar provisions can also be found in the OPA.[172] Compensation mechanisms are mentioned under the title of 'financial guarantees' or 'financial security' in the ELD.[173] The ELD does not elaborate the formulations of financial guarantees: it just mentioned insurance and other types of financial security. A detailed exploration of individual mechanisms is discussed in following chapter. This section focuses on how these mechanisms can act as remedies to liability rules to prevent and compensate for ecological damage.

3.2.1. Compensation

As discussed in last section, compensation is not a goal of liability rules. Other compensation mechanisms can provide a more secure access to compensation or at a cheaper price.

Firstly, with the burden of proving causation, restrictions on *locus standi* and difficulties in damage evaluation, liability for ecological damage is difficult to establish. These requirements restrict the application of the tort system and the compensation awarded to victims. Compensation mechanisms can help to overcome these difficulties: some mechanisms are independent of liabilities, such as first-party or direct insurance and a general environmental fund. Under first-party insurance, the victims seek insurance coverage and pay for the premiums. Under direct insurance, the potential injurer can act as policyholder on behalf of the injured parties. However, different from liability insurance, the trigger to compensate is no longer liability but the existence of damage.[174] A general

171 42 U.S.C. §9608(a)(1).
172 33 U.S.C. §2716(a).
173 Arts. 8(2), 14 of the ELD.
174 Faure 2007a, p. 88.

environmental fund can operate as a substitute for liability and insurance.[175] Therefore, under those instruments, the impaired environment can be restored or compensated even when liability cannot be established.

Secondly, compensation instruments can also act as a solution for the insolvency of liable parties which has become a substantial restriction for actual compensation. To guarantee compensation in the case of insolvency has been identified as an important task of financial measures both in the U.S. and in the EU. It is explicitly incorporated in the ELD that 'Member States shall take measures to encourage the development of financial security instruments…, including financial mechanisms in case of insolvency.'[176] Although not containing the task literally in CERCLA, the provision places the obligation on the liable parties to establish and maintain certain financial responsibility to cover their potential liability under 107(a) (1) of CERCLA.[177] Through this legal design, a stable financial resource is established before damage actually happens, so that compensation in case of insolvency can be better guaranteed. Many types of compensation instruments can contribute to this task, such as sectoral public and private environmental guarantee funds and insolvency insurance for the benefit of third parties.[178] A guarantee fund compensates the injured party in the event of insolvency of its members, with the assets assembled through voluntary or compulsory contributions from operators of similar activities. The insolvency insurance only covers environmental liability in the event of insolvency of the operator. The Swedish and Finnish Environmental Damage Insurance are interesting examples.[179]

Compensation mechanisms can also foster timely, low-cost access to compensation. The litigation to establish liability for ecological damage is a sophisticated, time-consuming process. 'Even when liability is firmly established, the possibility of appeal, delay and uncertainties associated with penalty collection can complicate the actual transfer of funds from defendants to victims and resource trustees.'[180] Rather than established *ex post* as liability, many compensation mechanisms are set up before the damage happened, and also provide simpler claim procedures for victims or other claimants for ecological damage.

3.2.2. Prevention

Prevention is an important aim of liability rules for ecological damage, which however again play only a far from satisfactory role. Although called 'compensation mechanisms', they can also improve the prevention of ecological damage.

Firstly, since the legal structure of the tort system itself restricts its application and the probability of being sued for potential liable parties is low, the potential liable parties are to a large extent under-deterred. Compensation mechanisms can

175 Faure & Grimeaud 2003, p. 238.
176 Art. 14(1) of the ELD.
177 42 U.S.C. §9608(a)(1).
178 Bocken 2009, p. 162-164.
179 Bocken 2009, p.164.
180 Boyd 2002, p. 423-424.

ensure that 'the expected costs of environmental risks appear on a firm's balance sheets and in its business calculations.'[181] Without compensation mechanisms, only the costs of damage which are attributed to an injurer in the judgment are taken as productive costs and internalized by him. He does not have to bear other costs even though they are actually caused by him. Compensation mechanisms, if designed carefully, can help to relieve this deficiency. For example, an environmental fund can offset the losses or be used to restore the impaired environment for certain kinds of damage. The existence of liability is not a necessary condition for offering compensation. Thus, if the fund is financed by actual injurers and if the contribution is allocated according to their contributions to the damage, such a mechanism can make the costs of ecological damage better internalized by the injurers and thus promote prevention.

Furthermore, as mentioned above, insolvency is also a hindrance to providing optimal incentives to reduce damage under liability rules for ecological damage. Compensation mechanisms, which are usually the results of the joint action of multiple potential injurers, or based on larger risk pooling, can provide better guarantees in case of injurers' insolvency. Compensation mechanisms can also contribute to better prevention.

There may be fear that if potential injurers choose compensation mechanisms with financial assets from third parties, such as banks and insurers, they may then be relieved from potential liability and have fewer incentives to prevent ecological damage. However, under this situation, the third party concerned will have a strong incentive to monitor the environmental safety of the potential injurers they provide assurance. The third party can set up the costs of capital or premiums according to the attributes of the potential injurers' activities. He can also only compensate within a certain cap to expose the potential injurers partly to the risk. These measures can help to generate incentives to reduce ecological damage.

3.2.3. Risk Spreading and Costs

Prevention and compensation are two aims of legal rules concerning ecological damage. Compensation mechanisms can contribute to the realization of these aims. Furthermore, a compensation system also influences risk spreading of ecological damage and generates new costs which should be balanced with the benefits it brings.

Liability rules place the burden of the whole damage on the injurer if the liability can be established. Otherwise, the impaired environment and other victims will not be compensated at all. Ecological damage often turns out to be enormous. The injurers, who are usually industrials, may be risk averse. The utility of the parties who take the restoration measures for the impaired environment can also be improved through risk-spreading mechanisms. Except mechanisms based on the assets of the liable party himself, such as self-insurance, mortgages and deposits, most compensation mechanisms have risk-spreading effects. For the mechanisms

181 Boyd 2002, p. 423.

financed by many potential liable parties, such as liability insurance and environmental funds, they can shift at least part of their liability to a larger pool by paying a smaller amount of money. As far as first-party insurance is concerned, victims themselves have to bear the costs. Under those mechanisms, through the involvement of many potential injurers or victims, the risk of a large amount of damage can be spread and is thus removed from the victims.

In spite of the desirable effects mentioned above, as additional mechanisms to regulation and liability rules for ecological damage, they can also generate new costs, which should be taken into consideration before advocating developing new compensation mechanisms. This argument has been used to oppose new or strengthened financial assurance rules. For example, the OPA contains vessel assurance rules, which was predicted to increase the insurance cost by seven to nine times of pre-OPA rates.[182] Some even came to more dire predictions including the possibility of a total halt in maritime trade and the collapse of worldwide vessel insurance markets.[183] However, to consider the costs of compensation mechanisms, a distinction should be drawn between created and redistributed costs. Compensation mechanisms with financial assets attributable to potential injurers force the internalization of damage costs which are otherwise avoided. They are new and additional costs for the potential injurers. However, from the social perspective, only those costs are only redistributed from victims to injurers and not new social costs. The cost-benefit analysis of compensation mechanisms should only take into account the new costs rather than redistributed costs. Compensation mechanisms, as new instruments to deal with ecological damage, can create real costs during their operation. The operators of those mechanisms need to collect information in deciding the coverage, premiums or sharing costs and other relevant issues. Administration, compliance and possible dispute settlement also create costs. To weigh those new costs against the benefits compensation mechanisms bring needs further research under each specific mechanism and empirical research is also necessary. Thus, this issue is discussed in the next chapter and in the empirical part of this thesis.

4. Conclusion

Booming science and technology have brought ecological damage into view and lead to increasingly legal rules on the prevention and compensation of ecological damage.

Regulation is a traditional approach to prevent ecological damage. Environmental standards are set to provide minimum protection for the environment; permits and licenses keep out the unqualified potential polluters; administrative guidance promotes the development and application of more environment-friendly technologies; market-based methods give actors flexibility in choosing the way to meet environmental protection obligations. However,

182 Shogren, Herrige & Govindasamy 1993, p. 109-133.
183 Boyd 2002, p. 435.

regulation cannot always provide the optimal incentives to reduce damage, with the restrictions of high information costs and administrative costs, and possibility of being captured.[184]

The deficiency of regulatory instruments makes liability rules a beneficial complement to provide incentives for prevention. Liability rules ask the injurers to bear the damage costs they create. The cost internalization gives potential liable parties incentives to reduce accidents to the extent that the marginal benefits of reduction in expected damage equals the costs of accident avoidance. Furthermore, liability rules may also promote compensation to victims. Liability itself intends to compensate the damage suffered by the victims with the resources or efforts of the liable parties. Thus, unlike regulatory instruments which are mainly used to prevent ecological damage, liability rules also have positive effects on compensation. However, as mentioned earlier in this chapter, from an economic perspective, compensation itself should not be treated as the primary goal of liability rules for ecological damage.[185] Many other compensation mechanisms can offer timely and much cheaper access to compensation. Liability is limited in offering victim compensation due to the difficulties in proving causation, existence of various defenses, and the needs to establish a duty of care under the negligence rule.

This chapter tried to examine to what extent liability rules for ecological damage are capable of providing optimal incentives to prevent damage. In the discussion, both economic analysis and legal analysis are used. From a law and economics perspective, this chapter tries to find the liability rules which can minimize total social costs of industrial activities. A comparison between different alternatives showed to what extent they can achieve the optimal balance between costs of accident avoidance and expected damage. The influence of various legal designs on risk allocation and risk spreading, as well as of the relevant administrative and enforcement costs, was also taken into account. In this chapter, some legislation and cases in the U.S. and the EU were also explored briefly, because liability rules should be designed within the limitations of a certain legal framework. Moreover, the variation of specific legal rules may help to find out different alternatives actually available in legal systems to certain legal issues.

Granting standing to sue is a major difficulty in establishing liability for ecological damage. The design of *locus standi* relates to the tradeoff between incentives to sue and relevant costs. If an explicit obligation to restore the environment is imposed on public authorities, and they have the capacity to perform environmental monitoring and management, public authorities can enjoy advantages of incentives and lower costs. However, public authorities are susceptible to influence from lobby groups. On the contrary, NGOs are more indifferent to lobbying but information costs and enforcement costs are high for them. They are eligible for filing litigation on behalf of the environment when they are equipped with necessary knowledge and expertise. Hence it is beneficial to

184 See Shavell 1984a.

185 See Chapter 3, Section 1.

grant the NGOs and individuals at least secondary standing to bring a claim for ecological damage.

A liable party is the party who should bear the losses brought by ecological damage and on whom the preventive incentives should be imposed. Economic analysis shows that the liable party should be the one who actually caused or contributed to the damage, thus channeling of liability is inefficient. To avoid the channeling rules, the liable parties should not be defined too restrictively, and recourse to a contributable third party should not be prohibited.

The debate about liability standards concerns the choice between strict liability and negligence rules. The choice depends on many conditions and presumptions, and the results can change dramatically when a certain presumption is relaxed. If judges can determine the due care level appropriately and assess the damage accurately, both rules can give incentives to adopt an optimal care level. In a unilateral case, strict liability is preferable to negligence since the former can give injurers incentives to reduce the activity level. In bilateral cases, since neither of the alternatives can give both parties incentives to adopt an optimal activity level, the choice relies on who can prevent the damage better. Moreover, strict liability is sensitive to the accurate assessment of damage; and negligence to an accurate assessment of the due care level. Thus the information problem should also be borne in mind when making a decision. Compared to negligence, strict liability is more vunerable to insolvency of injurers. Furthermore, when deciding the influence of the activity level, it is also important to examine whether an activity is so dangerous that its activity level should be reduced. As far as ecological damage is concerned, both dangerous activities and other industries can contribute to the damage. The damage evaluation and determination of the due care level are significant difficulties for judges. Elaborate damage assessment rules such as the DOI and NOAA regulations in the U.S. can help to make accurate assessments. Financial security is also beneficial to solve the problem of insolvency. If those conditions are satisfied, strict liability is more favorable for dangerous activities, while for other activities, negligence rules may apply.

Uncertainties in causation and damage caused by multiple tortfeasors are also difficulties in the establishment of liability rules for ecological damage. For uncertain causation, threshold probability rules and proportional liability are the main alternatives. From the deterrent aspect, proportional liability is more favorable since it adopts an *ex ante* perspective and can give the injurers better incentives than under threshold probability rules where the injurers are either over-deterred or under-deterred. However, when it comes to ecological damage, the proportion that an activity contributes to a result can hardly be determined precisely, which makes proportional liability impractical. Although not a first-best choice exists for ecological damage, assessment regulations with the information about how to indentify damage, finds pathway and quantify the damage can help to reduce the uncertainties and establish causation. As far as the problem of multiple tortfeasors is concerned, there is no dominant preference for either joint and several liability or non-joint liability.

To create efficient incentives, awarding damages correctly is also essential. For the injury to replaceable goods, full compensation is necessary in providing optimal preventive incentives under strict liability but not under negligence. For the damage to irreplaceable goods, as can be drawn from economic analysis, an *ex ante* system is preferable, which means the *ex post* damages awarded are determined by the *ex ante* value of the injured objects. Ecological damage can be treated as the combination of damage to both replaceable goods and irreplaceable goods. The injured environment can be deemed as replaceable as long as it can be restored or replaced.

The part of damage which cannot be restored and interim losses should be compensated as irreplaceable goods. Thus the willingness to pay can be used as an evaluation method. Models can also be established for relatively small and fixed types of damage.

The design of liability rules tries to give optimal incentives to prevent ecological damage. However, liability rules can hardly realize this goal by themselves. Insolvency of injurers is a substantial obstacle to ensuring efficient incentives as far as ecological damage is concerned. In view of the long time lapse between activities and the emergence of damage, difficulties in the establishment of liability and high litigation costs, the probability of being sued may be low for potential injurers. Thus they are largely under-deterred. Furthermore, liability also has high information requirements, such as accurate information about the magnitude of losses or due care level, which is difficult to reach for ecological damage. In addition, the economic analysis shows that although liability rules have positive effects on compensation, compensation is not the goal of liability rules. The difficulties in establishing liability for ecological damage leave much damage uncompensated. Although lowering the threshold of liability, such as lowering the burden of proof and broadening access to justice, may achieve better compensatory effects, the main goal of liability rules, that is deterrence, has to be compromised. Thus other compensation mechanisms are necessary to both prevent and compensate for the ecological damage. Compensation mechanisms can also promote risk spreading, but those benefits should be balanced with the costs they create. The individual mechanisms are examined in next chapter.

Chapter 4

ANALYSIS OF SPECIFIC COMPENSATION MECHANISMS

1. Introduction

Liability rules for ecological damage have been or are being established in many jurisdictions in response to the frequent and huge damage to the environment. As has been shown in the last chapter, subject to the difficulties of establishing liability, the large possibility of insolvency and the low probability of being sued, liability rules are far from satisfactory to either prevent or compensate for ecological damage. Hence, compensation mechanisms may be beneficial to overcome these deficiencies. As an introduction, a brief overview is provided for the available or possible compensation mechanisms. However, an exclusive and elaborate list is not provided here. After this overview, an analysis is made of the criteria to evaluate the specific compensation mechanisms.

1.1. *A Brief Overview of Compensation Mechanisms for Ecological Damage*

Insolvency has been a notable hindrance for liability rules to play an effective role in prevention and compensation of ecological damage. Not all the causers of ecological damage are big and wealthy firms. Even the wealthy ones are not immune from financial difficulties when facing possible catastrophic damage to the environment. One may easily come to liability insurance as a remedy to the insolvency problem. The tandem of liability rules and liability insurance can, on the one hand, leave the tort system untouched and, on the other hand, guarantee the compensation in case of the insolvency of the insured. Furthermore, liability insurance is also a risk-spreading mechanism which can improve the utility of the insured. Under this policy, a potential injurer can pay a certain and smaller amount of premium to protect himself from future liabilities. Thus the burden of compensation is shifted to the insurer, and the utility of a risk averse insured is increased.

Liability insurance has been introduced in many countries for ecological damage. In the U.S., 'environmental insurance developed from the 1980s, as consequences of the adoption of financial guarantee requirements in major federal environmental legislation.'[1] Under the U.S. legislation, the obligation of obtaining a financial guarantee is imposed on certain operations that are liable for natural resources damage under OPA and CERCLA.[2] Thus, financial responsibility is not exclusive for natural resources damage but also covers other liabilities under the legislation, such as response costs, removal costs, personal and property damage. Four mechanisms can be used to realize the financial responsibility for firms: insurance, surety bonds, self-insurance and financial guaranty.[3] These mechanisms are used by the potential injurers to realize their statutory liability, so the term 'insurance' mainly refers to liability insurance. The German Environmental Liability Act of 1990 even established an obligation of compulsory insurance, but only for particularly dangerous facilities listed in Annex 2.[4]

The risk should be insurable to make liability insurance function well. Different arrays of criteria have been proposed as to insurability.[5] Generally speaking, a risk should be accidental, determinable and measurable, independent and non-catastrophic.[6] When applied to liability for ecological damage, these criteria are not easily satisfied. The damage and its magnitude are often difficult to be indentified and measured even *ex post*, let alone to predict the risk of being held liable. The long-tail risks, difficulties in indentifying liable parties and uncertainty in causation combine to make the risk of liability for ecological damage less determinable and measurable for the insurance industry. Moreover, ecological damage sometimes turns out to be catastrophic. All those factors attenuate the insurability of liability insurance for ecological damage. In addition to the considerations about insurability, liability insurance is also criticized for the limited financial capacity of insurers, problems due to moral hazard and adverse selection, et cetera.[7] Given the deficiency of liability insurance, other alternative compensation mechanisms are designed to promote the prevention and compensation of ecological damage.

In the Netherlands, the environmental risks were previously mainly covered through liability insurance policies, of which the most important are: the liability insurance policy (AVB),[8] the environmental liability insurance (MAS)[9] and fire insurance for cleanup costs after fires.[10] The AVB covers sudden risks and occupational health risks related to the environment and the MAS covers risks of a

1 Bocken 2009, p. 147.
2 Boyd 2003, p. 259.
3 33 C.F.R. §138.80, 30 C.F.R. §253.28-31. These are the two financial assurance rules under OPA and CERCLA. The first rule governs waterborne vessels that carry oil or hazardous substances. The second rule governs offshore facilities used for oil exploration, drilling, production, or transport.
4 Taupitz 1993, p. 23.
5 See Wagner 2007, p. 87.
6 Wagner 2009, p. 399.
7 Faure 2007a, p. 76-78.
8 Aansprakelijkheidsverzekering bedrijven.
9 Milieu-aansprakelijkheidsverzekering Samenwerkingsverband.
10 Faure 2002a, p. 36.

more gradual nature. These arrangements have received criticism widely over the division of cover between the AVB and the MAS, the complicated procedure under the MAS, and the unclear definition in fire insurance.[11] These deficiencies are aggravated by the unpredictable expansion of liability by the judges, which lead to a shift towards first-party and direct insurance. The Dutch Insurers Association in 1998 presented the environmental damage insurance (MSV)[12] to take the place of the traditional liability insurance under the AVB and the MAS. The MSV provided integrated cover of all environmental damage occurring on or from the insured site if it concerns the pollution of soil or of water.[13] This policy provides direct insurance rather than liability insurance since the trigger for compensation is no longer liability but the damage itself. It is the insured's site that is insured. Thus once damage happens to or from the insured site, the cleanup costs are compensated through the insurance policy. The trigger of coverage is no longer liability but damage itself.[14]

In addition to insurance, many other alternative compensation mechanisms are also developed, such as risk sharing agreements, funds and self-insurance. Risk pooling by potential liable parties is a useful mechanism to increase the capacity to cover ecological damage. In some industries such as nuclear and oil transportation, the risk is so technical and complicated that the operators themselves rather than the insurance industries have the better information to predict the risks, make risk differentiation and monitor the operation. Under this situation, a risk-sharing agreement is a beneficial substitute for insurance. For example, Protection and Indemnity Clubs (P&I Clubs) provide insurance for marine oil pollution liability for the members. At the beginning of each year, a 'call' is made from the members — tank owners to cover the claims and administrative costs.[15] Other examples are the risk retention groups (RRG) in the U.S. The Federal Liability Risk Retention Act[16] stimulated the creation of professional mutual insurance companies under the head of RRG. The number of RRG rose from 60 to 254 from 1998 to 2007.[17]

Compensation funds are also mechanisms to promote the capacity to compensate for ecological damage and can operate in tandem with or independent from liability rules. The use of the term 'funds' can be misleading since it is widely used for a variety of financial arrangements. This research differentiates three types of funds: the funds which can be used to cover liability; the ones intervening only when liability cannot be established and the ones providing coverage irrespective of liability.

The aforementioned mechanisms all play a role in risk spreading — the potential injurers can either shift the risk of compensation to an insurance company or funds by a certain and smaller contribution to the risk pool or share the risks by

11 A summary of the criticisms can be found in Faure 2002a, p. 37.

12 Milieuschadeverzekering.

13 Faure 2007a, p. 91.

14 For the details on the Dutch direct insurance, see Chapter 8, Section 2.2.

15 See Bongaerts & Debièvre 1987, p. 145-187.

16 15 U.S.C. §§3901-3906.

17 Bocken 2009, p. 166.

gathering and building a larger risk pool. In addition to those risk-spreading mechanisms, there are still some other guarantee mechanisms which are based on the assets of the liable party himself or a third party guarantor. Under those mechanisms, risks are still borne by a few parties, thus they have a limited function in increasing the capacity to compensate. However, they can still play an important role to guarantee the availability of the assets when damage happens. A number of mechanisms fall into this category, for instance: a declaration of self-insurance, mortgages, liens and other traditional security interests in immovable or movable property, deposits, and guarantees by parent companies and bank guarantees.[18]

In addition, the capital market is also proposed as an alternative to increase the capacity to compensate for catastrophic damage. However, the capital market is used more to cover natural disasters and it has its own limited practical application as far as manmade environmental damage is concerned.

1.2. *Criteria for Compensation Mechanisms: Compensation, Prevention and Costs*

The aforementioned mechanisms provide a variety of methods to compensate for ecological damage. This section tries to identify a series of criteria in evaluating the compensation mechanisms. Some of those criteria have been discussed in the literature. In his research on the financial guarantees under the ELD, Bocken tries to find out the meaning and purpose of the financial provision through grammatical and textual interpretation, historical interpretation and systematic interpretation.[19] Based on the interpretation of the text and the purpose of the ELD, as well as Article 174 EC (now Article 191 TEU), he proposed three factors as the criteria for the evaluation of financial schemes: 1. the polluter pays principle and the effectiveness of the security mechanisms, 2. the proportionality principle and the cost of the security mechanism, and 3. the preventive effect of the mechanism.[20] Those criteria are derived from the basic principles of European environmental law contained in the recital of the ELD as well as from the provisions of the EC treaty (now the treaty on the Functioning of the European Union). Boyd analyzes the financial assurance rules in the U.S. from another perspective. Although he does not explicitly refers to formulations of 'principles' or 'criteria', he focuses his study on three aspects: the cost of assurance, the scope of the assurance rules (the liabilities and obligations that are covered by assurance, the dollar value of coverage that must be demonstrated) and the security of the assurance mechanism (the ability to deliver compensation when the compensation obligation comes due).[21] In this research, Boyd developed his discussion from the perspectives of both costs and effectiveness. However, as far as effectiveness was concerned, he laid his emphasis on compensation only, but not on the prevention of the damage.

18 Bocken 2009, p. 160-162.
19 Bocken 2006, p. 19.
20 Bocken 2009, p. 156-158.
21 Boyd 2002, p. 417-485.

As shown in the previous chapter, ecological damage needs to be prevented to the extent that total social costs are minimized. Moreover, compensation (restoration) can contribute to the deterrent effect and to corrective justice.[22] Even though liability rules have a positive effect on compensation, their aim is mainly focused on prevention. Thus the compensation mechanisms need to supplement the liability rules in the function of compensation: that is to provide the victims with swift and ample compensation, or restore the injured environment effectively. Since prevention can be regarded as the best way of victim protection, the compensation mechanisms should not counteract the effects of the tort system to provide incentives to prevent the damage. Furthermore, tort itself cannot guarantee efficient prevention, largely due to the insolvency problems, the low possibility to be sued and difficulties to establish liability. Therefore, a desirable compensation mechanism helps to promote the prevention effects by providing a remedy for those problems. Even when a mechanism is favorable from the perspectives of both prevention and compensation, it is still impractical if it is unaffordable or the costs are so high to offset its benefits in improving prevention and compensation. Thus three criteria should be considered in evaluating the compensation mechanisms: their effects on compensation, prevention and the costs of the mechanism. Admittedly, those criteria are only theoretical considerations which determine whether a mechanism is inherently suitable as a financial guarantee for ecological damage. However, it is noteworthy that the actual success of individual mechanisms is to a large extent 'determined by the political weight of the interests groups involved and by the economic development and the sophistication of the financial systems and of the environmental authorities of the countries in which they are used.'[23]Those issues, nevertheless, are outside of the scope here, but form part of the topic covered in the empirical part.

The criteria discussed above are also in line with Calabresi's well-known total social cost theory. According to Calabresi, the principal function of accident law is to reduce the costs of accidents and the costs of avoiding accidents, which can be divided into three sub-goals.[24] The first is reduction of the number and severity of accidents, focusing on the deterrent function; compensation is termed secondary, which concerns reducing the societal costs resulting from accidents. It can be accomplished in two ways: the risk-spreading method and the deep-pocket method. The third sub-goal involves reducing administrative costs.[25] Those three costs can be integrated with the three criteria to evaluate compensation mechanisms. Prevention is in line with reducing primary costs, which is offering optimal deterrent incentives to minimize the sum of accidents avoidance costs plus the costs of damage. Reducing secondary costs concentrates on increasing the utility of the property of the same value by requiring the one that is less risk averse to bear the losses or

22 Noticing that compensation here does not mean that all ecological damage needs to be fully compensated. The economic criteria require compensation only to the extent to minimize total social cost. See *supra* Chapter 3, Section 1.

23 Bocken 2009, p. 156.

24 Calabresi 1970, p. 26-31.

25 *Ibid.*

spread the losses. Secondary costs reduction is similar to the compensation function in the sense that they both concern allocation of the losses. However, those two concepts are not identical. Compensation focuses on making up the losses and redressing the victims with the resources of injurers or others. While secondary costs reduction involves the efficient allocation of losses. The victims can be cost bearers if they are less risk averse than the injurers. Another distinction between the two concepts is that compensation places emphasis on the victims' interest, thus containing the notions of ample and timely payment, while the other focuses on efficiency. The tertiary costs fit the criterion of reducing administrative costs. Therefore the evaluation of compensation mechanisms can also be explained by the total social costs theory, with some supplements to the secondary costs reduction.

To minimize the primary costs requires providing optimal preventive incentives. Under some mechanisms, such as insurance and compensation funds, the risks of serious damage are removed from the potential injurers after joining those policies. Therefore, the question arises whether they will still have incentives to behave carefully and diligently to prevent damage. Two factors are essential to determine the preventive effects of a compensation mechanism: the contribution of the guarantee and the possibility of recourse.[26] To achieve the optimal preventive incentives, the potential injurers should internalize all the costs they create. Full internalization requires that the contribution to a funding should be in proportion to one's own contribution to the risk. Furthermore, some mechanisms may allow the recourse from the real injurers, which may also help to improve the preventive effects.

The secondary costs reduction, or the concept of compensation, is of fundamental importance for compensation mechanisms. Different from traditional personal and property damage, ecological damage refers to damage to the environment, which has not only market value, but also ecological, and esthetic values. To make both the use and non-use value[27] of the environment whole, compensation here requires first to take timely and effective restoration measures; and if (full) restoration is impossible, to provide monetary damages to the party who can represent the damaged environment.[28] Those characteristics affect the functions of the compensation mechanisms: they should be helpful to the restoration of the damaged environment; the victims of ecological damage are less likely to be a large number of similar victims as the usual case of personal damage or property damage via the environment. Although the claim for, on the one hand, ecological damage and, on the other hand, relevant traditional damage, such as the personal and property damage, pure economic losses may lead to different legal procedures, they can be caused by the same activities. The financial guarantees for them, therefore, may not be differentiated or independent. For example, the

26 See Bocken 2009, p. 158. When discussing the evaluation criteria for financial guarantees under the ELD, he lists the preventive effect as one criterion. The contribution and recourse are important factors in judging the preventive effect.

27 For the discussion on the use and non-use value of the environment, see Cross 1989, p. 281-85.

28 Faure & Liu 2012, p. 263-264.

financial assurance established under CERCLA and OPA in the U.S. is used to guarantee the liability in the statutes, including not only natural resources damage but also response costs, real and personal property damage, and lost government revenue. A policy can also only focus on ecological damage. A case in point is the environmental damage insurance in the Netherlands, which covers only 'cleaning-up' costs.[29] Thus, when analyzing the effects of a compensation mechanism on compensation, it should be borne in mind whether it provides compensation only for ecological damage or also for other damage caused via the environment.

To award effective compensation, several issues may be taken into account: the coverage of the mechanism, the time to establish the guarantee, and the capacity to provide ample compensation. Ecological damage is defined distinctly in different countries: sometimes a broader approach is adopted to contain damage to various kinds of natural resources, such as the concept of 'natural resources damage' in the U.S.; under the ELD in Europe, however, 'environmental damage' refers only to damage to the protected species and habitats, soil and waters.[30] Correspondingly, the scope of compensation mechanisms may vary. The coverage of a compensation mechanism can either be sector based or more comprehensive. Given the advantages in technology and mutual monitoring, industries may establish a risk pool together through a risk-sharing agreement, which is often sector based, such as the practice in nuclear and marine oil transportation industries. However, when a first-party or direct insurance approach is used, the coverage can be more general. Under this policy, the risk is evaluated on the individuals or specific sites (as the example of direct insurance in the Netherlands),[31] which is not closely linked to the type of industries. By adopting a comprehensive policy, the insurance company can establish a bigger risk pool and lower its corresponding administrative costs.

Furthermore, a compensation mechanism can come into existence at different moments.[32] An *ex ante* intervention can ensure that the costs of the damage be internalized *ex ante*. Under this situation, financial guarantees can be the conditions to start an operation. A second possibility is to set up the mechanism as soon as environmental damage is caused, such as the advancement funds, which offer compensation before the liability is established. The least invasive solution is to provide a guarantee only when the recovery is undertaken by the competent authority. This solution provides the least guarantee since whether the claim can be satisfied depends on the assets available at the moment of foreclosure.

Another importance issue is the capacity of a compensation mechanism to provide ample compensation. Compensation mechanisms can act as a remedy to the insolvency problem under the tort system. Thus how a mechanism can help to increase the capacity of potential injurers to realize his liability, or to compensate when liability cannot be established, is highly relevant here. A financial guarantee can be funded by the injurer himself or by a third party. In this situation, although no risk spreading exists, it can still provide a higher guarantee for the availability of

29 Faure 2007a, p. 90.

30 See *supra* Chapter 2, Section 2.1.1.

31 See Chapter 8, Section 2.2.

32 Bocken 2006, p. 29-30.

assets at the moment of recovery. The risk-spreading mechanisms, for example the insurance and risk-sharing agreements, are less vulnerable in the case of insolvency, since it can establish a bigger risk pool, and the risks are more predictable when the subjects are expanded. However, the mechanisms themselves, such as the insurers or funds can also go bankrupt. The capacity of each mechanism should be examined specifically. Moreover, compensation mechanisms can also play a role when the liable party cannot be identified. For example, first-party insurance and environmental funds can be used to realize this aim.

A compensation mechanism may be desirable from the perspective of prevention and compensation; however, the benefits are not achieved without a price. All these advantages need to be balance with the administrative and enforcement costs of each mechanism. The term 'costs' itself deserves more elaboration here, since some opposition to assurance is made under the head of high 'costs'.[33] A distinction should be drawn between created and redistributed regulatory costs. A compensation mechanism which requires *ex ante* assurance 'can raise a regulated firm's costs by forcing the internalization of otherwise avoided obligations.'[34] The newly-internalized costs are an increase in production costs for the firms. However, it is only a shift of costs from the victims to the injurers from the social perspective. Only the created costs should be taken into account in the cost-benefit analysis of the compensation mechanisms. The administrative costs, such as the operation of such a mechanism, the compliance and coverage conflicts litigated needs to be discussed under specific mechanism.

Admittedly, prevention, compensation and minimization of administrative costs may not be achieved together. Different compensation instruments may achieve one of the goals better than the others. For example, prevention requires that the contributions to the compensation mechanisms come from the ones who actually contribute to the risk; which means the number of contributors may be limited. Hence the capacity to compensate for a catastrophe may also be limited. Environmental funds can be used to provide a higher level of compensation. However, it usually involves high administrative costs. Hence, when choosing compensation instruments, a delicate balance between the three criteria should be reached.

This chapter does not give an exhaustive list of available compensation mechanisms. Instead, it discusses several typical mechanisms which are already established in some countries or usually proposed by literature. They fall into different categories: they may be funded by the injurers themselves or risk-spreading mechanisms; they can come into existence in different stages; and they may be of either a private or public nature. The following mechanisms are analyzed here: liability insurance, first-party and direct insurance, risk-sharing agreements, compensation funds, other security mechanisms and the use of capital market. Under each mechanism, a short introduction or example of the practice is given;

33 This is the case with regard to the OPA vessel assurance rules. It was estimated that the cost of insurance will be increased dramatically relative to pre-OPA rates so that there is a possibility of a total halt in maritime trade. See Cropw & Knott 1994, p. 14.

34 Boyd 2002, p. 436.

thus a clear contour of each mechanism and its characteristics can be drawn. Then the theoretical analysis of the specific mechanism is made from the perspective of compensation, prevention and the costs it may produce. After the individual discussions of these instruments, they are compared, based on their capacity to reduce the three categories of social costs.

2. Liability Insurance

Liability insurance is a usual instrument to ensure the realization of liability. Under such a policy, the risks of being held liable for certain damage are removed from the insured parties, at the price of premiums paid by them. In other words, liability insurance is a contract that covers potential harm suffered by people other than the insured, which is the claim of victims against injurers.[35] Liability insurance can also be used as a compensation mechanism in the context of ecological damage. The parties who engage in environmentally hazardous activities pay premiums proportionate to their risks, and the risks of being held liable for ecological damage is transferred to their insurers. It is worth noting that ecological damage triggers not only civil liability of compensation but also administrative obligations to restore the damaged environment. Under a liability insurance policy, only the monetary burden is (partly) shifted to the insurers. The liable parties still have to fulfill their administrative obligation to report an accident and take cleanup and restoration measures.

2.1. *The Basic Rationale of Liability Insurance*

2.1.1. Risk Aversion and the Law of Large Numbers

Insurance is a risk transfer instrument, which allows the party initially bearing risk to shift the risk to another party less risk averse or more able to bear it. Insurers are the parties who take the risk from the parties initially bearing it, and the insured are the parties from whom the risk is removed.[36] The desirability of insurance can be explained from both the perspective of the insured and insurers.

'Utility' is the term used by economists to refer to an individuals' well-being. The prospect of the same monetary value can produce different utility effects in different contexts. A person or an agent may consider 'the utility of a certain prospect of money income to be higher than the expected utility of an uncertain prospect of equal expected monetary value.'[37] This person or agent is said to be risk averse. For a risk-averse party, he prefers to choose a certain loss of 1,000 rather than a lottery with a 10 per cent chance of losing 10,000. With the payment of a premium of 1,000, the loss of 10,000 in case of the materialization of the risk is removed from the insured. Thus from the perspective of the insured, their utility

35 Wagner 2009, p. 377.
36 *Ibid.*
37 Cooter & Ulen 2012, p. 45.

can be increased through liability insurance if they are risk averse. In spite of the variance among individuals, it is fair to assume risk aversion as a general attitude of human beings. This is because even modestly wealthy people would be ruined financially when confronted with an obligation to compensate for severe personal injuries.[38] Most firms are also risk averse although some are owned by a multitude of shareholders with diversified portfolios. This can be explained by the advantages of insurance companies in claims management expertise and the ability of insurance in saving administrative costs compared to the full diversification of investment portfolios.

In an optimal insurance market, the insurance company offers coverage with premiums equal to the expected accident costs – the probability 'p' multiplied by the possible magnitude of the damage once the accident has occurred 'D'. The payment p*D is the actuarially fair insurance premium. However, in the actual insurance market, the premium is higher than the actuarially fair premium. Running an insurance business causes administrative costs. The industry cannot thrive without a profit. All those costs should be added to the real premium. Thus, in addition to the actuarially fair premiums, the insurance premium also depends on the market structure, market situation of the insurer and profit margin.[39] For example, in the case shown in last paragraph, the actuarially fair premium is the probability of 0.1 multiplied by the magnitude of the loss which is 10,000, resulting in 1,000. However, the premium charged may amount to 1,200, considering the administrative costs and profits of the insurance company. Whether the premiums charged above the level of actuarially fair premium is acceptable for the insured relies on the extent of risk aversion and the associated willingness to pay to transform risk into certainty.[40]

The above analysis shows that it is desirable for a risk-averse insured to participate in an insurance policy, since the risk of a large amount of loss is removed from them with a certain, smaller amount of premium. However, it does not explain why taking on risk from the insured is also beneficial for insurers, for operating a business, as they are also risk averse. By pooling a large amount of similar but independent risks, the probability of the realization of risks and their magnitude become more predictable for the insurers, thus decreasing the risk for them. This is because of a mathematic theorem known as 'the law of large numbers'. According to the law of large numbers, 'in repeated, independent trials, with the same probability of a particular outcome in each trial, the actual incidence of that particular outcome will converge more closely with the probability of that outcome as the number of trials increase.'[41] A common example is the rolling of a dice. A single roll of a six-sided die produces one of the numbers 1 to 6, with equal probability. Thus the expected value of a single roll is (1+2+3+4+5+6)/6 = 3.5. With limited trials, the results can depart sharply from the expected value. However,

38 Wagner 2009, p. 378.
39 Faure & Hartlief 2003, p. 84-85.
40 Wagner 2009, p. 378.
41 Wagner 2009, p. 382.

repeated in large numbers, the average outcome approaches 3.5.[42] This theorem helps insurance to increase the predictability and control the risks effectively, given the capability of insurance to assemble lots of homogenous and independent risks.

2.1.2. Insurability

The foregoing analysis shows the desirability of insurance for both insurers and insured. But whether an insurance market can be established depends on the insurability of a certain risk. Insurability is a term of art in insurance economics and practice, which denotes the minimum requirements a certain category of risk must satisfy in order to be eligible for coverage. Scholars have developed arrays of criteria of insurability. According to Katzman, insurable risks have the following characteristics: exposures must be homogeneous and numerous enough to allow risk pooling; the loss occurs within a well-defined time period; and the frequency and severity of loss are calculable for premium setting.[43] Anderson gave a list of prerequisites of ideally insurable risks: a large number of units are independently and identically exposed to the risk, the loss is determinable and calculable, the loss is fortuitous or accidental and not subject to simultaneous destruction.[44] In spite of the variance in formulation, an insurable risk usually should be: accidental, determinable and measurable, independent and non-catastrophic.[45]

Accidentalness requires that the loss 'be contingent upon a move of nature and not the result of a deliberate choice by the insured.'[46] Otherwise, the insured can influence the insurer's obligation at their own will, which can lead to wrong incentives to the insured. Therefore, damage caused by an intentional tort of the insured is excluded from the coverage of insurance. The requirements of determinability and measurability are essential for setting the premiums of insurance. The premium should reflect the expected value of the loss, depending on both the probability of the risk materialization and the magnitude of the losses. This results in high information requirements on both the frequency and severity of losses. A well-defined risk, the existence of empirical data and explicit measurement methodology can help to obtain such information. Furthermore, an insurable risk must be independent rather than cumulative, which means that the materialization of one risk does not increase the probability of the realization of a significant portion of the other risks. Otherwise, simultaneous accidents may easily lead to the insolvency of the insurers. The limited financial capacity of insurers also imposes the requirement of non-catastrophic damage. The independence and non-catastrophic nature of losses are often interrelated. The modest and manageable single damage may pile up to be unmanageable if numerous people are negatively affected simultaneously.

42 *Ibid.*, p. 382-383.
43 Katzman 1986, p. 617.
44 Anderson 1998, p. 2.
45 Wagner 2007, p. 89.
46 Wagner 2009, p. 400.

It is misleading to deem insurability a hard concept which allows a clear and absolute distinction. It is 'an appreciation in terms of more-or-less instead of a decision in either/or mode.'[47] As has been said, 'A risk can be insured also when no statistics are available, and even when no theoretical analysis seems possible.'[48] The requirements of insurability can hardly be satisfied completely. Take accidentalness as an example: the intentional activity of the insured is excluded from the coverage, but harm caused by negligence is not. Though the insured does not pursue the negative consequence on purpose, the harm is not carefully considered or not in contradiction to his willingness, and thus it cannot be deemed to be utterly independent of his decisions. It can cause the problem of moral hazard, which can be cured by monitoring or risk-differentiation. The moral hazard problem is discussed in the next section. As far as the elements of determinability and measurability are concerned, additional premiums can be charged to expand the risk pool, thus to alleviate the negative effects of lack of predictability. To this extent, insurability is not only a matter of degree but also a matter of price. A risk is insurable as long as the premium is not beyond the affordability of insured and does not ruin the insurer's willingness to offer the insurance.

2.1.3. Problems of Insurance

In an ideal insurance market, the risk-averse individuals transfer their risk to insurers with the price of premiums setting as the actuarially fair premiums plus proper loading costs. Thus the utility of the insured parties is increased. At the same time, the incentives to prevent the damage remain for the insured, since they have to pay for the actuarial value of the risk. However, the insurance market is never ideal in practice in view of the information asymmetry between the insurers and the insured. The imperfect information gives rise to two problems haunting the insurance market: adverse selection and moral hazard.

2.1.3.1. Moral Hazard

Moral hazard means that the insured's motive to prevent loss tends to change once he has obtained coverage.[49] Since the risk of paying a large amount of damages is removed from the insured, changes happen to his incentives to take care.[50] Moral hazard is typical for insurance contracts for only one period. Under this situation, any investment in measures to reduce the risk adds new costs on his balance sheet for the benefit of the insurer.

Typically, two methods can be used to solve the problem of moral hazard: the insurer's monitoring of the care taken to prevent the loss and incomplete coverage.[51] Under the first approach, the insurer can monitor the insured's behavior and adapt

47 Wagner 2007, p. 91.
48 Borch, Aase & Sandmo 1990, p. 316.
49 Shavell 1979a, p. 541.
50 Wagner 2007, p. 95.
51 Shavell 1979a, p. 541.

the premium accordingly. In order to ensure the insured will not change his behavior so as to increase the risk insured, insurance creates incentives for insurers to take monitoring measures. If the monitoring shows that the behavior of the insured changes after the establishment of the insurance contract, the premium can be adapted to the new situation. Thus experience rating, which is 'an *ex ante* screening with a higher premium for certain high risk groups or an *ex post* premium increase of change of policy conditions based on previous loss experience'[52] helps to deal with the moral hazard problem.

Another solution exposes the insured partially to risk. Hence responsibility remains partially for the insured, which gives him incentives to prevent the damage. The partial exposure can be obtained either by deductibles, the introduction of an upper limit on coverage[53] or allowing certain exclusions (such as excluding coverage of liability for harm either expected or intended from the standpoint of the insured).

2.1.3.2. Adverse Selection

Another problem affecting the insurance market is adverse selection, which refers to the fact that 'the tendency of persons with relatively greater exposure to risk to seek insurance protection.'[54] In a well-functioning insurance market, the average premium should align with the risk profile of most members in a particular pool. However, if the risk different insured parties possess varies significantly, the good risks whose expected damage is less than the premium will leave the pool. Only the members with a higher risk have strong incentives to stay.

Both moral hazard and adverse selection are caused by information asymmetry. Risk differentiation has long been proposed as a solution to the information asymmetry.[55] The problem of adverse selection can be solved if the high risk insured are well discriminated from the low risk ones. Priest mentions narrowing the risk pool as a solution to the adverse selection problem. He also emphasizes the importance of the tradeoff between risk segregation and risk spreading.[56] In that case, the insured in a particular pool have similar risks level, which means that the premium reflects the expected risk of most members. With careful risk differentiation, the insured can shift his risk to the insurer for a premium corresponding to the actuarially fair premium plus reasonable loading costs. Not only is risk differentiation a remedy for adverse selection, it is also a solution for moral hazard. By classifying insured based on their expected losses, the insured will have preventive incentives even if coverage is offered. Usually, two forms of classification are available: feature rating and experience rating. The

52 Faure & Grimeaud 2003, p. 153.
53 Abraham 1988, p. 952.
54 Priest 1987, p. 1541.
55 See Crocker & Snow 1986.
56 Priest 1987, p. 1542-1543.

former bases premiums on objective features of the insured's operations, while the latter bases premiums on the loss experience during a previous period.[57]

According to the above analysis, insurance can promote the utility of the risk-averse insured by removing the burden of losses from him with a certain and smaller amount of premium. As for the insurer, under the law of the large numbers, the specific unpredictable risk becomes more predictable with large numbers of them assembling together. Thus from both the perspective of the insured and insurer, insurance is desirable. In spite of the desirability of insurance, whether an insurance market can be established relies on the insurability of a risk. An insurable risk needs to be accidental, determinable and measurable, independent and non-catastrophic. However, those criteria do not present a black-and-white distinction. Instead, insurability is more a question of degree and price. Moral hazard and adverse selection offer another angle to evaluate insurability. An insurable risk should be one for which the problems of moral hazard and adverse selection are reasonably under control.

2.2. *Liability Insurance for Ecological Damage*

After the brief discussion of the basic rationale of insurance, this section comes to the liability insurance for ecological damage: whether liability for ecological damage is insurable and the problems of moral hazard and adverse selection can be overcome.

2.2.1. Insurability of Liability for Ecological Damage

Under the policy of liability insurance for ecological damage, the insured risk is liability for ecological damage. Not only the nature and characteristics of the potential ecological damage the insured may cause, but also the features of the liability rules determine the insurability of such a risk.

2.2.1.1. Characteristics of Ecological Damage Influencing Insurability

Chapter 1 sketches some characteristics of ecological damage. Here only the factors influencing the insurability of ecological damage liability are briefly introduced.

Firstly, an insurable risk needs to be accidental. However, with regard to ecological damage, there is not always a clear-cut distinction between damage caused intentionally and unintentionally. A number of disputes may arise in determining 'accidentalness': whether the test used to determine accidentalness is subjective or objective; whether there is coverage if the policyholder expected another kind or magnitude of harm than that which happened.[58] Despite those uncertainties, 'there is a good deal of pre-existing case law developed in other kinds

57 Abraham 1988, p. 949.

58 Abraham 1993, p. 617.

of cases that can be drawn upon to guide decisions on these issues.'[59] Scholars also arrive at some methods to determine the accidentalness, such as comparing the records of a given company's action with the standard procedures practiced in that industry during that given time.[60]

The determinability and measurability are other problems ecological damage confronts. Sometimes, scientific uncertainty exists in the causation link between a certain activity and ecological damage. Under this condition, whether the causation is predictable from the insurer's perspective depends on the design of liability rules, which is discussed later. Even if the damage is indentified and the causation link is established, the measure of damage is another hindrance for insurability. Ecological damage is reflected not only in the reduction of the market value of the environment, but also in the losses of ecological and esthetic functions. The losses in various types of value create a challenge for measuring the ecological damage. Detailed rules about the assessment procedure and methods like the natural resources assessment regulations in the US[61] can contribute to measuring the ecological damage. Furthermore, additional premiums can be charged to alleviate the problem of less predictability. The long time lapse between the occurrence of an event and the appearance of damage also has a negative impact on the determination of ecological damage.

In addition, ecological damage sometimes turns out to be catastrophic, which constitutes another challenge for the capacity of insurers. Whether this obstacle is so serious that it precludes the establishment of an insurance market requires more empirical data, which is the topic in the empirical part.

2.2.1.2. *Liability Rules Influencing Insurability*

In addition to the features of ecological damage itself, the design of liability rules also has a significant influence on the establishment of a liability insurance market for ecological damage.

Financial Cap and Channeling of Liability

A limitation of compensation is an important feature of environmental liability, especially in many international conventions, such as the CLC and the Paris Convention.[62] Those limitations are often set as an argument to keep the risk insurable, since environmental damage can turn out to be catastrophic, which is an obstacle towards insurability. However, the financial cap poses problems from the deterrent perspective. If the victims are third parties who have no contractual relationship with the insured, they cannot negotiate on the division of risk as when the Coase theorem applies. Therefore the setting of the financial cap will affect the efficiency: if the liability is capped at an amount which is substantially lower than

59 *Ibid.*

60 Hollaender & Kaminsky 2000, p. 207-208.

61 43 C.F.R. §§11.10-11.93; 15 C.F.R. §§990.10-990.66.

62 See Chapter 6, Sections 1.1.1, 2.1.

the amount of damage the incident causes, the insured does not need to internalize fully the costs he produces and thus lacks the incentives to set an optimal care and activity level.[63] Moreover, the argument of insurability does not necessarily lead to a financial cap for liability. An insurance provision with an upper limit can also prevent the risk becoming catastrophic for the insurers.

As discussed in the last chapter, channeling of liability is another feature of some environmental liability regimes.[64] Under such regimes, the liability is channeled to a specific party, while the possibility to claim from other parties who actually caused the damage is denied. At first sight, channeling liability may promote insurability, since then only one party needs to purchase insurance coverage. Thus the insurance costs are reduced. However, under such regimes, the policyholder may have to bear the costs produced by other parties, which creates more uncertainties threatening the insurability of environmental liability.

Retrospective Liability

Retrospective liability refers to the situation that new liability rules apply to a tort committed earlier, when the activity was not regarded as wrongful. A well-known example is the CERCLA in the U.S., under which strict liability for the costs of cleaning up contaminated sites is imposed on the generators and transporters of the hazardous substances, and on the owners and operators of the sites, regardless of the time when the hazardous substances were deposited.[65] This rule supports the redistributive desire to compensate victims rather than the deterrent function of tort law.[66]

From the insured's point of view, the activities was not regarded as wrongful when they were committed, neither the damage nor the liability can be anticipated by him. Thus the *ex post* imposed liability has no impact on their preventive incentives. Retrospective liability also has a negative effect on the insurability. 'If the insurer were not aware that the behavior of this insured party might potentially have been considered wrongful, no premium would have been charged for the risk, no preventive measure would have been required in the policy conditions and no reserves against losses would have been set aside.'[67] Thus, the retrospective liability is highly unpredictable for insurers, which can lead to immensurability. The retrospective, joint and several liability set by statutes, accompanied by judicial strategies of interpreting insurance policy language from the consumer protection perspective and the common law expansions of liability results in the liability insurance crisis in mid 1980s in the U.S.[68]

63 Faure & Wang 2006a, p. 185.
64 See Chapter 3, Section 2.3.1.
65 42 U.S.C.A. §9607(a).
66 Bruggeman 2010b, p. 76.
67 Faure & Grimeaud 2003, p. 169.
68 Abraham 1988, p. 955-976.

Joint and Several Liability

Sometimes, several tortfeasors contribute to the same damage, which may cause a problem of apportioning the liability among them. Two types of liability rules can apply: non-joint liability as well as joint and several liability. The discussion in the last chapter shows that neither of the two rules dominates each other under the context of ecological damage.[69] Thus both non-joint liability and joint and several liability can apply here.

Joint and several liability is desirable from the victim's point of view, for he is then relieved from the burden to prove the injurers' contributions to the damage, and can get full compensation from any solvent injurers. However, this rule involves negative consequences for the injurer who is held liable. If he cannot prove other injurers' contributions or they are insolvent, the injurer who has offered compensation will have to bear losses that he has not caused. Under this uncertainty, his insurer cannot calculate *ex ante* the potential risks, since he is not in the position to take account of and monitor other possible injurers.[70] Moreover, it the other identified injurers are insolvent or not covered by insurance, the liability insurer has to bear the total costs.

Admittedly, joint and several liability increase the uncertainties for insurers. However, whether the uncertainties are so serious as to lead to uninsurability needs more empirical support. 'As long as joint and several liability represents a small percentage of total liabilities insured, it need not disrupt liability insurance premium calculations unduly, because the level of uncertainty it poses is minimal.'[71] As for the cases of the CERCLA, the scope of liability is potentially very large, including not only the generators and transporters of the waste, but also the owners and operators of the hazardous waste storage sites. Thus the uncertainty is magnified.[72] A non-faulty generator of parts of the waste might have to bear the cleanup costs contributed by other generators, owners or operators of the sites. So the extent of the uncertainty created by joint and several liability needs to be examined in a specific context.

Shifting the Risk of Causal Uncertainty

Unlike traditional damage, the identification of ecological damage and its causer is highly technical and the causation between them is sometime uncertain. The 'proportionate liability rule' has been introduced by some scholars as a remedy to the uncertain causation and to promote deterrence.[73] However, as shown in the last chapter, a proportionate liability rule is not realistic for ecological damage, for it is difficult to find out the precise probability that the damage is caused by an injurer.[74]

69 See Chapter 3, Section 2.5.2.
70 Monti 2001, p. 58.
71 Abraham 1988, p. 959.
72 *Ibid.*
73 Shavell 1985, p. 587-609; Faure & Bruggeman 2007, p. 105-121.
74 See Chapter 3, Section 2.5.

Since a proportionate liability rule does not apply, it is usually the threshold liability rule that is in place. Under the latter, a conclusion needs to be made on whether the damage is caused by a certain activity. The proof of causation between the damage and the activity of an injurer originally lies with the victim. However, the victim is inferior in information and position than the injurer to bear the burden of proof. Based on the perspective of victim protection, the proof of causation is increasingly shifted to the injurer.

Under the context of ecological damage, the causal uncertainty is relatively high, which makes the proof of causation a heavy burden on the injurers. Once he fails to prove that the damage is not caused by him, the injurer has to compensate fully for the damage. With the shift of the risk of causal uncertainty, the potential liability is more unpredictable for the insurers. The insurability is further attenuated if accompanied by the 'deep pocket policy' of the insurance industry, which is considered to serve the purpose of financing a social program rather than spreading risks.[75]

2.2.2. Moral Hazard and Adverse Selection

As discussed above, moral hazard and adverse selection are two problems affecting the insurance market. The same difficulties also arise in the case of liability insurance for ecological damage. Whether the insurance market is disrupted by those problems relies on how the remedies work. The above analysis reveals that moral hazard and adverse selection are caused by information asymmetry, which has three possible solutions: monitoring, exposing the insured partially to liability and risk differentiation.

The involvement of insurance gives the insurer incentives to monitor the insured's behavior to make him behave in a safe way. Safety regulation is usually the task of public authorities, through standards setting, license authorization and other methods. By operating an insurance industry, the insurer needs to set a premium reflecting the expected value of the damage the insured makes, which keeps him behaving in a safe way so that risk is not increased significantly after the offering of coverage. This involves both risk assessment and risk management.[76] The risk assessment evaluates the safety and integrity of the insured's operation at each specified site, and it is used by the insurer to decide whether to accept the risk, what the scope of coverage he will offer and how to set the premium.[77] The risk management after the provision of coverage is also important to combat the moral hazard problem. The growth of corporate subscription to environmental management systems, such as the ISO 14001 standard prescribed by the International Organization for Standardization or the E.U.'s Eco-Management and Audit Scheme can help to control the risk.[78]

75 Katzman 1986, p. 619.
76 Abraham 1988, p. 954.
77 Turner & Smith 1983, p. 341.
78 Richardson 2001, p. 304.

Insurers can also control the problem of moral hazard by exposing the insured partially to the liability. They can set either a threshold or an upper limit for the coverage. Thus the insured may still be partly responsible for the ecological damage they cause, and the incentives to prevent the damage remains for them.

Risk differentiation is a solution for both moral hazard and adverse selection. By discriminating between parties posing different levels of risks, the potential liable parties are asked to internalize the costs they may create. Risk differentiation may pose a challenge to the insurance industry with respect to ecological damage. Unlike health insurance or property insurance, where there are many individual insured parties without specialized knowledge about the risks they are exposed to, in the context of ecological damage, the insured are industries. They are specialized in their field and usually have better information on the risks they create. This situation requires insurers to invest in procuring information on the specific risk, which may trigger a need for specialization. This specialization gives the insurer adequate information on the features of a specific risk and the appropriate way to reduce the risk, thus to attract good risks with lower premiums.[79]

After years of development of environmental insurance, insurers have developed some methods of risk differentiation. Insurers can require specific preventive measures through the policy conditions and adapt the premiums to the performance of the insured. Two approaches can be used to set the premiums: feature rating and experience rating. If the feature rating premiums apply to liability insurance for ecological damage, the insurer needs to find the information about features of the insured's operations which may influence the risks—total production, the volume of waste generated, received or transported, the pollution prevention measures adopted and so on. For this highly technical issue, the insurer does not always possess the necessary information to differentiate the risks. Moreover, the risk is not only determined by the features of the operations, but also shaped by the liability rules and the courts. This uncertainty in the establishment of liability poses a further difficulty in risk differentiation. In spite of the fact that the insurers are sometimes inferior to the large and technical insured, they are more experienced in risk differentiation and monitoring under other conditions. Thus the feature rating 'can serve an informational function by indicating to the insured the features of its activities that affect the level of risk those activities pose.'[80]

Experience rating can create risk-reduction incentives for a reduction in losses that leads to a reduction in future premiums. It is easier in operation for insurers since it relies on the past performance of the insured rather than on a set of sophisticated standards. Its effectiveness depends on the availability of the past data and the length of the policy period.

Risks need to be differentiated so that the average premiums go well with most members expected losses. However, risk differentiation is not costless. Thus there is a trade-off between the benefits in solving the information asymmetry

79 Faure & Grimeaud 2003, p. 159.

80 Abraham 1988, p. 950.

problem and its costs. A further differentiation of the risk is only efficient when the marginal benefits of the further differentiation outweigh its marginal costs.[81]

2.3. *The Evaluation of Liability Insurance*

After the discussion of the insurability of liability for ecological damage, this research evaluates the desirability of liability insurance from a law and economics perspective. Before the examination of the influence of liability insurance on specific kinds of costs, the scope of liability insurance needs to be made clear: does it provide coverage only for ecological damage or other damage caused via the environment also? Is the instrument established on the basis of sectors such as insurance for nuclear liability or does it offer more comprehensive coverage? These questions are relevant since, on the one hand, they delimitate the application of liability insurance and depict a shape of such an instrument and, on the other hand, they also have an impact on the reduction of social costs. A policy with narrower coverage may enjoy an advantage in reducing primary costs since they may perform better in risk differentiation. However, a narrow-scoped, sector-based policy cannot offer compensation for all types of ecological damage. Thus lots of policies should be combined to give a complete coverage. Nevertheless, an enumeration of policies may still leave loopholes. As far as tertiary costs are concerned, comprehensive coverage can save administrative costs in the sense that fewer policies are required.

2.3.1. Primary Costs (Prevention)

The minimization of primary costs requires that the accident costs are internalized so that the marginal costs of accident avoidance equals marginal benefits in the reduction of damage. This requirement has two implications: firstly, it is the polluter that should bear the costs of accidents (to the extent that the total sum of prevention costs and damage is minimized); secondly, the costs are shared among polluters in such a way that the individuals' payment corresponds to their contributions.

From the perspective of the first implication, liability insurance is efficient. The premiums paid by polluters comprise the risk pool, which provides compensation when risks are materialized. Whether insurance can create optimal preventive incentives therefore depends on how the premiums are set. Risk related premiums do not mean an individual tariff to be employed in each case.[82] As discussed earlier, risk differentiation is a useful technique of insurance. The insurers can differentiate the risks of insured to different levels and associate the premiums with their risk levels. The aforementioned analysis gave a brief introduction on two forms of rating: feature rating and experience rating. In practice, those two forms are often combined in use. For example, a program developed by ECS Underwriting in the

81 Faure & Grimeaud 2003, p. 155.

82 Faure & Hartlief 2003, p. 110.

U.S. for pricing its third-party liability coverage gave many evaluation parameters for rating and pricing, including: total locations, exposures, sales, surrounding environment, current site use, material and waste, site history and so on.[83] The method of rating and pricing environmental risks is proprietary and variable with each insurer. Although the methods may not be perfect, they offer an approach to differentiate risks and ensure at least rough proportionality of premiums with their contributions to the risk.

Liability insurance can promote the deterrent effect of liability by relieving insolvency problems. However, it cannot eradicate such a problem totally, since insurers do not have unlimited capacity. Insolvency can still arise in the case of a catastrophe. Furthermore, a reasonable injurer, under the protection of limited liability, only has the incentives to take the insurance up to his total assets.

As discussed above, both scientific and legal uncertainties make the risks of liability for ecological damage less predictable. The insurance companies, as profit-pursuing entities, are also risk averse. Thus they often charge an additional cost in the premium, a risk premium, which reflects their aversion to the risks.[84] Early research shows that actuaries and underwriters tend to charge higher premiums in case of insurance ambiguity (on either probability or losses).[85] This may even sometimes lead to the reluctance of insurers to provide some kind of insurance policies, such as environmental liability policies in the 1980s.[86] Some techniques have been proposed as a solution to such a phenomenon, such as a revision of liability rules, better definition of contract terms, joint private-government programs, risk retention groups and subrogation.[87]

2.3.2. Secondary Costs (Compensation)

Secondary costs reduction is based on the fact that accident losses will be less burdensome if they are spread broadly among people and overtime. Two approaches may be used to reduce the secondary costs: spreading the losses among many people and the deep pocket policy, which means placing the losses on people less likely to suffer substantial social or economic dislocations when bearing them.[88]

Liability insurance can be used to reduce secondary accident costs. On the one hand, it plays a role in interpersonal loss spreading. The risk of being held liable for a catastrophe is removed from an insured with the payment of premiums. Thus once the risk is materialized, the cost is paid from the insurance pool which is financed by many insured. On the other hand, intertemporal spreading can also be achieved by insurance. Without a compensation mechanism, the damages usually need to be paid in a short period once liability is established; whereas under an

83 Kronenberg 1995, p. 338-339.
84 Faure & Fiore 2008a, p. 233-237.
85 Kunreuther & Hogarth 1992, p. 318.
86 *Ibid.*; Kunreuther 1987, p. 180.
87 Kunreuther 1987, p. 191-194.
88 Calabresi 1970, p. 39-40.

insurance policy, the damages are covered by premiums paid through the whole policy period.

As discussed earlier, there is some interplay between secondary costs reduction and the function of compensation. The coverage of an insurance policy, the time to intervene and capacity of insurers are explored to show how the aim of compensation can be achieved by liability insurance for ecological damage.

The coverage of liability insurance is quite diversified. Some policies cover only liability for ecological damage, while others also involve relevant traditional damages. The insurance policy can be either sector based or more comprehensive. However, whatever form a policy takes, the coverage can only be offered if liability is established. The previous chapter has shown that there are many obstacles to the finding of liability for ecological damage. With awards based on the establishment of liability, liability insurance can do little to broaden the scope of compensation.

A compensation mechanism can be established either in anticipation of any incident or only post factum. To ensure the effective coverage of compensation by the guarantee, it should be available from the start of the operations.[89] A post factum mechanism is only effective if the risk of damage is clearly limited. Otherwise, if a guarantee is only required at the moment of an incident, the liable parties may not have enough assets by then. Under a liability insurance policy, the payment of premiums and the establishment of a risk pool occur before the materialization of risks, which provides a comparatively high level of security for liability insurance.

In spite of the advantages in security, liability insurance is not an ideal instrument to award swift compensation. Compensation is only awarded after the establishment of liability. The cooperation between the public authority and liable parties in determining liability or the litigation often take months or even years.[90] The long gap between the appearance of damage and the establishment of liability means that the damage of victims and the restoration costs of the environment cannot always be covered by the policy in advance.

Insurance can increase the capacity to provide a higher amount of compensation. By establishing a pool[91] with the premiums paid by a large number of insured, an insurer has enhanced capacity to provide compensation for ecological

89 Bocken 2009, p. 157.

90 For example, the oil tanker Exxon Valdez spilled hundreds of thousands of barrels of crude oil in Alaska in 1989. $287 million in compensatory damages and $5 billion in punitive damages were awarded in the first trial in 1994. However, Exxon appealed and the litigation lasted decades. It was only in 2008 that Exxon Mobil agreed to pay 75% of the US$507.5 million damages (the cap of damages appealed earlier by Exxon) ruling to settle the case. See: <www.reuters.com/article/idUKN2641081120080827>, last accessed on 8 May 2013.

91 The pool here means that the insured aggregate their risk together through paying premiums to the insurers. Though paying a certain amount of premium to the insurance pool, the risk of a larger loss is relieved from him. This is different from the pooling under a risk-sharing agreement discussed in *infra* Chapter 4, Section 4. Under a risk-sharing agreement, the insured and the insurer are the same. The insured pool their resources together to cover their own risks, either through a contract arrangement or a legal requirement. For the details about pooling under a risk-sharing agreement, see Chapter 4, Section 4.

damage. A multitude of participants in the pool is helpful for the capacity. However, this benefit should be balanced with the needs of risk differentiation. Moreover, the size of a pool also depends on the willingness of the insured to take insurance coverage. Several factors have negative impacts on their incentives to seek insurance coverage: firstly, 'if the expected damage largely exceeds the injurer's assets the injurer will only have incentives to purchase insurance up to the amount of his own assets.'[92] Secondly, the incentives can be further reduced by the long time lapse problems. Given the long time lapse between the involvement of activities and the manifestation of damage, there is a strong tendency for the insured to commit to the low standard of safety and generate high profits for shareholders until the materialization of the damage. By then they can escape liability through bankruptcy. In such a context, 'the shareholders have no incentives to insure their company against the future liabilities looming beyond the horizon.'[93]

Insurance capacity can be promoted by using some traditional insurance techniques, such as co-insurance (insuring a similar risk jointly with a few insurers) and re-insurance. Another way to enhance capacity is pooling by insurers, which means that the insurers share risks in mutual pools on a non-competitive basis to provide coverage.[94]Although this is an effective way to enhance insurance capacity, pooling is also subject to the regulation of competition policy. Cooperation between insurers, aiming at increasing the capacity to cover large risks, may also result in the concentration of market, and subsequently, increased premiums. People who are in favor of exempting insurance from competition policy argued that pooling of risks can promote competition by involving a larger number of market participants, since it allows the access of small insurers who can otherwise not provide the large coverage.[95] However, pooling remains inefficient if it does not lead to increased competition. For example, an investigation about the liability insurance in France shows that the gap between the actuarial fair premiums for nuclear risks and the current premium charged is large. One important explanation for this situation is the current non-competitive market. The nuclear insurance market has been organized on a national and non-competitive basis since the beginning of the civil nuclear industry. There are 28 nuclear pools in the world and no competition exists between the various pools. This means the insurers can bear very high loading costs which will make it difficult for the insurers to survive on a competitive basis. Moreover, the nuclear insurer can determine the premium as he wishes and include an additional cost to pursue monopolistic profit.[96]

Another way to increase the capacity of liability insurance is to make it compulsory. When potential liable parties are obliged to buy the insurance coverage, the problem of a lack of incentives to seek an efficient level of insurance can be solved. With a large number of members, the insurance pool is substantially enlarged and its capacity to compensate can be significantly promoted. However,

92 Faure & Hartlief 2003, p. 213.
93 Wagner 2009, p. 396.
94 Faure & Hartlief 2003, p. 89.
95 Faure & Hartlief 2003, p. 93.
96 Faure & Fiore 2008a, p. 236-237.

liability insurance can also lead to competition and some other problems. The function of liability insurance as regards ecological damage is explored later in this section.

2.3.3. Tertiary Costs

In addition to the consideration of primary costs and secondary costs, the cost of the compensation mechanisms themselves should be kept reasonable. The examination of tertiary costs reveals whether 'an attempt to reduce accident costs, either by reducing accidents themselves or by reducing their secondary effects, costs more than it saves.'[97] A set of factors can be taken into account when deciding the tertiary costs: for example, whether the coverage is limited, if low-risk activities are excluded from the obligation to provide financial guarantees, and whether the mechanisms intervene only in case of insolvency.[98]

Loading costs and profits obtained by the insurance companies are two important parts of the tertiary costs. Loading costs refer to the overheads generated by the normal functioning of the insurance, including administration charges, management costs and so on.[99] To operate an insurance policy, an insurer needs information to determine and monitor the risk an insured has. Both factual and legal uncertainty exists with respect to environmental risks.[100] These uncertainties make the information costs and monitoring costs of liability insurance relatively high. In addition to the loading costs, the premiums charged by insurers also contain the profits pursued by them. The market structure is an essential factor to keep both loading costs and profits reasonable. On a competitive insurance market, the insurer who bears very high loading costs or a very high profit level cannot survive in the long run. However, on a monopolistic or non-competitive insurance market, the management is often inefficient and the waste significant due to a lack of incentives from competitors. The insurers can also pursue a very high profit.[101]

2.3.4. Interaction between the Different Costs Reduction Goals

Primary, secondary and tertiary costs all contribute to the reduction of total social costs. However, those three aspects are not fully consistent with each other. For example, a perfect system of secondary costs reduction with broad loss spreading may not be so ideal regarding primary costs reduction. Thus the aim is to find the

97 Calabresi 1970, p. 28.

98 Bocken 2009, p. 158.

99 Faure & Fiore 2008a, p. 234.

100 Faure 2007a, p. 76. As a relatively new risk, the necessary information to assess both the probability of accidents and possible magnitude of ecological damage may be missing. The uncertainty in finding liability further weakens the predictability of the risk that an insured may be held liable for ecological damage. Those problems make the information costs and monitoring costs of liability insurance relatively high.

101 Faure & Fiore 2008a, 236-237.

best combination of the three types of costs reduction, so that total social costs are minimized.

Before examining the discordance among the three reduction goals, we should bear in mind that the distinctions among those categories are not absolute. As Calabresi has clarified, the boundary between those categories is not always clear, which makes the division of accidents cost reduction, to some extent, an arbitrary one.[102] For instance, the dislocation and unemployment resulted from banning motorcycles can be treated either as primary (costs to avoid accidents) to secondary costs (economic and social dislocation costs 'resulting from primary accidents costs and their avoidance').[103] In spite of the vagueness in fringe areas, the categorization is useful for analytical purposes: 'the methods available for reducing costs, which in a loose sense fit each of the categories, are very different from one another.'[104]

As far as liability insurance for ecological damage is concerned, there are also some inconsistencies among the three reduction goals. For example, secondary costs reduction asks that the risks be spread as widely as possible or be allocated to the less risk averse parties. However, insurance should be financed on the basis of ones' contribution to the risk rather than the extent of risk aversion according to the primary costs reduction goal. Inconsistency also exists with respect to the partial exposure of the insured to risks, which is favorable in controlling primary costs. However, the exposure of the insured to risk also implies a risk of under compensation for victims, which does not correspond with the goal of compensation. To combine those three goals is not an easy task. Liability rules give priority to the reduction of primary costs. Therefore the deficiency in reducing secondary costs needs to be addressed by compensation mechanisms. The missed compensation function should be realized by those supplementary mechanisms. However, this does not mean that the other two aspects can be neglected. It is not desirable if the goal of secondary costs reduction is so far reaching that the preventive effects of liability rules are compromised.

In sum, liability insurance plays a positive role in reducing total social costs. It promotes the primary costs reduction in the sense that it cuts down the risk of insolvency. Moreover, the application of insurance techniques keeps the premiums proportionate to the contribution of the insured to the damage. Furthermore, the monitoring of insurers gives the insured additional incentives to reduce damage. As for the compensation role, liability insurance gives a relative securable guarantee since it is established before the accidents occur. It also increases the capacity to provide compensation.

However, liability insurance also encounters some problems and limitations. Firstly, liability insurance is based on the establishment of liability, which means that it retains most of the deficiencies of liability rules as well. Since coverage is offered on the condition that an insured is held liable, neither the timeliness nor the completeness of compensation can be guaranteed. The injured environment may

102 Calabresi 1970, p. 29.

103 *Ibid.*

104 *Ibid.*, p. 30.

not be restored or compensated in light of the difficulties in establishing liability; cleanup and restoration may not be funded on time because of cumbersome litigation.

Secondly, although significantly increasing the compensation capacity, the financial assets of the insurers are still limited, which imposes restrictions on both the primary cost and secondary costs reduction. Environmental harm sometimes has a catastrophic effect, even exceeding the capacity of the pooling insurers.

Information asymmetry is another daunting problem of liability insurance. To maintain the insurability of a risk, both the probability and possible magnitude of the damage should be predictable. However, both factual and legal uncertainty exists with respect to environmental risks.[105] On the one hand, as a relatively new risk, environmental damage may be difficult to predict due to the lack of information and scientific uncertainties. On the other hand, there may be uncertainty as to the scope of environmental liability with a view to the complexities in liability rules and judicial policy.

Fourthly, insurers need to deal with problems of moral hazard and adverse selection. They can be averted though deductibles, exclusions, risk differentiation and monitoring. However, risk differentiation and effective monitoring cannot be achieved without a price. Furthermore, risk differentiation places a high requirement on information, another challenge as regards ecological damage.

Fifthly, even if insurance is an efficient instrument with respect to prevention and compensation, the practical effect still relies on the incentives potential injurers have to insure and the willingness of the insurance market to cover a certain risk. In light of the long tail nature of environmental risk, the possible underestimation of the risk, the companies may have no incentives to insure against 'future liabilities looming beyond the horizon.'[106] Bearing in mind the low predictability of environmental liability and the lack of experience, whether the insurers are willing to underwrite such risks is also uncertain.[107]

2.4. *Compulsory Insurance*

With a view to the limitations liability insurance has, the question arises how the existing limits of liability insurance can be overcome. Two approaches may provide solutions: alternative compensation mechanisms and compulsory insurance.[108] The alternative compensation systems are discussed in the following sections. This

105 Monti 2001, p. 59-62.

106 Wagner 2009, p. 396.

107 A case in point is the insurance crisis in the mid 1980s in the U.S., when the pollution insurance market collapsed into a single insurer, the American International Group. Many insurers and reinsurers withdrew from the market. This was largely attributed to the unpredictability created by liability rules and the policy to favor the victims in judicial practice. See Katzman 1986; Anderson 1998.

108 For example, Wagner favored a private but mandatory insurance, rather than private voluntary insurance and public compensation schemes. See Wagner 2007, p. 87-112.

section focuses on the role of compulsory insurance as a solution for the deficiencies of voluntary liability insurance.

Compulsory insurance is one instrument to cure some of the limitations liability insurance has. Literature provides many arguments for compulsory insurance, which concern increasing the expected utility, curing the information problems, insolvency and adverse selection.[109]

A first argument for compulsory insurance is based on the theory that insurance removes risk from risk-averse individuals and thus increases their utility. However, the degree of risk aversion varies. Therefore a generalized duty to insure might force people to purchase insurance for which they normally have no demand.[110]

Another argument concerns the incentives to insure. As discussed above, the potential injurers may lack incentives to insure as a result of either underestimation of the risk or the long latency issue. Compulsory insurance gives a legal obligation to take out liability insurance, which is beneficial with regard to long latency issues. On the one hand, the potential injurers are forced to internalize the costs of the damage which may otherwise be escaped through bankruptcy procedure in future. On the other hand, the broader risk pool can increase the capacity of insurers. The argument of information problems is based on the deduction that there is an underestimation of risk. Thus two questions are relevant: is there such an underestimation that leads to the decision not to purchase insurance? Why will the decision not to insure create under-deterrence?[111] The latter is related to insolvency issue.

Insolvency is a third consideration for compulsory insurance. 'If the expected damage largely exceeds the injurer's assets the injurer will only have incentives to purchase insurance up to the amount of his own assets.'[112] Thus the injurer is only exposed to the risk up to his assets, which can lead to under-deterrence. Compulsory insurance offers a solution by compelling the injurers to purchase insurance for risks in excess of his assets.

Furthermore, compulsory insurance 'eliminates the dynamic workings of adverse selection and thus prevents the failure of private insurance.'[113] In a voluntary market, it is the person with a high propensity to cause damage that needs the insurance most, and a good risk may tend to leave the risk pool, leading to increasing actuarially fair premiums. Compulsory insurance offers a solution by forcing the good risks to stay. However, there the question arises why the good risks should stay to give a subsidy to the bad risks.

Despite the positive role compulsory insurance can play to ease the problem of adverse selection, scholars have pointed out that the introduction of compulsory insurance should be done with caution. In an inefficient market, the introduction of

109 Wagner 2007, p. 110; Faure & Grimeaud 2003, p. 181-185.
110 Faure & Grimeaud 2003, p. 181.
111 Faure & Grimeaud 2003, p. 182-183.
112 Faure & Grimeaud 2003, p. 183.
113 Wagner 2007, p. 110.

compulsory insurance may lead to less than efficient risk and premium differentiation.[114]

Moral hazard is a byproduct of insurance. An insured will change his behavior once the burden of compensation is removed from him. The market has developed some measures to deal with this problem, such as deductibles, monitoring and risk differentiation. However, under compulsory insurance, the coverage is often equal to the amount of liability, and deductibles are not allowed.[115] Thus to what extent the moral hazard problem can be controlled depends on the monitoring by insurers and how the insurers differentiate the risks and set and adapt the premiums according to risks. If those measures are difficult or very costly, compulsory insurance may create more problems than it solves.

Another limitation of compulsory insurance is the concentration on the insurance market. The above analysis is based on the assumption that the insurance market is perfectly competitive. However, 'if monopolistic premiums can be set an insurer will have fewer incentives to align his premiums to the individual behaviour of the insured and thus there is less control of the moral hazard problem.'[116] Moreover, in a concentrated market, compulsory insurance makes the insurers in fact become the licensor of the industry. The insurance coverage is the prerequisite for the operation of a company under a compulsory insurance policy. But in a monopolistic market, the industrials have few choices about insurers and cannot operate if denied of coverage by the insurance undertakings.

3. First-party and Direct Insurance

3.1. *Introduction: First-party Insurance and Direct Insurance*

As discussed above, insurance is an instrument to increase utility since the small probability of a large loss is taken away from the insured at the price of the certainty of a small loss.[117] The aforementioned liability insurance is one approach to realize this function. Moreover, first-party insurance presents another option, under which insurance coverage is provided and compensation is awarded directly by the insurer to the victim. Unlike liability insurance, it is the victim that buys insurance coverage to remove the risk of potential large damage to him. The compensation is awarded as soon as the damage occurs, provided that the specific damage is an insured risk under the insurance policy and irrespective of whether there is liability.[118]

Despite the variations in the insurance situation between European countries, first party insurance for environmental damage becomes available in several countries.[119] In the Netherlands, there is a movement towards first-party insurance

114 Van Boom 2008, p. 265.
115 Faure & Grimeaud 2003, p. 185.
116 Faure & Grimeaud 2003, p. 186.
117 Borch 1961, p. 245-255.
118 Faure & Bruggeman 2008, p. 11.
119 Faure & Hartlief 2003, p. 154.

with respect to environmental insurance.[120] Within the sphere of environmental damage, the first-party insurance can take two forms: first, the industrial operators can buy insurance for the damage they suffer themselves, for example, damage to the operator's site. Hence, injurer and victim will be the same and it is the polluter who finances the insurance.[121] In addition to this approach, other victims can also finance the insurance, which, however, will be vulnerable to criticism from the perspective of the polluter pays principle. The shift towards first party insurance can also be found in the sphere of medical malpractice insurance[122] and traffic accident victims' compensation.[123]

Many scholars are in favor of first-party insurance rather than liability insurance in the sense that the transaction costs are lower and risk differentiation might be easier.[124] First-party insurance directly covers the risk of damage with a particular victim or a particular site. Hence it is easier to indentify the circumstances influencing the risk.[125] For example, with regard to the fire insurance, distinctions can be made among houses with or without smoke alarms or houses closer or farther from fire hydrants. Furthermore, under first-party insurance, there are more forms of experience ratings which can serve effective risk differentiation.[126] On the contrary, under liability insurance, the insured risk is that the injurer will harm a victim whose properties are unknown *ex ante* to the insurer. Moreover, the legal uncertainties, such as how the judge will interpret the specific liability aggravate the difficulties in risk differentiation.[127]

Direct insurance is an alternative which looks like first-party insurance. In a direct insurance policy, the potential injurer who possesses a particular site seeks insurance coverage also for the benefits of third parties who could suffer damage resulting from that particular site.[128] Thus unlike the pure first-party insurance funded by victims, in a direct insurance policy, it is still the polluters that pay for the damage. An example of a direct insurance is the employer-provided health insurance, under which employers take insurance coverage on behalf of their employees in the event of work-related injuries and illnesses.[129]

Although similar to liability insurance, the potential injurers finance the insurance pool in a direct insurance scheme; the trigger is no longer liability but the mere existence of damage.[130] Under environmental liability insurance, the trigger of liability can be a substantial obstacle either because of the difficulties in establishing

120 Faure 2002b, p. 297-298.
121 *Ibid.*, p. 295.
122 See Wendel 2004, p. 367.
123 Faure 2001, p. 177-179.
124 Priest 1987; Bishop1983, p. 241; Epstein 1985, p. 645.
125 Faure 2002b, p. 292-293.
126 Priest 1987, p. 1545.
127 Faure 2002b, p. 293.
128 Faure 2002b, p. 294.
129 Gruber & Krueger 1991, p. 114.
130 Faure 2002b, p. 294.

liability themselves[131] or because of the uncertainties in judicial policy.[132] In a direct insurance scheme, those uncertainties can be alleviated since compensation is awarded as long as the damage is caused by an insured risk.

3.2. *The Feasibility of First-party Insurance and Direct Insurance in the Environmental Context*

The above section has briefly discussed the rationale of first-party insurance and direct insurance. Then it comes to the question whether the advantages mentioned above fit into the environmental context. Before discussing the two policies respectively, the scope of those policies should be made clear.

The focus of this thesis is on designing a compensation system for ecological damage, to fund restoration measures, and to award compensation for the damaged environmental elements when restoration is not available. However, as mentioned earlier, a compensation mechanism may cover both ecological damage and traditional damage caused via the environment. Hence, the role played by first-party and direct insurance in the compensation of traditional damage is also explored briefly here.

3.2.1. First-party Insurance

First-party insurance is financed by victims to cover their own damage. As for ecological damage, it may involve damage to biodiversity and other environmental elements where no individual rights are directly harmed and no individuals are able or willing to assert losses.[133] Thus, except the cases that industrial operators seeking coverage for the possible cleanup costs on their own plant, a pure first-party insurance scheme would not be very practical for ecological damage.[134] However, in a scenario of accidents leading to ecological damage, individual rights may be also infringed in another way than through ecological damage. Victims may suffer personal injury and property damage. This creates an opportunity for first-party insurance to intervene, despite that its efficiency in the environmental context is debated.[135]

To evaluate the suitability of first-party insurance as an instrument to provide compensation for traditional damage in the environmental context, the insurability of such a risk needs to be addressed briefly here. As discussed in the section on liability insurance, to be insurable, a risk needs to be accidental, determinable,

131 For example, the uncertain causation, multiple potential injurers, and long time lapse between the harmful activities and the manifestation of the damage are all obstacles in establishing liability. See Chapter 2, Section 3.2.

132 Sometimes, the judges use deep pocket policy to favor the compensation, by interpreting the policy language to expand the scope of coverage. See Abraham 1988, p. 960-969.

133 Richardson 2001, p. 323.

134 Faure 2002b, p. 294.

135 See Leiter 2001, p. 313-314.

independent and non-catastrophic.[136] Accidentalness requires that damage is not the result of a deliberate choice of the insured.[137] This would be less of a problem in the first-party insurance scheme, since most environmental accidents are of a unilateral nature. Even in some cases where victims can take some measures to prevent or reduce the damage, such as moving away from a chemical factory, their influence is, after all, much less than the influence of potential injurers. There are more uncertainties with regard to the requirement of determinability and measurability of risks. It is often difficult to estimate both the probability and magnitude of damage suffered via the environment. But some clues still exist for helping to estimate the expected losses, such as the distance of an insured's house (or property) from the potential pollution sources and the career an insured has. The damage is insurable if it is caused by 'an insured risk'. This notion implies that the causation between a risk and the damage still needs to be proved. However, the difficulties in indentifying a specific injurer and the uncertainties of judgments are less compared to a liability insurance policy. As for the independence and non-catastrophic effects, although environmental accidents themselves may be unrelated, an environmental disaster can cause damage to numerous victims simultaneously. Those losses are correlated and probably catastrophic. This problem is even more serious with respect to gradual pollution. If one kind of chemical is discovered to cause latent environmental harm, lots of claims may be made in the same period.

Furthermore, the insurability of a risk also requires that the problems of moral hazard and adverse selection be under control. Risk differentiation, which is mentioned as an advantage of first-party insurance compared to third party insurance, is a solution for both problems. However, the argument for first-party insurance with respect to risk differentiation is made in specific contexts. For example, Shavell developed a model to address questions of liability for personal injuries;[138] Priest analyzed commercial casualty insurance, and especially, product liability.[139] For the losses caused by product liability, risk differentiation is easier with a first-party insurance policy. On the one hand, a consumer knows more about his own consumptive choice and the extent of risk aversion; on the other hand, it is difficult for a producer to know who (his age, income and other relevant characteristics) would be his consumer *ex ante*.[140] However, the conclusion made with insurance in the product liability context is not necessarily applicable to the environmental context. On the one hand, victims may not have enough information about their risks of exposure to pollution; they may also have insufficient incentives to insure for the low probability risks.[141] On the other hand, insurers can ask

136 See Chapter 4, Section 2.1.2.
137 Wagner 2009, p. 400.
138 Shavell 2000, p. 166-179.
139 Priest 1987, p. 1521-1590.
140 There are also some scholars opposing the advantages of first-party insurance in risk differentiation, for example, Hanson & Logue 1990, p. 145-152.
141 Faure & Bruggeman 2008, p. 21.

information from potential polluters in a liability insurance policy. However, this is not the case for first-party insurance.

3.2.2. Direct Insurance

Under a direct insurance policy, the insured and the insurer can negotiate to provide coverage to traditional damage suffered by individuals, ecological damage or both. It is noteworthy here that the application of direct insurance is usually limited: the Dutch environmental damage insurance is such an example which provides coverage to damage occurring on or from that insured site.[142] This coverage is related to insured sites. But ecological damage and related traditional damage can also be created by events not related to specific sites, such as commercial trade activities, transport of hazardous waste and chemicals and maintenance or cleanup services.[143] In those contexts, both the damage to a potential polluter's property and the potential victims are more unpredictable. Therefore it is more difficult for a direct insurance policy to offer coverage under those conditions. This section discusses the use of direct insurance to cover ecological damage and traditional damage respectively.

3.2.2.1. *Ecological Damage*

According to the insurance contract, a direct insurance policy can provide protection against both ecological damage and traditional damage. As an instrument to offer compensation for ecological damage, the coordination between an insurance arrangement and the administrative measures concerning restoration and assessment of damage deserves more elaboration here. The non-economic value of the environmental elements determines that a restoration approach is primary in evaluating ecological damage. The restoration measures are taken either by public authorities or by liable parties with the approval of public authorities.[144] If a public authority decides to take restoration measures himself, he can later claim against the liable parties for the corresponding costs; if there is a dispute over the damages, the public authority can resort to justice. In such procedures, the public authority has power to decide the restoration measures, evaluate the losses and subsequently determine the amount of claims *ex post*.[145]

142 Faure 2002b, p. 300. For details of the Dutch environmental damage insurance, see Chapter 8, Section 2.2.

143 CEA, Navigating the Environmental Liability Directive: a practical guide for insurance underwriters and claims handlers, 2009, p. 26.

144 This is the case in both U.S. federal legislation and under the ELD. See, 42 U.S.C. §9604(a)(1), Arts. 5, 6 of the ELD.

145 According to the assessment regulations under CERCLA and OPA, the assessment taken according to the regulation leads to a rebuttable presumption. See, 43 C.F.R. §11.10, 15 C.F.R. §990.13. Thus, the decisions of a public authority have important effects on the determination of damages, although they are still subject to challenges in the court.

A direct insurance policy usually provides compensation for damage arising on or from an insured site. The trigger for compensation under this condition is no longer liability but the damage itself. Thus whether a loss is covered is mainly decided by the insurance contract. For example, in the Dutch environmental damage insurance policy, damage is covered if the concentration of pollutant exceeds the standards required by law.[146] These standards are, however, regulatory standards and exceeding such standards does not necessarily lead to the cleanups/restorations required by public authorities. Hence the coverage under a direct insurance policy and restorations taken by public authorities may diverge. If the insurance policy defines the covered damage more narrowly than required restoration by law, the public authorities cannot necessarily get the compensation from the insurance policies. Thus public authorities may still have to resort to liability rules and liability insurance to cover the losses. A major advantage of direct insurance is allowing easier and instant coverage. However, the coexistence of liability insurance and direct insurance for the same case will increase the complexities.

3.2.2.2. *Traditional Damage*

Direct insurance can also provide compensation to individual victims. The contract can stipulate that it extends coverage to damage caused from the insured sites. A direct insurance policy is often linked with specific sites, which, on the one hand, limits the application of the policy and, on the other hand, increases the predictability of the risks. There exists in theory a lot of off-site damage, such as the damage caused by transportation of hazardous chemicals or wastes, contract services and so on.[147] For those kinds of damage, where the potential injurers will take their activities and the one who may be influenced by their activities is unknown *ex ante*. Thus both the probability and severity of the potential damage is less predictable for either insurers or insured.

Under a direct insurance policy, the difficult road of liability law is avoided. According to the contract clauses concerned, individual victims may claim directly from the insurance company as long as they suffer damage from the insured sites. When there is such a direct action clause, victims do not have to claim against the liable parties first. The direct action of victims precludes the uncertainties caused by judges, which is stated as a significant reason for the shift from liability insurance to direct insurance in the Netherlands.[148] However, some uncertainties still remain: to be covered by a direct insurance policy, the claimant needs to prove his damage is caused from the insured sites. Proving causation continues to be a difficult job for victims. Instead of judges deciding, in a direct insurance policy, whether specific damage is covered is decided by the insurance contract signed between the insurers

146 Interview with Mr. Kremers.

147 CEA 2009, p. 26.

148 Faure 2002b, p. 298. Noting that although direct action was allowed in Dutch environmental damage insurance in the beginning, such a clause is abrogated later, see Chapter 8, Section 2.2.

and insured and, to a large extent, by the interpretation of that insurance policy. A victim can still go to court if a dispute over the coverage arises, thus the uncertainties caused by judges are not completely removed. Moreover, whether the expected losses are more predictable and reflect the actual costs made by the insured relies largely on the definition of the insured risks in the contract and the explanation of its provisions.

3.3. *The Evaluation of First-party Insurance and Direct Insurance*

In this section, the effects of first-party insurance and direct insurance in dealing with ecological damage and traditional damage are analyzed from the perspective of the total social costs theory.

3.3.1. First-party Insurance

As discussed above, ecological damage often involves damage to public natural resources and no individual rights are directly infringed. In many jurisdictions, it is the public authority that is eligible to take cleanup and restoration measures and claims for the damages.[149] Thus first-party insurance has limited implications for the restoration or compensation of ecological damage, with the exception that polluters buy insurance for their own sites. But as for the compensation of traditional damage caused via the environment, first-party insurance can still act as an alternative.

3.3.1.1. *Primary Costs (Prevention)*

First-party insurance is financed by victims as an exchange to take away the risks of bigger losses from them. Under such an insurance policy, it is not the injurers that provide the compensation, which leaves the costs of their activities externalized. If the insurer can get subrogation from the victims and recourse from the injurers, the problem of externalization is alleviated. However, one can hardly expect an insurer to recover all or much of the compensation he offers, recalling all the obstacles in establishing liability in the environmental context.

3.3.1.2. *Secondary Costs (Compensation)*

The aim of secondary costs reduction can be achieved by either risk-spreading instruments or allocating losses to the less risk averse party. First-party insurance is an instrument to spread the losses among a large number of victims at smaller and certain amounts of premiums. As discussed in the introduction to this chapter, three criteria are important to evaluate the effectiveness of the compensation function of

149 This is the case in both U.S. federal legislation and the ELD. See, 42 U.S.C. §9607(f)(2)(b), 33 U.S.C. §9607(f)(2)(b); Art. 8 of the ELD.

individual mechanisms: the coverage of the mechanism, the time of the establishment and intervention and the capacity to provide ample compensation.[150]

Recall that first-party insurance covers mainly traditional damage and only limited ecological damage. How broad its application would be depends on the definition of specific insurance contract. A first-party insurance policy can cover either personal injury or property damage or both. It can be either an independent policy providing coverage to losses caused via the environment or one part of a comprehensive personal or property insurance. Unlike liability insurance, which can be sector based, such as specific policy established for nuclear installments or hazardous chemicals factories, the first-party insurance can be more comprehensive. Since it is the victims who buy the insurance, they tend to remove all the risks to their bodies and property, rather than to discriminate from risks caused by different sectors.

A first-party insurance policy is established before risks materialize and offers the compensation once damage caused by an insured risk happens. In such a policy, the victims do not need to resort to the court and the difficult road of the tort system is avoided. One can get the compensation before the liability is established. Thus more instant compensation is guaranteed compared to the case of liability insurance.

The capacity to provide compensation to a large extent depends on the scope of a risk pool, which is determined by both the demand and supply of an insurance policy. Law and economics literature shows that victims will purchase first-party insurance if the premium is sufficiently low in comparison to the risks since then the expected utility can be increased.[151] However, some behavioral law and economics scholars argue that there may be a lack of insurance demand even if coverage is offered at actuarially fair premiums in competitive markets.[152] In some industrial sectors, such as in the nuclear sector, the probability of accidents and damage is low. Thus the potential victims may still take an 'it will not happen to me' attitude. For chemical factories with a higher possibility of producing pollution and damage, the people living in or whose property are in surrounding areas may have better incentives to insure. However, in this situation the premiums would be relatively high and the risk pool would be limited.

3.3.2. Direct Insurance

3.3.2.1. For Ecological Damage

As discussed above, direct insurance can be applicable to ecological damage. Direct insurance is established before losses happen, or even before the installations start to operate. Compensation is provided as long as the damage is caused from insured

150 See above Section 1.2.

151 Faure & Bruggeman 2008, p. 15.

152 See Slovic, Finucane, Peters & MacGregor 2004; Harrington 2000, p. 43-44; Kahneman & Tversky 1979, p. 268-269.

sites and irrelevant of the establishment of liability. Thus compared to liability insurance, it can offer earlier and easier access to compensation. However, how broadly direct insurance can cover depends on the definition of 'insured risk' in the contract and how the notion 'damage' from an insured site is interpreted. The potential consistency between an insurance contract and the restoration measures and damage assessment decided by public authorities deserves more attention.

As for the role of prevention, under a direct insurance policy, the potential injurers purchase coverage for cleanup costs of their own property and the damage caused from the insured sites. Such an insurance policy has deterrent functions since then it is the polluters that finance the risk pool. The easier access to compensation under direct insurance is also desirable from deterrence perspective. As discussed earlier, the probability of escaping liability maybe high in ecological damage cases, considering unidentifiable injurers, difficult damage assessment and remarkable litigation costs. Therefore the problem of inadequate deterrence will arise since injurers will base their care level on their expected liability, which is less than the actual expected losses caused by their activities.[153] This argument is used by some scholars to advocate mandatory liability insurance, provided the insurer is able to monitor an insured's conduct.[154] Under a mandatory liability scheme, every regulated injurer is forced to internalize at least parts of the costs it caused. However, such a scheme may still be imperfect since the premium of liability insurance is set according to the expected liability that might be traced to the insured parties, rather than the actual losses they may produce. Thus in a liability insurance policy, the costs of damage for which liability cannot be established are still externalized.

In this regard, a direct insurance might be more desirable by making the claim for damage easier. In a direct insurance policy, the trigger is no longer liability but damage itself. The victims or public authority do not need to engage in cumbersome litigation to get the compensation and can receive a direct action on compensation according to the insurance policy.[155] Thus they can get the coverage more rapidly and at lower transaction costs. Those advantages may promote the probability to claim for the damage and act as a remedy to the under-deterrence caused by low probability of being sued under liability rules. However, this plausible solution may propose another problem: will the potential injurers have incentives to buy this insurance since now they have to take into account more damage than the expected liability facing them? Without insurance, a potential injurer faces the burden to the extent that he is adjudicated as liable for the damage. In a liability insurance policy, according to Shavell, if insurers can monitor insured's level of care and set the premium at a reasonable level, risk-averse individuals will buy full coverage.[156] However, under this condition, the level of care adopted by injurers will remain suboptimal since insurers will set premiums based on expected liability rather than

153 Shavell 1987, p. 7.

154 Richardson 2001, p. 309.

155 Such as the case of Dutch environmental insurance policy. See Faure 2002b, p. 300-302.

156 Shavell 2000, p. 168.

expected loss.[157] However, under a direct insurance policy, if it is designed to combat the problem of low probability of being sued and thus allows more claims as described above, then it comes to the question why a potential injurer would choose to take such a policy rather than just facing the liability himself or seeking coverage under liability insurance.

3.3.2.2. *For Traditional Damage*

Potential polluters can also buy insurance for personal injury and property damage caused from their sites. This policy is desirable from the perspective of primary cost reduction, for the losses are internalized by polluters in the form of premiums. Whether the premiums they pay can reflect the real costs produced by polluters depends on the design of insurance contracts, such as the definition of causation between damage and pollution from the insured sites, the assessment of damage and probability of victims to claim from the insurance policy. Literature shows that the proof of causation under a direct insurance policy is easier than under a liability insurance policy.[158] To prove that the harm is caused by an insured risk is sufficient under the former, without the further need to establish the links to specific injurers. As far as the probability of claims being made is concerned, a direct insurance may allow more claims since there is no need to establish liability before the award of compensation. However, if a direct insurance policy allows more compensation than liability insurance, the problem arises whether polluters have incentives to buy such a policy.

Similar to first-party insurance, direct insurance is established before the damage happens. Thus it allows a more secure guarantee, for the risk of insolvency when losses happen is removed. Furthermore, direct insurance allows timely compensation before the liability is established.

4. Risk-sharing Agreements

4.1. *Introduction to Risk-sharing Agreements*

The earlier parts discussed commercial insurance as a compensation mechanism. Both liability insurance and first-party (direct) insurance offer avenues to shift the risk of compensation to a third party (the insurer) at the price of a premium. This section explores another method which, instead of shifting risks to commercial insurers, spreads the risks among polluters by establishing a mutual pool.

157 Leiter 2001, p. 276.
158 Faure 2002b, p. 295.

Under such a pooling system,[159] it is the polluters, but not commercial insurers, that aggregate resources to provide coverage. Nuclear pools[160] and pools under Protection and Indemnity Clubs (hereinafter as P&I clubs)[161] are such examples. It is the injurers themselves that finance this risk pool: they can either make advance payment or make an *ex ante* risk-sharing agreement (it can be called risk sharing, risk-sharing institutions[162] or mutuals[163] in the literature). A significant feature of this mechanism is that the members of this pool are both insurers and insured. Both the insurers and insured can influence the environmental accidents and its losses. But their interests are not fully congruent.[164] Rather than minimizing the risk, the insured tends to externalize its costs and get lower premiums; the insurer, a profit-maximizing firm itself, may limit and deny its coverage, and overcharge premiums. In a risk-sharing agreement, the potential injurers are both the providers and recipients of insurance. Thus the conflicts of interest between insurers and insured are avoided.

In a risk-sharing agreement, mutuality is formed, where the insurance premium paid by one member depends on the claims made by all other members.[165] It is in the interests of all for all other members' claims to be as low as possible and thus a mutual interest of risk minimization is created.[166] To reduce risks, the members of such a group have incentives to differentiate risks to align a member's contribution to the risk it proposes and monitor each other. Mutuality is established

159 Noticing again that a pool under a risk-sharing agreement is different from an insurance pool. This is also different from the pooling of the insurers to enhance the insurance capacity. The pool under a risk-sharing agreement is the aggregation of the resources of the insured themselves to cover their own risks. The pooling of insurers is one common technique used by insurers to work together to underwrite certain types of risks.

160 In the nuclear industry, insurance is often supplied by a monopolist in a jurisdiction, which leads to high premiums and limited coverage. Thus an effort has been made by nuclear operators to establish risk pools to cover their third party liability or their own property damage (onsite damage). The retrospective premium system under Price-Anderson Act and the pool established by NEIL are some examples. The Price-Anderson Act was originally enacted by Congress in 1957 to amend the Atomic Energy Act of 1954. Public Law 85-256. A retrospective premium system was introduced in its 1975 amendment. Public Law 94-197. NEIL is a nuclear insurer called Nuclear Electric Insurance Limited, which insures nuclear plants and their generating units for some first-party damage. Details see Chapter 7, Section 1.3.

161 A P&I Club is a non-profit mutual insurance association which is established by ship owners and charterers to cover their third-party liabilities to the use of operation of ships. Now thirteen separate and independent principle clubs cooperate together to comprise the International Group of P&I Clubs (hereinafter the Group), accounting for approximately 90 per cent of the world's ocean-going tonnage. See the website of the Group: <www.-igpandi.org/Home>, last accessed on 8 May 2013.

162 Faure and Skogh's discussed risk-sharing agreements as a compensation alternative for environmental damage and nuclear damage. See: Faure 2004; Faure & Skogh 1992; Skogh 1999, p. 505-515; Skogh 2008, p. 274-287.

163 See Bocken 2009, p. 165-166.

164 Boyer & Porrini 2008, p. 346-347.

165 Bennet 2001, p. 15.

166 *Ibid.*

usually when the members are subject to similar safety regulation and are exposed to comparable levels of liability. The members are faced with the same type of risk and have a more expert and precise knowledge compared to a third party insurer.[167] Therefore they can evaluate the risks each member creates and can monitor each others' behavior better.

Unlike the commercial insurance, under which *ex ante* information about the probability of a certain risk and its magnitude should be available to allow the calculation of an *ex ante* charged premium, each member's contribution can be agreed on beforehand and only actually paid *ex post* in a risk-sharing agreement policy. This characteristic makes it possible for a risk-sharing agreement to deal with uncertain risk, for which the statistical data about the occurrence are rare or the probability and size is less predictable. As long as a risk differentiation can be made among the members, a risk-sharing agreement can be feasible, since an *ex ante* charging of premiums is no longer necessary.

Another difference between a risk-sharing agreement and insurance concerns the costs. In an insurance policy, the risk is shifted to the insurer at the price of a premium. The premium is not recoverable by the insured no matter whether the insured risk happens or not. While according to a risk-sharing agreement, a member's contribution is lost only if an accident happens; the contribution can be carried over for the following year if there is no accident. A member can also recover his contribution by stopping creating the risk and leaving the pool.[168]

4.2. *Specific Institutional Arrangements of a Risk-sharing Agreement*

4.2.1. Examples of Risk-sharing Agreements

The previous section briefly discussed the risk-sharing agreement as an alternative to insurance to compensate ecological damage. The specific arrangement of such a policy under a risk-sharing agreement may vary: does the risk pool cover damage of members' own property or only the damage to a third party? Does this policy only apply to ecological damage or also extend to traditional damage? Furthermore, how members' contributions are determined and collected also needs to be clarified. Whether the contributions are paid *ex ante*, reserved *ex ante* or collected through retrospective premiums is of relevance to determine the security and costs of a risk-sharing policy. Moreover, risk-sharing policies can also differ in possibility of direct action by victims and recourse to the injurers.

As discussed above, a risk-sharing agreement is used to gather different members subject to the same risk within a single pool. The feature of the same type of risk results in such an agreement often being formed in a specific industrial sector, and thus sector-based rather than via a comprehensive policy. It is usually employed to deal with large and uncertain risks,[169] such as in nuclear and maritime

167 Faure & Fiore 2008b, p. 302.

168 *Ibid.*

169 *Ibid.*, p. 301.

transportation sectors. The nuclear risk is a typical catastrophic risk with low probability of happening but immense losses once materialized. Considering the low occurrence, reliable data on the probability and magnitude of an accident are often lacking. The nuclear mutual pools established in the U.S. are such examples. In the U.S., nuclear liability is governed by the Price-Anderson Act of 1957,[170] which has been revised approximately every decade.[171] According to the Price-Anderson Act, each license issued should have and maintain a financial guarantee to cover public liability claims.[172] The term 'public liability' is defined as 'any liability arising out of or resulting from a nuclear incident', with the exceptions of claims arising out of an act of war, workmen's compensation claims, and claims for damage to on-site property at a licensed nuclear facility.[173] The existing capacity of the insurance market (provided by a monopolistic national pool, American Nuclear Insurers, ANI) to provide is $300 million. If an accident creates damage in excess of $300 million, a retrospective premium needs to be called upon all American nuclear operators licensed by the regulatory party Nuclear Regulatory Commission (NRC). This premium is payable in annual installments up to a certain maximum amount per incident per power plant and is determined according to the size and number of reactors each plant has.[174] The amount of the second tier is now set at $111.9 million, with a maximum annual retrospective premium of $17.5 million per reactor per year.[175] If a catastrophic accident happens, which needs the collection of retrospective premiums over the years, the victims do not need to wait until the operators have paid all the premiums. The NRC guarantees those retrospective premiums or, in other words, advances the compensation in the second tier and later collects this from the operators.[176]

The Price-Anderson Act established only a risk-sharing pool for third party liability. However, a nuclear accident can also create serious onsite damage. The Three Mile Island accident has shown that cleanup of the onsite damage would be far more important than the third-party liability coverage of the nuclear operator.[177] This has led to the introduction of the so-called property rule by the NRC, which requires each power reactor licensee to obtain a financial guarantee to stabilize and decontaminate the reactor and the reactor station site.[178]

A national mutual pool has been established under this rule, Nuclear Electric Insurance Limited (NEIL). The risks covered in NEIL include costs associated with interruptions of electric generation due to accidental physical damage to insured sites, decontamination expenses, and other risks of direct physical loss at insured

170 42 U.S.C.A. §2210.
171 The Price-Anderson Act has been revised in 1967, 1975, 1988 and 2005.
172 Pub. L. No. 85-256, 71 Stat. 576, 577 (1957).
173 Pub. L. No. 85-256, 71 Stat. 576, 576 (1957).
174 Faure & Vanden Borre 2008, p. 243.
175 10 C.F.R. §140.11 (4); 75 FR 16646, 2 April 2010.
176 Faure & Vanden Borre 2008, p. 260.
177 *Ibid.*, p. 245.
178 Conditions of Licenses, 10 C.F.R. 50.54(W) (2008).

sites.[179] To cover those risks, NEIL established three insurance programs: the accidental outage program, the primary program and the excess program. The interruption risk is covered under outage and the primary program and other risks are covered under the primary and excess programs.[180] To reduce the exposing risks and guarantee the capacity, NEIL seeks reinsurance from the ANI.

In the area of marine oil transportation, the technical uncertainties with regard to the occurrences of oil spills, combined with the legal uncertainties about finding liability, make it less predictable to operate an insurance policy. The P&I Clubs appeared as a response to commercial insurers' reluctance to underwrite marine risks.[181] P&I policies cover the liabilities specifically enumerated in the agreement – the Club's rulebook. P&I coverage usually includes 'unlimited' reimbursement for claims arising from: liabilities in respect of persons, liability in respect of cargo, collision with ships, or with fixed and floating objects, salvage, compulsory wreck removal, fines imposed by government agencies, quarantine expenses, towage liabilities, 'sue and labor' and legal costs, any other liabilities which the club's directors deem appropriate to cover as well as limited reimbursement for oil pollution claims which arise from the entered vessels.[182] The oil pollution claim means 'a liability, cost, loss or expense, howsoever incurred, in respect of or relating to an escape or discharge of oil or any threat or consequence of such escape or discharge, but excluding liability for loss of or damage to such oil.'[183] The term 'liability' in this expression is determined by international conventions and national legislation applicable in each case. A major difference between private insurance and a risk-sharing agreement is that the insurers and insured are the same. As a typical risk-sharing arrangement, in a P&I Club, the ship owners make contributions to a risk pool according to an *ex ante* agreement. Instead of paying fixed *ex ante* premiums, at the beginning of each year, an 'advance call' is made to cover the claims and administrative costs.[184] The amount of each advance is based on the ship owners' claims history, the size of the fleet, and the Club's anticipated needs and the strength or profitability of the insurance market.[185] If the pool of the advance call is insufficient to cover the losses, a supplementary call can be asked from the members.

179 Nuclear Electric Insurance Limited, 2009 Annual Report, 20 (2009), available at: <www.nmlneil.com/media/1724/ar2009.pdf>, last accessed on 8 May 2013.

180 *Ibid.*

181 Ronnerberg 1990, p. 2.

182 Ronnerberg 1990, p. 7-9. Ronnerberg's analysis was based on the Swedish Club's 1990 rulebook. The similar coverage can also be found in the 2010 rulebook of the United Kingdom Mutual Steam Ship Assurance Association (Bermuda) Limited (Bermuda Rulebook). In the rulebooks, the 'unlimited' reimbursement does not mean that the Club should pay the full costs which fall into the categories. Instead, the reimbursement is subject to the limitation of liability set by law. While for oil pollution claims, the compensable sums are determined by the Directors of the Club. See Rule 5,B of the Bermuda Rulebook, <www.ukpandi.-com/ukpandi/resource.nsf/Files/2010Rules/$FILE/2010Rules.pdf>, last accessed on 8 May 2013.

183 See Rule 5,B, the Bermuda Rulebook.

184 Faure & Grimeaud 2003, p. 232.

185 Ronnerberg 1990, p. 29.

When discussing the institutional arrangement of a risk-sharing agreement, the section uses P&I Clubs and nuclear pools as examples to clarify some issues. Details of those pools are discussed in the empirical part.

4.2.2. Institutional Arrangements

The coverage of a risk-sharing agreement may vary significantly. A risk-sharing agreement can either cover member's third-party liability or cover damage to the polluters' own property. It can also intervene only when the members are insolvent.[186] Those pools can cover only ecological damage (cleaning up costs), relevant traditional damage, or even extend to damage irrelevant to environmental issues. The coverage of a risk-sharing agreement depends on the definition under specific pooling agreements. The P&I Clubs choose a broad approach to contain both environment-related and unrelated damage, but the coverage is limited to the liability a vessel owner assumes, precluding his own property losses.[187] The nuclear pools show both examples of mutual insurance for liability and operators' own property.[188]

Another interesting issue is the payment of pool members' contributions. In a risk-sharing agreement, members can just make an *ex ante* agreement on how to share losses and only actually pay the contributions when an insured risk happens. Besides these retrospective contributions, the financing methods of a risk-sharing agreement can have some other manifestations: the operators can make the full contribution *ex ante*, or contributions are not paid, but reserved by operators on an *ex ante* basis. The Price-Anderson Act takes the form of retrospective contributions, while the P&I Clubs combine an *ex ante* payment with retrospective payment. In a P&I policy, an advance is called at the beginning of each year, and an additional call is needed to supplement if the pool established earlier is not sufficient to cover the heavy claims. Those financing methods are relevant with respect to the security of compensation when risk have materialized and may lead to varying costs of a risk-sharing policy. Firstly, if contributions are not paid *ex ante*, the problem will arise whether the required amount will be available in the event of an accident. Secondly, considering the mobility of assets and corresponding monitoring and regulation, the costs created by each method are different. Those issues will be analyzed later in this section.

186 The Dutch COFIZE foundation is such an example. The Dutch legislation imposes an obligation on gas stations to provide financial security. By passing the safety assessment and making contributions, a gas station can participate in the COFIZE foundation to fulfill the financial requirement. The foundation only intervenes when the licensee is insolvent. See Faure 2007a, p. 82-84.

187 See the 2010 rulebook of the United Kingdom Mutual Steam Ship Assurance Association (Bermuda) Limited. (Bermuda Rulebook), Rule 5,B, available at: <www.ukpandi.-com/ukpandi/resource.nsf/Files/2010Rules/$FILE/2010Rules.pdf>, last accessed on 8 May 2013.

188 The retrospective pooling system under the PAA is an example of the former and the NEIL provides an example of the latter.

A risk-sharing agreement is established to protect its members' interests, while the protection of victims' welfare, although promoted by such a policy, is not usually the direct aim of it. Thus the question whether victims have a right of direct action against the pooling group arises. Direct action may lead to an increase in administrative costs with regard to the expense of defending and dealing with claims. But denial of such a right may be undesirable from the perspective of timely and securable compensation, especially in the case of a member's bankruptcy.

4.3. *Evaluation of Risk-sharing Agreements*

4.3.1. Primary Costs

To minimize the primary costs requires that the potential injurers are given optimal incentives to reduce the accidents. This means the contributions potential injurers have to pay should be aligned to the risks they create. Under a risk-sharing policy, it is the potential injurers who funded the risk pool. But whether the contributions of members in a risk-sharing agreement reflect the actual costs they produce deserves more analysis here.

To base the contribution on each member's actual risk requires information on both the probability of a certain risk and the magnitude of losses in the case of occurrence. However, under the background of ecological damage, such information is not always available. This poses a challenge to both the setting of contributions (premiums) and the capacity of a risk pool. Insurance policies try to deal with these problems by charging risk premiums for the uncertainties (insurer ambiguity),[189] which, however, can lead to high premiums and make insurance a less attractive policy. The lack of information is less of a problem under risk-sharing agreements. An arrangement can be made in the agreements for the contributions only actually to be paid when risks materialize. By then, the uncertain risks have been transformed into actual losses. Thus as long as a risk-sharing policy can differentiate between different levels of risks, accurate *ex ante* information on the probability and magnitude of the accidents would be no longer necessary. Risk-sharing policies are usually used to cover highly technical and complicated risks, such as nuclear liability and marine oil pollution liability. In specific sectors, members creating the same type of risks can cooperate together to tackle the potential losses. Different methods have been developed to differentiate risks among members in risk-sharing agreements. In the nuclear sector, each member's contribution is to a large extent dependant on the size and number of the nuclear reactors being operated.[190] This rough differentiation may be effective in the nuclear

189 Kunreuther, Hogarth & Meszaros 1993, p. 71-78.

190 This is the criterion to charge retrospective premiums under the Price-Anderson Act in the U.S. Stating Concerning the Price-Anderson Act: Hearing Before the Subcomm. on Capital Markets, Insurance and Government Sponsored Enterprises of the H. Com. on Financial Services, 107th Cong. (2001) (statement of Maijorie S. Nordlinger, Senior Att'y, U.S. Nuclear Regulatory Commission), available at: <www.nrc.gov/reading-rm/doc-collections/congress-docs/congress-testimony/2001/>, last accessed on 8 May 2013.

case. The risk a member creates depends on both his level of care and on the level of activity. As far as the level of care is concerned, the members of the same nuclear pool usually have to comply with the same safety regulations, which creates some consensus with regard to their risk levels.[191] However, it should be borne in mind that the safety regulation is offering just a minimal requirement, and the safe status of each nuclear operator may still vary. Risks can be further differentiated on the basis of other criteria such as the location of the nuclear power plant, construction type and operational features of the plant.[192] Theoretically speaking, the pooling also provides members with incentives for mutual monitoring, which contributes to maintaining the risks on a comparable level on the one hand, and alleviating the moral hazard problem on the other hand. However, whether effective mutual monitoring will happen in practice depends on specific situations. As far as the activity level is concerned, it can be largely attributed to the number of reactors a member possesses. Therefore, in a community where nuclear operators are exposed to the same level of safety regulation and create similar level of risks, and mutual monitoring works effectively, a risk differentiation based on the number of reactors in the nuclear sector is efficient. Otherwise, further risk differentiation is necessary. The P&I Clubs try to set premiums commensurate with the riskiness of each member, using a combination of empirical, objective knowledge and tacit, subjective knowledge.[193] At the beginning of the year, an advance called is made, based on the tonnage insured and the riskiness of their operations (expressed as a 'rate; per tonne).[194] With the renewal of coverage, the premium of a member can be adjusted according to his past claim experience.

It has been argued that a competitive market is a prerequisite for liability insurance to work efficiently to spread risks and at the same time to sustain the appropriate preventive incentives.[195] The risk-sharing agreement, as one type of mutual insurance, concerns the pooling of members and sometimes the co-insurance between different pools. For such an instrument, a competitive market is also essential for an efficient risk differentiation and maintaining a reasonable premium. The compatibility of risk-sharing agreements and competition law deserves more discussion here. The pooling of various members, on the one hand, can provide the otherwise unavailable large amount of coverage and, on the other hand, can decrease the competition by some arrangements such as agreements on premiums, standard form policies and so on.[196] Whether the P&I Clubs should be exempted from EU competition policy has long been a disputed issue. The International Group of P&I Clubs, composed of 13 major clubs, accounts for 93 per

191 For example, in the U.S., nuclear safety is governed by federal regulations, which guarantees that reactors propose comparable risks.

192 Faure & Fiore 2008b, p. 304.

193 Bennet 2001, p. 15.

194 *Ibid.*

195 Faure & Van den Bergh 2002, p. 282-291.

196 For example, the clubs do not allow competition by cutting rates. They argued that rate cutting would decrease stability and derogate from an effective claims records, based on which the premiums could be determined.

cent of the worldwide market for P&I insurance. The Group operates on the basis of its International Group Agreement and Pooling agreement, containing rules on the sharing of insurance claims and joint reinsurance. Those rules, especially the limitation on price cutting, have been hotly debated on their compatibility with EU competition policy. Despite the argument that it is doubtful that the conditions for exemption for the Group were fulfilled, the European Commission has granted the Group a 10-year exemption under the former Article 85(3) of the EC Treaty (Article 101(3), TFEU) in 1999. With the expiration of this exemption, this issue starts to attract attention again. The agreements between the members of the Group are not automatically covered by the new competition block exemption for the insurance sector.[197] The market share of the parties involved is substantially above the relevant thresholds provided by the block exemption (20 per cent for co-insurance pools and 25 per cent for co-reinsurance pools). Thus the European Commission opened formal proceedings to investigate whether the agreements between the P&I Clubs might infringe European Union antitrust rules on 26 August 2010.[198] This investigation closed in 2012, with the conclusion that the investigation cannot confirm the concerns over antitrust issues.[199]

As discussed above, members of a risk-sharing pool can pay their contribution through *ex ante* payments, *ex ante* reserves or *ex post* payments. The difference in forms of payment can also lead to variance in corresponding costs. An obvious advantage of the model to ask members to make a full contribution *ex ante* is that the funds would be available in the event of an accident.[200] However, it is costly in the sense that the assets paid would be immobilized during the policy period. The substantial contribution in some industries, such as the nuclear sector may make some smaller members insolvent.[201] Instead of actually making the payment *ex ante*, the contributions can also be reserved, but not paid on an *ex ante* basis. Without actual payment, arrangements should exist to ensure that the amount set aside is still available in the event of an accident, which creates high administrative costs on the control of the reserves. Furthermore, 'some priority regulation needs to be made in order to ensure that the trustee in bankruptcy cannot seize the funds.'[202] In the case of a retrospective contribution, there is no requirement of *ex ante* payment or reserves, which means that no assets are immobilized or lost until an accident occurs. This costs saving, however, has to be balanced with the security of such a financial guarantee and the risk of insolvency of members. Many other factors need to be considered in determining the insolvent risk. Other regulation may contribute

197 Commission Regulation (EU) No. 267/2010 of 24 March 2010 on the application of Art. 101(3) of the Treaty on the Functioning of the European Union to certain categories of agreements, decisions and concerted practices in the insurance sector.

198 <http://europa.eu/rapid/pressReleasesAction.do?reference=IP/10/1072&format=HTML&aged=0&language=EN&guiLanguage=en>, last accessed on 8 May 2013.

199 Antitrust: Commission closes investigation in P&I Clubs case, available at: <http://europa.eu/rapid/press-release_IP-12-873_en.htm>, last accessed on 8 May 2013.

200 Faure & Fiore 2008b, p. 306.

201 Faure & Fiore 2008b, p. 306-307.

202 Faure & Fiore 2008b, p. 307.

to the control of insolvency, as in the case of the nuclear sector.[203] The insolvency of one member will lead to increasing shares of other members, which creates high incentives of mutual monitoring with respect to the solvency status. Moreover, sometimes, guarantees are required to ensure the availability of funds in the event of an accident. The costs of the guarantee may also form a substantial part of the costs of the *ex post* payment method.[204]

4.3.2. Secondary Costs

A risk-sharing agreement also plays a role in the reduction of secondary costs and the provision of compensation to ecological damage and some other damage. A risk-sharing agreement can offer coverage for ecological damage, third-party liability and even the members' own property damage. A risk-sharing agreement is usually established to provide protection to its members. In addition to ecological damage, a member also faces risks of other third-party liability and his own property losses. Hence the coverage of a risk-sharing agreement is often quite comprehensive rather than specifically focusing on ecological damage.

It has been discussed above that the method of financing a risk-sharing pool influences the primary costs. The financing method or, in the other words, the time to establish the pool actually also has an impact on the security of a risk-sharing agreement as a compensation mechanism for ecological damage. The *ex ante* payment approach enjoys the merit of available funds when the obligation to compensate is due. As for the *ex ante* reserves approach, regulation and monitoring should be used to ensure the amount is actually set aside by each member and is still available in the event an insured risk materializes. Furthermore, the bankruptcy institution is also essential in the sense that other debtors should not have a priority over the claimers of the risk-sharing policy to be compensated by the reserved assets. Under the retrospective premium approach of the PAA, the *ex ante* reserves or payment is not required, thus the question of the availability of agreed contributions in the event of an accident arises.[205] The solvency regulation and financial guarantees of payment for shares of the members are then an important factor to determine the security of a risk-sharing agreement.

Except for the part of damage which can be regarded as damage to a member's own property, the damage is only compensated under a risk-sharing agreement when the liability is established for one of the members. In some arrangements,

203 Faure & Fiore 2008b, p. 308.

204 For example, the risk-sharing arrangement under Price-Anderson Act adopts the third approach. To guarantee the payment of retrospective premiums, the licensees are required to prove their solvency by: surety bond, letter of credit, revolving credit, maintenance of escrow deposits of government securities, annual certified financial statements and other approved types. See United States General Accounting Office, Report to Congressional Requesters, Nuclear Regulation: NRC's Liability Insurance Requirements for Nuclear Power Plants Owned by Limited Liability Companies, 2004, p. 7-8. As for the first four types, the cost of guarantee would be positive.

205 See Chapter 6, Section 1.3.2.

such as the P&I policies, an even more stringent condition is required: the members should first pay for the damage and suffer actual losses to get the reimbursement. Those arrangements, from the victims' perspective may be inferior, given the difficulties in establishing liability with respect to ecological damage.

Another factor to be considered here to evaluate the secondary costs of a risk-sharing agreement is its capacity to offer ample compensation. The risk-sharing agreement can provide coverage for the otherwise uninsurable risks given their unpredictability. Furthermore, it has high potential to give a large amount of compensation due to its function in both interpersonal and intertemporal risk spreading. On the one hand, with the contributions to a risk pool by many members, the risk of a large amount of compensation is spread among those members. On the other hand, in the case of catastrophes, it is not necessary to make the full contribution at once. Take the system of retrospective premiums in the Price-Anderson Act as an example. A risk pool is established with the participation of 104 nuclear operators in the U.S.,[206] making the capacity to compensate for damage amount up to $12.2 billion (including the amount available from insurance market). The limitation for each reactor's contributions is $111.9 million. This amount, however, is not required to be financed at once. Instead, it can be financed over some years, with a maximum annul retrospective premium of $17.5 million per reactor per year.[207] With this arrangement, the capacity of compensation by a risk-sharing agreement is substantially increased. As far as the P&I Clubs are concerned, the pool provides a mechanism for sharing all claims in excess of $8 million up to about $6.9 billion, with a retention of $8 million by each Club, a layer of claims between $8 million to $2.06 billion arranged by the Group, and the upper layer falls back onto each Club.[208]

4.3.3. Tertiary Costs

In addition to the potential to reduce primary costs and secondary costs, a risk-sharing agreement also creates tertiary costs. The agreement is usually established among members of the same sector and producing the same types of risks. Compared to a third-party insurer, the potential injurers themselves possess better knowledge about the potential risks, methods about risk assessment and controlling, as well as mutual monitoring. In this sense, the risk-sharing agreement can help to keep the costs to operate such a compensation mechanisms lower compared to the insurance. However, this benefit needs to be balanced against the advantages of the insurance, since insurers are specialized in dealing with risk.

A risk-sharing agreement can operate on the basis of an existing organization or through a specifically created one. The system of retrospective premiums in the Price-Anderson Act works through ANI, which is the insurer who provides

206 U.S. NRC, Power Reactors, available at: <www.nrc.gov/reactors/power.html>, last accessed on 8 May 2013.

207 10 C.F.R. §140.11 (4); 75 FR 16646, 2 April 2010.

208 See the website of The American Club, at: <www.american-club.com/page/international-group-of-pi-associations>, last accessed on 8 May 2013.

insurance for nuclear operators for the liability of the first layer under the Price-Anderson Act. There are no separate entities established specifically for the operation of the nuclear pool. The ANI is also responsible for collecting the retrospective premiums when the claims exceed the amount of the first layer. This design, on the one hand, keeps the two layers of compensation consistent; and on the other hand, saves the costs of establishing another organization and keeping the functioning of two separate systems. In the marine transport area, a system of mutual associations has been established. Unlike nuclear insurance, which contains limited insured parties, and is often operated on a monopolistic basis, the insurance market of marine liability is composed of a large number of participants on both sides of insured parties and insurers. Thus is it not feasible to rely on a specific insurer or public authority to run the risk-sharing pools. Specific entities, the P&I clubs, have to be established to run the risk-sharing pools. The operation of those clubs can lead to substantial administrative costs. However, the Clubs play more roles than providing insurance: they can provide correspondents and representatives to give on-the-spot assistance; Clubs also provide Letters of Undertaking when members' vessels are arrested; they also assist members in claims handling and settlement.[209] Those additional uses beyond providing insurance can be regarded as a counterweight to the high costs of running such a pooling system.

5. Environmental Funds

5.1. Introduction: the Various Types of Environmental Funds

The tandem structure of liability rules and insurance tries to improve the capacity to provide compensation for ecological damage and at the same time maintain the deterrent level to prevent damage. However, as discussed above, many problems may arise with the insurance of ecological damage.[210] Compensation funds are proposed as an alternative. The Hazardous Substance Superfund (superfund) under the CERCLA[211] and the Oil Spill Liability Trust Fund (OSLTF) under the Oil Pollution Act[212] are some of the well-known examples of environmental funds. In Europe, opinions in favor of the establishment of environmental funds can also be found in the literature. For example, The Interuniversity Commission for the Revision of Environmental Law in the Flemish Region proposed the introduction of a compensation fund as an optional second layer of compensation.[213] There are also some practices of environmental funds in Europe, such as the Dutch Air Pollution

[209] Ronnerberg 1990, p. 25-28.

[210] For example, the lower predictability of both the possibility and magnitude of ecological damage may weaken the insurability of liability insurance. Precise risk differentiation may be difficult or costly. The award of compensation is based on the establishment of liability. The insolvency problem prevents potential injurers to take sufficient insurance. See Section 4.2.

[211] 42 U.S.C. §9601.

[212] 33 U.S.C. §2701.

[213] Bocken, Lambrechts, Boes, De Nauw, Faure & Lavrysen 1996, p. 31-32.

Fund, the Walloon *Fonds pour la gestion des déchets* and the German *KlärschlammEntschädigungsfonds*.[214] In the marine pollution area, funds are also established at the international level, such as the International Oil Pollution Compensation Fund (IOPCF) established under the Fund Convention[215] and the International Hazardous and Noxious Substances Fund established under the International Convention on Liability and Compensation in Connection with Carriage of Hazardous and Noxious Substances by Sea (HNS Convention).[216] These examples show that the environmental funds have been used widely at both international and national levels. However, it is too early to conclude that the environmental fund is a panacea for all the problems ecological damage creates. In both literature and practice, the term fund has been used to refer to quite a variety of public and private financial instruments. Thus, to understand and evaluate the efficiency of environmental funds in dealing with ecological damage, it is worthwhile to explore their structures and different categories.

Scholars have classified environmental funds in different ways. For example, Faure analyzed four kinds of environmental funds: the limitation fund, advancement fund, guarantee fund and a general environmental fund. The limitation fund is a certain sum offered by the enterprise to compensate all the victims with a final settlement for damage caused by a specific tort. The advancement fund allows for compensation before the conclusion of civil procedures concerning liability and insurance coverage. Guarantee funds intervene in the case of the insolvency of a liable injurer or his insurer. The fourth type, a general environmental fund operates as a substitute for liability and insurance.[217] Bocken differentiates environmental compensation funds in different ways: the guarantee funds, complementary funds, autonomous funds and funds holding the polluter harmless. He defines the complementary funds as the funds that intervene when specific substantive requirements of tort law are not met. Autonomous funds offer compensation for victims that suffered damage of a certain type and origin, irrespective of the possibility of a tort claim. A fourth type is a bit special: it compensates 'the owner of the facility from which the pollution results for part of the cost of preventive measures or liability incurred by him.'[218] It can be seen that the term fund can have very distinct implications and the classifications may also

214 The Dutch Air Pollution Fund provided compensation for damages resulting from a sudden air pollution occurring above Dutch territory, if those damages should not reasonably be borne or not be borne in full by the injured party. The Walloon *Fonds pour la gestion des déchets* provides a limited compensation for damage caused by environmental effects of waster, and is intended to fill gaps of the liability mechanisms. The German *Klärschlamment-schädigungsfonds* compensates for personal and property damages and consequential losses caused by the agricultural use of sewage sludge. See Bocken 2002, p. 35-38.

215 Art. IV, International Convention on the Establishment of an International Fund for Compensation for Oil Pollution Damage, RMC I.7.111, II,1,7,111, Misc 37.

216 Art. XIII, International Transport Treaties, Suppl. 20 (October 1996), I-573–600/Cm. 3580 RMC I, II.7.125.

217 Faure & Grimeaud 2003, p. 236-238.

218 Bocken 1988, p. 3-6.

vary. Thus, it would be relevant here to discuss briefly the criteria to classify the funds.

5.1.1. Funds with or without Risk-spreading Functions and Ways to Finance the Funds

When talking about environmental funds, one may easily think of the financial arrangements which are funded through contributions by a group of potential polluters, other relevant groups or even more general tax arrangements. However, sometimes the term fund is also used to refer to a certain amount of money provided by an injurer himself to settle the claims in a certain tort. The limitation fund discussed by Faure is such an example.[219] In the cases of serial damage, liable parties may be willing to establish a fund to offer a certain sum to all the victims so that he can reach a final settlement for the damage caused by the specific tort. Examples can be found in the field of marine pollution. A limitation fund is established under the CLC convention. To avail the benefit of the limitation of liability, the ship owner shall create a fund for the total sum.[220] A similar provision can also be found in the HNS Convention.[221] The potential injurer may want to finance such a fund because he can then limit his liability to the amount he brought into the fund. Such a fund has its own limitations. Due to the intricacies and uncertainties about ecological damage, a long time lapse exists not only between the conduct of the harmful activities and the appearance of the damage, but also between the first time that the damage appears and the time that the damage is fully discovered. Thus if the limitation fund is not large enough to fully compensate victims, the latecomers are excluded from the accessibility of compensation.[222] Furthermore, there is no risk spreading in such an arrangement, so that the capacity of providing ample compensation is also limited. Bearing in mind that compensation is the primary goal of compensation instruments discussed in this research, this section will focus on the environmental funds involving risk spreading instead of on limitation funds.

Leaving aside the limitation fund, the term 'fund' is more frequently used to denote the financial instrument which is composed of the contributions from various actors, being potential injurers themselves or other relevant groups. The way to finance a fund deserves more explanation here since it is an essential criterion to determine whether the fund complies with the polluter pays principle and creates an efficient level of deterrence.[223] The financing of the fund can take

219 Faure & Grimeaud 2003, p. 236-237.

220 Art. V, 3 of the 1992 CLC.

221 Art. IX, 3 of the HNS Convention.

222 Faure & Grimeaud 2003, p. 237.

223 In addition to the way a fund can be financed, some other issues are also relevant with regards to the deterrent effect. Even if a fund is initially financed by parties other than the contributors to the risks themselves, its deterrent effect can be remedied by allowing for recourse. Whether the deterrence created by the tort system and liability insurance remains is another relevant issue.

different forms: taxes, charges from specific parties, recourse from responsible parties and et cetera.[224] In spite of those various funds collecting methods, from the perspective of financing parties, the funds can be categorized into three types: those financed by potential polluters, by other relevant groups and by the general public. The Superfund, before its depletion in 2003,[225] fell into the first category. It was then funded through a tax on petroleum and chemical industries.[226] An example of the second type is the IOPCF established under the Fund Convention. The annual contributions to the IOPCF are made by persons who received the oil in total quantities exceeding 150,000 tons, rather than the ship owners who transport the oil (the potential injurers). The third type, usually criticized as a subsidy to the pollution industries, is to finance the fund through a general tax. A case on point is, again, the superfund. Since its depletion by the end of financial year 2003, funding for orphan sites cleanups has been appropriated by Congress out of general revenues.[227]

5.1.2. Funds Established for the Benefits of Victims and Injurers

A compensation fund is often established by the potential injurers or other groups to compensate the victims. It usually operates to promote the welfare of victims. For example, the guarantee fund, which intervenes in case of injurers' insolvency, can increase the amount paid to the victims. A more general environmental fund established *ex ante* can guarantee a certain amount of compensation when damage actually happens.

However, the compensation fund can also be established for the benefit of injurers. The benefits of injurers are sometimes regarded as a goal to develop financial securities. For example, the ELD stipulates that enabling operators to use financial guarantees to cover their responsibility under the Directive is the aim of financial security mechanisms.[228] A few examples of environmental funds exist which protect the polluters against losses resulting from the pollution. There are, basically, two different approaches. Firstly, the funds reimburse the owners or

224 For example, the Dutch Air Pollution Fund was previously financed by a fuel tax and later by appropriations from the budget. See Bocken 2002, p. 37. These methods mentioned above are often combined to finance a fund. The OSLTF established under OPA is such an example. It was originally financed by a tax on imported and domestically-produced oil, amounts recovered by the Fund and those transferred from the previously related funds, penalties and interest on investment. 26 U.S.C.A. §9509(b). The tax was suspended in 1994 and was reinstated from 2006. See <www.uscg.mil/npfc/About_NPFC/osltf.asp>, last accessed on 8 May 2013.

225 US Government Accountability Office (GAO), Superfund Program: Updated Appropriation and Expenditure Data, GAO-04-475R Superfund Program, available at: <www.gao.gov/new.items/d04475r.pdf>, last accessed on 8 May 2013.

226 According to CERCLA, the superfund consists of the amounts appropriated, transferred to the fund, and the amount recovered, penalties and punitive damages. SEC 517 CERCLA creates three essential taxes to finance the trust: the excise taxes on crude oil, primary petrochemicals and certain inorganic substances. See Cartwright 2005, p. 308.

227 Cartwright 2005, p. 308.

228 Art. 14, 1 of the ELD.

operators of the polluting facility for the cleanup costs when defenses exist and liability is not established. The Deepwater Port Liability Fund and Offshore Oil Pollution Compensation Fund are such examples. The former compensated port licensees and operators of vessels for cleanup costs caused solely by an act of war or the negligence of the federal government in establishing or maintaining aids to navigation;[229] while the latter offered compensation for spills caused by an act of God, an act of war, hostilities, civil war or insurrection or negligent act of a third party.[230] Both funds have been transferred to the OSLTF with the coming into force of OPA.[231] Those funds can promote the protection of victims and cleanup of ecological damage in the sense that it allows the effective protective measures to be taken as soon as possible.[232] Otherwise, the owners or operators of a polluting facility will have little incentive to take those measures since they are not responsible. The second type compensates the polluter for (part of) the compensation paid to third parties. A case on point is the IOPCF under the 1971 Fund Convention. In addition to compensating the victims, the fund by then also played a role in indemnifying the owner and his guarantor for that portion of the aggregate amount of liability under the CLC.[233] This provision was later deleted in 1992 Protocol of Fund Convention. Those funds have a function other than compensating victims or cleaning up pollution. For the cases where no liability insurance is available, the fund financed by the operators themselves can be regarded as a mutual pool. But where liability insurance is available, the fund comprises a shift of burden from the specific liable parties and its insurer to other entities. Whether this shift is efficient in the sense of its deterrent effect depends on the way the fund is financed.[234]

5.1.3. Relationship to Liability Rules and Insurance

The third chapter of this research has discussed liability rules as an important instrument to both deter the potential injurers from preventing ecological damage and compensating for the damage. The inadequacies of liability rules in fully realizing those goals make the compensation mechanisms necessary. Those mechanisms can either complement liability rules to promote their function of deterrence and compensation or replace liability rules.[235] Liability insurance is a typical and frequently used instrument to complement liability rules. It can increase

229 33 U.S.C. §1517(h)(5).
230 43 U.S.C. §1813(b)(1)(A).
231 26 U.S.C.A. §9509(b).
232 Bocken 1988, p. 6.
233 Art. 5(1) of the 1971 Fund Convention.
234 Bocken 1988, p. 6.
235 Liability insurance and risk-sharing agreements fall into the first category. They can help to overcome some of liability rules' disadvantages such as the insolvency problems and promote monitoring from insurers or other members from the same risk pools. Direct insurance operates independently from liability rules. Compensation is awarded according to the prescribed conditions rather than establishment of liability. Thus it can avoid the inefficiencies of liability rules.

the capacity to compensate and at the same time maintain an effective level of deterrence. Thus the relationship of environmental funds with liability rules and other financial instruments (especially liability insurance) is another criterion to classify the various environmental funds.

The first type of environmental fund provides liable parties with a method to realize its liability. Many different instruments are established as financial security for the realization of liability, such as insurance, risk-sharing agreements and other market instruments. Compensation funds can also be used as one of such instruments to provide financial security. A fund can either provide the first layer of compensation, in parallel with other financial security mechanisms, or provide a second layer of compensation (for example, a fund can intervene when the limitation of liability and corresponding financial guarantee is exceeded, or when the injurers and their guarantors are insolvent).

A second type of environmental fund can complement liability rules by intervening in cases where no liability arises from ecological damage. There are some circumstances where polluters for ecological damage can be identified. Polluters sometimes can be exonerated from liability by invoking a valid defense. A fund can provide compensation in those instances to accelerate the restoration of ecological damage. It is called a complementary fund by Bocken.[236]

The above mentioned 1971 IOPCF is a hybrid of these two types. Under the CLC convention, a mandatory financial security system is established: the ship owners registered in a contracting state and carrying more than 2,000 tons of oil are asked to maintain insurance or other financial security.[237] The fund not only provides compensation when no liability arises, the liable parties and their insurers are insolvent, and damage exceeds the limitation of liability,[238] but can also indemnify the owners and their guarantors for a portion of the aggregate amount of liability under the CLC.[239] Treating the environmental fund as an upper layer can promote the compensation capacity, without weakening the deterrent effect of other instruments. But as a paralleling alternative of other instruments, problems, such as which instrument is more efficient, whether the intervention of the environmental fund comprises a shift of the burden from polluters to others or the public, may arise.

A third type of compensation funds provides compensation for victims having suffered damage of a certain type and origin, or cleanup and restoration costs for such damage, regardless whether liability is established or not. This is because liability for ecological damage cannot always be established. Even when liability can be established, the proof of damage and causation may constitute a substantive burden, or the intricate and long-lasting judicial procedure may prevent a timely remedy. Thus an environmental fund offering compensation irrespective of the tort system can be an attractive solution. Such an instrument makes compensation available when the tort system does not work, and can provide a timely and less

236 Bocken 1988, p. 4-5.

237 Art. VII of the 1969 CLC.

238 Art. 4(1) of the 1971 Fund Convention.

239 Art. 5(1) of the 1971 Fund Convention.

burdensome remedy. Usually such a category of funds still allows the tort system and the recourse from liable parties after payment from the funds. Though awarding of compensation is independent from the establishment of liability, the claimants still have the choice to resort to the tort system. After the payment from the fund, recourse against liable parties is allowed. An example of this type is the OSLTF under the OPA. According to the OPA, the OSLTF shall be available for the costs for removal actions consistent with the National Contingency Plan and the natural resources damages.[240] The Public authorities who have taken removal actions can present to the fund directly. Other claims have to be made from the liable parties first, and then present to the fund for the uncompensated damages. The Fund also provides compensation for a responsible party if he is entitled to make a defense against liability or a limitation of liability.[241] Thus the OSLTF is a hybrid of the first type and third type of fund. Usually a claim should be made against the liable party first, which makes finding liability the precondition for awarding compensation by the fund. However, the Fund also provides compensation directly for the removal costs paid by public authorities. Finding liability here is irrelevant for the awarding of compensation. A more typical example is the compensation for pollution-related health damage in Japan, where a fund is financed through the tax paid by polluters. Individuals who contract designated diseases and reside in designated polluted areas for a fixed period of time are eligible for compensation. The awarding of compensation is independent of liabilities, although the fund can later recover from polluters.[242]

Another variant of compensation funds replaces the liability system completely, thus resorting to or having recourse fromto liable parties is impossible. The environmental fund in place of the liability system may enjoy the benefits of better compensation and lower transaction costs.[243] However problems may arise concerning deterrence, which is much determined by how the fund is financed. A comprehensive compensation fund might be useful for compensating the individual victims who suffered personal injury or property damage, such as the example of the no-fault scheme for accidents in New Zealand.[244] In such a scheme, the damaged individual's rights can be recovered monetarily via a compensation fund. However, as far as ecological damage is concerned, where physical cleanup and restoration are important, this type of compensation fund is of less relevance. Ecological damage gives rise not only to civil liability but also to administrative consequences.

240 33 U.S.C. §2712(a).

241 33 U.S.C. §2712(a), 2708.

242 Aronson 1988, p. 1046.

243 The fund offers compensation also on the condition that no fault can be proved, which allows better compensation. Moreover, the procedure is faster and there are fewer lawsuits, leading to lower transaction costs. See Fiore 2009, p. 409-410.

244 New Zealand introduced a universal no-fault system for all accidents that happen in its jurisdiction. The system is governed by a public monopoly (the Accident Compensation Commission), and is financed through tax and revenues. The system covers work-related injuries, non-work-related injuries, non-earners' injuries, traffic accidents, subsequent work injuries and medical injuries. People who have suffered personal injury do not have the right to sue an at-fault party, except for exemplary damages. See Fiore 2009, p. 413-415.

The deteriorated environment cannot be simply recovered by monetary compensation from a fund without cleanup and restoration measures. To place all the burden of taking such measures on public authorities would exceed the capacity of authorities and would not be feasible. When an accident happens, legislation often imposes an obligation on the polluter to report to the relevant public authority and take necessary actions to prevent and restore the damage. A polluter's administrative obligation to report and restore cannot be entirely and clearly separate from its civil liability to compensate for ecological damage. Thus one can hardly imagine a compensation system for ecological damage which would completely exclude the liability system.

Liability rules and insurance are important mechanisms in tackling ecological damage from both the perspective of compensation and deterrence. The relationship of an environmental fund with them is an essential factor for understanding the function of the fund itself. Thus this criterion is used to discuss the various compensation funds in this section.

5.2. *Evaluation of the Various Funds*

Before the analysis on specific environmental funds, some common issues will be explored first.

Bocken has studied various forms of environmental compensation funds. According to him, the funds can go from separate legal entities to mere accounting devices.[245] Faure further analyzed that compensation funds could be run by a regulatory authority or could be administered privately on a competitive basis.[246] How a compensation fund is operated is relevant for both the primary costs and tertiary costs issues. For example, competition can give the operators of a private fund incentives to conduct efficient risk differentiation. Ecological damage compensation usually involves public authorities (in the procedure of cleanup/restoration). This makes compensation funds for ecological damage usually publicly operated. Another factor influencing the functioning of compensation funds is which specific authority operates the funds. When ecological damage happens, usually some public authority will intervene in the cleaning up or restoration measures or take the measures themselves. Sometimes, these authorities can also act as the operators of compensation funds. The Hazardous Substance Superfund is such an example. It is operated by the Office of Solid Waste and Emergency Response, a division of EPA.[247] According to CERCLA, whenever any hazardous substance is released or there is a substantial threat of such a release into the environment; or there is a release or substantial threat which may present an imminent and substantial danger to the public health or welfare, the president is authorized to take response actions or allows the response actions of responsible parties.[248] In general, it is the EPA that selects response actions to protect health and

245 Bocken 1987, p. 84.

246 Faure & Grimeaud 2003, p. 241.

247 See <www.epa.gov/superfund/about.htm>, last accessed on 8 May 2013.

248 42 U.S.C. §9604(a).

the environment at EPA lead sites (CERCLA sites).[249] The Superfund provides compensation for both costs of response actions and uncompensated damages.[250] Thus the EPA plays the dual role of the claimant and the payer of the compensation. This structure, on the one hand allows for instant response actions and promotes the availability of data and, on the other hand, provides fewer incentives in efficient risk differentiation and reducing administrative costs.

As far as the primary costs reduction is concerned, the method to finance a compensation fund is of great importance. Leaving aside the issue of recourse from injurers, the financing method determines the allocation of the costs of pollution. A fund can be financed by potential polluters themselves, other relevant groups or even general taxes. The Superfund and OSLTF fall into the first category, to which taxation on polluters are an important source of income.[251] A typical example of the second category is the fund established under the IOPCF convention, which is financed by the contributions by persons who received oil in excess of a certain amount.[252] Under such circumstances, it is the beneficiaries, but not the potential polluters, ship-owners, that bear the costs of pollution. The shift in costs burden may lead to less efficient deterrent incentives. Even when a compensation fund is financed by polluters themselves, whether the fund can create efficient incentives depends on whether the contributions of polluters are proportionate to the risks they create. The contributions are often collected in the form of taxes. However, as for ecological damage, the actors who cause the damage are sometimes difficult to identify, let alone their individual contributions. Thus it would be difficult to determine the marginal tax for specific actors. Moreover, even if further risk differentiation and connecting tax rates with their risk rates is possible, the differentiation is not costless. The benefits of further differentiation should be balanced with its corresponding costs. Thus to finance a compensation fund, the tax on potential injurers is usually based on their activity level. For example, the tax on crude oil and petroleum products was an important part of financing for both superfund and the OSLTF. The tax rate for the superfund was 9.7 cents a barrel. For the OSLTF, the rate is 8 or 9 cent a barrel depending on the date of entering of the crude oil or petroleum products.[253] Under such compensation funds, the contributions of potential injurers are mainly decided by their activity levels. However, the care level of actors becomes less relevant. Such a structure can create inefficient incentives to take optimal care levels to prevent ecological damage.

Another deterrence-related issue is whether recourse is allowed or feasible in a fund scheme. Even when the finance of a fund failed to produce efficient preventive

249 United States Environmental Protection Agency, OSWER Directive No. 9200.4-22A, p. 3. See: <www.epa.gov/compliance/resources/policies/cleanup/superfund/coord-nrd-trust.pdf>, last accessed on 8 May 2013.

250 42 U.S.C. §9611.

251 The barrel tax for OSLTF expired in 1994, and later was reinstated from 2006. See: <www.uscg.mil/npfc/About_NPFC/osltf.asp>, last accessed 8 May 2013. The tax for superfund expired in 1995. After that congress refused to renew the tax, which leads to the fund running out of money in 2003.

252 Art. X of the 1992 Fund Convention.

253 26 U.S.C.A. §4611(c).

incentives, this inefficiency can be remedied if the operator of the fund can recover their payment from the actual injurers. The recourse procedure will make the injurers bear the costs they create. The recourse issue is especially relevant to compensation funds irrespective of liability. The claimants can choose to claim against the liable parties or the fund for their damages. The fund often provides a swifter and easier procedure than liability litigation, which results in many claims without first resort to liability rules. Under those circumstances, the fund, from which compensation has been paid, can recover from the liable parties.

The above analysis has discussed the function of environmental funds on primary costs reduction. A compensation fund, however, attaches more importance to secondary costs reduction. As analyzed in the introduction to this chapter, three criteria are essential in understanding a compensation instrument from the perspective of minimizing secondary costs: the scope, security and capacity. Unlike liability insurance, which provides compensation only when liability is established, a compensation fund can also intervene when no one can be found liable, or the limitation of liability and ceiling of insurance are exceeded (the first category of compensation funds). Except for the limitation funds, compensation funds are financed *ex ante*, which makes compensation available when risks materialize. ALthough established *ex ante*, the compensation funds can intervene (compensate) at different times. For funds which provide compensation on the condition that injurers can invoke a valid defense or limitation of liability is exceeded, the procedure of finding liability should be before the payment from the funds. While for funds, regardless of liability, compensation can be offered before liability is established. The instant and less burdensome compensation is essential for ecological damage, since the easy access to funds makes timely and efficient cleanup and restoration available. As for the capacity of compensation funds, it can be unlimited in theory since funds are often operated by public authorities, and can be financed by specific taxes or even general revenues. However, the political resistance should be taken into account in understanding the capacity of a compensation fund in reality.[254]

Compensation funds also give rise to substantial tertiary costs. A specific entity may need to be established to ensure the operation of a compensation fund. Furthermore, in a competitive insurance market, the insurer has strong incentives to

254 How the tax to finance superfund was created and extended is such an example. CERCLA created three taxes to finance the trust fund with a comparatively low rate in 1980. After is expiration in 1985, there was much debate on the approval of a new tax. President Reagan refused two bills about funding for superfund and even vowed to veto any bill creating new taxes. After fierce lobbying by a number of legislators, SARA was signed and the rate of tax was significantly increased. However, after the second expiration of the superfund tax, no new tax had been approved yet. This led to the depletion of the superfund in 2003, which now depends mainly on revenues. For a detailed introduction of the development of superfund from the first Reagan Administration until the Clinton Administration, see Cartwright, 2005, p. 309-316. Now a bill on reinstating the superfund is under discussion. See: <www.washingtonpost.com/wp-dyn/content/article/2010/06/20/-AR2010062001789.html>, last accessed on 8 May 2013.

reduce administrative costs, which is not the case in a publicly operated fund system.

5.2.1. Environmental Funds to Cover Liability

5.2.1.1. Environmental Funds as One Type of Financial Guarantees

An environmental fund can be used as a financial instrument to guarantee the realization of liability. The fund provides compensation at the same level as other guarantees, such as insurance, rather than offering an additional layer of compensation.

To understand the deterrent effect of such kinds of environmental funds, it is worthwhile to see if liability insurance or another *ex ante* guarantee mechanism is available. When such mechanisms are unavailable, a fund financed by polluters can operate in a similar way to mutual insurance. It then guarantees at least some amounts when damage happens out of the pockets of polluters as a group. However, where liability insurance is available, compensation by a fund may comprise a shift of the losses from the polluter and its insurer to other parties.[255] To answer the question whether such a shift is acceptable needs a comparison of funds and insurance from the viewpoint of deterrence. An important issue in determining the deterrent effect is under which instrument better risk differentiation can be undertaken so that the polluters can contribute according to the risk they pose. Usually insurers are more specialized in risk differentiation and risk spreading than the public authorities which operate funds.[256] If the market is competitive, insurers also have stronger incentives to improve risk differentiation and to monitor the insured. However, such incentives are often weaker in a publicly-operated fund system. Furthermore, there is also no clear reason why a public authority running a compensation fund would have better information on risks than an insurer.[257] According to the above analysis, if insurance is available, there is no strong argument to favor a compensation fund over insurance from the perspective of deterrence.

5.2.1.2. Funds Offering an Upper Layer of Compensation

Rather than an alternative of financial security, a compensation fund can also provide another layer of compensation. Usually there is a cap for the financial security of liability of ecological damage (the cap is often equal to the limitation of liability[258]). A compensation fund can intervene when the cap is exceeded. The fund

255 Bocken 1988, p. 6

256 Faure & Grimeaud 2003, p. 240.

257 *Ibid.*

258 For example, under the OPA, the specific responsible parties are imposed an obligation to establish and maintain evidence of financial responsibility sufficient to meet the maximum amount of liability they subject to under that act. 33 U.S.C. §2716(a).

can also offer compensation when the injurers and their insurers (guarantors) are insolvent.

It is argued that the combined use of the liability system, insurance and a fund can keep the incentives of the liability system untouched, since the fund only intervenes when liable parties and insurers fail to fully cover the damage.[259] But Faure doubted the principle of using a guarantee fund as the solution for insolvency problem. According to him, mandatory insurance and some other *ex ante* guarantee mechanisms can also play a role in the event of insolvency.[260] Admittedly, as he discussed, mandatory insurance also has many deficiencies, such as a high concentration on market, actually acting as licensors and the duty to insure.[261] Thus it would be hard and too early to conclude which instrument is more desirable as far as the capacity problem is concerned. On the contrary, it is worth at least taking different alternatives into account when facing the problem of capacity.

5.2.2. Complementary Funds

A complementary fund provides compensation when no liability arises. For example, sometimes liable parties cannot be indentified for specific ecological damage, or injurers can invoke a valid defense.

Under a complementary fund, the deterrent effect of liability rules is untouched, since the fund only intervenes when no liability arises. If the fund is well designed, so that the potential injurers make the contributions, another preventive incentive is added. Under this situation, although no specific party is identified as liable, the costs of damage are collectively borne by polluters. This extra burden also creates deterrent incentives for the polluters.

From the perspective of secondary costs reduction, a complementary fund is also desirable. It provides the resources for cleanup and restoring the damaged environment which are unavailable under a liability scheme. Established *ex ante*, the amount is guaranteed when damage happens. Examined from the perspective of timely compensation, the effect of complementary funds differs according to different scenarios. When no liable parties can be identified, one can make a claim against the funds directly. However, in case of valid defenses, a procedure to find liability may still be followed before the claimants can make a claim from the fund.

The latter case is also relevant in discussing tertiary costs. In the situation that establishing liability is necessary to claim against a fund, the costs of two systems (tort litigation and claims under funds) arise.

5.2.3. Environmental Funds Irrespective of Liability

An environmental fund can also provide compensation irrespective of liability. A fund then compensates the victims who suffer damages of a certain type and origin.

259 Faure & Grimeaud 2003, p. 242; Monti 2001, p. 65-68.

260 Faure & Grimeaud 2003, p. 242.

261 See Section 2.4.

Whether the tort system can be used becomes irrelevant for the granting of compensation by the fund.

In addition to the method to finance such a fund, as discussed above, another deterrence-related issue is whether recourse from injurers is allowed. As a more comprehensive compensation system, this kind of fund offers compensation before liability is established. Thus recourse is an important approach to let the injurers bear the costs themselves. However, one needs to bear in mind that a theoretical possibility of recourse does not necessarily mean that recourse practically happens. An example was the Dutch Air Pollution Fund, which compensated victims suffering damage as a result of air pollution in so far as reasonable compensation had not or could not be obtained from other sources.[262] The fund could subrogate the victims to claim against injurers once it provided compensation. However, 'it appears that the fund has never found it sufficiently practicable to exercise recourse.'[263] Since recourse is costly and places a high requirement on both legal and technical issues, sufficient expertise and financial resources are essential for effective recourse. This would be more of a problem for funds compensating numerous individual victims suffering personal injury or property damage, since the fund may immerse in a large quantity of trivial recourse procedures. As far as ecological damage is concerned, it is usually the public authorities who have taken cleanup or restoration measures that claim from the fund. Those authorities should have the expertise to take such measures according to administrative procedures or assessment regulations. Thus it might be more practicable for the fund to take recourse from liable parties with the aid of authorities who have taken measures and obtained compensation from it.

This comprehensive compensation fund is desirable from the viewpoint of secondary costs reduction. It allows compensation before liability is established. A comparable easier and swifter claim procedure is established under the fund scheme. The burden of proof is also often relieved compared to a litigation procedure. Easy access to funds is especially essential to deal with ecological damage. This is because timely cleanup measures can help to prevent or reduce damage and speed up the restoration of the damaged environment.

5.3. *Conclusion*

Funds are an often advocated solution for the problems of environmental damage. However, before arguing for or against funds as one financial instrument to compensate for ecological damage, one needs to notice the differences between various types of funds. Compensation funds can either be used to cover liability, to intervene when liability cannot be established, or to compensate victims regardless of whether liability has been found or not.

The previous analysis has shown that when liability insurance is available, there is no clear reason why funds would be preferred. If the insurance market is

262 Bocken 1987, p. 86; Bocken 1988, p. 5.

263 Bocken 1988, p. 5.

competitive, it is the insurer rather than the public authority operating the fund that can better differentiate risks and reduce administrative costs. However, this comparison does not mean environmental funds have no added value.

Firstly, some other financial instruments, such as liability insurance and risk-sharing agreements, only work when liability is established. Bearing in mind the difficulties to identify injurers and the heavy burden of proof, there is quite some ecological damage for which liability cannot be established. A compensation fund can provide resources to cleanup and restore the damaged environment on this condition. It is also desirable from the perspective of deterrence if the fund is financed by the polluters. Since in that case, the polluters bear the losses they create collectively, which are otherwise externalized.

Secondly, when an accident happens, it is important to react at once to prevent the occurance and expansion of damage and to recover the damaged environment as soon as possible. However, the tort system often leads to a long and burdensome procedure. Thus it is beneficial to make resources available to take emergent actions before liability is established. A compensation fund can play such a role. Although it is impossible to pre-finance all the cleanup and restoration measures for ecological damage, a design that would allow claims against the fund when the potential responsible parties deny their liability would be beneficial.

Thirdly, one can note the comparable higher capacity of compensation funds. Financed by taxes and fees, a public fund has theoretically unlimited capacity, which in practice is subject to political will and resistance. Environmental funds can be used to alleviate the capacity problems other financial instruments have, by adding another layer of compensation. However, attention should be attached to whether the existence of environmental funds reduces the incentives of potential injurers to seek the coverage of other *ex ante* financial instruments

6. Security Mechanisms Provided by the Liable Party or by a Third-party Guarantor

In addition to the risk-spreading instruments discussed above, financial security mechanisms can also be based on the assets of the potential responsible parties themselves or some third parties. A potential responsible party can prove its financial capacity or provide compensation by using self-insurance, captive, mortgages, liens, deposits, trust funds and escrow agreement. He can also pursue a guarantee from the parent corporation, from banks or from other financial institutions.

6.1. The Various Security Mechanisms

6.1.1. Self-insurance and Captives

By resorting to self-insurance, the potential responsible party (usually the operator of environmentally harmful installations) can submit a financial statement providing evidence of a satisfactory financial status to satisfy its financial responsibility.[264] Although using the nice word 'self-insurance', this instrument cannot be considered insurance in the traditional sense, for there is no risk-spreading or risk distribution and thus no loss-spreading when damage happens.[265] In fact, self-insurance is 'a system in which a company sets up a loss reserve account and depending on whether the account is funded or unfunded, the company may segregate an asset account to match the reserve.'[266] To establish a valid self-insurance policy, a company has to pass the financial test. For example, rules may require that a company's capital and net worth are both greater than the coverage requirement or a bonding rate test.[267] Those tests make sure that only companies with relatively deep pockets can satisfy their financial responsibility by demonstrating sufficient financial strength.[268]

One can also retain the risks but meanwhile satisfy the financial responsibility requirement by establishing a captive. A captive is 'a special purpose entity established as a licensed insurance or reinsurance company, owned by a corporation or group of companies which are not active in the insurance business themselves.'[269] Sometimes, the insurance market is reluctant to offer coverage, or only provides the coverage at high prices. Under this condition, a large company can choose to make a reserve and cover its own risk by establishing an insurance company itself. Captives are a popular way to cover the primary layer risks which are somehow predictable. They can also be tailored to meet the needs of various companies. There are three general forms of captives: those related by ownership, those related by a specific area of commerce or coverage and those with heterogeneous membership. Pure captives and group captives fall into the first category, differentiated by the number of their owners. The second category focuses on association and risk retention via a group of captives. Rent-a-captives are typical examples of the third category.[270] Group captives are established by various industries posing the same risk, and thus amount to pooling.[271] Pooling as a compensation instrument has been discussed earlier in this thesis. Thus the focus of this section is the pure captive established by one company as a financial security instrument. The captives operate in the same way as other insurance or reinsurance companies, by 'establishing unearned premium reserves and loss reserves, adhering to minimum capital/surplus level, actively managing the risk portfolio through diversification and cessions to the reinsurance market, managing the investment portfolio, and so on.'[272]

264 Bocken 2009, p. 160.
265 Faure 2004, p. 457-458.
266 Bruggeman 2010b, p. 144.
267 Boyd 2002, p. 434.
268 *Ibid.*
269 Bruggeman 2010b, p. 144.
270 Elliott 2005.
271 Faure 2004, p. 458.
272 Bruggeman 2010b, p. 145.

6.1.2. Mortgages and Liens

The traditional security interest in immovable and movable property can also be used as a financial security instrument. Here the potential responsible party can set aside his assets by establishing security interest for the potential claimants (victims). It is worth noting that these instruments such as mortgages and liens are of little relevance as a guarantee for future liabilities resulting from unforeseen occurrences,[273] since, on this occasion, the identity of victims, the creditors, is unknown *ex ante*. This means that security interests *in rem* are not an appropriate instrument to cover the individual damages caused via the environment. However, for the ecological damage, for which the claimants are mainly public authorities, these instruments may be more promising. A mortgage or lien for the benefit of a public authority can be established either at the occasion of the delivery of a permit or after an accident happens.[274]

6.1.3. Deposits, Trust Funds and Escrow Agreements

One can deposit a sum of money with the creditor and use it in the event of default by the depositor. The deposits can also be made with a third party under an escrow agreement,[275] which allows the release of the sum under the conditions determined in the agreement. The secondary category is more relevant with ecological damage, for which the public authorities are claimants. In the event of default by responsible parties, the authority can liquidate the account of deposits for the payment of covered damage. Having the accounts managed by independent financial institutions or other third parties is essential to guarantee the security of the deposits.

Another related instrument is trust funds, from which compensation is not conditioned upon the default by responsible parties, but is provided more comprehensively. A trust fund is a vehicle to collect money dedicated to a specific purpose. It is administrated by an independent trustee in charge of collecting, investing, and disbursing funds.[276] Compensation is provided when the specified condition is due, such as the establishment of liability in the court, being agreed by the sponsor of the fund or being recognized through an agreed procedure. Due to the comprehensive coverage and usually the large amount involved, the fund is typically financed over some period of time.

6.1.4. Corporate Guarantees, Bank Guarantees and Surety Bonds

A company can also seek the guarantee from a third party, be it the parent corporate or financial institutions. The third party may provide compensation

[273] Bocken 2002, p. 16.

[274] Bocken 2009, p. 160-161.

[275] Fogleman 2005, p. 118.

[276] Boyd 2002, p. 434.

simultaneously with the debtor or only after the responsible party is in default or insolvent, depending on the specific contract and national law.[277] The previous discussion mentioned self-insurance as an instrument to satisfy a large company's financial responsibility. However, self-insurance is not such a useful instrument for small and medium enterprises (SMEs), since it is difficult for them to pass the financial test. However, their capacity to cover the potential liability can be guaranteed by their parent corporation who can possibly pass such a test.

One can also think of bank guarantees and surety bonds as financial security instruments, under which banks and other financial institutions (especially insurance companies) assume the compensation obligation in the event of certain conditions being fulfilled. The potential responsible party has to pay a bank fee or premium in exchange for the banks' or financial institutions' coverage.

6.2. *Analysis: From the Perspective of Compensation and Prevention*

Different from the various compensation instruments discussed earlier, the mechanisms provided by the liable party himself or a third-party guarantor involve no risk spreading. The capacity to cover potential liability is not obtained by aggregating a large number of similar risks together. Instead, a company proves his capacity by either passing a financial test, providing assets or money as assurance or seeking a guarantee from a third party with stronger financial capacity.

Lack of risk spreading, on the one hand, has the advantage of keeping preventive incentives by potential liable parties, since it is still the polluters who pay for the damage. Even in the case of a third-party guarantee, where a third party may have to compensate in the event of default by the liable party. The potential liable parties have to pay a bank fee or premium in exchange for the coverage. On the other hand, however, without spreading the risk among a large number of members, the capacity to compensate for damage is still limited. This not only creates a problem of compensation, but also relates to the preventive effect. The problem of insolvency may indeed still arise, which leads to a lack of efficient preventive incentives.

The use of security mechanisms provided by a liable party or a third party is limited. Except the cooperate guarantee given by a parent company, most mechanisms mentioned here are only available for large companies. To resort to those mechanisms as a financial security, one has to pass a financial test to prove its deep pocket, provide sufficient and unencumbered assets or cash as assurance, or seek guarantee from banks or financial institutions by showing a good financial status and credit rating. Those requirements allow larger companies to qualify but keep out the SMEs. For SMEs, its capacity to compensate can be assured when a parent company, which itself meets the financial standards, agrees to satisfy the coverage requirements.[278]

277 Bocken 2009, p. 162.
278 Bocken 2009, p. 160.

In spite of the common characteristics mentioned above, these various security mechanisms also have many different features. For example, the intervention obligation of a security mechanism may come into play on different conditions: when damage happens or a liable party defaults. The level of security provided by each mechanism also differs. Passing the financial test cannot guarantee the availability of assets when damage occurs. While an irrevocable bank guarantee is a highly reliable security instrument, leaving the bank crisis aside.[279] A big range of variance also exists as regards to the tertiary costs each mechanism causes. Self-insurance creates no substantial costs since no prepayment is needed. Cost-saving is a frequently used argument for captives compared to insurance, but a bank guarantee may not be cheap or readily available. The following part discusses these mechanisms respectively.

6.2.1. Self-insurance and Captives

In the U.S., self-insurance is an acceptable security mechanism. By passing a financial test required by regulatory authorities, a large company can prove his capacity to cover the potential liability it may create. Without resorting to a third party or collective mechanisms, the risks still remain with the potential liable parties. Thus the preventive incentives remain untouched. However, the security interest provided is minimal. Passing the financial test at the beginning of an operation does not necessarily mean the assets are still available when risks are materialized. Moreover, even if the reserve is still available, the question will arise whether the amount will be used to cover the specific losses or is accessible to all creditors. Without a regulation of the ultimate goal of these reserves through self-insurance, the reserves will be considered one of the companies' assets and will be subject to execution by all creditors.[280]

Captives are often used to cover risks with some extent of predictability or risks that might otherwise be hard to insure. They are attractive instruments for companies considering the possible tax advantages related to investment income, premiums, and incurred losses, direct access to reinsurers, allowing the retention of risks according to the companies' needs and improved cash.[281]

From the perspective of deterrence, a captive is desirable considering its potential to control moral hazard of the potential responsible party who set up the captive. In practice, a captive or self-insurance can be regarded as the retention of risks: many potentially responsible parties self-insure a certain amount of risks, and only purchase 'excess' coverage from insurance market. The captive or self-insurance is then a deductible, which means a potential responsible party is still exposed partially to risks. Thus the potential responsible parties still have incentives to prevent the risks, and the problem of moral hazard is alleviated. However, the sole use of captives provides limited coverage, which may pass parts of the losses

279 *Ibid.*, p. 162.

280 Faure 2004, p. 459.

281 Bawcutt 1991, p. 25-27.

on to the taxpayer in case of the liable party's insolvency. As for the security of compensation, the regulation of the goal of the reserves discussed above for self-insurance is also important for captives. Furthermore, compared to insurance, a captive also has the advantage of cost-effectiveness, in the sense that it can avoid agent and broker commissions and insurance overhead/profit loadings.[282] This, however, does not mean a captive creates no substantial costs, since the organization of such a legal entity is also costly. Many companies may lack expertise and assets to establish and sustain such an entity. A choice for them under this condition is resort to the rent-a-captive.

Another issue which needs to be mentioned here is that captives focus primarily on high-frequency low-severity risks, which are statistically highly predictable.[283] However, the less predictable risks, for example, the low-frequency high-severity risks are still transferred to reinsurance market or covered by other compensation mechanisms.

6.2.2. Mortgages and Liens

The previous brief introduction reveals that mortgages and liens are not a plausible assurance for individual damage caused via the environment. However, they may play a role for ecological damage, for which the public authorities are the representatives of the injured environment and thus claimants. Bocken has discussed a mortgage for the benefit of a government agency established at the occasion of the delivery of a permit. He argued that these mechanisms can be used to satisfy the ELD requirement.[284] It is worth noting that this deduction only applies under certain conditions. In various jurisdictions, the competent authorities to represent the environment and claim for the damage often differ and many different authorities may be involved. For the noise damage and soil damage caused by hazardous installations, at least parts of the victims and concerned public authorities can be identified *ex ante*. However, there are also many occasions, such as the oil pollution, for which the concerned authorities are unknown *ex ante*. Thus it would be difficult to determine for which specific public authority these mechanisms can set a security interest. A security arrangement can also be made *ex post*, after the occurrence of an incident. But then the level of security is attenuated, since it is not sure whether there are still sufficient and unencumbered assets available.

Mortgages and liens do not create substantial costs since prepayment is not required and neither is a specific institution. However, by setting a security interest, a company reduces the collateral that he can offer to pursue other commercial interests.[285] Furthermore, the size of the security interest depends on the value of encumbered assets at the time that damage is caused.

282 Bruggeman 2010b, p. 145.
283 Banks 2005, p. 89.
284 Bocken 2009, p. 160.
285 *Ibid.*, p. 161.

6.2.3. Deposits, Trust Funds and Escrow Agreements

The same as mortgages and liens, deposits are only used when liable parties (the depositors) fail to satisfy their obligations. An example is Article 514-1 of the French *Code de l'environnement*, which requires the licensee, who does not abide by an injunction, to take safety measures to provide a cash deposit.[286] The potential responsible parties are required to pay such a deposit *ex ante*, which constitutes an immediate financial burden. However, they do not need to pay the insurance premiums or bank fees as a consideration for the guarantee provided by third parties. The deposits can be recoverable if no damage occurs, which gives the potential responsible parties the incentive to prevent the damage. Institutional arrangements can make the deposits attractive to potential responsible parties. For example, in the Flemish Draft Decree on Environmental Policy prepared by Interuniversity Commission, the environmental guarantee account (EGA) can be used to satisfy ones' financial responsibility. Potential responsible parties can put deposits in this account in order to guarantee their obligations. The deposits are tax deductible and the interest can be added into the account.[287]

A trust fund provides more comprehensive coverage. However it is often established after a specific incident, such as the establishment of the BP Oil Spill Liability Trust Fund after the Deepwater Horizon oil spill.[288] Since it is established *ex post*, little preventive incentives are created, except that it gives a reminder to the potential responsible parties that they may be held responsible for the damage they cause in the future. However, a trust fund can play a role in terms of secondary costs reduction. It can promote instant and easy access to compensation and efficient cleanup and restorations. An inadequacy that may still remain regarding compensation is that the trust funds may not be fully funded at the time of the claim. They are often financed over some period of time, which may lead to retardation in compensation. Furthermore, less security is provided, since there is no guarantee that a sufficient amount of assets will be available to set up such a fund when damage occurs. It is also costly since a trust fund is usually established for a specific incident. All the administrative loadings, costs spent on identification of damage and dealing with claims should be taken into account.

6.2.4. Corporate Guarantees, Bank Guarantees and Surety Bonds

Instead of providing assurance by himself, a liable party can also seek guarantees from a third party. A party providing a guarantee can be held liable simultaneously with the debtor or after the latter defaults, according to the contracts and national

286 An English copy of the French Environmental Code can be found at: <www.wipo.int/wipolex/en/text.jsp?file_id=180787> last accessed on 8 May 2013.

287 Bocken, Lambrechts, Boes, De Nauw, Faure & Lavrysen 1996, p. 31.

288 After the Deepwater Horizon drilling rig explosion, an agreement was made between the major responsible party British Petroleum (BP) and the American administration to establish a trust fund to compensate for the natural resources damage, removal costs and individual damage. See: <www.thebpclaimsfund.com/>, last accessed on 8 May 2013.

laws.[289] In the context of environmental liability, it is more common for the provider of a third-party guarantee to intervene when the debtor is in default.

To ensure the security of a guarantee, it should be irrevocable or only revocable with the approval of the regulatory authorities. Otherwise, a debtor or provider can revoke the guarantee after a potential responsible party proves his capacity to the regulatory authority and leads to no assurance when damage happens. As far as the security is concerned, some other issues in contract law are also relevant. For example, whether the purpose of surety becomes nugatory because of the debtor's default; whether the obligations of a guarantee provider are independent from those of the debtor and from the defenses he can invoke.

A bank guarantee or a surety bond is often issued on an annual basis. The debtor and the provider can extend the guarantee every year based on the past damage and claim experience. However, as for the ecological damage, a long time lapse often exists between the occurrence of accident, the first appearance of damage and the extensive appearance of damage and claims. Thus a short-term based guarantee may not reflect the real costs a debtor creates.

As far as the tertiary costs are concerned, a potential responsible party does not have to tie up a large amount of money or encumber other assets. The payment of bank fees or premiums may also be tax deductible. However, the bank guarantees and surety bonds are not always cheap or readily available. It may be only achievable by the larger companies with good credit rating.

The security mechanisms provided by the liable party or a third-party guarantee can be a complement for other risk-spreading mechanisms to prevent and compensate for ecological damage, bearing in mind that it is usually only available for larger companies. The SMEs, except the ones that can seek enough coverage from their financially-sound parent companies, may still resort to other risk-spreading mechanisms. Some of these mechanisms have advantages in costs saving, such as self-insurance and captives. But some are not cheaply available. Furthermore, more governmental oversight is needed for these mechanisms. Regulation of the purpose of reserves is of essential importance for self-insurance and captives. More monitoring should be done on the financial status of potential responsible parties or their corporate parent as guarantors. The guarantees provided by third parties also need to be irrevocable.

7. The Use of the Capital Market to Provide Coverage

Ecological damage sometimes turns out to be catastrophic. For example, after the Deepwater Horizon oil spill, BP established a liability trust fund of $20 billion to compensate for the removal costs, natural resources damages and individual victims. Similar incidents which lead to millions billions of dollars damage are not rare in the history. To improve the capacity of compensating for catastrophes, one may think of the capital market as a possible vehicle.

289 Bocken 2009, p. 162.

The capital markets are often discussed as one type of alternative risk transfer (ART). ART is a pliable concept about which no precise definition has been agreed upon.[290] It is usually used to denote a variety of solutions which 'can assist insurance companies and other corporations in the financial management of their business by drawing upon methodologies from the insurance and banking sectors.'[291] ART comprises three general types of mechanisms. Self-insurance, captives and risk-retention groups fall into the first category, which have been discussed in last section. The other two types concern the use of the capital market, containing hybrid reinsurance-financial products and financial instruments. The hybrid products combine characteristics of financial instruments and traditional reinsurance, but they do not access capital markets directly. The financial instruments can go beyond reinsurance industry capacity to access capital markets directly.[292] Those financial instruments are often insurance-linked. When talking about capital markets, scholars often regard them as an approach to increase the capacity of insurance. Among the various capital markets instruments, catastrophe bonds are regarded as the most successful ones.[293]

In addition to acting as instruments to increase insurance capacity, the capital markets are also proposed by some scholars, especially Tyran and Zweifel as instruments which can be used directly by operators of environmental hazardous installations (nuclear operators) to cover their potential liability.[294] According to them, the existing liability and compensation systems for nuclear damage in Europe failed to provide sufficient compensation and create efficient deterrence. This was due to the low cap on the operators' liability combined with the inefficiency in the insurance sector because of the lack of competition. Consequently, governments subsidized the nuclear sector substantially and exacerbated externalization. To cure those inefficiencies, Tyran and Zweifel suggested using capital markets to provide coverage (to obtain environmental risk internalization through capital markets, ERICAM). The operators could float risk participation shares on capital markets, where the investors could choose to sign such a share to create a warrant. The investors could earn a risk premium and interest or dividend payments as long as there were no claims. When the liability claim in excess of a certain threshold was presented, the warrant could be exercised.[295] They mentioned the advantage of competition between both operators and investors. The operators needed to compete to get the coverage from investors, which created incentives for them to reduce risks. The individual investors and traditional insurers would also compete with each other which then led to reasonable prices for their coverage. A promising an assumption as it sounds, this ERICAM system deserves more exploration here. Tyran and Zweifel suggested that the individual operators could issue their own capital markets instruments. Then the questions arise whether they are capable and

290 See Banks 2004, p. 49.
291 Bruggeman 2010b, p. 141.
292 Cummins & Weiss 2009, p. 506.
293 Cummins 2008, p. 23.
294 Tyran & Zweifel 1993, p. 434-437.
295 *Ibid.*, p. 437.

have incentives to do so. This firstly keeps out the SMEs which lack expertise and assets to tackle such a sophisticated and technical financial issue. For the nuclear sector, this may be a less problem since most operators have a comparatively deep pocket. However, in many sectors the operators' liability is limited, and they are not obliged to seek financial security high enough that they have to resort to the capital markets. Without such an obligation, operators may not have incentives to seek coverage from the capital markets, because in case of the catastrophes, many operators tend to take an 'it will not happen to me' attitude or because they can escape liability by being insolvent. Of course one can imagine a system where the individual operators are obliged to seek a high-enough coverage. However, the catastrophes are often low-frequency high-severity risks. It will be unrealistic and wasteful to ask every individual operator to seek such a high coverage. A more realistic solution may be to ask the groups of operators or the sector as a whole to seek the broad coverage from the capital market.

Marcus Radetzki and Marian Razetzki have made a more modest suggestion.[296] They proposed financial instruments, such as hedge funds, pension funds, options and futures as methods to tackle the uncovered liabilities of industrial catastrophes.[297] Among these instruments, catastrophe bonds are mentioned by them as the most promising ones. According to them, $100 billion of catastrophe bonds can be issued by the nuclear sector through a group of insurers of nuclear operations, or a pool of nuclear operators or an intergovernmental institution.[298] This might be a practical suggestion, however, several problems still need consideration. Firstly, from the perspective of deterrence, the interest rate of a capital market instrument should reflect the real risks responsible parties are exposed to. However, as for the catastrophe, the accidents are a rare event. Thus the rating of risks may be impossible. Secondly, since only limited entities can be qualified as an issuer for capital market instruments, the market is still not competitive. Thus the actual and reasonable price for the instruments may not appear. Thirdly, the deterrent effect on responsible parties is direct. There are different triggers for capital market instruments, such as the indemnity trigger, index trigger and parametric trigger.[299] The last two triggers are based on some subjective conditions, and have better ability to control moral hazard and adverse selection of securities buyers.[300] To what extent a capital market instrument can contribute to the deterrence of operators then depends on the measures taken by instrument issuers.

[296] Radetzki & Radetzki 2000, p. 180-195.

[297] Radetzki & Radetzki 2000, p. 188-190.

[298] Radetzki & Radetzki 2000, p. 190.

[299] A capital market instrument with an 'indemnity trigger' provides compensation according to an issuer's actual exposure to a particular pre-defined event; an instrument with a 'parametric trigger' according to the occurrence of an event with one or more certain defined physical parameters; an instrument with an 'index trigger' according to a recognized industry loss index. See Bruggeman 2010b, p. 138.

[300] Bruggeman 2010b, p. 170-171.

Using capital markets to cover catastrophes has happened in practice. However, they are mainly linked to natural disasters. A strong link between capital market instruments and man-made accidents is still rare. According to a report published by Swiss Re, among the risks securitized by the capital market from 1997 to July 2010, only one per cent ($405 million) of the covered risks concerns industrial accidents.[301] This coverage is created by the first and only successful securitization of third-party liability risk until 2010. Oil Casualty Insurance, Ltd (OCIL), a captive insurer for the petrochemical industry, successfully issued a $405 million catastrophe bond through Avalon Re Ltd. (Avalon) in 2005. It provided OCIL with coverage against losses and most associated loss adjustment costs under its portfolio of excess general liability policies. The Avalon transaction was made practicable by operating on a claims-made basis, capping the payment for each event and attaching a relatively high return period. Besides, the limited number of OCIL insured parties (79 petrochemical companies at the time of issuance) allows the availability of historical data from the insured, which adds to the predictability of risks. Since the success of Avalon is due to the unique exposure profile and the coverage needs of OCIL's client base, the replication potential of this transaction may be limited.[302]

8. Conclusion: Comparison between Individual Instruments and Policy Recommendations

Different compensation instruments can be developed to compensate for ecological damage, each having different influences on the reduction of social costs. This section tries to compare those instruments, considering the evaluation criteria mentioned in the section above. The use of the capital market, strictly speaking, cannot be classified as an independent category of compensation instruments. This chapter reveals that rather than being used directly by operators to cover their liability, a more promising approach is to use the capital market as an instrument to increase the capacity of insurance pools or risk sharing pools. Admittedly, the actual practice to use the capital market in the context of man-made accidents is still rare. Hence it is not included in the comparison. As discussed earlier, the compensation mechanisms can either cover only ecological damage or also traditional damage. The first element to be compared is the scope of coverage. Then instruments are compared according to the three categories of social costs. From the perspective of primary costs, three factors are compared: the relationship with liability, the setting of contributions (who finances the instruments and how the contributions are allocated) and their ability to control moral hazard and adverse selection. Then the potential in reducing secondary costs of those instruments is compared, considering the time necessary to establish the compensation instruments, the time necessary to

301 Swiss Reinsurance, Insurance-linked Securities Market Update, 2010, p. 7. See: <www.artemis.bm/articles/swissre%20update%20h12010.pdf>, last accessed on 8 May 2013.

302 Marsh & McLennan Companies, The Catastrophe Bond Market at Year-End 2005: Ripple Effects from Record Storms, 2006, p. 5. See: <www.guycarp.com/portal/extranet/-pdf_2007/GCPub/CatBond_yr_end05.pdf>, last accessed on 8 May 2013.

provide compensation, whether there is risk spreading, and capacity. The last issue to be compared is the tertiary costs created by each instrument. After those comparisons, policy recommendations are given for specific instruments.

8.1. *Liability Insurance*

Liability insurance is the most popular instrument used by potential injurers. The deterrent effect of liability insurance remains as long as the premium an insured pays corresponds with his actual risk.[303] Setting a risk-based premium for natural resources liability insurance is not an easy task, considering factual and legal uncertainties.[304] Insurers are experts in valuing uncertainties, managing risks, and in handling claims. Both feature rating and experience rating are used by insurers to set premiums.[305] Liability insurance can create moral hazard and adverse selection problems, which can be controlled through deductibles, monitoring, and risk differentiation.[306] However, even if risk-based premiums are used and moral hazard and adverse selection are controlled, efficient deterrence may still be unachievable because potential injurers are only incentivized to seek insurance coverage up to the amount of their own assets. In spite of this inefficiency, liability insurance is generally thought to encourage safety, as it alleviates insolvency problems and controls risks.[307] Additionally, empirical analysis supports the use of insurance. A recent study compared accidental underground fuel tank leaks in two states over a fourteen-year period.[308] The research shows that the shift from a state-level government assurance programs to private insurance markets reduced the frequency of underground fuel tank leaks by more than 20 per cent.[309]

Liability insurance is established before damage happens, and compensation is only awarded when liability is established. Thus the difficulties in finding liability will still prevent liability insurance from providing sufficient compensation. Under a liability insurance policy, risks are spread among potential injurers participating in the same pool. Re-insurance and capital markets can be used by insurers to increase capacity.

As far as tertiary costs are concerned, a competitive market will give insurers incentives to reduce loading costs and provide the products at reasonable costs.[310] Compulsory insurance is a frequently proposed alternative to overcome some

303 See Abraham 1988, p. 949-950.

304 For example, uncertain causation, multiple potential injurers, and long time lapse between the harmful activities and the manifestation of the damage are all factual uncertainties in establishing liability for natural resources damage. There are also legal uncertainties about liability for natural resources damage. For example, sometimes judges use a deep pocket policy to favor the compensation, by interpreting the policy language to expand the scope of coverage. See Abraham 1988, p. 960-969.

305 Abraham 1988, p. 949-950.

306 *Ibid.*

307 See Kehne 1986, p. 405-412.

308 Yin, Kunreuther & White 2011.

309 Yin, Kunreuther & White 2011, p. 327.

310 See Calabresi 1970, p. 26-30.

drawbacks of voluntary liability insurance, as it can incentivize operators to seek insurance and reduces problems such as adverse selection and insolvency.[311] However, compulsory insurance is only efficient if it is provided to operators seeking liability insurance; if moral hazard is controlled; if risk-based premiums are used; and if a competitive market exists.[312]

8.2. *First-party Insurance and Direct Insurance*

In addition to liability insurance, first-party insurance and direct insurance can also be used to cover ecological damage. First-party insurance directly covers risk of damage to a particular victim or specific to a particular site, making it easier to identify circumstances influencing risk.[313] Individuals can use first-party insurance to cover their personal injury or property damage. However, ecological damage often concerns public natural resources, which are owned collectively by all. Public authorities often act as trustees of public resources, thus having standing to bring ecological damage claims.[314] Nonetheless, without individual victims first-party insurance is not as effective for compensating public natural resources as it is for compensating traditional damage. First-party insurance does effectively cover on-site damage and cleanup costs of the polluters themselves.

Alternatively, under a direct insurance policy the potential injurer is covered for any damages suffered by third parties damaged on the insured party's particular site. Thus, unlike the pure first-party insurance funded by innocent victims, in a direct insurance policy, it is the polluters that pay for damages. Direct insurance may be desirable from the perspectives of both deterrence and compensation. On the one hand, the unpredictability of the tort system is removed; and on the other hand, compensation can be awarded before the liability is established. However, why would potential injurers seek coverage for damage suffered by other parties or the public, if they cannot be held liable? Questions arise as to whether the insurers are able to identify the covered damage and execute an efficient risk differentiation. Another essential issue is the compatibility of insurance coverage with administrative procedures of damage assessment and restoration.

8.3. *Risk-sharing Agreement*

Sometimes risks are unpredictable, so much so that the insurance market may be reluctant to provide coverage. In these circumstances a risk-sharing agreement is a useful alternative, because the establishment of an *ex ante* risk pool is no longer necessary. The highly technical and complicated risks some industrial activities possess make them favor the use of risk-sharing policies. Such policies are especially useful in covering nuclear liability and liability for marine pollution by

311 Wagner 2009, p. 397-398; Faure & Grimeaud 2003, p. 181-182.
312 Bruggeman 2010b, p. 222.
313 Faure 2002b, p. 292-293.
314 See Boyd 2003, p. 266-267.

oil.[315] The problems of moral hazard and adverse selection can be effectively controlled through mutual monitoring.[316] Three approaches can be used to finance the risk pool: an *ex ante* payment, reserves, or an *ex ante* agreement to pay if damage happens.[317] The *ex ante* payment approach has security advantages, as the availability of funds, but may lead to immobilized assets.[318] The *ex ante* agreement needs to be guaranteed by some financial instruments.[319] Monitoring is necessary for both reserves for *ex ante* agreements, to ensure the amount is still available in case of damage.[320] Implicit risk sharing is available for the nuclear risk under the Price-Anderson Act, since the majority of compensation will be collected through retrospective premiums paid by nuclear operators collectively.[321] Risk-sharing agreements have great cost-saving potential, since these pools are not profit seeking, and payment or reserves are not actually lost if no damage occurs.

8.4. *Environmental Funds*

There are many kinds of environmental funds, which can be used to help liable parties to cover their liability for ecological damage, can intervene when no liability is established, or provide timely compensation irrespective of liability. In other words, compensation funds have added value in comparison to other compensation instruments, because they can intervene when no liability arises and provide an upper layer of compensation in addition to other instruments where insufficient. As timely response and restoration is essential for ecological damage, funds providing finances for cleanup and restoration measures before the liability is established is especially desirable.

In spite of the desirable roles played by funds, when instruments adequately cover liability there is no clear reason why funds should be preferred to liability insurance. Funds are disadvantageous as they poorly differentiate risks and administrative costs are often high.[322] To ensure that environmental funds function efficiently, clearly defined coverage and compensation procedures are necessary. Funds should be financed by potential injurers rather than via general revenues. General revenues funding shifts the burden from polluters to the general public, thus reducing prevention and deterrence incentives. Funding environmental funds by polluters is an important criterion for effective and efficient natural resources compensation. However there is an exception. If a fund covers only historical damage and is of a temporary nature, a shift in costs bearing is not necessarily inefficient.

315 Faure & Fiore 2008b, p. 301.
316 See Faure & Vanden Borre 2008, p. 274.
317 Faure & Fiore 2008, p. 306-307.
318 Faure & Fiore 2008, p. 306.
319 Faure & Fiore 2008, p. 306-307.
320 Faure & Fiore 2008, p. 307-308.
321 See Faure & Vanden Borre 2008, p. 242-244.
322 See Faure & Hartlief 2003, p. 323-324.

8.5. *Guarantees by the Polluters*

Security mechanisms provided by the liable parties or a third-party guarantor can also be used by (potentially) liable parties to cover their liabilities. This includes a variety of instruments including self-provided insurance instruments such as mortgages, liens, deposits, trusts funds and escrow agreements; as well as instruments provided by third parties, such as bank guarantees, sureties, and corporation guarantees. For example, in the U.S. many instruments can be used to achieve financial responsibility for hazardous waste storage facilities to ensure closure, post-closure care, and third-party liability. These include trust funds, surety bonds, letters of credit, insurance, financial tests and corporate guarantees.[323] However, except for guarantees provided by a parent corporation, these instruments are usually only available for large companies. Small companies are typically unable to pass the financial tests necessary to prove deep pockets, are unable to provide sufficient unencumbered assets or cash assurance, or cannot seek a guarantee from financial institutions by showing good financial status and credit rating. Larger companies can more easily satisfy these requirements, but small and medium enterprises often cannot. Where possible, the capacity of small and medium enterprises to compensate damages can be assured when a parent company meeting financial standards agrees to satisfy coverage requirements.[324] Some instruments have advantages in costs saving, such as self-insurance and captives. Other instruments, like bank guarantees or sureties, are not cheaply available. Additional governmental oversight is needed for these mechanisms. Regulation of reserves is essential for self-insurance and captives.[325] The financial status of potentially responsible parties, and their parent corporations as guarantors, should be monitored. Guarantees provided by third parties also need to be irrevocable.

The table below gives an analytical overview of the potential of the above-mentioned compensation instruments to reduce Calabresi's social costs, and makes policy recommendations accordingly.

323 40 C.F.R. §264.146-147 (2006); 40 C.F.R. §265.146-147 (2006).

324 See Bocken 2009, p. 160-162.

325 Faure 2004, p. 459.

Table 1: Comparison of Various Compensation Instruments

Compensation Instrument	Comparison of Instruments				Policy Recommendation
	Secondary costs			Tertiary Costs	
	Time to Provide Compensation	Risk Spreading	Capacity		
Liability Insurance	When liability is established	Spread among potential injurers taking the insurance coverage	Depends on the size of the pool, incentives to seek insurance and whether the market is competitive	Profit seeking; competition creates incentives to reduce loading costs; not recoverable if no damage happens	1) can be used to cover both ecological damage and traditional damage 2) adequate risk differentiation 3) control of moral hazard and adverse selection 4) reinsurance and capital market can be used to promote capacity 5) competition on the insurance market
First-party Insurance and Direct Insurance	When damage happens	Spread among potential victims in first-party insurance, and among potential injurers in direct insurance	Maybe limited considering the incentives to seek coverage	Profit seeking; competition creates incentives to reduce loading costs; not recoverable if no damage happens	1) first-party insurance can be used to cover traditional damage; direct insurance can be used to cover both ecological damage and traditional damage 2) direct insurance can be used to preclude the unpredictability of liability 3) clear definition of insured risk and compatibility with administrative procedure in direct insurance 4) tort should still be allowed, but double compensation is prohibited 5) adequate risk differentiation 6) control of moral hazard and adverse selection

Compensation Instrument	Comparison of Instruments				Policy Recommendation
	Secondary costs			Tertiary Costs	
	Time to Provide Compensation	Risk Spreading	Capacity		
Risk-Sharing Agreement	Depending on types of agreement: compensate when liability is established (used to cover liability); compensate when it is proved that the injurers are insolvent (used to provide coverage in case of insolvency); compensate when damage occurs (used to cover losses of injurers)	Spread among potential injurers	Depends on size of the pool; can be big since the payment for a catastrophe can be spread over many years		
Environmental Funds	Depends on the specific kinds of funds; can provide compensation when damage happens or after liability fails	May be spread among potential injurers or general public	Non-profit seeking; saving costs since *ex ante* payment is no longer necessary; Recoverable if no damage happens	1) can be used to cover both ecological damage and traditional damage; especially useful for highly technical sectors, or sectors that insurance market is reluctant to provide coverage 2) different methods of financing can be used, *ex ante* payment, reserve and *ex ante* agreement.	

Compensation Instrument	Comparison of Instruments				Policy Recommendation
	Secondary costs			Tertiary Costs	
	Time to Provide Compensation	Risk Spreading	Capacity		
				Other financial guarantees should be used to guarantee the fulfilling of *ex ante* agreement. Monitoring is important for both reserve and *ex ante* agreement. 3) re-insurance and capital market can be used to promote capacity 4) risk-based contributions 5) whether competition is encouraged depends on specific situations	
Guarantees Provided by Operators Themselves or Third Parties	When liability is established	No risk spreading	Non-profit seeking, but public authorities usually have less incentives to reduce loading costs	1) can be used to cover both ecological damage and traditional damage; preferably intervene when no liability arises or liable parties are insolvent; can be used to fund the restoration of public authorities before liability is found. 2) financed by potential injurers 3) promote risk differentiation 4) clear compensation procedure	1) can be used to cover both ecological damage and traditional damage; useful for big companies rather than for small/medium enterprises 2) regulation of the ultimate goal of reserves when self-insurance or captives are used 3) monitoring of the financial status of potential injurers or their parent corporate as guarantors 4) guarantees provided by third parties are irrevocable

Chapter 5

COMPENSATION MODELS FOR ECOLOGICAL DAMAGE

1. Introduction

Serious and extensive damage to the environment has come about alongside modernization and industrialization. The damage has led to not only personal sufferings and economic losses, but also losses in the sense of ecological, esthetic and cultural values the environment has. Therefore, on the one hand, the ecological damage should be minimized and, on the other hand, the damaged environment needs to be restored.

The previous chapters have shown that regulation and liability rules can contribute to both the function of prevention and compensation (restoration of) for ecological damage. Regulation is a primary instrument to prevent ecological damage. The deterrent function can be achieved by setting environment and emission standards, requiring licenses for specific operations, environmental taxes, and administrative guides and so on. However, as far as environment risks are concerned, which are full of uncertainties and highly technical issues, government cannot always have the information to come to efficient regulation. Having prevention as its primary goal, liability rules can also not come to optimal deterrence. This is due to the difficulties in establishing liability, information asymmetries barring efficient setting of care levels or accurate measurement of damage, and low probability of being sued with respect to ecological damage.[1] Liability rules also have a positive function as regards compensation (restoration). However, the difficulties in establishing liability may also leave much damage to natural resources unrestored.

Carefully designed compensation mechanisms can help to overcome the inefficiencies of regulation and liability rules with regard to both prevention and restoration. Firstly, both regulation and liability rules have prevention as their primary goal. While compensation instruments place their emphasis on restoration. They are instruments used to help potential liable parties to cover their liability or finance the restoration when no liability can be found, or limitation of

1 Chapter 3, Section 3.1.

liability/limited liability can be revoked by liable parties. Furthermore, those instruments can also play a role in promoting prevention, by alleviating the problem of insolvency or avoiding the difficult road of liability.[2] Chapter 4 discusses some compensation instruments individually: liability insurance, first-party and direct insurance, risk-sharing agreements, environmental funds, guarantees provided by the liable party themselves or third parties, and the potential use of the financial market. A comparison is also made to show their difference in coverage and potential to reduce three types of social costs.

This chapter tries to design some models, with the interaction between regulation, liability rules and compensation mechanisms to prevent and compensate for ecological damage (Section 2). In all those sections, different compensation mechanisms can be used. However, this does not mean all compensation mechanisms mentioned above should be established in all scenarios. This would be too costly an approach to follow. Based on the comparison discussed in Chapter 4, this chapter identifies some indicators in choosing among the compensation mechanisms, and proposes two multi-layer compensation systems (Section 3).

2. Proposed Models to Prevent and Compensate for Ecological Damage

2.1. *Compensation, Liability Rules, and Insolvency*

The section discusses models that provide efficient deterrent incentives to potential injurers and promote compensation (restoration) of the damaged environment. Regulation, liability rules, and compensation instruments can be combined to achieve this goal.

2.1.1. Recalling Characteristics of Ecological Damage

Before examining the respective roles of regulation, liability rules, and compensation instruments, some characteristics influencing the design of compensation models are briefly recalled.

Ecological damage concerns both public and privately-owned natural resources. Various values of the environment make restoration a more important way of restoring than monetary compensation. Thus compensation mechanisms for ecological damage should be explained in such a way that they finance and promote restoration rather than merely provide monetary compensation. Emphasis on restoration means public authorities play an important role in this system. The mix of administrative and civil elements may influence the operation of compensation instruments, and requires cooperation between public authorities and the providers of compensation instruments. Another issue worth noting is that an environmentally harmful activity usually does not only produce ecological damage.

2 Chapter 3, Section 3.2.

Ecological damage is often accompanied by personal injuries and property damage. Many compensation instruments are then developed to cover both types of damage. When a specific instrument is analyzed, attention must be paid to its coverage: whether it covers only ecological damage or both.

As discussed earlier, ecological damage can be caused in the different stages of activities: construction, operation and post-closure phrase. During the construction period, an *ex ante* approach is taken to predict the potential damage caused by the construction. In the other two stages, damage is assessed *ex post*. In the post-closure phase, how financial resources can be guaranteed after the closure of a certain activity creates a challenge. A full discussion of compensation systems in all these stages is out of the scope of this research. This thesis focuses on the compensation system for ecological damage caused during the operation of activities.

2.1.2. Mandatory Financial Security

Various scholars propose the imposition of financial requirements on specific environmentally harmful enterprises. For example, Alberto Monti proposed an institutional model incorporating regulation, liability, funds and insurance in response to environmental harm.[3] According to his model, an enterprise wanting to enter or remain on the market must satisfy certain financial requirements.[4] Compliance with these requirements can be achieved by the adoption of several financial products, including environmental insurance.[5] Mandatory financial security was also proposed by the Interuniversity Commission for the revision of environmental law in the Flemish region in 1995 (Draft Decree on Environmental Policy).[6] According to Part Nine of this Draft Decree, the permit-holder or holder of a notification certificate of the classified installations and activities designated by Flemish government is bound to provide financial guarantees where liable for damage to or impairment of the environment.[7] Under this policy there is no general duty of financial security on all the classified installations.[8] Instead, the government or the administrative authorities determine the requirement type and amount of financial security.[9] A variety of compensation instruments can be used to satisfy the financial security requirement, including: an insurance policy, a guarantee provided by a financial institution, any other form of personal or collateral security and a deposit. The introduction of a mandatory financial responsibility system is widely favored by many scholars.[10] Literature shows that financial responsibility plays an

3 Monti 2001.
4 *Ibid.*
5 *Ibid.*
6 Bocken, Lambrechts, Boes, De Nauw, Faure & Lavrysen 1996, p. 31-32; Faure & Grimeaud 2003, p. 226.
7 *Ibid.*
8 *Ibid.*
9 *Ibid.*
10 Many other scholars are also in favor of a system with financial responsibility, although they may differ in the concrete design of regimes. For example, Kehne mentioned the advantages of a financial responsibility system with respect to both deterrence and compensation. He →

integral role in promoting deterrence and compensation, and saving administrative costs[11] A financial responsibility system can strengthen the deterrent effects of liability rules and supplement regulatory standards.[12] It is well established that insolvency risks result in inefficient deterrence under strict liability regimes.[13] Environmental accidents are often followed by a high magnitude of claims from victims and expensive restoration, greatly reducing a firm's wealth. Given this reality, firms must demonstrate the availability of assets to cover potential liability to alleviate insolvency risks.[14] Law and economics literature supports the use of other instruments, especially extended liability, to resolve insolvency problems.[15] Extended liability holds third parties contractually connected with the directly liable party liable.[16] Other literature proposes that managers should additionally be held liable where corporations are undercapitalized.[17] Further, literature also suggests that a firm's lenders should be held liable for ecological damage.[18] While extending liability reduces insolvency risks, some scholars argue that financial responsibility is preferable to extended liability since the latter does not guarantee costs internalization and can distort production decisions.[19] A detailed analysis of and a comparison between extended liability and financial responsibility is beyond the scope of this research. This research focuses on whether financial responsibility is necessary and how it should be arranged to promote the prevention and compensation of ecological damage.

Another advantage of financial responsibility is that firms are then more likely to monitor activities, inducing efficient care. Financial responsibility establishes a principal-agent relationship between the assurance providers and the firm engaged in environmentally risky activities.[20] The principal is paid a fee, and in return bears the risk of liability for ecological damage.[21] Under such an arrangement, the

also indentified the conflicts between these goals and tried to come up with a system (mainly insurance policies) that can balance those goals. See Kehne 1986. Feess and Hege favored a financial responsibility system consisting of both insurance and lender guarantees rather than pure strict liability and extending liability. See Feess & Hege 2000; Boyer and Porrini focused on proposing a model with efficient interactions and liability sharing between governments, firms and insurance companies. See Boyer & Porrini 2008. Kambia-Chopin analyzed financial responsibility from the perspective to solve the judgment-proof problem. Instead of the traditional insurance policies, he favored a particular form of contract which is close to an alternative risk transfer product. See Kambia-Chopin 2010.

11 See Kehne 1986, p. 403; Boyer & Porrini 1986, p. 344.

12 Kehne 1986, p. 403.

13 See Summers 1983, p. 155; Shavell 1986, p. 45.

14 Kambia-Chopin 2010, p. 78.

15 *Ibid.*

16 Feess & Hege 2000, p. 220.

17 See Kraakman 1984, p. 872-877.

18 See Balkenborg 2001; Boyer & Laffont 1997; Dionne & Spaeter 2003; Hutchison & Van't Veld 2005; Pitchford 1995.

19 See Boyd 2002, p. 424-425; Feess & Hege 2000, p. 222.

20 Kehne 1986, p. 408.

21 See Shavell 1979b, p. 55.

principal has strong incentives to monitor the environmental safety of the agents.[22] Through regulatory instruments, public authorities can also monitor and influence a firm's activity. However, financial responsibility also has added value. It both induces private party involvement and encourages firms to seek additional information about hazardous activities. Further, if the environmental risk is long term there is an added incentive for the intervention of insurers and other assurance providers. Assurance providers have more incentives to assess risks accurately to operate profitably, as compared to regulators who are more likely to be influenced by powerful political constituencies.[23]

Financial responsibility can also promote timely, low-cost public access to compensation.[24] Alleviation of insolvency risks provides a certain level of guarantee for the availability of assets in case of damage. Assurance may also save administrative costs. Delay and great information requirements of rulemaking procedures are obstacles to the specificity and adaptability of regulatory standards. In a competitive market, assurance providers have strong incentives to control administrative costs and can amend the policy conditions more easily and cheaply.[25]

2.2. *Models of Compensation*

This thesis presents three models of compensation, incorporating a combination of the compensation mechanisms discussed above. The models are distinguished based on whether financial security is mandated. These models illustrate that different strategies are necessary, based on whether or not the liable party can or cannot be identified. Additionally, different approaches are necessary dependent upon whether financial coverage can be required.

2.2.1. Model 1: Mandatory Financial Security

In a mature and competitive market, financial requirements should be imposed on specified parties before operations begin. A ceiling should be set for the financial requirement, as no market is able to provide unlimited coverage. It is imperative that the efficient amount of coverage be determined. James Boyd argues that optimal coverage should be just enough to internalize future liabilities.[26] Coverage beyond this level ties up capital and is wasteful, while coverage below this level results in externalities and under-deterrence.[27] He further clarifies this level to be the maximum of realistic environmental costs.[28]

22 Boyd 2002, p. 423.
23 Kehne 1986, p. 410-411.
24 See Boyd 2002, p. 423-424.
25 Kehne 1986, p. 411-412.
26 See Boyd 2002, p. 441-442.
27 *Ibid.*
28 *Ibid.*

Figure 1: Model I: Mandatory Financial Security

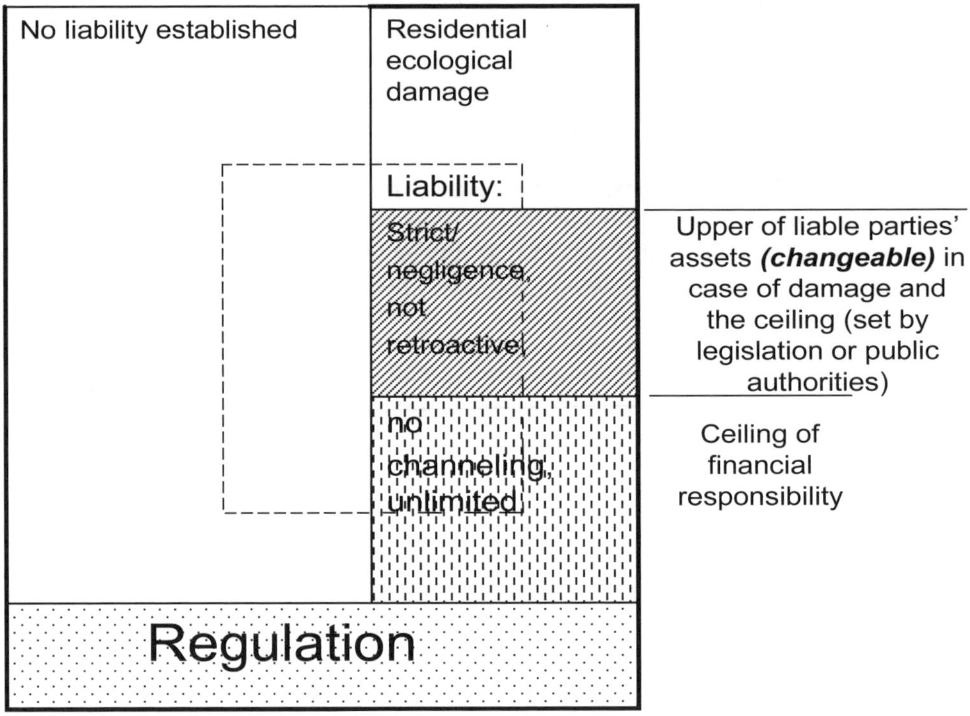

Compensation instruments to fulfil the financial requirement

Personal liability borne by liable parties themselves (maybe zero if the liable parties' assets in case of damage are less than the ceiling)

Collective compensation instruments

Compensation instruments irrespective of liability

This theoretically sound requirement, however, is difficult to satisfy in practice. It is difficult to estimate environmental costs, as variables including the scope of damage, costs of restoration, climate change, hydrology, and geology are all uncertain. Given the uncertainty, financial requirements are often deemed insufficient ecological damage compensation instruments. For example, in the U.S. under the Surface Coal Mining and Reclamation Act (SMCRA), first enacted in the

1970s, mining bonds are required to cover reclamation costs.[29] Despite improvements to bonding requirements over the years, bonds are not always sufficient to cover the full cost of reclamation.[30] Similar financial requirements also exist in the area of nuclear damage and oil pollution regulation.[31] Given difficulties in estimating potential environmental costs, this thesis does not identify the reasonable amount of financial responsibility. Instead, the thesis focuses on who can best determine financial responsibility ceilings, and how the ceiling should be determined.

Three approaches can be used to determine the ceiling of financial responsibility. The ceiling can be determined either on a case-by-case basis, via established estimation methodologies, or according to a relatively fixed industry-wide 'schedule'.[32] Under the first approach, the ceiling can be tailored to the specific risk an operation creates. Tailoring promotes better risk differentiation and cost internalization. However, individual determinations create high information and enforcement costs, and these costs may outweigh the benefits. Established methodologies can also be used to determine the ceiling. This approach allows for a variety of cost estimations, and saves administrative costs as compared to the case-by-case approach. The third methodology is a sort of middle ground, as it is somewhat tailored to the specific activity, but only to an extent. The administration and enforcement costs are not as extreme as in the case-by-case method, but at the same time they are more tailored than the second approach.

Additionally, three approaches may be used in determining who has the right to set the ceiling. The ceiling can be stipulated to by statutes, the government can be

29 30 U.S.C. §1259(2012).

30 Boyd 2002, p. 445.

31 For example, the Price-Anderson Act (hereinafter PAA) determines the cap of the financial guarantee of the nuclear power plants. The limit has increased from $560 million to $12.2 billion since the enforcement of the PAA. This amount is also higher compared to the requirement in other jurisdictions. In spite of the uncertainty in estimating the potential damage caused by a nuclear accident, the caps in liability and the corresponding financial guarantees have long been criticized as a subsidy to the nuclear industry. See U.S. Nuclear Regulatory Committee, The Price-Anderson Act – The Third Decade: Report to Congress, NUREG-0957, 1983, available at: <http://pbadupws.nrc.gov/docs/ML0727/ML072760026.-pdf>, last accessed on 8 May 2013; Berkovitz 1989; Dubin & Rothwell 1990; U.S. Department of Energy, Energy Information Administration, Federal Energy Subsidies: Direct and Indirect Interventions in Energy Markets, SR/EMEU/92-02, 1992, available at: <ftp://ftp.eia.doe.gov/service/emeu9202.pdf>, last accessed on 8 May 2013. This is also true for financial requirements under OPA, which imposes limited liability and financial responsibility on tank vessels and offshore facilities. Until 2009, the financial requirement for tankers is the greater of $3200 per gross ton or $22 million. For offshore facilities the requirement is no more than $150 million. However, the Deepwater Horizon explosion in April of 2010 largely dwarfed the prescribed liability limit and financial requirements. The released oil from the drilling platform is estimated at approximately 206 million gallons before its containment on July 15, 2010. Although there is no official estimation of total costs yet, it is reasonably regarded to be far more than the liability limit. To date the administration has sent eleven bills to responsible parties for the removal costs, totaling $711.8 million dollars. See Restore the Gulf, <www.restorethegulf.gov>, last accessed on 8 May 2013.

32 Boyd 2002, p. at 442-443.

authorized to decide the ceiling for specified operators on a case-by-case basis, or a third possibility is to allow operators to estimate their own costs and set a ceiling accordingly, then seeking government or third-party approval.

By statutorily prescribing a financial requirement ceiling, the first approach makes risks more predictable for both operators and for the financial market as a whole. This approach results in fewer administrative costs, as ceilings are set according to a fixed schedule, not on a case-by-case basis. However, statutorily established ceilings suffer from the same disadvantages as limited liability. The financial requirements for operators are not tailored to individual characteristics and needs and, because financial requirements are based on general factors, inefficiencies may result.

The second solution is to authorize government to set financial security that is somewhat tailored to a party based on information provided by that party. Under such a scheme, the statute classifies certain activities for which the actors must provide financial security. It is then up to the government to decide which specific parties and to what extent proof of financial security is necessary. This scheme is advantageous in that government can tailor the ceiling for each party to satisfy specific conditions, including the type and size of installations, locations, potential risks, and preventive measures. However the advantage is only realized where government is capable of differentiating parties where accurate information is available. Further, ceilings may be inefficiently set where lobbying groups influence government decision making. This is especially true for developing countries that lack reliable administrations, but it is also a challenge for developed countries.[33] The limited governance capacities, corruption and external influences are several usual problems perplexing developing countries.[34] The standard-based model provides additional discretionary powers to the environmental regulatory authority, but discretion without a solid and reliable administrative structure is risky, and abuses more prevalent. On the contrary, a rule-based system concentrates more power at the legislative or executive level, limiting the power of civil servants (who often have greater expertise) to execute precise rules.[35] In a country with an unreliable administrative structure it is advisable to set minimal statutory financial requirements. However, it may be beneficial to allow for some flexibility by allowing governments to ask for greater financial security when appropriate for a specific party. Under such a regime, minimal financial coverage is guaranteed and at the same time it is flexible enough to set more efficient standards when necessary.

Under the third approach, theoretically the individual operators can more accurately estimate costs and establish efficient financial ceilings. Realistically, relying on self-estimates alone can easily lead to inadequate coverage levels, as owners and operators tend to minimize their costs of assurance. Hence information provided by operators suffers from moral hazard problems. This approach provides

33 Faure, Goodwin & Weber 2010, p. 108-123.

34 *Ibid.*

35 Faure, Goodwin & Weber 2010, p. 110-111.

sufficient coverage levels only if public authorities conduct frequent, highly critical reviews of closed plans.[36]

The traditional instrument used to establish financial responsibility is liability insurance. Law and economic literature provides that insurance is advantageous in that it flattens out risk.[37] The risk-averse operator can instead pay insurance premiums, thus the risk of a catastrophic event resulting in natural resources damage is shifted to the insurance company.[38] Where insurance policies are sold in a competitive market, they will be priced effectively.[39] However, this is only accurate if one assumes that insurers are capable of monitoring the insured party's activity, that accidents are unilateral, that there are no administrative costs, and that no errors or uncertainty exists as to legal outcomes.[40] This is rarely, if ever, the case. When these assumptions are relaxed, as they must be realistically, liability insurance alone is less attractive, and alternative instruments may be used to satisfy financial requirements. The operator can choose between the various alternative instruments discussed above, which include liability insurance, direct insurance, risk-sharing agreements, environmental funds, and self-provided or third-party provided guarantees. These alternative instruments alleviate some of the problems related to insurability, and induce capital market involvement in the internalization of environmental costs.[41] The use of various instruments allows greater flexibility and competition. Competition gives providers an incentive to promote the efficiency of their products and to reduce loading costs.

Collective compensation instruments are effective when liability is unclear, or when the financial coverage sought is insufficient to cover the damage and the injurer is insolvent. When collective instruments are used, it is essential to make polluters finance instruments, thus creating additional deterrent incentives.

The above table illustrates that the threshold of collective compensation instruments in case of insolvency lies in the ceiling of the financial requirement or the assets of the liable party when damage occurs, whichever is larger. The ceiling may be specified in legislation or by the government. The only possibility an operator has to shift the burden of compensation through collective compensation instruments is to establish smaller companies and to use limited liability entities.

36 Kehne 1986, p. 417.
37 See Shavell 1982.
38 *Ibid.*
39 *Ibid.*
40 Shavell 1982, p. 131.
41 Boyer & Porrini 2008, p. 344.

2.2.2. Model II: Voluntary Financial Security

Figure 2: Model II: Voluntary Financial Security

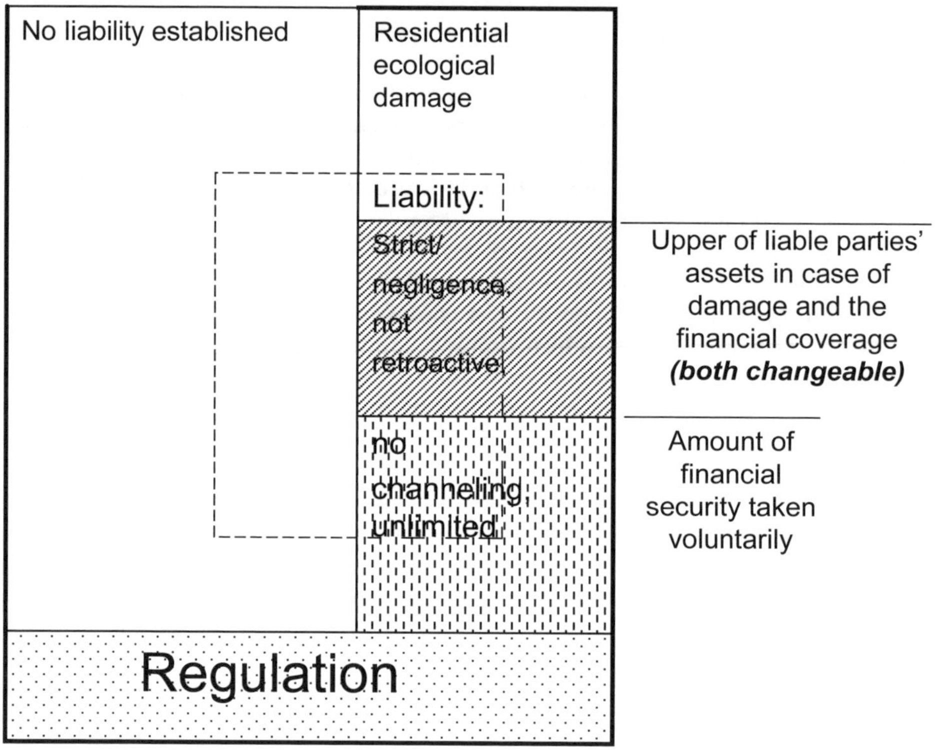

Compensation instruments voluntarily taken by liable parties

Personal liability borne by liable parties themselves (maybe zero if the liable parties' assets in case of damage are less than the amount of financial security taken voluntarily)

Collective compensation instruments

Compensation instruments irrespective of liability

Where the market providing financial products for ecological damage is immature or uncompetitive, voluntary financial security promoting the development of financial instruments with the gradual introduction of mandatory financial security is a feasible approach. However, the problem with this voluntary approach is whether, or to what extent, operators are incentivized to seek financial coverage. Resorting to a compensation instrument can increase the utility of risk-averse operators. When seeking insurance, for example, operators choose to make certain specified payments, instead of risking larger, uncertain liabilities. Thus the incentives for seeking financial coverage depend on the probability that the operator will be held liable, and on his personal attitude towards risk. If liability rules are efficiently set and if operators are always held liable for their ecological damages, reasonable operators will seek insurance coverage up to the probable amount of damage, or up to the value of their assets, whichever is less. If operators are unlikely to be held liable for natural resources damage, they will have no incentive to seek an efficient level of financial coverage. Additionally, government may take measures to make financial coverage attractive to potential injurers. For example, favorable tax policy makes captive insurance more attractive. Norbert Pelzer argues that identifying the obligation of members to pay a premium in a pooling system is desirable.[42] In the Flemish Draft Decree on Environmental Policy (Draft Decree), an environmental guarantee account is proposed as one financial guarantee, by which potentially liable parties can make deposits to guarantee certain obligations.[43] According to the Draft Decree, deposits are tax deductible and interest is added to individual accounts.[44] These designs incentivize potentially liable parties to seek financial coverage.

Operators can shift the burden of compensation to collective instruments by seeking a lower coverage of financial security. Further, collective funds may also be used to dilute incentives for operators to seek individual financial security to cover liability.

Under a voluntary scheme, the government is no longer responsible for setting the ceiling of the financial requirement. Instead, the government's role is only to promote the predictability of environmental liability by improving monitoring, establishing databases, exchanging information, and making assessment guidance for ecological damage. A predictable liability system is essential to financial security and insurability. Promotion of the financial market to cover ecological damage is also a role for government. With the development of financial markets, mandatory financial requirement may be introduced gradually.

42 Pelzer 2007, p. 52.

43 Bocken, Lambrechts, Boes, De Nauw, Faure & Lavrysen 1996, p. 31.

44 *Ibid.*

2.2.3. Model III: Optimal Financial Cap

Figure 3: Model III: Optional Financial Cap

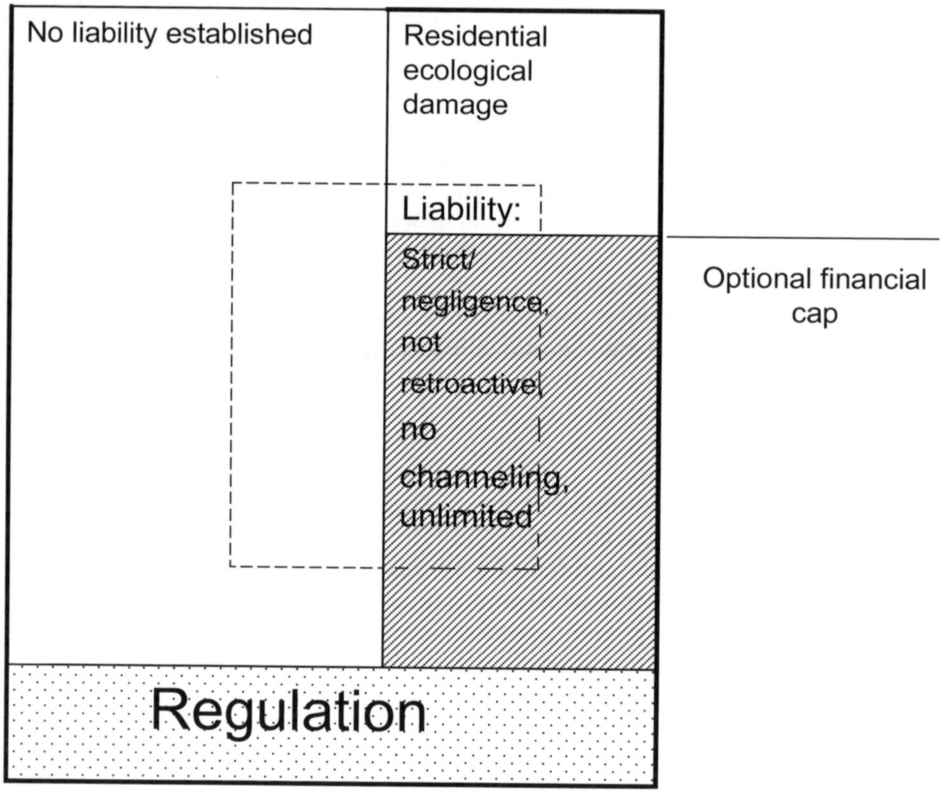

Compenation instruments taken voluntarily and used to satisfy optional finanical caps

Collective compensation instruments

Compensation instruments irrespective of liability

In addition to increasing the predictability of environmental liability and promoting markets to provide financial coverage, government may also introduce a system with optional caps for financial security. After all, the previous two systems may reduce an operator's duty to seek insurance coverage, but do not necessarily limit liability itself. A recommendation to this end was made in the Flemish Draft Decree on Environmental Policy.[45] The Draft Decree adopts unlimited liability and imposes financial responsibility on the permit holder or the holder of a notification certificate of classified installations.[46] However, operators may limit liability for ecological damage by providing a guarantee fund up to a regulated amount.[47] Hence to enjoy the limitation of liability, operators must provide a guarantee fund, guaranteeing that injured parties will be compensated for natural resources damage above the amount for which operators are liable to pay. If operators can prove the availability of a financial guarantee fund, their liability can be limited to a specified amount. Financial caps incentivize operators to establish sizeable 'optional limitation funds', providing greater certainty for risk-averse injurers. Individualized optional caps provide greater differentiation of risk as compared to generalized statutory caps, as they are tailored to the operator's specific risks and possible natural resources damage.

The design of optional financial caps in the Draft Decree can shed some light on the model incorporating voluntary financial security.[48] In case of a lack of capacity and high concentration on the insurance and financial market, it is unfeasible to impose an obligation to seek a certain amount of financial coverage. However, in these cases government may impose an optional cap. By establishing guarantees up to the cap, the injurer limits his liability up to that specified amount. This approach gives injurers incentives to seek a financial coverage to a certain level in the undeveloped market. At the same time, since operators are not obliged to seek coverage, possible negative influences on market concentration under a mandatory scheme is avoided.

3. Towards a Multilayered Approach to Compensate for Ecological Damage

It has been advised in the above-presented models that a variety of compensation instruments should be developed and utilized to promote competition. However, this does not mean that each instrument is fit for all types of operators and for all types of damage. Each instrument has its own specific strengths and weakness when analyzed in the context of a specific situation.[49] Thus, it is desirable to identify which instruments best suit specific situations. Chapter 3 compares different compensation mechanisms on their coverage and potential to reduce three categories of social costs. Based on the comparison, this chapter identifies a few factors to take into consideration in choosing between these mechanisms. Then this section tries to use these factors to design a compensation system for ecological damage.

45 Bocken, Lambrechts, Boes, De Nauw, Faure & Lavrysen 1996, p. 29-32.

46 *Ibid.*, p. 31; Faure & Grimeaud 2003, p. 227.

47 Bocken, Lambrechts, Boes, De Nauw, Faure & Lavrysen 1996, p. 31.

48 Faure & Grimeaud 2003, p. 227.

49 See generally Ad-Hoc Industry, Natural Resource Damage Group, Report: Survey of Industrial Companies, Insurance and Other Financial Security Instruments and Remediation of Environmental Damage under the EU Environmental Liability Directive, 2010.

3.1. *Factors Influencing the Choice of Instruments*

When deciding which instrument is more suitable in specific conditions, many factors need to be considered. Most importantly, it must be determined if liable parties are identifiable. When liable parties can be identified, liability rules can be used as an instrument to prevent and compensate for ecological damage. Liability insurance is then the most popular instrument. If the potential injurer can transfer risk to insurers with a premium reflecting actuarial costs plus administrative costs, liability insurance can guarantee compensation without reducing the preventive incentives of potentially liable parties, making this solution most desirable.[50] When liable parties cannot be identified, alternative instruments must be used, such as direct insurance or environmental funds.

A related issue is the type of damage. For example, soil pollution often has localized influence, with identifiable liable parties. In these cases liability insurance can be a useful compensation instrument. Air pollution is of a more diffuse nature and identifying liable parties is more difficult. In that case environmental funds may be a more promising instrument than liability insurance.

Another factor is the nature of damage: is it general or catastrophic damage? A catastrophe means that there is damage with low frequency and high severity.[51] Self-insurance may be better suited for predictable risks, while risk-sharing pools are able to cover less predictable risks as well since the contribution can be paid *ex post*.[52] High severity imposes high requirements on the capacity. The use of the financial market is an instrument with high potential to increase the capacity of insurance or risk sharing agreements. Financial instruments have been developed to cover natural disasters but have only limited use in the field of man-made disasters.[53]

The size of enterprises creating environmental risks also plays a role in the choice of instruments. The guarantees provided by potential injurers themselves and bank guarantees may be more readily available to larger enterprises than they are to small or medium enterprises (SME). An *ex ante* established instrument is desirable as it can provide a securable guarantee. Sometimes big companies may have the resources to provide *ex post* and ad hoc instruments to compensate for ecological damage.[54] However, for SMEs with greater risk of insolvency, an *ex post*

50 See Wagner 2009.

51 Bruggeman 2010b.

52 Skogh 2008, p. 282.

53 Using capital markets to cover catastrophe happens in practice. However, they are mainly linked to natural disasters. A strong link between capital market instruments and man-made accidents is still rare. According to a report by Swiss Re, among the risks securitized by the capital market from 1997 to July 2010, industrial accidents account for only one per cent ($405 million) of the covered risks. See M. Wittman *et al.*, *Insurance-linked Securities Market Update*, 2010, p. 7, available at: <www.artemis.bm/articles/swissre%20update%20h12010.pdf>, last accessed on 8 May 2013; Bruggeman 2010a.

54 The compensation trust fund established by BP after the Deepwater Horizon Oil Spill is an example of this. Instead of seeking insurance coverage from the market, BP has chosen to be self insured by its captive insurance company with coverage of $700 million. This amount is →

instrument is not reliable and may lack the assets to establish ad hoc compensation instruments.

The sector in which the potential injurers operate is also a factor to be considered. For example, in highly technical sectors, such as the nuclear industry, it is the operators that have the best information about risks created and how to control such risks. Thus cooperation between operators via a risk-sharing agreement, for example, would be an effective compensation instrument. For more traditional sectors, where insurers have developed corresponding products based on statistical risk and claim experience, the expertise of insurers can make insurance a competitive choice.

The primary criterion in choosing instruments and establishing a compensation system is whether the liable party is identifiable. Taking into account this distinction, other indicators are also briefly examined.

3.2. *When the Liable Party is Identifiable*

When a liable party can be identified, tort law provides the polluter with an incentive to prevent damage. As the law and economics literature proves, if the magnitude of damage can be accurately measured (under strict liability), or the due care level can be efficiently set (under negligence rules), a fully solvent party will have efficient incentives to prevent the damage.[55] However, the potential magnitude of ecological damage is great, and may push a potentially liable party into insolvency. Liability insurance is thus a useful compensation instrument: by providing an *ex ante* guarantee ensuring availability of assets in case of damage, liability insurance can both promote compensation for ecological damage and alleviate preventive inefficiencies caused by insolvency.

Not only operators, but also financial institutions, public authorities, and insurers are experts in risk spreading and risk differentiation. By spreading risks among a group of members exposed to the same type of risks, the utility of risk-averse parties can be increased and the capacity of compensation raised. By establishing premiums that correspond to an insured party's contribution to risks, the deterrent incentives created by liability rules remain in effect. Considering information asymmetry problems, liability insurance may also result in moral hazard and adverse selection issues. However, insurers are specialized in dealing with those problems: both experience rating and feature rating can be used to differentiate risks, and monitoring and limited coverage can be used to control moral hazard. These Insurers are better skilled than operators, financial institutions, and public authorities in contributing risk to injurers, thus providing better

far from enough to cover all damage and costs that arise after the accident. Thus BP established a trust fund of $20 billion to compensate for the removal costs, natural resources damages and individual victims. See G. Gonzales, *BP Oil Spill Damages to Stretch Insurance Coverage*, available at: <http://oilprice.com/Environment/Oil-Spills/BP-Oil-Spill-Damages-to-Stretch-Insurance-Coverage.html>, last accessed on 8 May 2013.

55 Shavell 1980, p. 4.

deterrent incentives. If the market is competitive, insurers are incentivized to reduce loading costs.

Despite insurer's specialized skills in risk spreading and differentiation, liability insurance may not always create efficient incentives. Potentially liable parties are only incentivized to seek insurance up to the value of their assets owned, meaning that adverse selection may result, limiting the capacity of insurance.[56] Thus compulsory insurance is an often proposed remedy to these problems.[57] However, as discussed above, compulsory insurance can aggravate problems of moral hazard and market concentration.[58] Therefore, the introduction of compulsory insurance should be done with caution: it is only efficient if the scheme applies exclusively to those demanding liability insurance, if moral hazard is controlled, if risk-based premiums are used, and where competitive markets exist.[59]

A preference for liability insurance where liable parties are identifiable does not mean that liability insurance is the only necessary compensation instrument. There are still some scenarios under which other instruments may play a role. Sometimes the insurance market is highly concentrated, thus insurers are either too reluctant to provide coverage or only provide policies at very high prices.[60] Under these conditions, the potentially liable parties must resort to other compensation instruments. Risk-sharing agreements, captives, and guarantees provided by operators themselves or by third parties are often alternatives used.

As mentioned above, the sector or industry to which a potentially liable party belongs to is also a factor influencing choice of instruments. For highly technical sectors in which operators have better access to information concerning differentiation and control risks, such as the nuclear and oil transportation sectors, a risk-sharing agreement is a useful alternative.[61] For example, in the nuclear sector, if there is an equal or comparable level of nuclear safety and security, and equal or comparable economic conditions and a legal framework, the pooling system can be efficient.[62] A risk-sharing agreement has also an advantage in dealing with less predictable risks, because an *ex ante* payment of the share is no longer necessary. It can also reduce costs in the sense that contributions can be paid *ex post*, or payment

56 See Faure & Grimeaud 2003, p. 153-155.

57 For example, Wagner favored a private but mandatory insurance, rather than private voluntary insurance and public compensation schemes. See Wagner 2007, p. 87-112.

58 Faure & Grimeaud 2003, p. 157-161.

59 Bruggeman 2010b, p. 222.

60 For example, the nuclear insurance market has been organized on a national and non-competitive basis, since the development of the civil nuclear industry. This has led to high nuclear insurance premiums. During the 17th and 18th centuries, English commercial vessels were repeatedly seized or destroyed by the enemy navies. Those catastrophic losses made private insurers unable or unwilling to underwrite to write such risks. The Protection and Indemnity Clubs began to develop and cover the risks created by marine transportation in response to the market needs since the mid-19th centuries. Ronnerberg 1990, p. 3-4.

61 See Radetzki & Radetzki 2000.

62 Pelzer 2007, p. 50-53.

is recoverable if it is paid *ex ante*, and less damage happened during the policy time.[63]

Larger companies can also use captives or other guarantees if liability insurance is unavailable or too expensive. For example, nuclear companies can use captive companies to cover their first layer of liability.[64] Captives are attractive to these companies because they adhere to minimum capital or surplus levels, actively managing the risk portfolio through diversification and cessions to the re-insurance market and favorable tax policies.[65] A bank guarantee is also a possible alternative. Bank guarantees also have the disadvantage of being expensive, as banks are usually not specialized in risk spreading and are themselves risk averse.

Even where liability insurance is available, its capacity is not unlimited. Considering the potentially catastrophic effects of ecological damage, other instruments can provide an upper layer of compensation. The risk sharing agreement discussed above can act as an upper layer instrument.[66] Environmental funds are frequently used instruments to intervene when others fail. In addition to providing an upper layer of compensation, environmental funds can also be used to finance the cleanup of ecological damage when a threat is imminent, or when other financial resources are not immediately available.[67] It is worth noting that environmental funds should be organized in such a way that potential injurers, instead of the public at large, finance the fund.

3.3. *When the Liable Party is not Identifiable*

As far as ecological damage is concerned, it is possible that the liable party cannot be identified, or causal links cannot be established. In these situations liability cannot be established, and consequently liability insurance is no longer a useful instrument. The alternatives providing compensation irrespective of liability can then come into play, such as first-party and direct insurance, and environmental funds.

In the context of ecological damage, first-party insurance has only a limited role to play.[68] Irrespective of personal injury and property damage, the victims of ecological damage are the public, whose interests are minor and indirect. Collective action problems exist, thus disincentivizing operators to seek insurance coverage for ecological damage to the fullest extent. In some jurisdictions, public authorities are given standing to sue for ecological damage. First-party insurance can be used as an

63 Faure 2004, p. 471.

64 Faure & Grimeaud 2003, p. 215.

65 *Ibid.*

66 For example, 10 C.F.R. §140.11(a)(4) (2010) (explaining the requirement that nuclear power plants of a given capacity must seek insurance up to $375 million, and in the event damage exceeds this amount, licensees will pool together to cover the second layer of compensation).

67 See, for example, 33 U.S.C. §2712(a)(d) (2012) (granting State officials immediate access to funds if necessary for the immediate removal of an oil discharge, or the mitigation or prevention of a threat of discharge).

68 Leiter 2001, p. 313-314.

instrument to cover on-site remediation costs. In these circumstances, polluters themselves suffer damages and they can insure against the costs of remediation and restoration with their own resources. Direct insurance is a more promising alternative, by which owners of potentially injurious sites seek insurance coverage benefitting third parties.[69]

Direct insurance is advantageous in that liability rules may then be avoided. However, a causal link between damage and an insured site must still be established. When the liable party cannot be identified, environmental funds may prove to be the superior compensation scheme. While insurers are more specialized in risk differentiation, environmental funds are often operated by public authorities who have little incentive to making sufficient risk differentiation and control tertiary costs. Ideally funds are financed by potential injurers, rather than by the public at large. The operators of funds need to be incentivized to differentiate risk and reduce administrative costs. If environmental funds are being financed by potential injurers, these instruments can also promote deterrence. Since liable parties cannot be identified, the environmental costs cannot be borne by liable parties and thus are externalized. Environmental funds can still make otherwise externalized costs internalized by the injurers as a whole. Hence, an additional level of deterrence can be achieved.

[69] See Faure & Grimeaud 2003, p. 217-224.

PART II

EMPIRICAL PART

EMPIRICAL PART

The first part of this research proposes three models to prevent and compensate for ecological damage. In all those three models, regulation, liability rules and compensation instruments interact to compose an efficient system. This research also tries to indentify a few indicators in choosing among the different compensation instruments. In the second part, those theoretical models and indicators are tested through the practice in a few jurisdictions. The practice in international regimes, the U.S. and the EU are discussed in this part. The international compensation systems for oil pollution and nuclear damage are introduced briefly. Three systems in the U.S. are chosen here: the systems established under the OPA, the CERLCA and the PAA. The ELD framework is also addressed briefly, followed by a few examples in some EU Member States. This research chooses quite a variety of systems, hence many detailed information cannot be addressed. The empirical study does not aim at providing a tentative comparison of all the systems. It, actually, tries to show whether the practice is in line with the theoretical models; and, if not, why there is such a divergence.

Among the systems discussed in this part, some are of particular importance to ecological damage. For example, under the ELD, only three types of damage are covered and individual damage caused via the environment does not play a role. The OPA and the CERCLA are also closely related to ecological damage (under the headings of response action and natural resources damage).[1] Some other systems, however, have only limited relevance for ecological damage, such as the international systems, which cover mainly personal injury and property damage. Ecological damage plays a limited role only when it comes to compensate for preventive measures and restorations measures.[2] One important reason for such a choice is because ecological damage and other damage caused via the environment are usually treated in the same framework. For example, the CLC establishes a liability regime for both parts of ecological damage (restoration costs) and other

1 In both systems, natural resources damage is explicitly included as being compensable. Response actions and liability rules have been established for natural resources damage. The two acts also authorize the promulgation of assessment regulations. See Chapter 7, Section 2.

2 See the definition of oil pollution under the CLC. Details see Chapter 6, Section 1.1.

individual damage. Some compensation instruments, such as liability insurance, environmental funds are also, quite often, used to cover both types of damage. Actually, the liability system for ecological damage only started to develop recently. Leaving aside the U.S., there are not many experiences in this field. However, many jurisdictions have gained experience in compensating for general environmental damage. In such systems, regulation, liability rules and many different compensation instruments have been developed to address the damage. Those examples can provide inspiration for the establishment of a compensation system for ecological damage. In addition, the international and the U.S. nuclear compensation systems are also discussed in this research. Nuclear damage usually focuses more on individual suffering than on environmental damage. However, the expansion of the compensable damage under the second generation of international nuclear liability conventions allows compensation to ecological damage to some extent. Moreover, the nuclear compensation systems provide examples of how a potential catastrophe can be compensated.

Those systems are chosen in this part for the following reasons. Firstly, those systems are in different stages in developing a compensation system for ecological damage. The international conventions cover only limited parts of ecological damage (prevention/restoration measures).[3] In the EU, the ELD introduced a comprehensive liability framework for biodiversity damage, land damage and water damage. However, the transposition has only been finished by 2010[4] and there has been only limited practice in this area yet. Given the limited experience, a full discussion of the EU Member States in compensating for ELD damage is not feasible here. Instead, this research chooses a few examples in the Member States to cover environmental liability more generally. The U.S. has gained more experience to compensate for natural resources damage, which started to develop since the 1980s, under the CERCLA. Secondly, different liability rules have been established under those systems. They possess a lot of different characteristics (strict/negligent rules; channeled/not, capped/not, retroactive/not). It is interesting to find out whether those liability regimes are in line with/divergent from the theoretical models. Thirdly, in those systems, different combinations of compensation instruments are available. Those varieties provide an opportunity to figure out whether the theoretical indicators in choosing compensation instruments work in practice.

After this introduction, the practice in the international regime, the U.S. and the EU are analyzed respectively (chapters 6-8). Followed by individual descriptions, a short conclusion reviews these systems according to the theoretical models and indicators. When there is a divergence between the theoretical models and practice, this research also tries to find out the reasons for this.

3 Such as the coverage of 'impairment of the environment' under the CLC and the second generation of the Paris Convention. See Chapter 6, Sections 1.1, 2.1.

4 Environmental Liability, available at: <http://ec.europa.eu/environment/legal/liability/-index.htm>, last accessed on 8 May 2013.

Chapter 6

COMPENSATION SYSTEMS UNDER THE INTERNATIONAL REGIME

1. International Compensation System for Oil Pollution

The international compensation regime for oil pollution provides an interesting example how the combination of liability rules and financial security mechanisms can prevent and compensate for ecological damage. Two regimes are introduced briefly here: the first is composed of the International Convention on Civil Liability for Oil Pollution Damage (the CLC and the International Convention on the Establishment of an International Fund for Compensation for Oil Pollution Damage (the Fund Convention). Those two conventions are important early conventions on oil pollution liability, and apply to ships carrying oil in bulk as cargo.[1] But they still leave gaps: when the pollution results from the bunker oil from other ships, rather than that covered under the CLC, another convention starts to fill the gap. The International Convention on Civil Liability for Bunker Oil Pollution Damage (the Bunker Convention)[2] was adopted in 2001. Both regimes adopt some similar principles, such as strict liability, channeling liability to ship owners and compulsory financial security. Furthermore, international funds are also established under the Fund Convention. This section first gives a short summary of the liability rules under the conventions and then a brief introduction of the financial security mechanisms is given. This is followed by a law and economics analysis.

1.1. *Liability Rules under the International Regimes on Compensation for Oil Pollution*

As shown above, there are two international liability regimes for oil pollution. This part introduces liability under them respectively.

1 Art. I 1 of the 1992 CLC; Art. I 2 of the 1992 Fund Convention.

2 The Bunker Convention, RMC I.7.130, II 7.130.

1.1.1. Liability Rules under the CLC Regime

1.1.1.1. The Adoption of the 1969 CLC and the 1971 Fund Convention

The international liability regime for oil pollution started to develop in the 1960s. The Torrey Canyon oil spill raised the broad public awareness of the oil pollution risks. Before the coming into being of the international conventions, a voluntary scheme was adopted: the Tanker Owners Voluntary Agreement concerning Liability for Oil Pollution (the TOVALOP),[3] under which fault liability with a reversal of the burden of proof is imposed on the tanker owner or the bareboat charter. The TOVALOP is complemented by another voluntary agreement by the oil industry: the Contract Regarding an Interim Supplement to Tanker Liability for Oil Pollution (the CRISTAL). The CRISTAL established a fund to cover oil pollution damage in addition to the TOVALOP with the contributions from the oil industry.

Shortly after the adoption of the voluntary agreement, an international convention was adopted under the auspices of the International Maritime Organization (IMO). In 1969, the 1969 CLC was passed, together with a Resolution on the Establishment of an International Compensation Fund for Oil Pollution Damage. This resulted in the adoption of the 1971 Fund Convention.

According to the 1969 CLC, the convention applies to pollution damage that happened on the territory (including territorial sea) of a Contracting Party and applies to the preventive measures.[4] Oil pollution is defined as:

> Loss or damage caused outside the ship carrying oil by contamination resulting from the escape or discharge of oil from the ship, wherever such escape or discharge may occur, and includes the costs of preventive measures and further loss or damage caused by the preventive measures.[5]

This definition is constricted by the word 'contamination', which means that damage caused by fire or an explosion following a discharge is not covered. The definition is still vague and its concrete scope only became clear over the years. Personal injury is eligible for compensation, but not including exposure to health risks, anxiety and upset. Damage to property should be 'real' and not speculative. Pure economic loss is eligible if the loss is quantifiable in economic terms.[6] A more problematic category is damage to the environment. Preventive measures contain cleanup and restoration. They are compensable if they are reasonable and the loss is quantifiable in economic terms. The claims based on abstract methods of calculation are not admissible.[7]

Strict liability is adopted in the 1969 Convention. Hot debates took place during the negotiation of the Convention on whom the liability should rest. Under

[3] TOVALOP, 8 International Legal Materials, 497-501, 1969.
[4] Art. II of the 1969 CLC.
[5] Art. I 6 of the 1969 CLC.
[6] Verheij 2007a, p. 138-139.
[7] Verheij 2007a, p. 139.

the influence of the international regimes for nuclear liability, no doubts have been formulated on the reasonability of the channeling of liability. The debates focused on whether it should be the ship owner or the oil industry that bears the liability.[8] In the end, a compromise was reached: liability under the CLC fell on the shoulders of the ship owner. As a price, the oil industry also needed to contribute to the compensation through a compensation fund. At the conference to pass the 1969 CLC, it was agreed that an international compensation fund would be established in the near future. Therefore, the channeling of liability is achieved under the 1969 CLC. The ship owner is defined as 'the person or persons registered as the owner of the ship or, in the absence of registration, the person or persons owning the ship.'[9] The 1969 CLC preempts other legislations: no other claims are eligible other than those under the convention. It shows explicitly that no claims are made against the servants or agents of the owner.[10] Scholars deduce from the explicit list that claims against other parties, for example, the cargo owner and the operator are still possible according to applicable national laws.[11] Recourse against third parties is allowed under the 1969 CLC.[12]

The liability established under the 1969 CLC is capped at 210 million francs or 2,000 francs for each ton of the ship's tonnage. The amount is higher than that under the 1957 Convention on the Limitation of Ship-owners' Liability, but still far from covering the whole potential damage that can be caused by oil pollution. The ship owner's right of limitation cannot be used if the incident occurred as a resulted of his actual fault or privity.[13]

Two years after the adoption of the 1969 CLC, the 1971 Fund Convention was passed. The 1971 Fund Convention has two aims: to provide compensation when the protection available from the 1969 CLC is inadequate and to relieve the ship owners from additional financial burden.[14] The details of the 1971 Fund Convention are discussed in the following section on compensation instruments.

1.1.1.2. The 1984 and 1992 Protocols

Several serious oil spills that happened after the adoption of the original CLC and the Fund Convention, for example, the Amoco Cadiz in 1978 and the Tanio in 1980, triggered the revisions to the original conventions. The first Protocols to revise the conventions were reached in 1984. Since the U.S. did not ratify the protocols, the protocols could not come into force. Nevertheless, the changes in the 1984 protocols are largely incorporated in the 1992 Protocols.

8 Wang 2007, p. 212-219.
9 Art. I 3 of the 1969 CLC.
10 Art. III 4 of the 1969 CLC.
11 Verheij 2007a, p. 141.
12 Art. III 5 of the 1969 CLC.
13 Art. V 2 of the 1969 CLC.
14 Art. 2(1) of the 1971 Fund Convention.

In 1992, two protocols were adopted to revise the original conventions: the 1992 CLC and the 1992 Fund Convention. Under the 1992 CLC, the definition of pollution damage is broadened. It includes:

> (a) loss or damage caused outside the ship by contamination resulting from the escape or discharge of oil from the ship, wherever such escape or discharge may occur, provided that compensation for impairment of the environment other than loss of profit from such impairment shall be limited to costs of reasonable measures of reinstatement actually undertaken or to be undertaken;
> (b) the costs of preventive measures and further loss or damage caused by preventive measures.[15]

This definition further delimitates the scope of compensable environmental damage. The loss of profit from the impairment of the environment is compensable, including both consequential loss and pure economic loss. In addition to that, the term 'impairment of the environment' is limited to the costs of reasonable measures of reinstatement actually taken or to be undertaken. Even with this definition, its clear implication has to be understood in practice. Disputes have taken place on the issues of damage quantification, the state as environmental trustee and ecological restoration.[16] It is often up to the national courts, where the lawsuits are brought, to interpret the meaning of vague terms such as 'reasonable measures'.

Although there is no authoritative interpretation of the international conventions, the IOPC Fund has developed, through years of experience with handling compensation claims, a Claims Manual which contains specific criteria for the assessment of environmental damage.[17] The Claims Manual is not designed as an authoritative interpretation of the international conventions, but it is rather meant to assist claimants by giving a general overview of the Fund's obligation to pay contributions.[18] However, the Claims Manual may serve as a useful guideline.

As far as compensation for environmental damage is concerned, the Claims Manual provides that 'Compensation is payable for the costs of reasonable reinstatement measures aimed at accelerating natural recovery of environmental damage. Contributions may be made to the costs of post-spill studies provided that they relate to damage which falls within the definition of pollution damage under the Conventions, including studies to establish the nature and extent of environmental damage caused by an oil spill and to determine whether or not reinstatement measures are necessary and feasible.'[19] The Claims Manual provides

15 Art. I 6 of the 1992 CLC.

16 Mason 2003, p. 3-5.

17 The Claims Manual was first adopted by the Assembly of the Fund in October 2004, and later revised in 2007. The most recent version is 2008. The goal of the Claims Manual is to assist claimants with marine oil pollution compensation, and hence not an authoritative interpretation of the relevant conventions. However, it is of great use in practice for claims processing and proves to be a useful guideline.

18 Claims Manual, December 2008 edition, p. 5.

19 Claims Manual, December 2008 edition, p. 13.

that 'reasonable costs associated with the capture, dealing and rehabilitation of wildlife, in particular birds, mammals and reptiles' are to be compensated.[20]

For the criteria of 'reasonableness', there should be a sufficiently close link of causation between the loss or damage and the contamination. Such a close link may be considered in the light of such factors as geographic proximity of the claimant's business activity to the contaminated area, the degree to which a claimant's business is economically dependent on an affected resource, the extent to which a claimant had alternative sources of supply or business opportunities, and the extent to which a claimant's business forms an integral part of the economic activity within the area affected by the spill.[21]

Moreover, claims for compensation for environmental damage, calculated by theoretical models in accordance with abstract models, are inadmissible.[22] Punitive damage based on the degree of fault of the wrong-doer is also not compensable.[23]

Furthermore, there are also some other changes in the 1992 CLC. The limit of liability is increased to 4.51 million SDRs or 89.77 million SDRs, depending on the size of the ships. As a compromise to the increase of the liability limit, the criteria when the ship owners lose their right to limit liability are further restricted: damage should result from their willful misconduct.[24]

Liability is still imposed on ship owners. However, the broadened scope of the list that can be protected from the liability strengthens the effect of channeling. According to the 1992 CLC, not only servants or agents of the owner, but also some other parties, such as the pilot or any other person who is not a member of the crew and performs services for the ship, any charterer, any person performing salvage with the consent of the owner or on the instructions of a competent public authority; any person taking preventive measures, and their agents or servants are also exonerated from oil pollution liability.

1.1.2. Liability Rules under the Bunker Convention

The Bunker Convention was adopted in 2001. It fills some gaps left by the 1992 CLC. The 1992 CLC constricts its application through the definition of 'ship' and 'oil'. A vessel can be regarded as a ship under the 1992 CLC only when it carries oil in bulk as cargo.[25] 'Oil' means 'any persistent hydrocarbon mineral oil such as crude oil, fuel oil, heavy diesel oil and lubricating oil, whether carried on board a ship as cargo or in the bunkers of such a ship.'[26] In other words, the 1992 CLC applies not only to the discharge of oil as cargo, but also to the discharge of oil in bunkers, but then only when the ships carries oil as cargo as well. Thus questions can be asked

20 Claims Manual, December 2008 edition, p. 12, under the heading of 'Clean-up and preventive measures'.
21 Claims Manual, December 2008 edition, p. 29.
22 Claims Manual, December 2008 edition, p. 36.
23 *Ibid.*
24 Art. V, 1.2 of the 1992 CLC.
25 Art. I 1 of the 1992 CLC.
26 Art. I 5 of the 1992 CLC.

how damage from the discharge of oil in bunkers is compensated if the ships do not carry oil as cargo. Under this situation, the Bunker Convention can apply. But it is worth noting that the definition of 'bunker oil' under the Bunker Convention is different from that of 'oil' under the 1992 CLC. 'Bunker oil' contains both persistent oil and non-persistent oil.[27] To avoid double application, the Bunker Convention does not apply to the damage under the 1992 CLC.[28]

The Bunker Convention also adopts some similar liability rules as the 1992 CLC, such as strict liability and exclusive liability.[29] One important difference between the Bunker Convention and the 1992 CLC is that the former does not introduce a cap on liability itself but leaves the possibility to applicable national laws or international conventions.[30]

1.2. *Compensation Instruments for Oil Pollution Damage*

The above description shows the structure of liability rules under the CLC and Bunker Convention regimes. To guarantee the availability of compensation, both regimes introduce compulsory financial security. The specific ship owners are required to seek insurance or other instruments to cover their potential liability.

The 1969 CLC requires the owner of a ship registered in a contracting state and carrying more than 2,000 tons of oil in bulk as cargo to maintain insurance or other financial security up to his limits of liability. In addition to insurance, the financial security can also be in the form of a bank guarantee or a certificate delivered by an international compensation fund.[31] This requirement remains in the 1992 CLC.[32] Similar provisions can also be found in the Bunker Convention.[33] Although both regimes allow insurance and other instruments to be used to cover the liability, the most commonly used instruments are insurance, especially Protection and Indemnity Policies (P&I Policies).

In addition to the financial security, funds are also established to complement the compensation available from the 1969 and the 1992 CLC. Here lies another important distinction between the Bunker Convention and the CLC/the Fund Convention regimes: there is no fund established under the former. The following section describes the P&I policies and compensation funds respectively.

27 Art. I 5 of the Bunker Convention. For the detailed analysis about the oil under the two conventions, see Han & Wang 2010, p. 149-151.
28 Art. IV 1 of the Bunker Convention.
29 Art. III of the Bunker Convention.
30 Art. VI of the Bunker Convention.
31 Art. VII 1 of the 1969 CLC.
32 Art. VII of the 1992 CLC.
33 Art. VII of the Bunker Convention.

1.2.1. Protection and Indemnity Policies

The theoretical part discussed Protection and Indemnity Clubs (P&I Clubs) briefly as an example of a risk-sharing agreement. This part elaborates its function more in detail.

P&I Clubs began to develop with the reluctance of traditional insurance companies to provide sufficient insurance for marine vessels. The first English Protection Associations were founded in the mid-19th century as a reaction to a lack of capacity and unwillingness to underwrite the marine risks by private insurers. This lack of capacity and interest is the consequence of the increasing risk of English commercial vessels being seized or destroyed by the enemy navies of France, Spain or the United States during the previous two centuries.[34] Following the development of those associations in Britain, P&I Clubs began to grow in late the 19th and early 20th century in Scandinavia and America.[35] The International Groups of P&I Clubs (the P&I Group) has formed on the bases of individual P&I Clubs. Its thirteen member clubs provide liability cover for approximately 90 per cent of the world's ocean-going tonnage.

As discussed earlier, P&I coverage usually includes both pollution damage liability and other liability caused by ship owners. When defining the term 'liability' under the P&I policies, one has to refer to the applicable international conventions and national legislation. The CLC and the OPA are important examples. The CLC involves traditional damage and ecological damage, which is evaluated by the costs of reasonable measures of reinstatement. The OPA in the U.S. defines liability even broader, under which the responsible parties are liable for removal costs and six kinds of damages: natural resources damage, real or personal property, subsistence use, revenues, profits and earning capacity and public services.[36] Thus the coverage of a P&I policy can be rather broad: not only does it provide a coverage of the liability for ecological damage, the relevant personal injury and property damage as well as other non-environment related losses are also covered. A P&I Club provides services more than a pure insurer and operates as a mixture of an insurance company, a law firm and a loss adjuster. In addition to offering an insurance coverage, a P&I Club can also provide a worldwide network of correspondents and representatives to give on-the-spot assistance to the ship owner when required, give Letters of Undertaking to offer a security when members' vessels are arrested and assist in claims handling and settlement.[37]

Under the P&I policies, the insured parties must have suffered actual monetary losses before they can seek reimbursement from the insurers. That means a member is only entitled to seek compensation for the amount he has in fact lost due to the occurrence of a covered incident. This is called the 'pay to be paid' rule, which is usually incorporated in the Club's Rule Book. In a P&I policy, the Club is only obliged to assist his contractual counterpart, the Club's Member, in case of

34 See Ronnerberg 1990, p. 3.

35 See *ibid.*; Libby 1952, p. 685.

36 33 U.S.C. §2702.

37 Ronnerberg 1990, p. 25-29.

losses Thus, usually, the injured cannot bring a direct action against a P&I Club and can only get the compensation by a claim, litigation against or settlement with the injurer. This policy can create problems in the case of the insured's bankruptcy, since then the injurer may have insufficient assets to realize his liability. Clubs vary in their attitudes towards this issue. In English law, the injured party ranks as an ordinary creditor, and thus does not have a general direct right against the insurer in case of bankruptcy of the insured. However, according to the 1930 Third Parties Act and some case law, an injured party is provided a limited right of direct action against an insurer in particular circumstances.[38] In the U.S., there is no right of a third party of a 'direct action' against the insurer under the common law. But if the federal state enacts a 'direct action statute', an injured party can be provided with such a right.[39] The Louisiana Direct Action Statute[40] is the first legislation in the U.S. to stipulate a right of direct action, and it goes so far as to provide such a right independently from the assured's insolvency. The New York Statute is more restrictive than the Louisiana one, which allows direct action where the insured is bankrupt and if the assured's liability is already determined by an actual court decision.[41] According to the above analysis, the Clubs vary with regard to the extent of the protection they award for the injured.

The P&I Group arranges reinsurance together for each Club. At this moment, for the ship owners' policies, each Club retains the first $80 million as their retentions. The amount between $8 million and $60 million is divided among all the Clubs. The captive insurer of the Group, Hydra Insurance Company, and re-insurance with the international insurance market also play an important role in providing reinsurance for the upper layers. This brings the upper limit of its reinsurance program to $3060 million. Within this amount, the limit for compensation for oil pollution is limited to $1060 million.[42]

1.2.2. Compensation Funds

1.2.2.1. The 1971 Fund

The 1971 Fund Convention plays two roles: to compensate the victims and to indemnify the ship owners. Firstly, it complements the compensation from the 1969 CLC under these situations: no liability can be established under the 1969 CLC; owners and his financial guarantors are financially incapable of compensation and the damage exceeds the owners' liability. To encourage preventive measures, their costs raised from the voluntary activities of the owners are also treated as pollution damage.[43] The Fund has no obligation to pay if it can prove the damage resulted

38 Holstein-Childress 2002, p. 205.
39 Dougherty 1985, p. 1478-1481.
40 LSA-R.S. 22:655.
41 Foster 1998, p. 281.
42 <www.igpandi.org/Group+Agreements/Pool+reinsurance+programme>, last accessed on 8 May 2013.
43 Art. 4(1) of the 1971 Fund Convention.

from an act of war, hostilities, civil war or insurrection or oil from a warship or a state-owned/ operated ship; or if the claimant cannot prove that damage resulted from a ship-related incident. Contributory negligence can also be used as a defense for the fund.[44] The compensation available from the 1971 Fund is not unlimited: the total sum of the amount from the 1969 CLC and the 1971 Fund Convention is capped at 450 million francs; for the damage caused from a natural disaster of an exceptional, inevitable and irresistible character, the amount payable from the Fund is capped at 450 million francs. The Assembly of the Fund has the right to increase the amount up to 900 million francs. When claims exceed the amount payable from the fund, payment should be reduced proportionally for each claimant.[45]

In addition to providing complementary compensation to victims, the 1971 Fund also indemnifies the ship owners. Indemnification is available for the payment between (1) the amount in excess of 1,500 francs for each ton of the ship's tonnage or 125 million francs, whichever is less; and (2) the amount not in excess of 2,000 francs for each ton of the said tonnage or 210 million francs, whichever is less. But the indemnification is not available if the damage is caused due to the willful misconduct of the ship owners.[46] Claims for compensation or indemnification should be made within three years from the occurrence of the damage and six years from the occurrence of the incident.[47]

The Fund is financed by big oil importers in contracting states. The eligible importers need to make initial contributions as the working capital of the fund, and the annual contributions to cover the administrative expenses and claims.[48] The calculation of contributions is based on a fixed sum for each ton of contributing oil received.[49] The Contracting State has the obligation to ensure that the eligible contributing importers appear on a list and to communicate it to the Fund.[50] The Contracting State can also declare that it assumes the obligation that lies on the importers within its territory to make contributions.[51]

1.2.2.2. The 1992 Fund and the winding up of the 1971 Fund

The original CLC and Fund Convention proved to be insufficient to cover the potential catastrophic oil pollution damage. In 1992 a new compensation fund was established. The 1992 Fund Convention removed the function of the Fund to alleviate the burden of liability of ship owners. Therefore the only function of the 1992 Fund is to provide additional protection to the victims of oil pollution. The conditions of the application of the 1992 Fund to compensate victims are the same

44 Art. 4(2)(3) of the 1971 Fund Convention.
45 Art. 4(4)(5)(6) of the 1971 Fund Convention.
46 Art. 5(1) of the 1971 Fund Convention.
47 Art. 6(1) of the 1971 Fund Convention.
48 Wu 1996, p. 98.
49 Arts. 11(1), 12(2) of the 1971 Fund Convention.
50 Art. 15(1) of the 1971 Fund Convention.
51 Art. 14(1) of the 1971 Fund Convention.

as that of the 1971 Fund.[52]The available compensation from the 1992 CLC and the 1992 Fund Convention is increased to 203 million SDRs.[53]

The 1992 Fund is also financed by the oil industry. To remove an obstacle for ratification, the requirement of the considerable initial contributions is removed. The 1992 Fund is solely financed by annual contributions. The calculation of the contributions is also based on the amount of oil received.

The 1992 Fund Convention came into force in 1996, when the 1971 Fund Convention was still in force. The co-existence of two funds was problematic since that diluted the capacity of each fund to provide sufficient compensation. In 2000, a protocol was passed to allow for an early winding up of the 1971 Fund. According to this protocol, the 1971 Fund ceases to be in force when the number of contracting states fall bellow 25, or the total quantity of contributing oil fall below 100 million tons, whichever is earlier. The 1971 Fund ceased to be in force on 24 May 2002, but it still continues to compensate for damage from incidents before that day.[54] As of January 2012, the 1992 Fund Convention has 105 Member States.

1.2.2.3. The 2003 Supplementary Fund

Although the limitation of compensation under the 1992 Fund Convention had been considerably increased, shortly after its adoption, the amount was dwarfed again by yet more catastrophic oil pollution cases.[55] Under this background, a Supplementary Fund was established in a 2003 Protocol.[56] The Supplementary Fund provides an additional layer of compensation for oil pollution victims under the 1992 CLC and the 1992 Fund Convention. In other words, a condition for the payment from the Supplementary Fund is that the victim is entitled to compensation under the 1992 CLC and the 1992 Fund Convention, and is unable to obtain full and adequate compensation from them.[57] The Supplementary Fund Convention increases the aggregated amount of compensation up to 750 million SDRs.[58]

The Supplementary Fund Convention imposes an additional burden on oil importers to compensate for pollution damage. After the adoption of this convention, two voluntary arrangements were made to alleviate the additional burden imposed on the oil importers and to balance the interests between ship owners and the oil industry. Under the Small Tanker Oil Pollution Indemnification Agreement (STOPIA) and the Tanker Oil Pollution Indemnification Agreement

52 Art. IV 1 of the 1992 Fund Convention.

53 Art. IV 4 of the 1992 Fund Convention.

54 Website of IOPCF, <www.iopcfunds.org/about-us/>, last accessed on 8 May 2013.

55 The Nakhodka accident near Japan in 1997 and the Erika disaster in France in 1999 are two examples.

56 Protocol of 2003 to the International Convention on the Establishment of an International Fund for Compensation for Oil Pollution Damage, 1992. (The Supplementary Fund Protocol), RMC II. 7.115.

57 Art. IV 1 of the Supplementary Fund Convention.

58 Art. IV 2 of the Supplementary Fund Convention.

(TOPIA), the tanker owners indemnify parts of the additional burden under the Supplementary Fund Convention.[59]

1.3. *Critical Analysis*

The liability regime concerning oil pollution provides an example on how the combination of liability and compensation instruments can contribute to the prevention and compensation of environmental damage. Two international schemes have been established: the CLC/Fund Convention scheme and the Bunker Convention Scheme.

The two schemes both create a liability framework for oil pollution and there are many similarities between them. The CLC allows compensation for ecological damage. However, it defines the impairment of the environment other than loss of profits narrowly to incorporate only costs of 'reasonable' restoration undertaken or to be undertaken. Hence the irreparable damage is excluded. The Claim Manuals of the IOPCF also denies compensation for damage calculated via theoretical models. Several efforts have been made to broaden the scope of compensable environmental damage under the CLC/IOPCF. However, none of them have been accepted, since there are worries that allowing broad compensation for environmental damage will lead to many arbitrary decisions of national courts and jeopardize the capacity to compensate for private victims.[60] This limited scope results in a large part of ecological damage uncompensated and may lead to under-deterrence. Admittedly, when considering the internalization of the whole costs caused by ecological damage, one has to take notice of the difficulties in evaluating ecological damage.

Strict liability is established under both regimes. This is an important shift. Before that, the existing voluntary arrangement (TOVALOP) was based on fault liability.[61] Strict liability has advantages since ship owners are usually better informed concerning their potential risks and preventive methods than the legislators and judges. Thus strict liability can give better incentives to ship owners to take efficient care. Furthermore, strict liability not only creates incentives concerning the care level but also concerning the activity level that ship owners take.

As under the international regimes of nuclear liability, channeling of liability is also used in establishing oil pollution liability. As discussed earlier, during the negotiation for the 1969 CLC, there was a hot debate on who should be liable for oil pollution, but there was little doubt on the principle of channeling itself. In the end, under both the CLC and the Bunker Convention, liability is imposed exclusively on ship owners. The CLC also has a list of other parties who are exonerated from liability. A further channeling is achieved under the 1992 CLC, which broadens the scope of the exonerating list under the 1969 CLC. Under this situation, other parties, who may equally contribute to oil pollution risk, have no incentives to take

59 Wang 2007, p. 210.
60 Wang 2011, p. 154-158.
61 Wang 2011, p. 210.

sufficient care. This situation may be alleviated if the ship owners make use of the recourse provision to claim back their payment from other parties according to the applicable national law. But then one may still question whether, in practice, this provision can be used that often. The channeling of liability to ship owners can also be explained from the private interest approach. During the negotiation of the CLC, various proposals have been made by delegates from different states. For example, Ireland, the Netherlands and Denmark advocated strict liability on the cargo; U.S., UK, Australia and Germany were in favor of strict liability on the ship; other proposals included fault liability on the ship owner and joint liability. Those debates are understandable, the countries with a big ship industry preferred liability on cargo; while, on the other hand, countries with a powerful oil industry proposed liability on ships.[62] Finally, the consensus of imposing liability on ship owners was achieved with the promise of a compensation fund financed by the oil industry. The balance of interest between the ship and oil industry can also be found during the establishment of STOPIA and TOPIA, which aim at alleviating the additional burden brought by the Supplementary Fund Convention.

Although the CLC has increased the limitation of liability significantly compared to the earlier conventions, and the amount has been increased over the years, the limits established under the CLC are still far from sufficient in covering potential catastrophic losses. The Bunker Convention does not introduce a limit itself but allows the application of the national law and other international regimes. Limited liability, combined with strict liability gives less than efficient incentives for potential liable parties to prevent damage. One may immediately come with the question whether unlimited liability is economically feasible. However, one early empirical study has shown that by increasing the amount of compensation from $52 million (the limit of that moment) to $250 million, the effect on the price of oil is quite limited.[63] Even when some small oil companies are driven out of business due to unlimited liability, it is not necessarily undesirable since it is only when facing the whole liability they create, they will take optimal care.[64]

Both the CLC and the Bunker Convention require bigger ship owners to provide financial security up to the liability limit. Insurance, bank guarantees or guarantees from a compensation fund can be used as financial security. Economic analysis shows that strict liability is only efficient when the injurer is fully solvent. Compulsory financial security can alleviate the insolvency problem. Insurance is the most commonly used method to provide financial security. However, liability insurance may lead to the moral hazard problem.[65] This problem can be solved by monitoring and risk differentiation. In the field of marine pollution, insurance policies can mainly be provided by P&I Clubs, a mutual association of ship owners. Thus, rather than private insurance, the policy is a risk-sharing agreement. It has the advantage that ship owners, who have better information on their risks, can control

62 Wang 2011, p. 213-219.
63 Smets 1983, p. 31-43.
64 Wang & Faure 2010, p. 19.
65 Cummins & Tennyson 1996, p. 29-50.

the risks through mutual monitoring and the *ex ante* information burden is alleviated.

One major difference between the CLC/Fund Convention regime and the Bunker Convention regime is that compensation funds are only established under the former. Two International Compensation Funds for Oil Pollution were established in 1971 and 1992. Both funds provide additional compensation for the victims of oil pollution. The function of indemnifying the ship owners under the 1971 Fund is removed in the 1992 Fund. One may question the efficiency of the fund since it is the oil industry, rather than the liable party, the ship owners, who finance the compensation. However, this needs to be understood from the background of the balance of interests between ship owners and the oil industry. However, the efficiency of such funds can still be questionable since the contributions of the funds are based on the amount of oil received. Therefore, the oil importers still lack incentives to choose safer ships.[66]

2. The International Compensation System for Nuclear Damage

This section tries to depict and analyze the international compensation system for nuclear damage. Many Western countries started to develop the nuclear industry to meet their increasing energy needs since the 1950s. However, a few barriers to the development existed, especially the serious public concern about the sufficient protection against potential damage and the hesitation of investors to step into the highly risky industry.[67] To guarantee a certain level of compensation for potential victims, relieve the nuclear investors from the potential heavy claims and promote the peaceful use of nuclear energy, the governments tried to establish domestic nuclear liability acts and international conventions on nuclear liability since then. There are also some conventions addressing nuclear liability issues in case of transport.[68] However, in this research, only damage caused by nuclear accidents in land-based nuclear facilities is discussed. In the 1960s, two international compensation regimes were established for nuclear damage: the OECD regime and the International Atomic Energy Agency (IAEA) regime. Under the auspices of the OECD Nuclear Energy Agency (NEA), the Paris Convention and the Brussels Supplementary Convention to the Paris Convention on Third Party Liability in the Field of Nuclear Energy of 31 January 1963 (Brussels Supplementary Convention) have been developed.[69] The second regime was developed under the aegis of IAEA: the Vienna Convention on Civil Liability for Nuclear Damage of 21 May 1963 (Vienna Convention).[70] Those two regimes are usually called the first generation of

66 This has been criticized as soon as the 1971 Fund Convention was adopted. See Wood 1975, p. 46-47.

67 Schwartz 2006, p. 38-39.

68 For example, the 1971 Brussels Convention Relating to Civil liability in the Field of Maritime Carriage of Nuclear Material, UNE 944 UNTS 255.

69 Convention Supplementary to the Paris Convention on Third Party Liability in the Field of Nuclear Energy (Brussels Complementary Convention).

70 Vienna Convention on Civil Liability for Nuclear Damage (Vienna Convention).

nuclear liability conventions.[71] Those two regimes have, on the one hand, made an effort to establish international/regional regimes for nuclear liability and, on the other hand, have obvious limitations in terms of restricted geographical scope, a narrow definition of nuclear damage and an insufficient amount of available compensation.[72] The Chernobyl accident in 1986 triggered an intensive discussion about those limitations and later a revision process of the existing regimes. The so-called second generation of nuclear liability conventions was established thereafter. Those conventions consist of the Joint Protocol Relating to the Application of the Vienna Convention and the Paris Convention (Joint Protocol),[73] the Protocol to Amend the 1963 Vienna Convention on Civil Liability for Nuclear Damage (the Protocol to the Vienna Convention),[74] the Convention on Supplementary Compensation for Nuclear Damage (CSC),[75] the Protocol to amend the Convention on Third Party Liability in the Field of Nuclear Energy of 29 July 1960 (the Protocol to the Paris Convention)[76] and the Protocol to amend the Convention of 31 January 1963 supplementary to the Convention of 29 July 1960 on Third Party Liability in the Field of Nuclear Energy (the Protocol to the Brussels Supplementary Convention).[77]

Before the introduction to the content and development of the international nuclear liability conventions, the relationship between nuclear damage and environmental damage needs to be clarified here. This thesis focuses on the compensation system for damage to the environment, instead of personal injury and property damage caused via the environment. Even though how 'nuclear damage' is defined depends on the specific instrument that stipulates the liability for it, more attention has been paid to potential personal injury/death, and property damage resulting from a nuclear incident instead of to potential harm to the environment. However, the limited definition of 'nuclear damage' is subject to change. Under the revised Paris Convention and Vienna conventions, three new headings which are environment-concerned are added to the compensable damage.[78] The Protocol to the Vienna Convention came into force on 4 October

71 See Faure & Vanden Borre 2008, p. 220-232.

72 See Vanden Borre 2010, p. 192.

73 Joint Protocol relating to the Application of the Vienna Convention and the Paris Convention (Joint Protocol).

74 Vienna Convention as Amended by the Protocol of 12 September 1997 to Amend the Vienna Convention on Civil Liability for Nuclear Damage (the Protocol to Vienna Convention), 1063 UNTS 358.

75 Convention on Supplementary Compensation for Nuclear Damage (CSC).

76 Paris Convention on Third Party Liability in the Field of Nuclear Energy as amended by the Additional Protocol of 28 January 1964 and by the Protocol of 16 November 1982, and by the Protocol of 12 February 2004 (the Protocol to the Paris Convention), 956 UNTS 251.

77 Convention Supplementary to the Paris Convention on Third Party Liability in the Field of Nuclear Energy (as Amended by the Additional Protocol of 28 January 1964 and by the Protocol of 16 November 1982, and by the Protocol of 12 February 2004) (the Protocol to the Brussels Supplementary Convention), 1041 UNTS 358.

78 Under both revised regimes, the concept of 'nuclear damage' is expanded to include 'the costs of measures of reinstatement of impaired environment', 'loss of income deriving from a direct economic interest in any use or enjoyment of the environment' and 'the costs of →

2003, and the ratification of the Protocol to the Paris Convention is still under discussion in the Member States, especially in the EU.[79] If those two new regimes can be broadly accepted, they will also establish a compensation system for environmental damage caused by nuclear events. Moreover, the compensation regimes for nuclear damage are especially interesting since a multi-tiered compensation system is established with the intervention of public funds. There have also been hot debates on the capacity of insurance to cover the increased liability and broadened scope of compensable damage under the revisions[80] and the possibility to establish an international pooling system.[81]

This section firstly sketches the development and framework of the international nuclear liability conventions. The introduction to the system is followed by a critical analysis based on both the liability regimes and the financial security system.

2.1. *International Regimes for Compensation of Nuclear Damage*

2.1.1. Origins of International Regimes

At the beginning of the development of the nuclear industry, the Western-European market had to rely on American suppliers and technology. The American nuclear industry, however, was unwilling to bear liability for possible nuclear accidents in Europe.[82] Therefore, a 'hold-harmless' clause was introduced in the first bilateral agreements between the U.S. and the Europe, which required the European nuclear operators to indemnify the American suppliers for all claims resulting from their activities.[83] Even with those 'hold-harmless' clauses, the American industry was still unsure whether it could provide sufficient protection and the U.S. 'Atomic Industrial Forum' conducted a compressive study on the possibility of European victims claiming against American suppliers in the case of a nuclear incident. Two reports were published later by the Atomic Industrial Forum: the 'Preliminary Report on Financial Protection against Atomic Hazards' ('the Preliminary Report')[84]

preventive measures'. See the Protocol to Paris Convention, Art. 1(vii), the Protocol to Vienna Convention, Art. 1(k). Those headings can also be the components of 'environmental damage'. The implications of those headings are analyzed in the following section.

79 See Vanden Borre 2010, p. 200-205. How to harmonize nuclear liability has been hotly debated in Europe, and a ratification of the Protocol to the Paris Convention is proposed as one alternative to achieve this goal. Other alternatives under discussion include a partial harmonization by ratification of the Protocol to the Paris Convention and the Protocol to the Vienna Convention respectively by Member States; Euratom as a contracting party to the Protocol to the Paris Convention and issuing a European Nuclear Liability Directive. See Hanrilica 2009.

80 See Froggatt & Carroll 2010, p. 555-614.

81 See Carroll 2009, available at: <http://papers.ssrn.com/sol3/papers.cfm?abstract_id-=1505373>, last accessed on 8 May 2013; Pelzer 2007.

82 See Vanden Borre 2007, p. 262.

83 Vanden Borre 2010, p. 180.

84 Prepared by the experts of Columbia University and published in March 1956.

and the 'International Problems of Financial Protection against Nuclear Risk' ('the Harvard Report').[85] Those two reports developed some principles for nuclear liability, which were later incorporated into the international conventions.

The Preliminary Report identified four parties who can be affected by nuclear risks: the nuclear industry, private insurers, the government and the victims of a a potential nuclear accident. In this report, the liability of the nuclear industry is linked with insurance capacity; for the part of damage that cannot be compensated by insurance, the government needs to provide additional compensation to protect the victims. The Preliminary Report introduced a system of unlimited government intervention: any damage in excess of the limitation of liable parties' liability should be covered by the government in terms of indemnity.[86]

The channeling of liability to nuclear operators was advised in the Harvard Report. According to the Harvard Report, the suppliers and contractors were exempted from liability for the following reasons: the suppliers are afraid of being held liable instead of or jointly with the operators and then being burdened by the lengthy trials; suppliers will lose control after the delivery of goods and services, and operators are more capable of obtaining insurance.[87]

Those two reports proposed several principles for nuclear liability, which were later adopted in a few national legislations and international conventions.

2.1.2. First Generation of Nuclear Liability Conventions

As discussed earlier, two compensation regimes for nuclear damage were established in the 1960s under the auspices of the NEA and the IAEA. Since the Paris Convention and the Brussels Convention are established under the auspices of the OECD/NEA, they are regionally confined to Western Europe, Slovenia and Turkey. The Vienna Convention, under the aegis of the IAEA, is world-wide in scope. A few principles proposed in the Preliminary Report and in the Harvard Report were accepted under those two conventions. Before the description of those principles, this section first clarifies what constitutes 'nuclear damage' under those conventions, or in the other words, which categories of damage the nuclear industry is liable for according to the conventions.

According to the Paris Convention, the nuclear operator is liable for 'damage to or loss of life of any person' and 'damage to or loss of any property other than' onsite damage or property used in connection with the installations.[88] The Convention does not explain the two headings of damage further. What should be considered as damage to persons or damage to property is left to the discretion of the competent court in accordance with the applicable national law.[89] Those two headings are also recognized as 'nuclear damage' under the Vienna Convention. In addition to those two headings, 'nuclear damage' under the Vienna Convention also

85 Prepared by Harvard Law Schools and published in 1959.

86 Preliminary Report, p. 30.

87 See Vanden 2007, p. 265-266.

88 Art. III(a) of the Paris Convention.

89 The *exposé des motifs* of the Paris Convention, point 39.

includes 'any other loss or damage so arising or resulting if and to the extent that the law of the competent court so provides.'[90] Therefore, more discretion is given to the competent court.

As for the geographical scope, the Paris Convention does not apply to 'nuclear incidents occurring in the territory of non-Contracting States or to damage suffered in such a territory unless otherwise provided by the legislation of the Contracting Party in whose territory the nuclear installation of the operator liable is situated.'[91] The geographical scope of the Vienna Convention is less clear. Some argue that since the convention itself is silent on its geographical scope, the Vienna Convention on the Law of Treaties will apply. The convention applies to damage occurring in the territory of a State Party, on board aircraft registers in that state and on ships flying its flag.[92] While others deduce from the article about the competent court[93] that the geographical scope depends on the competent court and thus damage resulting from non-Vienna states could be covered by Vienna Convention.[94] The geographical scope has been broadened by the Joint Protocol later, which is discussed in Section 2.1.3.

2.1.2.1. Strict Liability

In Western Europe, there is a long-established tradition of a presumption of liability for hazards resulting from a dangerous activity. Nuclear industry covered under the Paris Convention is obviously qualified as a dangerous activity and there is a serious difficulty in proving negligence of the nuclear operators. Therefore, a system of absolute liability is established under the Paris Convention.[95] According to the Paris Convention, the operator is liable for damage caused by a nuclear incident in a nuclear installation or involving nuclear substances coming from such installations.[96] To prove the fault of nuclear operators is no longer necessary to establish liability. Liability established under the Paris Convention is quite stringent since many classical exonerations, such as force majeure, Acts of Gods or intervening acts of third persons under general tort law are no longer applicable.[97] The available exonerations are an act of armed conflict hostilities, civil war, and insurrection. The operator is not liable for damage caused via a grave natural disaster of an exceptional character unless the legislation of the Contracting Party in whose territory his nuclear installation is situated provides to the contrary.[98] Similar stipulations about absolute liability and exonerations can also be found under the

90 Art. I(k) of the Vienna Convention.
91 Art. II of the Paris Convention.
92 Lamm 2006, p. 172.
93 Art. VIII of the Vienna Convention.
94 Blanchard 2000, p. 236.
95 The *exposé des motifs* of the Paris Convention, point 14.
96 Art. III(a) of the Paris Convention.
97 The *exposé des motifs* of the Paris Convention, point 48.
98 Art. IX of the Paris Convention.

Vienna Convention.[99] However, under the Vienna Convention, there is an additional possibility for operators to be relieved of this liability: the competent court can, according to the applicable law, relieve the operator wholly or partly from his obligation if the operator can prove that damage resulted from gross negligence or an act or the omission of the victims.[100]

2.1.2.2. *Channeling of Liability*

Under the Paris Convention, liability is channeled to the operators. In addition to the operators, no one else is liable for the damage caused by a nuclear incident.[101] The 'operator' is defined as 'the person designated or recognized by the competent public authority as the operator of that installation.'[102] These provisions can exonerate other parties engaged in nuclear activities from liability since the Paris Convention is the only legal basis for a claim against a nuclear operator in case of the identified incidents.[103] This concentration of liability is based on two reasons: to avoid the complicated legal procedures to identify the liable parties and to allow a concentration of insurance capacity.[104] Under the Paris Convention, the operators in principle do not have a right of recourse against the other parties. This is because it is argued that allowing recourse will make it necessary for suppliers to seek insurance coverage and will lead to costly duplication of insurance.[105] However, recourse is possible if the damage results from an act or omission done with intent to cause damage or if and to the extent provided by contract.[106] The Vienna Convention also has similar provisions.[107]

2.1.2.3. *Limited Liability*

Under both the Paris Convention and the Vienna Convention, the operator's liability is limited both in amount and in time. The Paris Convention sets the maximum liability of the operator at 15 million SDRs, but allows the Contracting Party to establish by legislation a greater or lesser amount considering the capacity of insurance and financial security. The Contracting Party can also require a lower amount according to the nature of the installation. The lower amount should be no less than 5 million SDRs.[108] The Paris Convention introduces a cap on liability, taking into account the difficulties of operators to seek financial security.[109] The flexible expression allows the Contracting Parties to set the limit higher than that set

99 Arts. I(1)(k), IV(1),(3) of the Vienna Convention.
100 Art. IV(2) of the Vienna Convention.
101 Art. VI(a),(b) of the Paris Convention.
102 Art. I(a)(vi) of the Paris Convention.
103 Art. VI(c)(ii) of the Paris Convention
104 The *exposé des motifs* of the Paris Convention, point 15.
105 The *exposé des motifs* of the Paris Convention, point 18.
106 Art. VI(f) of the Paris Convention.
107 Arts. II(5), X of the Vienna Convention.
108 Art. VII(b) of the Paris Convention.
109 The *exposé des motifs* of the Paris Convention, point 43.

in the Paris Convention. For example, in Sweden, the limit on liability is set at 300 million SDRs according to the Nuclear Liability Act (SFS 1968:45).[110] Germany even adopted a system with unlimited liability.[111] Under the Vienna Convention, the cap of liability should be no less than U.S. $5 million.[112] The statute of limitations under both the Paris Convention and the Vienna Convention is set at ten years from the date of the nuclear accident. They both allow the extension of the extinction periods by Contracting Parties if the operator is covered by insurance or other financial security.[113]

2.1.2.4. *Financial Security*

To seek financial security coverage for the operator's liability is important for international regimes of nuclear liability. Both conventions require the operator to have and maintain insurance or other financial security up to the cap of its liability.[114] Insurance is the most commonly used instrument for an operator to cover its liability. In fact, the cap on liability is usually set as the maximum available amount from the insurance market. Since insurance is available per installation for a fixed period rather than in respect of a single incident, the potential resources available for compensation may be reduced after the first incident. Under these conditions, the Contracting Parties may need to intervene.[115] It is for the Contracting Parties to decide the nature, form and extent of the compensation according to applicable national law.[116]

2.1.2.5. *Exclusive Jurisdiction*

Both the Paris Convention and the Vienna Convention have an 'exclusive jurisdiction' clause: the jurisdiction lies only with the courts of the Contracting Party within whose territory the nuclear incident occurred. If the nuclear accident happens outside the Contracting Parties' territory or the place cannot be determined, the jurisdiction lies with the courts of the Installation State of the liable operator.[117] Since nuclear damage may have a transboundary effect, the exclusive

110 OECD, Nuclear Legislation in OECD Countries, Regulatory and Institutional Framework for Nuclear Activities: Sweden, 13, available at: <www.oecd-nea.org/law/-legislation/sweden.pdf>, last accessed on 8 May 2013.

111 *Gesetz über die friedliche Verwendung der Kernenergie und den Schutz gegen ihre Gefahren (Atomgesetz – AtG) vom 23. Dezember 1959, Neufassung vom 15. Juli 1985, letzte Änderung vom 31. Juli 2011* (Act on the peaceful utilization of nuclear energy an the protection against its hazards (Atomic Energy Act) of 23 December 1959, as amended and promulgated on 15 July 1985, last amendment of 31 July 2011), §31.

112 Art. V of the Vienna Convention.

113 Art. VIII of the Paris Convention; Art. VI(1) of the Vienna Convention.

114 Art. X of the Paris Convention; Art. VII of the Vienna Convention.

115 The *exposé des motifs* of the Paris Convention, point 49.

116 Art. XI of the Paris Convention; Art. VIII of the Vienna Convention.

117 Art. VIII of the Paris Convention; Art. XI of the Vienna Convention.

jurisdiction can ensure that the cap of liability is not exceeded and that a fair distribution of compensation takes place, if the claims lead to insoluble problems.[118]

2.1.2.6. *Public Funding*

The liability limits established under the Paris Convention and the Vienna Convention are quite low compared to the potential catastrophic damage that a nuclear incident can cause. The liability is limited to promote the development of the peaceful use of nuclear liability. To provide the potential victims better protection, the Brussels Complementary Convention was established under the auspices of the NEA in 1963. Under the Brussels Complementary Convention, two additional layers of compensation are added in terms of public funds. Therefore, the total amount of money available for compensation is increased to 300 million SDR, and a third layer compensation system is established: firstly, the operator is liable up to an amount of at least 5 million SDRs via its financial security; between this amount and 175 million SDR, the Installation State needs to make public funds available; for the amount between 175 and 300 million SDR, compensation is provided out of public funds by all Contracting Parties according to a specific formula.[119] The Installation State can escape its obligation under the second layer by setting the liability limit at no less than 175 million SDRs. In that case the whole amount up to 175 million SDR needs to be made available by liable operators.

2.1.3. The Second Generation of Nuclear Liability Conventions

2.1.3.1. *General Issues*

The terrible nuclear accident in Chernobyl in 1986 triggered people to think about the insufficiency of the existing nuclear liability conventions in terms of the ability to protect the victims. The Chernobyl accident led to serious damage to human health, property and the environment with a broad transboundary impact. However, at that moment, the Soviet Union had no national legislation on nuclear liability. Nor was it a member of any international nuclear liability regimes. Even if the Soviet Union had joined the Vienna Convention, serious problems still prevent an effective compensation: the authorities in the Soviet Union questioned the necessity of the preventive measures;[120] the cap on liability under the Vienna Convention was dwarfed by the serious damage it caused; western European countries joined another regime which is independent of the Vienna Convention.

The lack of the acceptance of the international regimes and the insufficient resources available in case of nuclear damage triggered the effort to revise the conventions. Another five conventions have been passed since the Chernobyl accident. The first effort was made to link the geographical scope of the existing

118 The *exposé des motifs* of the Paris Convention, point 54.

119 Art. III(a)(b) of the Brussels Complementary Convention.

120 See Vanden Borre 2010, p. 192.

regimes. In 1988, the Joint Protocol was adopted, under which the operator of a nuclear installation situated in the territory of a party to one convention shall be liable for the nuclear damage suffered in the territory of a party to either the Paris Convention or the Vienna Convention.[121] It was believed that the Joint Protocol would give incentives for the Central and Eastern European countries to join the Vienna Convention and would broaden the application of the two regimes. However, only 12 out of 18 countries in these areas which have joined the Vienna Convention became a member of the Joint Protocol. Globally, of the 52 States Party to the Paris and Vienna Conventions, only 27 Parties have joined the Joint Protocol.

In addition to the geographic links, some substantive changes were also adopted in the following years. Under the auspices of the IAEA regimes, two nuclear conventions were open for signature in 1997: the Protocol to the Vienna Convention and a stand-alone convention – the CSC. The CSC was made to open to more countries: it maintains the essential principles established under the Paris Convention, and tries to establish two tiers of public funds in case of nuclear damage. It is open to any state, if it is a party to either the Paris Convention or the Vienna Convention, or if it has national legislation containing the same principles. An effort was also made to revise the NEA conventions. In 2004, the Protocol to the Paris Convention and the Protocol to the Brussels Supplementary Convention were open for signature.

Under the regimes of both NEA and IAEA, a few important changes have been made, especially the broadened scope of nuclear damage and the increased amount of the limitation of liability and public funds. Under the Protocol to the Paris Convention, in addition to personal injury and property damage, four new titles have been added to the concept of 'nuclear damage': the economic loss arising from personal injury and property damage; the costs of measures of reinstatement of the impaired environment; the loss of income deriving from a direct economic interest in any use or enjoyment of the environment, and the costs of preventive measures.[122] Similar provisions can also be found in the Protocol to the Vienna Convention. The difference is that, under the latter, the definition of nuclear damage is open: it includes any other economic loss which is permitted by the general civil liability law in the competent court.[123]

Since the health impact of nuclear radiation may not manifest itself after decades, the revised conventions also extend the statute of limitation for claims for personal injury and death. Those kinds of claims need to be brought within 30 years from the date of the accident.[124] In addition, the available defenses for the operators are further limited: natural disasters are no longer an applicable defense.[125] The Protocols are also designed to be attractive by broadening the applicable

121 Art. II of the Joint Protocol.

122 Art. I(vii) of the Protocol to the Paris Convention.

123 Art. I(k) of the Protocol to the Vienna Convention.

124 Art. VIII(a)(i) of the Protocol to the Paris Convention; Art. VI(1)(a)(i), the Protocol to the Vienna Convention.

125 Art. IX of the Protocol to the Paris Convention; Art. IV(3), the Protocol to the Vienna Convention.

geographical scope. The Protocol to the Vienna Convention applies to nuclear damage wherever suffered, with a permitted exclusion if a non-contract party has nuclear installations but does not offer reciprocal benefits.[126] The Protocol to the Paris Convention also covers damage suffered in some non-Contracting Parties which satisfy some specific requirements.[127]

Another important change made under the NEA and IAEA regimes is the increased liability limitation and public funds. Under the Paris Convention, a three-tier compensation system was established in 1960. The Protocol to the Paris Convention increases the limit for nuclear operators to be no less than 700 million Euro. The Contracting party can reduce the liability to be no less than 70 million Euro for an incident originating from a nuclear installation, or to be no less than 80 million Euro for the carriage of nuclear substances according to the reduced risks.[128] The Convention even allows the adoption of unlimited liability by the Contracting Parties, as long as the financial security required is no less than the amount mentioned above.[129] In addition to the liability of operators, the Protocol to the Brussels Complementary Convention provides another two tiers of compensation. From the maximum amount of operators' liability set by the Protocol to the Paris Convention up to 1,200 million Euro, the Installation State needs to pay in terms of public funds. Public funds need to be made available by all Contracting Parties according to a set formula of compensation between the amounts of 1,200 million Euro and 1,500 million Euro.[130] Thus the available resources for compensation in case of a nuclear accident are increased significantly: from 300 million SDRs (approximately 327 million Euro) to 1,500 million Euro.

The compensation capacity also increased under the IAEA regime. The Protocol to the Vienna Convention increases the liability limitation to no less than 300 million SDRs, or no less than 150 million SDRs, provided the Installation State will make public funds available to cover the amount between the set the limitation to 300 million SDRs.[131] Thus, the revision has introduced a second layer of compensation. Although the amount is also significantly increased, it is modest compared to the possible significant damage and it is just set at the same level as the original Paris Convention and the Brussels Supplementary Convention. The CSC Convention also provides two tiers of compensation: first 300 million SDRs is paid by public funds from the Installation State and another 300 million SDRs from the collective funds from the Contracting Parties.[132]

These new conventions and protocols are designed to overcome the deficiency of the first generation of nuclear liability conventions. However, the revisions have made eight international conventions available for nuclear liability, which lead to the so called a 'labyrinth of international conventions' dealing with nuclear liability

126 Art. I A of the Protocol to the Vienna Convention.
127 Art. II(a) of the Protocol to the Paris Convention.
128 Art. VII(a)(b) of the Protocol to the Paris Convention.
129 Art. X(b) of the Protocol to the Paris Convention.
130 Art. III(b) of the Protocol to the Brussels Complementary Convention.
131 Art. V(1) of the Protocol to the Vienna Convention.
132 Arts. III 1, IV 1 of the CSC Convention.

issues.[133] Moreover, among the five new conventions, only two have come into force: the Joint Protocol entered into force in 1992 and the Protocol to the Vienna Convention came into force in 2003. Until now, only five countries have ratified the protocol to the Vienna Convention: Argentina, Belarus, Latvia, Morocco and Romania, and none of them have a significant nuclear generating capacity.[134] There are still many large nuclear generating countries which are not members of any of those conventions, such as the U.S., Canada, Japan, China, Korea, Russia, South Africa and Switzerland.

There is no comprehensive study on why some countries are reluctant to join nuclear liability conventions, but a few hypotheses have been proposed. Firstly, some countries are geographically remote from the Contracting Parties to the international conventions. Thus they can benefit less from the available compensation in case of transboundary damage. This may be a concern for Australia, Japan and Canada. Another reason may be that unlimited liability has been adopted in the domestic legislation in Switzerland, Austria and Japan. However, the Protocols to the Paris and Vienna Conventions have allowed the participation of the states with unlimited liability. Some countries may hesitate because a discriminatory treatment between domestic and foreign victims is not allowed under the conventions and they may be reluctant to pay foreign victims out of public funds.[135]

2.1.3.2. The Coverage of Environmental Damage

The international conventions try to establish a compensation framework for nuclear damage. Before the discussion on detailed liability rules, the first question to be answered is what qualifies as 'nuclear damage'. The first generation of conventions focuses more on personal injury and property damage. Admittedly, those two headings are the two most obvious kinds of damage in case of a nuclear accident. However, the Chernobyl accident showed that the environment could also be seriously influenced by a nuclear accident. To leave the environmental damage uncompensated leads to an externality of costs by the nuclear industry and leads to lax safety measures adopted. Therefore the second generation of conventions tries to incorporate the environmental damage under three headings: the costs of measures of reinstatement of impaired environment; loss of income deriving from a direct economic interest in any use or enjoyment of the environment and the costs of preventive measures.

In this section, the analysis of nuclear damage uses the Protocol to the Paris Convention as an example since the provision under the Protocol to the Vienna Convention is similar. The Protocol to the Paris Convention applies mainly to

133 For the patchy and complicated approach under international nuclear liability conventions, see Duncan 2008.

134 Of these five countries, only Argentina and Romania have nuclear capacity. The net nuclear power generating capacity of Argentina is 935 MWe and that of Romania is 1,300 MWe in 2009. See Schwartz 2009, p. 50.

135 *Ibid.*

Western Europe. In Europe, in addition to the Protocol to the Paris Convention, there is also another instrument which stipulates the compensation for the environmental damage, specifically the ELD. However, compared to the ELD, the compensable environmental damage under the Protocol to the Paris Convention is still quite limited. The ELD, however, uses the term 'environmental liability' and is more of a public law nature.[136] The specific operators under the ELD are obliged to take preventive and remedial measures. Public authorities can also take the remedial measures by themselves and claim the costs from the liable operators. However, the Protocol to the Paris Convention is part of the tort law.[137] The liability framework covers reinstatement costs, 'if such measures are actually taken or to be taken'. However, the protocol does not establish an obligation for the liable parties to take reinstatement measures or to authorize other parties to do so but leaves it to the Member State. Thus whether the reinstatement costs will be compensated depends on whether there is someone to take the reinstatement measures. 'Measures of reinstatement' is defined as:

> any reasonable measures which have been approved by the competent authorities of the State where the measures were taken, and which aim to reinstate or restore damaged or destroyed components of the environment, or to introduce, where reasonable, the equivalent of these components into the environment.[138]

This definition uses a far-reaching concept of remediation. After comparing this concept with remediation under the ELD, Pelzer argues that the two terms have the same implications. In other words, the reinstatement measures cover primary remediation, complementary remediation and the compensatory remediation under the ELD. This leads to the concern of insurers who are reluctant to cover compensatory remediation.[139] The issues about insurability of environmental damage are discussed in the following section. There are also some scholars criticizing that the obscure words used in the definition do not give a hint on the standards of reinstatement, in other words: to what extent does the damaged environment need to be restored. Should it be restored to the baseline level that is the status before the accident happened or to the extent that the environmental components are present and are functioning normally?[140]

Another important heading, as far as environmental damage is concerned, is

> loss of income deriving from a direct economic interest in any use or enjoyment of the environment, incurred as a result of a significant impairment of that environment.[141]

Under this heading the individual loss as a result of the harm to the environment is covered. Both damages and remedial measures can be used as remedies under this

136 See Winter, Jans, Macrory & Krämer 2008, p. 163-165.

137 Pelzer 2010, p. 52.

138 Art. I(a)(viii) of the Protocol to the Paris Convention.

139 Pelzer 2010, p. 49, the accompanying text of note 1.

140 Emmerechts 2008, p. 101-102.

141 Art. I(a)(vii)(5) of the Protocol to the Paris Convention.

heading. If the remedial measure is chosen, the liable party needs to remediate the environment to the extent that the individual's interest is influenced. But the liable party has no obligation to remediate the whole polluted environment under this heading. One example would be that a nuclear accident leads to the pollution of a beach. The owner of the hotel at the beach can require the operator to replace the sand at the part of beach that is used by hotel clients. The liable operator has an obligation to reinstate the whole beach under this heading.[142] However, to incorporate the loss of income resulting from the enjoyment of the environment results in criticism due to the difficulty of making this assessment.[143]

Pelzer points out another limit of those two headings: they only apply when they are not covered as damage consequential to property damage.[144] Since most environmental components are subject to the ownership of the State or private persons, the application of those headings is quite limited.

A third heading relevant here is 'the costs of preventive measures, and the further loss or damage caused by such measures'. The preventive measures can be taken either after a nuclear incident or at an event which creates serious threat. Both reinstatement measures and preventive measures need to be reasonable measures; and to be reasonable, the measures should be 'appropriate and proportionate according to the competent court', depending on a few criteria, such as: the nature and extent of damage or risk; the possible effectiveness of the measures and relevant scientific and technical expertise.[145]

The revised conventions introduce three headings related to the harm to the environment. The sketchy expression, however, does not establish a harmonized standard in evaluating the damage. On the contrary, wide discretion has been given to the competent court. The protocols limit the recoverable costs 'to the extent determined by the law of the competent court.'[146] Under this formulation, to what extent the three headings of damage can be compensable depends on the applicable law in the Contracting States. A detailed analysis of the practice in all Member States is outside the scope here. This section uses the discussion in the UK as an example to show the potential function of the new headings under the revised conventions.

In the UK, nuclear liability rules are mainly incorporated in the Nuclear Installation Act of 1965. In light of the revised Paris Convention, the UK tries to incorporate the changes in the Protocol to the Paris Convention into its domestic legislation. In 2011, a paper on 'Implementation of Changes to the Paris and Brussels Conventions on Nuclear Third Party Liability' (consultation paper) and a 'Draft Order with the Proposed Amendments to the Nuclear Installations Act 1965 to Implement the Convention Changes' were published and made open to public

142 Pelzer 2010, p. 55.
143 Danzi 2009, p. 197.
144 Pelzer 2010, p. 54.
145 Art. I(a)(x) of the Protocol to the Paris Convention.
146 Art. I(a)(vii) of the Protocol to the Paris Convention.

consultation.[147] In the consultation paper, it is expressed that, as far as the new categories of damage are concerned, it will not go further, but will adopt the wording and definitions used in the Convention.[148] For the reinstatement costs, it notices that in the UK, there are various arrangements that require public authorities or other bodies to take steps to reinstate the environment. The consultation paper does not want to create a new freestanding regime for the reinstatement, but instead limits the liability to the extent covered by existing reinstatement arrangements.[149] One of the important arrangements is for the contaminated land under Part IIA of the Environmental Protection Act 1990. Under this regime, the eligible person to bear responsibility for remediation is the one 'who caused or knowingly permitted the substances ... to be in, on or under that land.'[150] However this arrangement is in contradiction to the channeling of liability under nuclear liability law. Thus how the existing domestic reinstatement arrangement can be compatible with the new categories under revised Paris Convention is a challenge. In the UK, pure economic loss is only recoverable in very limited circumstances. Therefore the heading loss of income deriving from a direct economic interest in the environment 'incurred as a result of a significant impairment of that environment' will provide a new entitlement to compensation. To set a threshold for 'significant impairment' is proposed in the consultation paper. It is the judges who will decide whether this threshold is passed, what 'the environment' is, and what constitutes 'a direct economic interest in any use or enjoyment of the environment.'[151] The Protocol to the Paris Convention defines 'preventive measures' as 'any reasonable measures taken by any person ... subject to any approval of the competent authorities required by the law of the State where the measures were taken.'[152] There are already existing national plans and powers to take preventive measures, thus the heading preventive measures and consequential losses do not create a new freestanding regime. In spite of the wording 'subject to any approval of the competent authorities' under the revised Paris Convention,[153] the consultation does not require the approval of preventive measures because of the likely urgency of the preventive measures. However, the consultation paper does create some limit on the 'consequential loss': the damage should be suffered by someone other than the person taking the preventive measures; claims need to be made for personal injury or property damage.[154]

147 Reference NO. 10D/974, available at: <www.decc.gov.uk/en/content/cms/-consultations/paris_brussels/paris_brussels.aspx>, last accessed on 8 May 2013.

148 Department of Energy & Climate Change 2011, Implementation of Changes to the Paris and Brussels Conventions on Nuclear Third Party Liability: A Public Consultation 2011, Paragraph 1.6.

149 Consultation Paper, Paragraph 4.21-4.24.

150 Environmental Protection Act of 1990, Part II A, 78F.

151 Consultation Paper, Paragraph 4.56.

152 Art. I(a)(ix) of the Protocol to the Paris Convention.

153 *Ibid.*

154 Consultation Paper, Paragraph 4.68-4.69.

2.2. *Evaluation from a Law and Economics Perspective*

The international conventions provide a framework to compensate for nuclear damage, with the following features: strict liability, channeling liability to nuclear operators; limited liability linked to insurance capacity and public intervention in case of damage in excess of the limitation. Those rules were designed at the beginning of the development of nuclear energy and were aiming at both promoting the development of the new industry and at the same time guaranteeing a certain level of protection for the public. However, after 50 years of development, it can be questioned whether the already developed nuclear industry still needs a subsidy in terms of special liability arrangements. This section tries to analyze the Conventions, mainly the liability rules and financial security mechanisms.

2.2.1. Liability

Nuclear liability can, on the one hand, provide protection to potential victims by compensation in case of damage, and on the other hand give the nuclear operators incentives to take precaution measures and prevent damage. The international conventions establish a strict liability system for nuclear operators. The victims only need to prove the damage they suffered is caused by a nuclear accident or nuclear substance, without proving the fault of the operators. This is efficient from an economic perspective. The nuclear accident can be categorized as a unilateral accident, where only the operators contribute to the risks. In such a highly technical industry, the operators are in a better position than the judges to determine their optimal care level. Moreover, only strict liability gives the operators incentives to adopt an efficient activity level.[155] The nuclear liability under the international conventions is even called absolute liability since less exoneration is available. Only damage resulting from an act of armed conflict, hostilities, civil war, insurrection can be used as a defense. A grave natural disaster of an exceptional character is a possible defense under the original Paris and Vienna Conventions, but is removed by the Protocols. Terrorism is also not included in the available defenses. Damage caused by terrorism and national disasters are less predictable, and thus is usually excluded from insurance policies. Other financial security mechanisms need to intervene in those cases to provide a guarantee for compensation.

For strict liability to work efficiently, the amount of compensation should be equal to the actual costs of the accidents. However, under the international nuclear liability conventions, strict liability is accepted at the price of the limitation of compensation. The cap of liability does not only put the potential victims in an unfavorable position. More importantly, under the strict liability regime, liable parties will only have incentives to prevent damage to the extent of the cap. Under-deterrence may lead to a less efficient care level. Under the original Conventions, the cap was set as 15 million SDR or $5 million. The amount was substantially increased in the Protocols, to be no less than 700 million Euro or 150 million SDRs.

155 Faure 1995, p. 27.

The Protocols even allow the Member States to adopt unlimited liability. In practice, only a limited number of countries adopted unlimited liability.[156] The amounts set under the protocols are still nowhere near to the real possible costs of a major nuclear accident. Although there is no consensus on the estimation of the worst scenario of nuclear accidents yet, the Chernobyl accident and Fukushima accident have caused damage much higher than the limit set in nuclear conventions. Some studies also estimate the possible damage to be as high as several billions of Euro: it is estimated that the total damage of a reactor meltdown in Germany would be over 5,000 billion Euro; a Greenpeace review of the costs cited some estimation varies from several billion to 6.8 trillion USD dollars.[157] Given the potential catastrophic losses the nuclear accidents can create, the limited liability creates a substantial subsidy to the nuclear industry. This subsidy may not only lead to a less than efficient care level adopted by operators but also gives nuclear energy in an advantageous status compared to other kinds of energy. This may also lead to too high a demand for nuclear energy. For example, research showed that with subsidies to nuclear sectors, the energy production costs for nuclear energy is lower than coal, gas and other renewable energy.[158] A recent German research shows that if the total damage should be borne by the nuclear industry, and all the liability is covered with insurance, it will lead to a net price increase for nuclear energy of 0.139 to 2.36 Euro per kWh or 3.93 to 67.3 per kWh, depending on the payout period.[159] If this holds true, it will, undoubtedly, make nuclear energy prohibitively expensive.

The international nuclear conventions also introduce a system of channeling the liability to nuclear operators. Other parties, such as suppliers, transporters, subcontractors, test operators, consultants, nuclear plant designers and constructors are released from liability. A right of recourse is only permitted if the nuclear incident is caused intentionally by others, or the contract expresses so. This leads to the so-called legal channeling problem: other parties who actually contribute to the risks are relieved from liability and thus have no incentives to take efficient care. This problem has been criticized by a lot of law and economics literature: after comparing the legal channeling under the international conventions and economic channeling in the U.S. system, Vanden Borre argues that the legal channeling leads to under-derterrence;[160] Trebilcock and Winter criticize the combination of channeling and limited liability.[161] The channeling is especially problematic since, generally, it does not allow recourse. One may argue that the operators can

156 For example, unlimited liability was adopted in Austria and Germany. In Denmark and Sweden, the unlimited liability is seriously considered or will be applicable when the revised Paris Convention comes into force. See Schwartz 2009, p. 54.

157 Cited in Currie 2008, p. 92.

158 Faure & Fiore 2009.

159 Versicherungsforen Leipzig, Study: Calculating a Risk-appropriate Insurance Premium to Cover Third-Party Liability Risks that Result from Operation of Nuclear Power Plants, 2011, available at: <www.laka.org/docu/boeken/pdf/6-01-0-30-34.pdf>, last accessed on 8 May 2013.

160 Vanden Borre 1999, p. 27.

161 Trebilcock & Winter 1997, p. 232-235.

negotiate with other parties and make arrangements to abolish channeling. However, this is not always possible for the nuclear industry. A survey shows that considering the different bargaining power, the contracts between operators and designers/constructors rarely make use of the exception to abolish channeling.[162] There are some reasons which are used quite often in justifying the channeling of liability: allowing lawsuits with all parties will launch a prolonged recovery procedure; if all parties need to buy insurance, it will lead to an escalation of nuclear insurance costs. However, it is doubtful whether the simplified claiming procedure outweighs the advantages in allowing the victims to sue more parties in terms of the protection of victims. Furthermore, alternative instruments such as pooling, bank guarantees, letters of credit and so on can also be used.[163]

2.2.2. Financial Security

Liability itself cannot guarantee compensation. Some financial security mechanisms, such as insurance, risk-sharing agreements or bank guarantees can be used to guarantee the availability of resources for compensation in case of damage and to give potentially liable parities incentives to take care. Under the international conventions, nuclear operators are required to seek financial coverage up to the limitation of liability. Insurance is the most commonly used financial security mechanism. Actually, in many cases, insurance capacity is an important factor in determining the limitation of liability. The limitations under the international conventions are set quite modestly, thus state intervention plays an important role in addition to liability.

2.2.2.1. Public Funds

Under the Paris Convention, a three-tiered compensation system is established. The Installation State needs to make public funds available to cover the second tier of damage and the contracting parties together should provide a collective fund to provide the third tier of compensation. The public funds intervention is even strengthened under the second generation of conventions. There was a shift towards public funding in those conventions.[164] The required amount of public funds from Installation States under the Paris Convention is increased from (up to) 175 million SDRs to 500 million Euro, and the collective fund is increased from 125 million SDRs to 300 million Euro. The shift is also the case in the conventions under the auspices of the IAEA. Under the original Vienna Convention, there is no requirement concerning public funds. The Protocol to the Vienna Convention adds a new tier: the public fund from the Installation State. A stand-alone instrument, the CSC Convention establishes a two-tier compensation system: 300 million SDRs from the Installation State or 300 million SDRs from the collective Fund. The shift

162 Ameye 2010, p. 56.
163 Ameye 2010, p. 36-39.
164 Vanden Borre 2007, p. 305.

towards public funding has been criticized by law and economics literature, since it constitutes a subsidy to operators and thus further dilutes the incentives for operators to take efficient care.[165] It is worth noting that to what extent the public funds will cause under-deterrence is also determined by the way the fund is financed. For example, the U.S. ratified the CSC in 2008, under which it has the obligation to make 300 million SDRs available in case of an accident happening in its territory. However, this ratification of CSC relieves the suppliers from potential liability for foreign victims; the suppliers are required to participate in a retrospective risk-pooling program to cover the contribution of the U.S. according to the CSC.[166] Thus, the polluters rather than the public actually finance the compensation.

2.2.2.2. *Insurance*

Insurance is a most commonly used instrument for nuclear operators to cover their liability. Considering the catastrophic nature of the nuclear accident, insurance is usually provided by nuclear insurance pools, which are bundling resources at national level. Under an insurance-pooling agreement, each insurer who is a member of the pool needs to contribute according to the ratio contractually agreed with the pool.[167] Pooling can promote competition in the sense that it allows the small insurers to participate without putting their financial capacity at risk. However, this only holds true if pooling is not combined with further restrictions on competition.[168] The nuclear pools are a monopoly that excludes all competition. The operators can only buy insurance from their own national pools; thus there is no competition between national pools. In this situation, the insurability of a certain nuclear risk and the premium are not based on an objective actuarial study but on the willingness of the insurers to cover such a risk.[169] One example of the monopolistic premium for nuclear liability is the case of France. Nuclear energy is operated monopolistically by the EDF in France. According to the legislation, the limitation of liability is set at 91 million Euro, of which the first 60 million Euro is covered through a reserve of EDF, while the other 31 million Euro is covered by the insurance. A study shows that the EDF has to pay 6.4 million Euro per year as a premium, which is 355 times higher than the actuarial premium.[170] There are some explanations for the high premium: firstly, the insurers have less incentives to control loading costs and can charge a monopolistic profit; secondly, the largely immeasurable risk makes the risk-averse insurer take into account a high risk ambiguity; a third reason is that nuclear risk is to some extent non-diversifiable.[171]

165 See Faure & Vanden Borre 2008, p. 265-269.

166 Energy Independence and Security Act of 2007, Pub. L. No. 110-140, § 934 (a)(1)(D), 121 Stat. 1492, 1744 (2007).

167 Vanden Borre 2010, p. 198.

168 Faure 1995, p. 31.

169 Faure 1995, p. 32.

170 Faure & Fiore 2008a, p. 236.

171 Faure & Fiore 2008a, p. 236-238.

Another factor influencing nuclear liability insurance is that nuclear insurance pools usually provide both coverage for third-party liability and first-party damage. The concerns of operators about the coverage for their own assets further decreases the capacity for liability insurance.[172] One example is that in the 1980s, the limits of liability and liability insurance coverage in Belgium for the nuclear licensee was 4 billion Belgian francs, while the available property damage coverage was 40 billion Belgian francs.[173]

The above analysis shows that insurance coverage for traditional nuclear damage is already problematic. The second generation of nuclear liability conventions further broadens the scope of compensable damage and increases liability limits. Whether and to what extent traditional nuclear pools can meet those new requirements still remains to be seen. In the first few years after the publishing of the revised conventions, insurers have shown their concerns about: the increased amount, extended prescription period for personal injury and the new heads of nuclear damage. According to the insurers, few insurers are prepared to offer coverage of more than 10 years. They are reluctant to provide coverage for reinstatement costs and are only willing to provide limited coverage for loss of income deriving from a direct economic interest in the environment as well as the preventative measures based on specific restrictions. They even argue that the new headings of coverage will lead to financial uncertainties and cause a reduction in insurance coverage.[174] Now, in some countries, the insurers are trying to examine whether and how the new headings, especially the environment-related ones, can be covered. According to a study from 2009, the insurers in the UK are willing to provide coverage for economic loss arising from personal injury and property damage. For reinstatement costs and environment-related loss of income, they are willing to provide coverage as long as the accident is classified as an accident of INES 5, 6, 7. This is because the accident of INES 1-4 has no or little off-site impact and is more difficult to quantify.[175] The coverage for preventative measures was still under discussion.[176]The study establishes a methodology to assess the potential risks under the new heading and tries to give some advice in calculating the premiums.[177] However, the coverage for personal injury beyond 10 years is still not coverable by insurers.

Admittedly, as mentioned earlier, the insurability argument is often made by insurers according to their willingness to cover rather than based on an objective actuarial study. It is possible that one day the insurers may be capable of providing wider coverage for nuclear liability. However, the question still remains why the insurers will bind their assets to the nuclear industry facing increasing risks and

172 Vanden Borre 2010, p. 198-199.

173 Faure 1995, p. 32.

174 Tetley 2006.

175 Hathlia 2009, p. 77.

176 According to the consultation paper published in 2011, the coverage for preventive measure is now accepted by insurers. Point 9.6.

177 Hathlia 2009.

costs. The introduction of alternative financial instruments may be beneficial not only for operators and victims but also the insurers.

2.2.2.3. *Pooling*

Some literature has proposed to establish a pooling system to cover nuclear liability.[178] Insurers are unwilling to provide full coverage for the nuclear liability facing operators or are only willing to do so at a very high price. The individual operator cannot provide sufficient self-guarantees for the potential catastrophic damage and, moreover, usually their only asset, the nuclear installation itself, will be damaged in the case of a nuclear accident. Under this situation, the pooling system may provide a better alternative. The operators can make an *ex ante* agreement to cover their nuclear liability together according to an agreed share of contribution.

Examples of pooling systems exist within the national level: in the U.S. system and in the German system. In the U.S., the 1975 amendment to the Price-Anderson Act introduced a so-called retrospective pooling system. Under the legislation, a two-layered compensation system is established. The operator needs to provide coverage through insurance for the first layer of $375 million in case of a nuclear accident. If this amount is exceeded, each operator needs to pay up to 111.9 million. Given the current number of reactors in the U.S., the total available compensation amounts to $12.2 billion.[179] It is worth noting that this is a retrospective system, since the obligation to pay only comes due when a major accident happens. In Germany, there is no legislative obligation for the operators to pool. Unlimited liability is established and the operators are required to provide financial security up to an amount of DEM 5 billion (2.5 billion Euro). The first layer of 255.6 million Euro is provided by the insurance market. To cover the remaining liability, the four parent companies of the 19 German nuclear power plants concluded a 'Solidarity Agreement' in 2001. According to this agreement, in case of damage in excess of 255.6 million Euro, the operators need to provide compensation for the remaining 2.24 billion. Each partner needs to contribute a certain percentage of the total amount according to a ratio plan based on the square foot of the thermal reactor power.[180] The details of the U.S. and German pooling system are discussed in the following chapters.[181]

Compared to insurance, the pooling system has the advantages that *ex ante* information about the possibility and extent of risks are no longer necessary. One needs only to know the proportion of each member's contribution to the risk. As a low frequency high magnitude risk, nuclear liability is less predictable. A

178 Pelzer 2007; Simon Carroll, European Challenges to Promoting International Pooling and Compensation for Nuclear Reactor Accidents, Nonproliferation Policy Education Center Working Paper Series, <http://papers.ssrn.com/sol3/papers.cfm?abstract_id=1505373>, last accessed on 8 May 2013; Faure & Fiore 2008b.

179 10 C.F.R. §140.11(4).

180 Pelzer 2007, p. 44.

181 See Chapter 7, Section 1.3.2; Chapter 8, Sections 2, 3.

retrospective pooling system can relieve the heavy information burden. In addition, a pooling system also has financial advantages. The above analysis has shown that the premium in a monopolistic insurance market is very high. A pooling system is not for profit and can offer a cost-effective way to cover nuclear liability, especially the less predictable new headings under the second generation of the conventions. The pooling system can also be beneficial for insurers, since they can be relieved from the burden of the new headings they do not feel able to cover.[182]

In spite of the promising features mentioned above, several requirements need to be satisfied for operators to pool effectively, especially at the international level. The states whose operators wish to participate in the pool should have comparable political, legal and economic backgrounds; the pooling members should have equal or comparable levels of nuclear safety and security.[183] The countries with nuclear installations have varying political, legal and economic backgrounds, which creates a challenge for international pooling. This may be less of a problem in the EU, where the members share more transboundary risks. However, with the enlargement of EU Member States, the development status of the members are more diverging now. It is even more problematic as far as nuclear liability legislation is concerned. The EU states have a so-called patchwork nuclear liability regime.[184] They are subject to different nuclear liability conventions or do not join any convention at all. The domestic legislations have different attitudes towards many important issues such as limited/unlimited liability; amounts required for financial security, and so on.[185] As for levels of nuclear safety and security, it is still problematic. Since pooling is based on trust and confidence, it should be possible for the operators to make mutual monitoring or an organization should exist to control the risks of its members. However, an EU-wide approach towards safety regulation and standards and a single regulatory body are still missing in Europe. In spite of those weaknesses, it is worth noting that the EU is now making efforts towards a more harmonized system of nuclear liability, and many scholars are working on that.[186] The current difficulties of insurance to cover new risks offers a chance to 'break new ground in providing financial security'. An attempt towards alternative financial security mechanisms, especially a pooling system, can still be desirable. Some of those attempts have already been made in Europe.

Two mutual pools are established in Europe: European Mutual Association for Nuclear Insurance (EMANI) and European Liability Insurance for the Nuclear Industry (ELINI). They are both mutual insurance associations and both provide not

182 Carroll 2008, p. 96.

183 Pelzer 2007, p. 50.

184 See Reyners 2009.

185 For example, in Austria and Germany, unlimited liability was adopted. But in most Member States, the liability is capped. Within those countries with limited liability, the required limitations also vary. For an overview of the liability limit in EU Member States, see European Commission, Legal Study for the Accession of Euratom to the Paris Convention on Third Party Liability in the Field of Nuclear Energy, TREN/CC/01-2005, 2005, p. 22-23, available at: <http://ec.europa.eu/energy/nuclear/studies/doc/2009_12_accession-_euratom.pdf>, last accessed on 8 May 2013.

186 For example, Ameye 2009; Hanrilica 2009.

only coverage to nuclear power stations but also to other nuclear facilities. EMANI was established in 1978, and provides coverage for material damage, business interruption, machinery breakdown, terror and erection all risk. The overall insurance capacity in 2010 is 600 million Euro and the capacity for terrorism is 500 million Euro.[187] The decontamination costs for installations in case of an accident are also covered in the martial damage policy within the limitation of 600 million Euro.[188] However, since there is no legal obligation for nuclear operators to seek financial coverage for their onsite decontamination, as is the case in the U.S., the coverage for decontamination costs by EMANI may be not significant. EMANI can either act as the leading underwriter or as a coinsurer to provide insurance directly, or act as a reinsurer. Since EMANI can act either as a leading underwriter or as a coinsurer with national pools, two different types of policies exist: the national pool policy and the EMANI policy. They may have different coverage and prices.

ELINI was created at the end of 2002 to provide insurance capacity for nuclear liability risks. It is the first worldwide nuclear pooling system aimed at providing nuclear third-party liability coverage.[189] ELINI is not designed as a competitor to the national pools, but instead to offer excess liability coverage: it provides coverage to all the headings under the revised Paris and Brussels Supplementary Conventions and also coverage for terrorism and the 30 years prescription period.[190] The capacity to provide coverage for the less predictable risk is because of the special structure of ELINI. There are three types of membership for ELINI: the insured members who have an insurance policy and have contributed to the constitution of the guarantee fund; non-insured members who take part in the constitution of the guarantee fund (supportive members) and non-insured members who do not take part in the constitution of the guarantee fund but pay an administration fee.[191] Both insured members and supportive members make annual contributions to the guarantee fund according to the agreed ratios. If the total losses exceed the contributions, an additional call can be made up to 20 times the annual contributions.[192] The availability of additional calls makes an *ex ante* calculation of premiums for the unpredictable risks less necessary. It is worth noting that ELINI is different from national mutual pools established in the U.S. and Germany. Under the national regimes, the insurers and insured are the same; the operators make contributions according to pre-set ratio when a major accident happens to one of the operators. However, under the system of ELINI, supportive members who are not insured parties can also make contributions to the guarantee fund. ELINI can either act as a leading underwriter or a coinsurer. Thus the individual insurance policies can be

187 EMANI Annual Report 2010; <www.emani.be/EMANI_AR_FINAL_2010.pdf>, last accessed on 8 May 2013.

188 Interview with Danny Van Welkenhuyzen, Managing Director of ELINI and Ann Geivaerts, Legal Manager of ELINI on 28 June 2011.

189 Faure & Vanden Borre 2008, p. 257.

190 Annual report ELINI 2010, at p. 16. <www.elini.net/ELINI_2010_(2011)_FINAL.pdf>, last accessed on 8 May 2013.

191 *Ibid.*, p. 17.

192 Interview with Danny Van Welkenhuyzen, Managing Director of ELINI and Ann Geivaerts, Legal Manager of ELINI on 28 June 2011.

varied from one to another; the extent of involvement of ELINI and the percentage of each contributor in different policies may also be different. The insured parties are also free to choose the amount of insurance. Through cooperation with national pools, ELINI works more like a combination of a traditional insurer and a mutual pool.

To guarantee sufficient compensation for future damage, ELINI makes reserves. The contributions made by members are put into a special bank account and can only been withdrawn after the dissolution of the pool or 30 years after the expiration of its policies. In 2010 the maximum insurance capacity per policy is 89 million Euro, of which 37.5 million Euro is provided through retention.

ELINI can provide insurance by itself, as a leading underwriter or coinsurer with national pools. The content of the contract and insured amount depend on national laws, since different countries have divergent liability limits and liability and insurance requirements. When it acts as a coinsurer to national pools, ELINI relies on the national pools to decide the content of the insurance contract but can decide its own premium. Usually, the basic premium among electricity producers is the same, which are subject to increasing or decreasing according to the grading made by ELINI. When accepting one as a member of ELINI, ELINI has to make a risk review and grades the risks according to technical, insurance and financial features. The grading can be used to adjust the premium later.[193]

ELINI makes an attempt at establishing an international pooling system for nuclear liability. If the revised Paris Convention comes into effect and is accepted by many countries, the additional capacity provided by ELINI may become more attractive. However, given the divergent liability requirements among different countries, and given the lack of consistent and harmonized standards on nuclear safety, nuclear operators may not have a strong interest in establishing a mutual pool. That may be one of the reasons why the mutual pools have developed so slowly since its birth. After several decades, the capacity of EMANI only increased from 150 million Euro in the first decade of existence to 600 million Euro in 2010,[194] while the capacity of ELINI is still only 89 million Euro in 2011.[195]

193 The pool in the U.S., NEIL can rely on some self-regulatory organizations, such as WANO and INPO to control the safety of members. WANO is the World Association of Nuclear Operators, INPO is Institute of Nuclear Power Operations in the U.S. They fufill important roles in checking the nuclear safety in nuclear power plants. In the U.S., NEIL cooperates with INPO. However, to be a member of WANO or INPO, one should be the operator of nuclear power plants. But the members of ELINI vary: they can be either operators for nuclear power plants or other nuclear installations. Thus, unlike NEIL, which can rely on INPO's information in setting premiums, ELINI are not affiliated with those institutions. Hence, ELINI cannot affiliate with a nuclear self-regulatory organization like WANO or INPO to control the nuclear risks.

194 Faure & Vanden Borre 2008, p. 257; Annual Report 2010 EMANI, at p. 16.

195 Annual report 2010 ELINI, p. 18.

Chapter 7

COMPENSATION SYSTEMS IN THE U.S.

1. Compensation System for Nuclear Damage in the U.S.

The Price-Anderson Act (PAA) and Nuclear Regulatory Commission (NRC) Regulations require the licensees of nuclear installations to seek financial protection for both the potential public liability caused from the installations and for property damage of the installations.[1] The compensation instruments used in the U.S. can provide compensation for nuclear incidents of a catastrophic nature.

The same as the international nuclear liability system, the U.S. regime has only limited relevance for ecological damage. There is no explicit definition of nuclear damage in the Price-Anderson Act, which sets out the liability and compensation for nuclear damage in the U.S. Instead, a related concept 'public liability' is defined in the Act, as 'any legal liability arising out of or resulting from a nuclear incident or precautionary evacuation.'[2] The analysis below shows that the term 'public liability' used in the Price-Anderson Act only refers to the liability arising from bodily injury and property damage but not to damage to the environment. The ecological damage is of more relevance with regard to on-site property damage which contains the costs of the decontamination and stabilization of the nuclear reactor.[3] The limited influence of the American nuclear compensation regime on ecological damage does not eliminate the need to examine the compensation scheme for nuclear damage. Admittedly, a nuclear incident can primarily have a significant influence on human health and property, which have already attracted much attention. However, the environment can also be seriously affected by a nuclear accident, as the Chernobyl

1 The Price-Anderson Act was originally adopted in 1957: Act to Amend the Atomic Energy Act of 1954, Pub. L. No. 85-256, 71 Stat. 576 (1957). It was recently revised by the Energy Policy Act o f 2005, Pub. L. No. 109-58,119 Stat. 779 (2005). The NRC also adopted many regulations about the nuclear industry. The parts relevant to financial protection and compensation are compiled under Title 10, Part 140 of Code of Federal Regulations. 10 C.F.R. §140.1-140.109.

2 42 U.S.C. §2014(w).

3 10 C.F.R. §50.54(w).

accident demonstrated.[4] The measures taken to decontaminate the polluted reactor site have also an influence on the environment. As shown in the second generation of international nuclear liability conventions, the limited interpretation of nuclear liability may be broadened to incorporate ecological damage one day.[5] Moreover, the compensation scheme for nuclear damage in the U.S. is interesting since it makes compensation for a catastrophe up to $12 billion dollars possible, which is incomparable to other areas of environmental liability. This scheme may then shed some light on the establishment of a compensation system for ecological damage.

1.1. *Liability for Nuclear Damage*

Both regulation and liability rules can be used to prevent nuclear damage. After World War II, the U.S. established the Atomic Energy Commission (AEC) to foster and control the peace time development of atomic industry. The AEC was abolished by the Energy Reorganization Act of 1974,[6] which assigned its functions to the Energy Research and Development Administration and the Nuclear Regulatory Commission (NRC). The latter is the agency in the U.S. to regulate the nuclear industry: it oversees reactor safety, reactor licensing and renewal, radioactive material safety and spent fuel management. Furthermore, there is also a self-regulation institution, the Institute of Nuclear Power Operations (INPO) that monitors and controls nuclear risks. A detailed examination of how the nuclear industry is regulated in the U.S. is outside the scope here. This section then introduces briefly the liability rules for nuclear damage in the U.S.

The nuclear sector may create damage during the operation of nuclear facilities, during the decommissioning and after the decommissioning of the facilities. The nuclear waste, if not disposed properly, can also lead to severe damage. Furthermore, cautious decommissioning and monitoring after the decommissioning period is also necessary to prevent nuclear damage.

Liability rules for nuclear damage during the operation of nuclear facilities are mainly prescribed in the Price-Anderson Act and in the Nuclear Regulatory Commission Regulations. When congress passed the Atomic Energy Act in 1954, it decided to allow private participation in nuclear power development. The thinking toward peaceful uses of nuclear energy brought with it the new problem of assigning liability caused by the industry. Without adequate insurance or financial coverage, utilities are reluctant to accept the risks.[7] The Price-Anderson Act was passed in 1957 to provide for a system of financial responsibility. As mentioned in the Act, it has a dual-purpose: 'to protect the public and to encourage the

4 The nuclear accident which happened on 26 April 1986 at the Chernobyl Nuclear Power Plant is the most serious nuclear accident so far, which lead to a wide spread of radioactive substances. It caused not only serious damage to human health, but also contamination of the aquatic and forest environment. See International Atomic Energy Agency 2006.

5 See Chapter 6, Section 2.

6 Energy Reorganization Act of 1974, Pub. L. No. 92-438, 42 U.S.C. §5801.

7 GAO, Analysis of the Price-Anderson Act, EMD-80-80, 1980, p. 3.

development of the atomic energy industry.'[8] The dual purpose can be seen in the provisions about limitation of liability and financial protection provided by the government.

A sole and exclusive federal cause of action for any property damage or personal injury from radiation exposure, the Public Liability Action (PLA), is established in the 1988 Amendment Act.[9] The term 'Public liability action' is defined broadly to contain 'any suit asserting public liability'. The Price-Anderson Act also defines the related concepts 'public liability' and 'nuclear incident' broadly.[10] Through those broad definitions, Congress preempted all state causes of action. There are two types of PLA causes of action: the cause of action for personal injury and that for property damage.[11] Hence the Price-Anderson Act does not intend to cover damage to the environment.[12]

As far as substantive rules are concerned, they 'shall be derived from the law of the State in which the nuclear incident involved occurs, unless such law is inconsistent with' the Price-Anderson Act.[13] Thus usually, the state tort law applies with the exception that the act has different prescriptions. The Act provides one such exception: if the nuclear accident comprises an 'extraordinary nuclear occurrence' (ENO), it requires the nuclear plant owner to waive certain defenses they might otherwise have under local tort law.[14] In another words, the operators are strictly liable if damage arises from an extraordinary nuclear occurrence. Whether an accident comprises an ENO is determined by the NRC. To be qualified as an ENO, a nuclear incident needs to satisfy the following criteria: the discharge or dispersal constitutes a substantial amount of source, special nuclear or byproduct material, or has caused substantial radiation levels offsite; and there have in fact been or will probably be substantial damages to people or property offsite.[15] The NRC regulations give detailed criteria in determining whether the above-mentioned conditions come due.[16] The legislation restricts the waivers to ENOs for two reasons: to avoid 'nuisance' suits and to protect the public from the consequences of catastrophic nuclear accidents.[17] However, the substantial criteria are very difficult to satisfy. Even the Three Mile Island accident was not classified as an ENO.[18] After the Three Mile Island accident, a petition has been proposed to the NRC which

8 Publ. L. No. 85-256, 68 Stat. 921.

9 Jose & Garza 2007, p. 13.

10 42 U.S.C. §2014(q)(w).

11 Jose & Garza 2007, p. 16.

12 However, this does not mean the compensation instruments established under the Price-Anderson Act have no relevance to environmental damage at all. The following discussion shows that case law has successfully explained the coverage clause of the insurance policy broadly to include clean-up costs.

13 42 U.S.C. §2014(h h).

14 42 U.S.C. §2210(n).

15 10 C.F.R. §140.81(b).

16 10 C.F.R. §140.84-140.85.

17 Tousley 1981, p. 617-618.

18 Metropolitan Edison Co., 11N.R.C.519 (1980).

requires the revision of the criteria of an ENO.[19] The NRC also indentified the difficulties in determining an ENO,[20] and published three proposed amendments to the criteria in 1985 and solicited public comment on these amendments.[21] Some scholars even advised to abolish the threshold of ENO to allow a better protection to victims.[22] However, after the procedure of public comments on the proposed amendments, the NRC decided that the current criteria for determining an ENO were adequate and were consistent with the intent of Congress, and thus should be sustained.[23] Under this situation, the majority of claims for nuclear damage may still not arise out of an ENO.[24] When an incident is not identified as an ENO, the claims are still subject to state tort laws. During the legislating the rules of waivers for an ENO, the Joint Committee on Atomic Energy has considered state law. It was expressed that, by then, there was no assurance that all State courts would impose a rule of strict liability in the event of a nuclear incident.[25] At that moment, there were some jurisdictions which purported to reject the doctrine of strict liability and a sizable number of others where the law relative to strict liability was unsettled.[26] The approach of minimal interfering with state law was kept in legislating on the waivers under an ENO. Thus many other issues are still subject to state tort law. Although the application of state tort law does not mean strict liability could not be used, there were at least some cases where the negligence rule or fault-based approach was still followed.[27]

Another feature of liability rules for nuclear damage arising from ENOs is that other statutes of limitations are waived as long as a suit is instituted within three years from the date on which the claimant first knew or reasonably could have known of his injury or damage and the cause thereof.[28] The requirement of within twenty years of the nuclear accident was abolished by the 1988 Amendment Act.[29] As far as nuclear damage is concerned, the damage may not manifest itself decades after the accident. Thus the relaxation of the statute of limitations can provide better protection to victims.

19 65 FR 61283-01, 2000 WL 1577628 (F.R.)

20 NRC 1983 Report.

21 50 FR 13978.

22 Tousley 1981, p. 609-630.

23 65 FR 61283-01, 2000 WL 1577628 (F.R.)

24 A. Roman *et al.*, *Canada and International Nuclear Liability*, available at: <www.millerthomson.com/assets/files/article_attachments/Canada_and_International_Nuclear_Liability.PDF>, last accessed on 8 May 2013.

25 Senate Report No. 89-1605, 89th Cong. 2nd sess. 1966.

26 *Ibid.*

27 Although under those cases, the courts disagree on the extent to which state tort law should be used or whether the federal law can preempt state tort law, they require the establishment of fault or identification of the fault-based standard of care. For example, *O'Connor v Commonwealth Edison Co.*, 13 F.3d 1090 (7 th Cir. 1994); re TMI Gen. Pub. Utils. Corp., 67 F.3d 1103 (3d Cir. 1995); *Merilyn Cook et al. v Rockwell International Corporation and the Dow Chemical Company*, 273 F. Supp. 2d 1175.

28 42 U.S.C. §2210(n).

29 Pub. L. 100-408, §§10.

Unlike the situation under the international regime, there is no legal channeling of liability to nuclear operators. Instead, the Price-Anderson Act creates a system of economic channeling.[30] The victims can still claim from other parties other than operators. But the financial protection clause has shifted the burden of compensation mainly to operators. For damage caused by the qualified operators of nuclear power plants, the total available amount of compensation is $12.2 billion since.[31] This amount is far higher than the amount of financial protection needed from contractors and other licensees determined in the Price-Anderson Act.[32] Moreover, the Price-Anderson Act provides an 'omnibus' coverage. It covers 'anyone liable' for 'public liability'. Thus victims may choose to claim from operators, the most identifiable and capable parties for their damages.

To encourage the development of the nuclear industry, the PAA exempts the licensees and contractors from unlimited liability. It adopts a limitation on aggregate public liability for a single nuclear incident. The amount of limitation is different among nuclear power plant owners with a capacity of 100,000 electrical kilowatts or more, contractors with the Department of Energy (DOE) and other licensees. This amount is closely linked to the amount of financial protection the PAA requires from those parties.[33] For a nuclear incident with damages in excess of the amount of aggregate public liability limitations, the Congress will 'take whatever action is determined to be necessary (including approval of appropriate compensation plans and appropriation of funds) to provide full and prompt compensation to the public.'[34]

To prevent nuclear damage does not only require actors to take care during the operation of nuclear facilities, but also asks for cautious behavior during the decommissioning of nuclear facilities and disposal and monitoring of nuclear waste. The licensees of byproduct material, production and utilization facilities as well as special nuclear material have the obligation to decommission their facilities for unrestricted use or for use under restricted conditions.[35] As far as nuclear fuel manufacturing facilities are concerned, since the federal government was responsible for the nuclear fuel production for several decades, the decontamination and decommissioning are done through the Uranium Enrichment Decontamination and Decommissioning Fund.[36] For the low-level radioactive waste, the States are responsible for the disposal.[37] The Federal Government is responsible for disposal of high-level radioactive waste and spent nuclear fuel underground in a deep geologic

30 Vanden Borre 2007, p. 300.
31 10 C.F.R. §140.11.
32 10 C.F.R. §140.11-12.
33 42 U.S.C. §2210(e)(1).
34 42 U.S.C. §2210(e)(2).
35 10 C.F.R. §30.35, 50.75, 70.38.
36 42 U.S.C. §2297.
37 42 U.S.C. §2021c.

repository. A Nuclear Waste Fund is established to bear the costs of carrying out the disposal activities.[38]

1.2. *The Requirement of Financial Protection*

Under the PAA, licensees and contractors are required to maintain certain levels of financial protection for the potential public liability. In addition to those third-party liabilities, the NRC regulation also requires power reactor licensees to provide financial assurance for their obligations, in the event of an accident, to stabilize and decontaminate the reactor and the reactor station site.[39] This section examines the financial requirements for third-party damage and property damage respectively.

1.2.1. Financial Requirement for Third-party Liability

Both nuclear reactor operators and contractors who conduct activities according to a contract with nuclear facilities may create risks of public liability. The PAA imposes different financial protection obligations to operators and contractors.

The PAA of 1957 required that the licensees of facilities having a rated capacity of 100,000 electrical kilowatts or more should provide financial protection up to the maximum amount available from private sources. The financial protection could be private insurance, private contractual indemnities, self insurance, other proof of financial responsibility, or a combination of such measures.[40] The most commonly used protection is private insurance, which was provided by a stock company, the Nuclear Energy Liability Insurance Association (NELIA) and a mutual pool, Mutual Atomic Energy Reinsurance Pool (MAERP). Those pools combined to provide coverage for nuclear liability up to $60 million.[41] Under the PAA of 1957, the government should agree to indemnify the licensees in case of an accident in excess of the required financial protection. The aggregated indemnity from each incident should not exceed $500 million.[42] This is because in the 1950s, the nuclear industry was still in its infancy and government wanted to encourage the development of the nuclear industry by exonerating the operators from the burden brought by potential catastrophes.

An important step which shifted the burden from government to operators was made in 1975. By then, the Joint Committee on Atomic Energy argued that the nuclear industry should have been able to bear the financial burden after years of development.[43] Although the total amount of compensation was kept the same as

38 42 U.S.C. §10222.

39 10 C.F.R. §50.54(w).

40 Act to Amend the Atomic Energy Act of 1954 (Price-Anderson), Pub. L. No. 85-256, 71 Stat. 576, 577 (1957).

41 McClure 1968, p. 256.

42 Act to Amend the Atomic Energy Act of 1954 (Price-Anderson), Pub. L. No. 85-256, 71 Stat. 576, 577 (1957).

43 Report No. 94-648, House of Representatives (1975); Senate Report No. 454, 99th Congress, 1sr Session (1975) 10.

that in 1957, a new tier of compensation, the retrospective premium, was introduced in place of the indemnity provided by the government. All operators should pay the retrospective premiums in the case of an incident in excess of the primary liability coverage up to the amount of $5 million.[44] The amount of primary financial protection and retrospective premiums has been increased over the years. Now the primary financial protection was set as $375 million, and the retrospective premium for each operator from each incident should not exceed $111.9 million since 2010.[45] Given the current number of operating reactors, the maximum compensation for a nuclear incident is up to $12.2 billion.

In addition to the power plant operators, the PAA and the NRC regulations also require other licensees and contractors with the DOE to provide financial protection. For example, the NRC requires the power plant operators at a lower thermal power level to provide a lower amount of financial protection from $1 million to $2.5 million according to their size. For the other licensees, the regulation provides a formula to calculate the necessary amount of protection, which would be in no event in excess of $4.5 million or $74 million. In addition, certain holders of construction permits and combined licenses, plutonium processing and fuel fabrication plants, and uranium enrichment facilities are also imposed an obligation to seek coverage of financial protection.[46] For an incident in excess of the required financial requirement, the licensees or other relevant persons can be indemnified by the NRC through the indemnification agreement with an amount up to $500 million.[47] As far as the contractors are concerned, the Secretary of Energy shall enter into an indemnification agreement with them. Under this agreement, the Secretary may require the contractor to provide and maintain financial protection. For the damage in excess of this amount, the Secretary shall provide indemnification up to $10 billion.[48] As a price of the indemnification agreement, the NRC or the Secretary can collect a fee from the licensees or contractors. The fee is set as $30 per year per thousand kilowatts of thermal energy capacity, subject to deduction by the NRC or the Secretary.[49]

For nuclear damage in excess of the aggregated liability discussed above, the Congress shall take 'whatever action is determined to be necessary' to provide full and prompt compensation.[50] The Secretary or the Commission shall make a survey of the causes and extent of damage, and submit a report to the Congress. The President shall submit an estimate of the aggregate dollar value of personal injuries and property damage, recommendations for additional sources of funds and compensation plans. The additional sources of funds include 'possible revenue

44 Pub. L. No. 94 -197, 1975 HR 8631, 89 Stat. 1111.
45 10 C.F.R. §140.11(4); 75 FR 16646, 2 April 2010.
46 10 C.F.R. §140.12-140.13b.
47 42 U.S.C. §2210(c).
48 42 U.S.C. §2210(d).
49 42 U.S.C. §2210(f).
50 42 U.S.C. §2210(e)(2).

measures on the sector of the economy, or on any other class, to which such revenue measures might be applied.'[51]

1.2.2. Financial Requirement for Property Damage

In addition to damage to third parties, nuclear incidents can also create serious on-site damages. The Three Mile Island accident shows that the onsite damage to the nuclear power plant can be more important than the third-party liability. The NRC thus requires the operator to maintain a minimum coverage for on-site property damage and gives a priority to decontamination.[52] The operators are asked to seek a 'coverage limit for each reactor station site of either $1.06 billion or whatever amount of insurance is generally available from private sources.' If the estimated costs of stabilization and decontamination exceed $100 million, the proceeds must be payable first for stabilization of the reactor and next for decontamination of the reactor and the reactor station site.[53] Accordingly, this rule requires a first-party insurance coverage for the property damage of the nuclear operators. The guarantee of funds for decontamination also provides protection to the potential victims and the environment.

1.2.3. Compensation under the Convention on Supplementary Compensation

In 2007, the Energy Independence and Security Act was passed.[54] It provides an instrument to implement the Convention on Supplementary Compensation for Nuclear Damage (CSC). The U.S. deposited this instrument of ratification of the CSC at the Vienna headquarters of the International Atomic Energy Agency in 2008.

Under the Energy Independence and Security Act, the U.S. Treasury will pay into the fund provided for by the CSC. But the costs can then be reimbursed. For a non PAA incident, costs can be reimbursed from a retrospective premium scheme the suppliers have to set up: 'each nuclear supplier shall participate in a retrospective risk pooling program ... to cover the contingent cost resulting from a covered incident outside the United States that is not a Price-Anderson incident.'[55] The contribution made by each supplier into the retrospective risk pool will be fixed according to a risk informed assessment formula determined by the Secretary of Energy.[56] In other words, in determining the contribution of each nuclear supplier, the particular risk it contributes to will be taken into account.[57] For the PAA

51 42 U.S.C. §2210(i).

52 Changes in Property Insurance Requirements of NRC Licensed Nuclear Power Plans, 49 Fed. Reg. 44645, 44649-50 (8 November 1984).

53 10 C.F.R. §50.54(w)(1),(4).

54 Energy Independence and Security Act of 2007, Pub. L. No. 110-140, 934.

55 Energy Independence and Security Act of 2007, Pub. L. No. 110-140, 934(e).

56 Energy Independence and Security Act of 2007, Pub. L. No. 110-140, § 934(e)(1), 121 Stat. 1492, 1744 (2007).

57 Faure & Vanden Borre 2008, p. 273-274.

incident, funds already available under the PAA should be used.[58] In this case, in spite that the country has an obligation to make public funds available under the CSC, it does not mean that taxpayers have to pay for it. This burden is actually shifted to nuclear suppliers (non-PAA incident) and operators (PAA incident). However, it is worth noting that the CSC has not come into force yet.

1.2.4. Financial Requirement for Decommissioning Nuclear Facilities and Financing for Nuclear Waste Disposal

Decommissioning the nuclear facilities and disposing of nuclear waste property can also help to prevent damage. Financial requirements in this stage can make assets available for decommissioning. For example, the NRC also requires the nuclear power reactor licensees to provide assurance that funds will be available for the decommissioning process (including decontamination). Prepayment, external sinking fund, a surety method, insurance or other guarantee method can be used to provide the assurance.[59] Similar requirements are also imposed on licensees of byproduct material, and special nuclear material.[60] Decommission is defined as 'means to remove a facility or site safely from service and reduce residual radioactivity to a level that permits release of the property for unrestricted use and termination of the license; or release of the property under restricted conditions and termination of the license.'[61] However, completing decommissioning does not mean the facilities will not produce any nuclear damage any more. This is because decommissioning activities do not include the removal and disposal of spent fuel. The Federal Government is responsible for disposing of the high-level radioactive waste and spent nuclear fuel underground in a deep geologic repository. Nuclear Waste Fund fees collected on nuclear power generation are set to fully recover the estimated costs of the disposing activities. However, until now, there is still no such operational repository. The GAO report shows that there may be an implicit subsidy in this program because the federal government has potentially assumed risks of cost overruns and schedule delays for the repository.[62] In other words, parts of the costs created by nuclear actors may still be externalized.

However, a full discussion of financial requirements and compensation instruments used in the nuclear industry is outside the scope here. The focus of this research is on the instruments used during the operation of nuclear facilities, especially by the operators of nuclear power reactors.

58 Energy Independence and Security Act of 2007, Pub. L. No. 110-140, 934 (a)(J).

59 10 C.F.R. §50.75.

60 10 C.F.R. §30.35, 70.38.

61 10. C.F.R. §20.1003.

62 GAO, Federal Electricity Subsidies: Information on Research Funding, Tax Expenditures, and Other Activities That Support Electricity Production, GAO-08-102, 2007, p. 55.

1.3. *The Instruments to Compensate for Nuclear Damage*

As discussed above, the PAA requires operators to provide primary financial protection up to the maximum amount available from private sources. The private sources which provide nuclear coverage are mainly insurance pools. Two nuclear insurance pools provided third-party liability coverage in earlier stages. Since 1998, American Nuclear Insurers (ANI) is the only remaining nuclear insurance pool.[63]ANI provides insurance coverage to both operators and suppliers. However, the on-site damage is not covered in ANI's policies. Instead, the property damage is covered by a mutual insurance company, Nuclear Electric Insurance Limited (NEIL). This section discusses liability insurance, the retrospective premiums scheme and property insurance respectively.

1.3.1. Nuclear Liability Insurance

1.3.1.1. Basic Content

ANI is the only provider of nuclear liability insurance in the U.S., which is organized as an unincorporated voluntary joint underwriting association comprised of 21 member companies.[64] It provides four types of liability insurance to either nuclear operators or suppliers. The Facility Form Policy is issued to owners or operators of nuclear facilities, which is used to cover the primary financial responsibility required by PAA. The losses that exceed the primary limit and need to be compensated through retrospective premiums are covered by the Secondary Financial Protection (SFP) Policy. The damage suffered by workers at nuclear facilities is excluded from the 'public liability' in the PAA. However, it is covered by the Facility Worker Form Policy provided by ANI. Moreover, ANI also provides a policy for other parties than operators: the Supplier's and Transporter's Policy. The Facility Form Policy provides an omnibus coverage: it covers not only the damage caused by power reactor operators, but also 'any other person or organization with respect to their legal responsibility for damages because of bodily injury, property damage or environmental damage caused by the nuclear energy hazard.'[65] Thus the liability of suppliers and transporters to the insured operators is also covered. But the coverage is limited to liability for the insured locations, and whether to purchase facility form insurance is at the discretion of the facility operator (he can also choose another instrument to provide financial security). Therefore not all the liability facing suppliers can be covered. Furthermore, when the suppliers and transporters do not provide services to power reactor facilities, there is not necessarily a requirement of financial protection from those facilities.[66] Thus the protection that

63 Faure & Vanden Borre 2008, p. 254.

64 See website of ANI: <www.amnucins.com/Membership.html>, last accessed on 8 May 2013.

65 See website of ANI: Need for Nuclear Liability Insurance: <www.amnucins.-com/library/Nuclear%20Liability%20in%20the%20US.pdf>, last accessed on 8 May 2013.

66 For example, the contractors with DOE can make an indemnity agreement with the government with the coverage of $12 billion. But the other facilities, such as the private →

suppliers and transporters can get from those facilities' financial coverage is much weaker. In those cases, the supplier's and transport's policy can provide additional protection. The SFP policy which provides coverage for the retrospective premiums is discussed in the following section. This section focuses on the Facility Form Policy.

The Facility Form Policy covers liability for bodily injury or off-site property damage caused by nuclear material at the defined location, or while in transit to or from the defined location. According to ANI, this policy is not designed to cover environmental cleanup costs. Off-site cleanup costs are excluded unless they result from clearly identifiable accidents. On-site property damage and on-site cleanup costs are specifically excluded too.[67] This may be because the Facility Form Policy is used to provide coverage for the primary financial requirement under the PAA. But liability for environmental damage is not explicitly covered under the 'public liability' under the PAA. This makes neither the PAA nor the Facility Form Policy closely linked with environmental damage. Furthermore, ANI worries that if it provides coverage to the environmental cleanup of primary site contamination, its ability to cover third-party claims in the event of a nuclear catastrophe would be jeopardized.[68] However, a recent case has interpreted the coverage broadly to cover on-site clean-up costs as well.[69] In this case, the insured parties were named as potential responsible parties under CERCLA by EPA. Thus they are asked to conduct a Remedial Investigation and Feasibility Study. Later the insured parties claimed from ANI for the response costs and the latter denied coverage on a number of grounds, among them that the Policy did not cover response costs.

According to the insurance policy, it should pay on behalf of the insured 'all sums which the insured shall become legally obligated to pay as damages because of ... property damage caused by the nuclear energy hazard'. The term 'property damage' was defined as:

> physical injury to or destruction or radioactive contamination of property, and loss of use of property so injured, destroyed, or contaminated, and loss of use of property while evacuated or withdrawn from use because possibly so contaminated or because of imminent danger of such contamination.[70]

According to the judge, a reasonable insured could not conclude from this language that the words 'destruction' or 'radioactive contamination' of property would not include environmental harms. The policy contains an endorsement which precludes environmental cleanup costs and on-site cleanup costs.[71] However, the court precludes the use of this endorsement according to state law since the issuance of the endorsement represented a reduction in coverage and does not satisfy the

research reactors, nuclear fuel fabricators, and university reactors are not asked to provide financial protection or can get the indemnity agreement up to a much lower amount.

67 <www.amnucins.com/Insurance.html>, last accessed on 8 May 2013.

68 *Whittaker Corp. v American Nuclear Insurers*, 671 F. Supp. 2d 242 (D. Maas. 2009).

69 *Ibid.*

70 *Ibid.*

71 *Ibid.*

procedural requirement.[72] By holding the pollution exclusion endorsement unenforceable, this case results in environmental damage being covered by the insurance policy.

The Facility Form Policy provides an omnibus coverage. Not only the operators who buy the policy but also any other person or organization (except the U.S. government and its agencies) who contributes to the insured risks are defined as 'insured'. ANI contends that all operating power reactors have satisfied their financial protection requirements by purchasing liability insurance from it.[73] Therefore, the liability of any party related to the insured facilities is also covered under this policy. Although different from the legal channeling under the international regime, the victims can still claim from other parties theoretically, the burden of compensation for nuclear damage is practically shifted to the operators.

Nuclear damage may have a long tail nature. A long time lapse may exist between the nuclear incident or exposure and the manifestation of damage (for example the diseases). Under this situation, which date is the trigger of the insurance policy is an important question. According to the ANI policy, it covers

> bodily injury or property damage (1) which is caused during the policy period by the nuclear energy hazard and (2) which is discovered and for which written claim is made against the insured, not later than two years after the end of the policy period.[74]

The term 'caused during the policy period' is explained by the court as an occurrence trigger. It means that coverage is provided if any of the three events: exposure, progression, or manifestation takes place during the policy period.[75] This interpretation is advantageous to victims since they are covered as long as one stage of the damage happens during the policy period. However, considering the long tail nature of the damage, the occurrence coverage may make nuclear risks more unpredictable to insurers. The occurrence policy has created many problems in insurance for liability under CERCLA in the 1980s. However, it is less of a problem as far as nuclear insurance is concerned. The Facility Form policies are written on a continuous basis and contain no expiration date. They remain in force until cancelled by the policyholder of ANI. Furthermore, ANI is the only insurer in the U.S. that provides nuclear liability insurance. Thus the trigger of policy will not lead to such a big problem in allocating the burden among different insurers. But the problem can still arise when the policy is cancelled. Once the policy is cancelled, the claims need to be made within 10 years.[76] Given the long tail nature of nuclear

72 *Ibid.*

73 See website of ANI, Need for Nuclear Liability Insurance: <www.-amnucins.com/library/Nuclear%20Liability%20in%20the%20US.pdf>, last accessed on 8 May 2013.

74 *Ibid.*

75 *Babcock & Wilcox Co. v American Nuclear Insurers*, 51 Pa. D.7 C.4TH 353 (Pa. Com.Pl. 2001).

76 See website of ANI, Need for Nuclear Liability Insurance: <www.-amnucins.com/library/Nuclear%20Liability%20in%20the%20US.pdf>, last accessed on 8 May 2013.

damage, the damage which happened during the policy period may still be uncompensated.

The coverage of primary nuclear liability insurance was set as $60 million when the PAA was passed. The catastrophic nature of nuclear damage makes its liability excluded from the traditional liability insurance policy. To cover this potential catastrophic damage, many insurance companies pool together to secure a large amount of capacity. Given the lack of experience and data about the risks, the insurers said that the coverage was based on the companies' willingness to invest in nuclear insurance rather than on any actuarial bases in the earlier stage.[77] The coverage has grown through the years to $375 million. To provide this coverage, the insurance companies do not only pool together but also use the re-insurance market. In 1998, only 31.1 per cent of the liability exposure under each policy was retained by ANI's members and 68.9 per cent was ceded to reinsurers around the world.[78]

The Facility Form Policy is capped at $375 million. Different from a conventional insurance policy, which is capped either by each accident or the aggregated limit in a certain policy year, this nuclear insurance is subject to a single lifetime aggregate limit.[79] The defense costs are also included within the policy limit. All indemnity payments and claimed expenses made under an ANI domestic liability policy reduce the policy's limit of liability. The limit can be reinstated if agreed by ANI.[80]

1.3.1.2. Risk Differentiation

The primary nuclear liability insurance in the U.S. is not compulsory. However, all the eligible operators are required to seek primary financial protection up to the amount available from private resources. ANI is the main party and the only insurer which provides nuclear liability insurance. Thus there is less of a problem of adverse selection. However, the problem of moral hazard still exists. To control moral hazard, the insurer needs to differentiate the risks of different operators and to set the responding premiums or make deductibles.

The Facility Form Policy provides coverage up to $375 million. It is written without deductibles or self-insured retention programs.[81] In this case, the preventive incentives cannot be achieved through retaining some risks. The main method that can be used to control moral hazard is risk differentiation. ANI operates on a monopolistic basis. Law and Economics literature shows that incentives for insurers to differentiate risks are reduced in a monopolistic insurance market.[82] However, the monopolist may still practice some extent of risk differentiation since it can raise

77 GAO 1980, p. 8.

78 NRC: NRC's 1998 Report to Congress on the Price-Anderson Act, SECY-98-160, 1998.

79 See ANI 2006, Insuring the Future of Nuclear Power Worldwide, 6, available at: <www.nuclearinsurance.com/library/Ani%20Brochure.pdf>, last accessed on 8 May 2013.

80 Jones 2009, p. 71.

81 *Ibid.*

82 See Faure & Van den Bergh 2002, p. 282-291.

profits by risk differentiation.[83] Both feature rating and experience rating can be used to differentiate risks and set premiums. Historically, the premium of ANI policies was based on five components: reactor type, use, size, location and type of containment. Since 1981, a sixth component, the operational performance of individual reactor plants, has been taken into account in deciding premiums. This is made through the so-called ANI Engineering Rating Factor (ERF), which is based on the operators' performance on environmental releases, regulatory performance, significant events, radioactive waste, safety system failures, unplanned automatic scrams, safety system actuations and ANI liability recommendations. According to the performance, the operators are eligible to receive a credit of up to 20 per cent or a surcharge of up to 30 per cent.[84] In this sixth component scheme, both the feature of nuclear operators and the experience of their past performance which may influence the nuclear risks are considered in setting premiums.

To control the risks created by operators, ANI can also execute onsite inspections. The inspection is a supplement to that conducted by the NRC and INPO. They communicate together to minimize redundant efforts.[85] The ERF program also gives incentives to operators to control risks. In addition, the Industry Credit Rating Plan program also creates such incentives. Under this program, approximately 75 per cent of each insured's liability premium is set aside in a reserve fund of which the purpose is to pay loss or loss expenses. The reserve premiums can be returned to policyholders in ten years based on historical losses experience.[86] In that case, the operators are given additional incentives to increase care level.

1.3.2. The Retrospective Premiums Scheme

A retrospective premiums scheme was introduced in 1975. Initially, the financial requirement was satisfied with the coverage from private insurance and a government indemnity agreement. This was because it was regarded that nuclear industry was not capable of bearing all the burdens at its infancy period and the PAA intended to encourage the development of the nuclear industry. However, after years of development, it was believed that the industry should shoulder its responsibilities.[87] This was achieved by phasing out federal indemnity and establishing a system of retrospective premiums paid by nuclear operators.

Under the retrospective premiums scheme, if an accident leads to the damage in excess of $375 million, all qualified nuclear operators are obliged to pay the retrospective premiums up to $111.9 million. This arrangement is a form of a risk-sharing agreement. However, different from the usual understanding about risk-

83 Faure & Vanden Borre 2008, p. 270.

84 W. Wendland, 2002, *Effluent Release Performance Indicators- An Insurer's Perspective, Presentation at 12th Annual RETS/REMP Workshop*, available at ANI's website: <www.amnucins.com/library/rets-rempr.pdf>, last accessed on 8 May 2013.

85 <www.nuclearinsurance.com/Loss%20Control.html>, last accessed on 8 May 2013.

86 See ANI 2006.

87 H.R.Rep.No. 94-648, 10.

sharing agreements, this arrangement is a mandatory system. Rather than the voluntary pooling of operators, the retrospective premiums scheme is established according to the legislative requirement of the amended PAA. This arrangement ensures that all nuclear power plant operators participate in the system and provide strong capacity. Furthermore, this system is used only when the primary instrument, the insurance market, fails to cover the full damage. In other words, the retrospective premiums scheme provides an upper layer of compensation.

The retrospective premiums are collected only when damage with a magnitude in excess of the primary coverage happens. The premiums can be charged over the years with an annual amount within $17.5 million per incident.[88] Since the obligation of payment only comes due after the occurrence of damage, a special arrangement is needed to ensure that the operators can pay their retrospective premiums in the case of damage. According to NRC regulations, the operators need to provide one of the following guarantees: surety bond, letter of credit, revolving credit/term loan arrangement, maintenance of escrow deposits of government securities, annul certified financial statement and other type of guarantee approved by NRC.[89] Since the late 1990s, the bond for payment of retrospective premiums an operator made with ANI was accepted as a guarantee complying with the NRC regulations. The bond is a contractual arrangement between ANI and operators (SFP policy), under which ANI will collect retrospective premiums in the event that claims exhaust primary coverage. If the operator fails to pay the retrospective premium, ANI would pay for up to $30 million in one year of the premiums and collect it later from the defaulting operator.[90] According to NRC regulations, if an operator fails to pay the retrospective premium, NRC reserves the right to pay those premiums on behalf of the licensee and recover the amount from the operator.[91] After the deregulation of the electricity industry, 31 out of the 103 operating commercial nuclear power plants in the U.S. were limited liability companies till 2004.[92] The NRC does not differentiate the limited liability companies from the others and does not conduct in-depth financial reviews to determine the operators' ability to pay retrospective premiums. However, ANI goes further than the NRC. It requires limited liability companies to provide a letter of guarantee from their parent or other affiliated companies with sufficient assets to cover retrospective premiums as a condition to issue the bond for payment of retrospective premiums.[93]

The existing SFP policy is a bonding made between ANI and operators. Although ANI has a contingent liability if retrospective premiums are not paid when due, its payment will be reimbursed from operators. Thus the burden of secondary financial protection still lies with the nuclear operators. However, ANI also shows its interest in developing a system on a guaranteed basis: with the

88 10 C.F.R. §140.11(4).

89 10 C.F.R. §140.21.

90 GAO 2004b, p. 8.

91 *Ibid.*

92 *Ibid.*, p. 6.

93 *Ibid.*, p. 9.

payment of an additional per reactor premium, ANI would cover the secondary financial protection without reimbursement afterwards from operators.[94] This coverage shifts the burden of payment after a severe nuclear accident from the industry to the insurance industry.

1.3.3. The Mutual Pool for Property Damage

NRC regulations require nuclear power plant operators to obtain insurance or demonstrate their financial status to cover property damage, especially to stabilize and decontaminate reactors up to an amount of $1.06 billion. The operators satisfy this requirement though a mutual insurance system, Nuclear Electric Insurance Limited (NEIL). The reason for creating this mutual insurance was to provide nuclear operators with an alternative to commercial insurance.[95]

NEIL provides two types of policy: on-site accidental property damage and decontamination coverage and accidental outage coverage.[96] This research focuses on the internalization of the costs created by nuclear risks, especially ecological damage. On-site property damage policy covers the costs to decontaminate and stabilize the reactor. Thus the focus of the research is the former policy.

On-site accidental property damage policy contains two programs: the primary program with the limit of $500 million and the excess program with a limit of $2.25 billion.[97] NEIL provides coverage different from ANI and is re-insured by the latter. The primary program provides coverage to expenses incurred in connection with the stabilization and decontamination of the reactor and direct physical damage to, or destruction of, the insured property as a result of an accident.[98] The excess program provides coverage to both property damage and decommissioning liability. The payment of expenses under the excess program is made in the following order: losses under nuclear liability coverage, losses under decommissioning liability coverage, losses under debris removal and decontamination coverage, property damage coverage and functional total loss coverage.[99] The property damage policies cover property damage caused by an 'accident'. The term 'accident' is defined broadly to incorporate more than nuclear related incidents: 'a sudden and fortuitous event, and event of the moment, which happens by chance, is unexpected, and unforeseeable.' However, 'it does not

94 Testimony of John L. Quattrocchi, Senior Vice President, Underwriting American Nuclear Insurers Before The United States Senate Energy and Natural Resources Committee, May 24, 2001, available at: <www.amnucins.com/library/johnQ5-01testimony.pdf>, last accessed on 8 May 2013.

95 Faure & Vanden Borre 2008, p. 255.

96 2009 Annual Report of NEIL, 20. available at: <www.nmlneil.com/media/1724/ar2009.pdf>, last accessed on 8 May 2013.

97 *Ibid.*

98 J. Gittus, 2006, *Introducing Nuclear Power to Australia: An Economic Comparison, A Report Prepared for the Australian Nuclear Science and Technology Organization*, 226, available at: <www.ansto.gov.au/__data/assets/pdf_file/0016/12445/nuclear_options_paper_Gittus_complete.pdf>, last accessed on 8 May 2013.

99 *Ibid.*, p. 233-234.

include any condition which develops, progresses, or changes over time, or which is inevitable.' In this case, damage that occurs over time, such as erosion, corrosion, cracking and gradual accumulation of radioactive contamination are excluded.[100] The property damage policy is related to ecological damage, since it covers on-site decontamination and cleanup, debris removal and disposal, damage repair and/or site restoration. However, NEIL believes that the likelihood of such an event that results in payment of those aspects is very remote, and the more likely insurance risk involves equipment failures that do not challenge nuclear safety.[101]

The premium is based on the amount of the coverage, the deductible and various rating criteria. One of the missions of NEIL is to insure Members' nuclear risks by 'providing value and equitable treatment'.[102] Thus Members are provided with the same coverage, subject to the same loss control standards and requirements.[103] But the policy allows deductibles. The NEIL policy is linked with the self-regulation under INPO. In the event that the site is determined by INPO as not meeting the industry standard of acceptable performance, the premium can be increased up to 25 per cent.[104] NEIL can make a call to the member insured for payment of a retrospective premium up to 10 times the annual premium to cover losses incurred by the company during the policy year.[105]

NEIL also developed methods to evaluate risks and control losses. It first assumes that each insured plant creates the same level of risk and minor differences are reflected through premium adjustment.[106] Insurance evaluations focus on balance-of-plant or non-nuclear safety areas. For the nuclear safety issues, NEIL relies on work done by regulators, such as IAEA, the World Association of Nuclear Operators (WANO) and INPO.[107] NEIL Members' Engineering Advisory Committee developed loss-control standards.

1.4. *Critical Analysis*

The compensation system for nuclear damage in the U.S. provides an interesting example which allows the availability of a large amount of compensation. In such a system, regulation, liability and compensation instruments are combined to prevent and compensate for nuclear damage. After the description of the existing system, this section analyzes the system according to the theoretical findings of Part I.

100 Wilks 2002, p. 331.

101 *Ibid.*, p. 6.

102 See website of NEIL, <www.nmlneil.com/about-us/company-information/about-neil-overview.aspx>, last accessed on 8 May 2013.

103 G. Wilks, 2009, *Managing Nuclear Risks: Nuclear Property Insurance, a presentation at Managing Nuclear Risks, Win 2009 Global Conference*, 22, available at: <www.winus.org/-Portals/0/2009win/5_Thursday.pdf>, last accessed on 8 May 2013.

104 Gittus 2006, p. 231.

105 *Ibid.*

106 Wilks 2009, p. 25.

107 Wilks 2001, p. 7, 8.

1.4.1. Combined Use of Regulation, Liability and Compensation Instruments

In the U.S., regulation, liability and compensation instruments are used together to tackle nuclear risks. A regulatory scheme is established for the whole process of nuclear activities: from the mining phase, to radioactive substances and nuclear fuel, nuclear installations, radioactive waste, export and transportation. Many regulatory and supervisory authorities, as well as semi-public agencies, are involved in the regulatory framework. The involvement of multiple actors in the whole regulatory process ensures that at least a certain level of nuclear security can be achieved and some nuclear damage can be prevented.

However, the nuclear sector involves many highly technical issues, which make the regulators sometimes lack sufficient information to prevent nuclear damage at an optimal level. In that case, the use of liability rules can be desirable. Liability can give nuclear actors additional incentives to take care if the regulatory standard is less than the socially optimal care levels.[108] In such a highly technical sector, the actors themselves sometimes have better information than regulators. Therefore a system of strict liability is beneficial in the sense that it gives actors incentives to take care tailored to their own conditions, especially in the dimensions that are unregulated. Under the PAA, a strict liability system is established for ENOs. However, as criticized by some scholars, the conditions to determine ENOs are so strict that very little damage can be traced to an ENO.[109] Thus in many cases, the state tort law can still be used. The consistent application of strict liability is not guaranteed. There are at least some cases where the negligence rules may be used, and where hence the care level followed by operators needs to be examined.

Under the PAA, a so-called scheme of economic channeling is established. The victims can still claim from the parties other than operators who actually contribute to the damage. However, the PAA requires power plant operators to provide financial protection up to $12.2 billion, which is far more than the amount required from other parties, if there is such a requirement at all. Thus the major burden is still borne by operators. This arrangement is more desirable than a system of legal channeling, under which the actual damage contributors rather than operators are not exposed to risks at all and thus may choose to externalize the nuclear damage.

It is worth noting that a system of limited liability is adopted under the PAA. For the operators of nuclear power plants, they have to make available primary financial protection of $375 million and the retrospective premium is $119.9 million per operator per accident. This results in a total available amount of compensation of $12.2 billion. However, this amount may still be insufficient to cover a catastrophic nuclear accident. Many scholars thus argued that this constitutes a subsidy to the nuclear industry.[110] Whether and to what extent the PAA provides a subsidy is analyzed specifically in the following section.

108 Trebilcock & Winter 1997, p. 226.

109 NRC 1983.

110 For example, Dubin & Rothwell 1990, p. 73-79; Heyes & Liston-Heyes 1998, p. 122-124.

The PAA stipulates the liability that occurs during the operation of nuclear power reactors. Since nuclear damage may turn out to be catastrophic, it can make the operators insolvent in case of such a catastrophe. Under this situation, compensation instruments, such as insurance and risk-sharing pools can be used to alleviate the insolvency problem and promote compensation.

1.4.2. Is there a Subsidy for the Nuclear Sector?

Under the PAA, the liability of the nuclear power plant operators and DOE contractors are limited to $12.2 billion. The other nuclear actors are subject to a much lower limit. A fierce debate has been taking place whether the limited liability under the PAA constitutes a subsidy to the nuclear industry. Some scholars argue that the PAA does not confer an actual subsidy, since no accident requiring payouts, beyond the PAA limit, actually happened.[111] However, even without such an accident, the nuclear industry can still benefit from limited liability. Without such protection, the nuclear industry will have to either bear the costs itself in case of such an accident or purchase full liability insurance. Many efforts have been made to estimate the scope of the subsidy.[112] The estimates vary from $403,000 ($317,000) per reactor per year (2001) to $32 million per reactor per year (2001). Although the existing studies differ in the extent of the subsidy created by the limited liability under the PAA, and the methodology they used is open to criticism,[113] many of the researches were carried out when the required financial security from operators were much lower, and even the government admits that no credible quantification of this value is available,[114] it is widely accepted that a substantial subsidy to the nuclear industry does exist.[115] This subsidy will, on the one hand, make the nuclear industry adopt an inefficiently low care level and, on the other hand, give the nuclear industry an advantageous competitive status and encourage the excessive installation of nuclear capacity.[116]

Arguments are also made that despite that limited liability does not expose the nuclear industry fully to the risks it creates, the PAA provides a second best solution. It is argued that without the PAA, the existing large amount of compensation will not be available from the individual nuclear operators.[117] Furthermore, the nuclear operators, suppliers and other related actors will not be

111 Zycher 1992; G. Rothwell, 2002, *Does the U.S. Subsidize Nuclear Power Insurance?*, *Stanford Institute for Economic Policy Research*, available at: <www.stanford.edu/group/siepr/cgi-bin/siepr/?q=system/files/shared/pubs/papers/briefs/policybriefjan02.pdf>, last accessed on 8 May 2013.

112 US Department of Energy, Energy Information Administration 1992, p. 77; Dubin & Rothwell 1990; Heyes & Liston-Heyes 1998.

113 See Rothwell 2002.

114 GAO 2007, p. 53.

115 Pacific Northwest National Laboratory, An Analysis of Federal Incentives Used to Stimulate Energy Production, prepared for U.S. Department of Energy, 1978, p. 116; NRC 1983; Berkovitz 1989, p. 1-68; U.S. Department of Energy, Energy Information Administration 1992.

116 Heyes & Liston-Heyes 2000, p. 197.

117 Rothwell 2002.

able to or be willing to participate in the nuclear electricity industry.[118] However, this is not accepted as a sound argument. Without the subsidy, the full insurance costs can be passed on to the consumer in higher electricity prices, then 'the utilities would have to decide whether nuclear power could be competitive and profitable in relation to other energy sources.'[119] The question then is whether a system of unlimited liability is feasible or can remove the subsidy in practice. If the financial protection under the PAA works together with unlimited liability, then the operator is still liable if that damage in excess of the financial protection required under the PAA happens. However, for nuclear operators, especially the limited liability companies, the main asset is the nuclear installation, which will be damaged in the event of a catastrophic accident. In that case, theoretical unlimited liability cannot guarantee that damage can be compensated in practice. The costs will still be borne by government or by the victims themselves.

One proposal has been made to introduce a system of unlimited liability with financial guarantees. Under the existing PAA regime, there are aggregated limits and annual limits on retrospective premiums. In the 1983 NRC report, an alternative was provided. The annual limit of retrospective premiums still applies. However, these premiums have to be paid until all public liability has been satisfied.[120] In such a system, on the one hand, the compensation is paid out of the pocket of the nuclear industry and, on the other hand, the limit on annual premiums makes this burden not too stressful for the industry. But problems may still exist, for example, the question can arise whether the operators will be able to pay the premiums and how they can guarantee their capacity. This proposal at least provides an alternative and makes an effort to cover the full liability by the nuclear industry. Congressional hearings on this report were first held in 1984 in the 98th Congress.[121] Because of the limited time remaining in that session, no legislative action was taken by Congress.[122] Various proposals aiming at eliminating the limitation of liability have been discussed in the 99th and 100th Congress.[123] Those proposals were strenuously opposed by the nuclear industry. After a long debate, the President signed into the PAA of 1988, under which the unlimited liability system was not adopted and instead, it increased the maximum amount of compensation by nearly ten-fold.

118 Testimony of Ralph Nader before the Committee on the Budget U.S. House of Representatives, 30 June 1999.

119 Pacific Northwest National Laboratory 1978.

120 NRC 1983.

121 Amendments to the Price-Anderson Act of 1954: Hearing on H.R.3277, and the Nuclear Regulatory Commission and the Department of Energy Reports on the Price-Anderson Act of 1954 Before the Subcomm. on Energy and the Environmental of the House Comm. On Interior and Insular Affairs, 98th Cong., 2d Sess. (1984).

122 Berkovitz 1989, p. 19.

123 For example, Senator Robert Stafford introduced a bill (Senate bill 1761) that embodied the system proposed in the NRC report in the 99th Congress. (S.1761, 99th Cong. 1st Sess., 131Cong. Rec. S 13, 127) in the 100th House discussions, Representative Dennis Eckart also offered a similar proposal in line with the 1983 NRC report. (133 Cong. Rec. H 6769-81.) See also Senate bill 44, and Senate bill 843 at the 100th Congress. (S. 44, 100th Cong. 1st Sess. 133 Cong. Rec. S 289; S. 843, 100th Cong., 1st Sess., 133Cong. Rec. s 3878.)

How a compensation system can be established depends not only on the economic criteria to prevent and compensate nuclear damage optimally, but also on the weighing and compromising of different interests.

In addition to the limited liability under the PAA, other factors can also reinforce the subsidy effect. For example, though limited liability is also adopted in some other areas, such as oil pollution, the limited liability does not apply at least in some situations, for example in case of from gross negligence or willful misconduct.[124] However, the limit on liability in the PAA is absolute. This can aggravate the distortive effect created by limited liability.

For nuclear actors other than power plant operators, the subsidy may be larger. They are required to provide much less financial protection. The government makes indemnification agreements with DOE contractors. The indemnification is provided at a rather low price,[125] and even in case that damage arises from gross negligence and willful misconduct.[126] In this case, the other actors have even less incentives to adopt optimal care levels.

The above analysis is based on the liability arising during the operation of nuclear facilities. However, liability and subsidies can also exist in other activities and stages. For example, the 2007 GAO report shows that subsidies also exist in decontaminating and decommissioning of nuclear fuel manufacturing facilities, as well as in the construction and operation of a geologic repository for the disposal of radioactive waste.[127]

Quantitative and in-depth analysis of the subsidy given to the nuclear industry, especially during other periods of nuclear activities rather than those covered by the PAA is still lacking. To internalize the costs caused in all process of nuclear activities can give incentives to the nuclear industry to adopt optimal care and activity levels (efficient installation capacity). Whether such a system can be established depends to a large extent on political willingness and feasibility. Moreover, stringent liability rules and financial requirements may create negative incentives sometimes. For example, the existing nuclear operators may be unable to decommission their facilities. Lots of resources will be spent by operators or insurers on defending the claims. Defending claims fits the interests of insurers and operators since it avoids numerous future claims and payments. However it constitutes a large amount of social costs. For example, between 1957 and 2008, 229 claims were defended by ANI, which lead to $64.4 million to be paid for damage.

124 For example, under OPA, the liability of the responsible party for an offshore facility is capped at 'all removal costs plus $75 million', and that for on-shore facilities and deepwater ports is limited to $350 million. However, the responsible party will lose the right to limit its liability if the incident is caused by gross negligence or willful misconduct or the violation of an applicable Federal safety, construction or operating regulation. See Wang & Faure 2010, p. 6-8.

125 42 U.S.C.A. §2210(f).

126 Opinion of Anna Aurilio, Hearing at United States Senate Committee on Energy & Natural Resources on 24 May 2001.

127 GAO 2007, p. 53-55.

However, $243 million has been spent on defending costs and other expenses, far more than the damage being paid.[128]

1.4.3. Compulsory Financial Requirements

In the U.S., nuclear operators and some other actors are required to provide financial protection under the PAA. For example, the qualified nuclear power plant operators are required to provide primary financial protection of the amount available from private sources and a second layer of retrospective premiums up to $111.9 million. The PAA also allows the NRC to assess annual retrospective premiums less than the requirement under the PAA according some specified criteria.[129] For the other licensees and contractors, the PAA authorizes the NRC to determine the amount of financial requirements. However, the amount of indemnification provided by the government is stipulated in the PAA.[130]

This compulsory scheme makes sure that at least some assets will be available in case of nuclear damage. Thus a certain level of costs internalization and preventive incentives can be achieved. In the infancy of the nuclear industry, it might be difficult for operators and other nuclear actors to satisfy their financial requirement by resorting to private sources. Therefore the 1957 PAA provided a scheme under which the government indemnifies the operators and other actors substantially. However, after decades of development, it was believed that the nuclear industry should be able to bear this burden itself, and the capacity of the insurance market was also improved. In that case, a retrospective premium system has replaced the indemnification agreement for nuclear operators, and the amount of the financial requirement has increased substantially.

The amount of financial protection is to a large extent determined by the PAA. For example, the amount needed for retrospective premiums and the indemnification agreement is prescribed in legislation. The regulatory authority, the NRC has also some discretion to shape the financial protection requirement. The annual retrospective premium is subject to reduction by the NRC. For other actors, the NRC has more discretion. According to the NRC regulations, the amount of financial requirement for other actors can be stipulated in regulations, determined based on some specified methodology or made on a case-by-case basis.[131] This is in line with the economic analysis: stipulating the ceiling of the financial requirement in legislation creates predictability and avoids the potential lobby influence to

128 M. Berger, 2009, *Managing Nuclear Risks in the United States, a presentation at INLA Inter Jura Congress 2009*, 8, available at ANI's website: <www.amnucins.com/library/INLA-2009.pdf>, last accessed on 8 May 2013.

129 42 U.S.C.A. §2210(b)(2).

130 42 U.S.C.A. §2210(c),(d).

131 For example, for nuclear reactors which operates at a thermal power level not exceeding 11,000 electrical kilowatts, the amount of financial protection is set in NRC regulations. 10 C.F.R. §140.11. For other reactors, the amount should be determined according to the formula and other provision of the NRC regulations. 10 C.F.R. §140.12. NRC regulations authorize NRC to determine the amount of liability insurance required for uranium enrichment facilities according to specific situations. 10 C.F.R. §140.13b.

public authorities. At the same time, the NRC also has some discretion in determining the ceiling. Thus more risk differentiation can be achieved. In other words, the amount of the financial requirement can be tailored to the specific situation of the nuclear actors.

1.4.4. Multiple Layers of Compensation Instruments

This section focuses on compensation instruments used during the operation of nuclear power reactors. Different from damage caused by diffuse pollution, such as air pollution, usually, nuclear damage can be traced to specific nuclear actors or activities. The economic analysis in the theoretical part shows that in case of identifiable liable parties, there is a preference for liability insurance. When liability insurance is not available, some alternatives such as risk-sharing agreements may be useful. The compensation instruments used by nuclear operators in the U.S. correspond with these theoretical criteria. In the U.S., two main compensation instruments available for nuclear operators are liability insurance and the risk-sharing agreements.

In the U.S. regime, third-party damage resulting from nuclear incidents is treated differently from property damage (including the contamination of reactors and sites) arising from the same incident. For damage to the third parties, a multi-layered compensation system is established. The primary instrument is an insurance policy provided by ANI. ANI is the only insurer that provides third-party nuclear insurance in the U.S. It is an insurance association with the cooperation of many insurance companies. To provide insurance coverage for the low frequency high consequence damage, insurers have to work together to increase their capacity. However, what amount of insurance coverage is available depends to a large extent on the willingness of insurers to invest in nuclear insurance. Furthermore, even with the cooperation of insurers, the capacity of the insurance market is still limited. In that case, an upper layer of compensation is necessary. This was initially provided by government indemnification. With the development of the nuclear industry, the burden has been shifted from the government (taxpayers) to the nuclear industry itself. A risk-sharing pool (retrospective premium system) provides the second layer of compensation. The use of a risk-sharing pool also corresponds with theoretical criteria. Nuclear damage is a type of damage with a low frequency and high magnitude. The unpredictability makes the risk-sharing pool an attractive alternative to insurance, since *ex ante* payment of contributions (premiums) is no longer necessary. Furthermore, existing regulation exposes the nuclear operators to similar levels of risks. Moreover, in such a highly technical sector, operators themselves may have more information than insurers. It is worth noting that the risk-sharing pool is also operated on the basis of ANI. In case of damage in excess of the insurance coverage, ANI will collect retrospective premiums from nuclear operators. If the operator fails to pay for retrospective premiums, ANI can pay it first and get reimbursement from the operator. This design is desirable in the sense of providing a smooth link between two layers of compensation. In case of a severe nuclear incident, lots of claims may be filed over time and dealing with those claims is also time consuming. It takes some time to determine whether the limit of primary insurance coverage is exceeded and a delay in compensation may happen.[132] ANI provides primary insurance and collects the retrospective premiums. Therefore the problem of linking the two layers is avoided. In case of a catastrophic nuclear damage, even the compensation provided by retrospective premiums may be insufficient. Then the third layer of compensation can be used. According to the PAA, in that case, the Congress will take whatever action is determined to be necessary to provide compensation.[133] Since the operators are protected by limited liability, it is not very likely that the compensation will be paid out of the pocket of the polluter. The costs of damage may then be externalized.

The property damage is insured by a risk-sharing pool: NEIL. ANI covers liability arising under the PAA and the property damage is excluded. NRC regulations require operators to provide a financial guarantee for the potential decontamination and stabilization costs. The risk-sharing pool, NEIL, is used to satisfy this rule. The advantages of the risk-sharing pool in dealing with nuclear liability also apply to the property damage suffered by nuclear operators.

1.4.5. Use of Insurance and Risk-sharing Pools

Insurance and risk-sharing pools are used as compensation instruments to cover nuclear operators' liability and property damage. This section analyzes these two instruments respectively.

Liability insurance for nuclear operators is exclusively provided by ANI, an association with the cooperation of many insurance companies. On the one hand, there is a monopoly in the insurance market, which may dilute the incentives of risk differentiation and lead to high prices. On the other hand, to cover the potential catastrophic nuclear damage, insurers need to cooperate together to ensure the capacity. In spite of the fact that there is still insufficient experience and data to support accurate actuarial premium calculation, ANI has identified a series of factors in determining the premiums. Not only the feature of the reactors, such as the size, location, and type of reactors, but also the past performance in the sphere of environmental releases, regulatory performance, and significant events, and so on, are considered in premium setting. Those rating methods give operators preventive incentives.

A risk-sharing pool is also used to cover the operators' liability and property damage. From a deterrence perspective, a risk-sharing pool is preferable to insurance in the sense that operators have more information than insurers to make an effective risk differentiation or a risk-sharing pool allows better mutual monitoring.[134] As far as risk differentiation is concerned, a pool for liability is operated through ANI, with the same amount of premiums from different operators. In that case, there is no risk differentiation among different operators. However, regulation and self-regulation can expose all operators to similar levels of

132 Pelzer 2007.

133 42 U.S.C.A. §2210(e)(2).

134 Faure & Grimeaud 2003, p. 234.

risks. Moreover, the substantial amount of potential retrospective premiums may make some operators insolvent and increase the risks of others. Thus the operators have incentives to monitor each others' solvency.[135] In that case, the risk-sharing pool can still operate efficiently. The pool for property damage is operated through NEIL, which can do on-site inspection, control losses and adapt the premiums according to each operator's risk. It is worth noting that NEIL's inspection focuses on non-nuclear safety areas. For the nuclear safety area, it relies much on the work of regulators and other parties, especially a self-regulatory organization, INPO. The members of NEIL are also members of INPO. Non-compliance with INPO standards can lead to an increase of premiums under NEIL.[136]

To participate in the pool for property damage, the operators have to pay premiums *ex ante*. In case of heavy claims in the policy year, NEIL can call for retrospective premiums up to 10 times the annual premium.[137] The pool for liability is financed *ex post* through retrospective premiums. This arrangement is cheap for operators, since no *ex ante* payment or reserve is necessary. Thus the assets are not immobilized during the policy period. The actual costs for operators are the possibility of losing the assets in the future and corresponding loading costs. The advantage of costs saving has to be balanced with the security of the pool and the risk of insolvency of members. To ensure the availability of resources in case of a severe nuclear incident, ANI requires that limited liability companies provide guarantees from their parent companies or affiliated companies.[138] Another problem created by the *ex post* payment system is that operators do not have to pay much *ex ante* in exchange of the secondary coverage and the possibility of damage to trigger the payment of retrospective premiums is very low. Thus to what extent operators will regard the potential loss of retrospective premiums as their productions costs and have corresponding preventive incentives depends much on how risk averse they are.

1.4.6. Private Interest Approach

The above analysis evaluates the compensation system for the nuclear sector in the U.S. according to the theoretical criteria established in Part I. However, how a compensation system is shaped is determined not only by the criteria to provide optimal prevention and compensation, but also by the involvement and interaction between relevant private interests.

Compensation for nuclear damage is more than a legal issue; it relies on the country's attitude towards nuclear industry and even on the whole energy policy. The 1957 PAA has mentioned that one aim of the legislation is 'to encourage the development of the atomic energy industry'. This is done by limiting the operators' liability for nuclear damage. Private parties fear facing the potential catastrophic losses it may create and are reluctant to participate in the nuclear industry without

135 Faure & Vanden Borre 2008, p. 271.

136 Gittus 2006, p. 234.

137 *Ibid.*

138 GAO 2004b, p, 3.

the protection under the PAA. Thus whether full liability is established relies to a large degree on the government's attitude towards the nuclear industry. This also leads to much criticism. Since after decades of development, it is believed that the nuclear industry should be able to face the full liability it creates. If the nuclear industry is as safe as it claims to be, to bear full liability should not be too burdensome. If the costs it may cause are too great and could overburden the industry, then the nuclear industry is worthwhile, compared to other energy industry sectors.[139] The government's attitude towards the whole energy sector also influences the shape of the compensation regime for nuclear industry. For example, nuclear power produces much less carbon dioxide during the production compared to traditional fuel and natural gases. Thus the hot debate of climate change makes the nuclear industry more attractive. The Energy Policy and Security Act of 2005 provides some incentives to encourage the development of the nuclear industry. If carbon charges come into force, the nuclear industry will be put in a more advantageous competitive position compared to other sectors[140] and may conduct too high an activity level.

The nuclear industry also plays an important role in shaping the compensation regime. Nuclear operators have successfully lobbied for a system of limited liability. The benefits of nuclear suppliers are protected through channeling the burden of liability economically to operators. The ratification of the CSC is also to protect the suppliers from potential open-ended liability.[141]

The public and potential victims, for which the members only become relevant if accidents happen, belong to a so-called shadow interest group, since the vague scope of this group, their interest in the early stage in nuclear legislation is weakly protected. The PAA then was regarded as a protection to the nuclear industry rather than for the victims. However, since politicians know that shadow interest groups have the potential to become an effective lobby group, the politicians may still respond to them in the same way as to normal interest groups.[142] The appearance of some large nuclear damage, such as Three Mile Island accident has made those shadow interest groups visible and the public more aware of the risks involved. The increasing groups and their demands might be one explanation of the increasing limits of liability and the PAA. Compared to the public, the nuclear industry is a well-organized small group, and it can carry out wealth transfers to the members because of low transaction costs.[143] On the contrary, the public and legislator are faced with high information costs. Those situations result in the nuclear industry having an advantageous position over the public in nuclear legislation

Insurers can also influence the structure of financial protection. As mentioned above, nuclear liability is excluded from the general liability insurance policies. Two insurance associations provided nuclear liability insurance in the early stage, and

139 Opinion of Anna Aurilio.

140 See U.S. Congressional Budget Office, Nuclear Power's Role in Generating Electricity, 2008.

141 Energy Independence and Security Act of 2007, Pub. L. No. 110-140, 42 U.S.C.A. §934(a)(1)(D), 121 Stat. 1492, 1741 (2007).

142 Keenan & Rubin 1988, p. 22.

143 Faure 1995, p. 34.

since 1998 ANI is the only insurer for nuclear liability in the U.S. The monopolistic insurer also has a strong power and incentives to lobby for favorable nuclear legislation. The capacity of the insurance market has long been an argument for limited liability. The amount of coverage, rather than basing it on an actuarial basis, depends much on the insurers' willingness to invest.

In conclusion, the design of a compensation system for nuclear damage is also the result of the weighing and compromise of those interests involved. The shape of such a system depends not only on legal and economic criteria but also on the interaction between those interests.

2. The Compensation System for Natural Resources Damage in the US

2.1. *Introduction*

As discussed in Chapter 1, the term 'natural resources damage' is used in the U.S. to denote the damage to the environment. This concept contains both damage to public natural resources and damage to the privately owned resources which have ecological value in excess of the owners' economic value. The compensation system for natural resources damage in the U.S. is especially interesting. A comprehensive liability and compensation regime is established with a combination of statutes and common law. The evaluation procedure and standards adopted in regulations make assessment more feasible and predictable. In addition, compared to other jurisdictions, claims for natural resources damage and related compensation instruments are not only possible on paper. The U.S. has witnessed an intensive practice in those aspects as well.

Originally, natural resources damage was compensated under common law to a limited extent. The common law provides different regimes according to the nature of the affected interest. When the damage to natural resources involves public interest, common law allows states to claim for damage based on the public trust doctrine or the doctrine of *parens patriae*.[144] If private interests are affected by damage to natural resources, claims can also be made based on negligence, nuisance, trespass or strict liability.[145] Although common law allows claims for natural resources, the recovery is still very limited. For example, only states have limited *locus standi* under the public trust doctrine; it does not give federal authorities such standing. The methodology used to assess damage and rules concerning causation also limit the compensation for natural resources damage.[146]

The rapid development of economic and technology after World War II have lead to increasingly serious natural resources damage in the U.S. However, traditional common law failed to provide sufficient protection to the environment. Against this background, federal statutes began to develop in the 1970s to provide

144 Peck 1999, p. 287.

145 Anderson 1997, p. 420-422.

146 For more details about the limitation of common law in compensating natural resources, see Schmid 2008, p. 488; Kanner & Ziegler 2006, p. 140-141.

compensation regimes for natural resources damage. Those early statutes include: the National Marine Sanctuaries Act, the Trans-Alaska Pipeline Authorization Act, the Deepwater Port Act, the Outer Continental Shelf Lands Act and the Clean Water Act.[147] The most important and more recent federal statutes allowing compensation for natural resources damage are the Comprehensive Response, Compensation, and Liability Act (CERCLA) and the Oil Pollution Act (OPA). The OPA provides a compensation regime for damage resulted from oil pollution. The compensation system for other hazardous substances is stipulated under the CERCLA. The CERCLA is also known as the superfund statute. Those two systems are examined respectively in the following sections. It is worth noting that the liability regimes established under the OPA cover not only natural resources damage but also other categories of damage caused via the environment. Correspondingly, the financial responsibility and compensation fund required under the OPA also provide coverage to other categories of damage. The CERLCA does not cover liability for private parties. But some compensation mechanisms, such as liability insurance may cover both CERCLA liability and liability for private parties. Section 2 and 3 introduce the liability rules and compensation instruments under the OPA and the CERCLA respectively. This description is followed by some empirical data about natural resources damage claims in the U.S. in Section 4. Section 5 provides a critical analysis of the compensation system.

2.2. *Compensation System for Oil Pollution*

Before the adoption of the OPA, an international compensation regime was established for oil pollution: the 1969 CLC and the 1971 Fund Convention. The U.S. sent delegations to the negotiation of those conventions. However, due to some characteristics of the conventions, such as the preemption of states laws, low-liability limits and channeling liability to ship owners, the U.S. failed to ratify the conventions.[148] In response to the serious accident of Exxon Valdez in 1989, the U.S. Congress quickly passed the OPA 1990. The OPA has some similarities with the international regime, such as strict liability, limited liability with compulsory financial guarantee. However, it has some substantial differences as well: the scope of compensable damage is much wider, liability is not channeled, and higher liability limits with higher possibilities for the potential responsible parties to lose their right to limit. The OPA does not preempt state laws, which means states can still impose additional liability or financial responsibility.[149]

147 For the discussion of those pre-OPA statutes, see Lewis 2001, p. 101-107.

148 Wang & Faure 2010, p. 3.

149 33 U.S.C. §2718.

2.2.1. Liability for Oil Pollution

2.2.1.1. *Liability under the OPA*

The OPA does not only adopt a liability regime for oil pollution, but also requires the establishment of response mechanisms for oil spills. The federal responders are required to prepare a National Contingency Plan outlining federal and state responsibilities, and responsible parties are also required to prepare private response plans.[150]

Removal Costs and Natural Resources Damage

Before the OPA, many other federal or state statutes, such as the CWA also authorized federal, state authorities and Indian tribes to remove or arrange removal of a discharge of oil.[151] The OPA allows compensation for removal costs. Under the OPA, removal is defined as 'containment and removal of oil or a hazardous substance from water and shorelines or the taking of other actions to minimize or mitigate damage.'[152] This definition shows that removal is the first step taken after oil pollution to clean up. Broadly speaking, removal cost is also one part of ecological damage defined under this research. But the U.S. treats removal costs differently from other damages, including natural resources damage. Under the OPA, responsible parties are liable for removal costs and damages.[153] The term 'damages' includes: natural resources, real or personal property, subsistence use, revenues, profits and earning capacity and public services.[154] It is clear from this definition that the term 'natural resources damage' does not include removal costs. This dichotomy is different from the structure in the ELD, under which the term 'environmental damage' includes both the emergency response and restoration after awards.[155] Hence, when discussing the compensation for ecological damage in the U.S., it should be borne in mind that both removal costs and natural resources damage are relevant. Removals are taken immediately after or in the case of a substantial threat of discharge to prevent/mitigate the damage. Only when there is still damage left after the removals, it is necessary to make natural resources damage assessment to quantify the remaining damage and to restore the environment. The distinction between removals (response actions) and natural resources damage can also be found in the CERCLA and is discussed in Section 3.

How to determine the scope and to quantify the damage has long been a problem perplexing people when compensating for damage to the environment. The OPA gives a definition to the term 'natural resource damage', which includes:

150 Sump 2011, p. 1102.
151 33 U.S.C. §1321(c).
152 33 U.S.C. §2701(30).
153 33 U.S.C. §2702(a).
154 33 U.S.C. §2702(b)(2).
155 The details of the term 'environmental damage' under the ELD, see Chapter 8, Section 1.

(A) the costs of restoring, rehabilitating, replacing, or acquiring the equivalent of, the damage natural resources;
(B) the diminution in value of those natural resources pending restoration; plus
(C) the reasonable cost of assessing those damages.[156]

This definition is quite broad. Restoration is used as the primary method to evaluate the loss of natural resources and the diminution of services.[157] Not only restoration costs, other alternatives and interim losses pending the restoration are also compensable. To quantify the damage, the OPA authorizes federal, state authorities and Indian tribes as trustees of natural resources to assess the damage. It also requires the National Oceanic and Atmospheric Administration (NOAA) to promulgate regulations for natural resources damage assessment.[158] The assessment made according to these regulations has the force and effect of a rebuttable presumption.[159] The final rules concerning natural resources damage assessment were promulgated in 1996 by the NOAA.[160] Under this regulation, a concrete procedure and available assessment methods are prescribed. Under the NOAA assessment rule, 'the costs of restoring, rehabilitating, replacing, or acquiring the equivalent of, the damage natural resources' is defined as 'primary restoration'.[161] To compensate for interim losses, the NOAA rule also adopts a restoration-based approach: compensatory restoration should be taken to compensate for the lost natural resources and service during primary restoration.[162] When determining compensatory restorations, trustees should use a resource-to-resource or service-to-service approach to compensate for the lost natural resources service or value.[163] If these approaches are not possible, trustees can use other evaluation techniques to estimate the dollar value of the lost services and select the scale of the restoration action that has a cost equivalent to the lost value.[164] A variety of valuation techniques are allowed to calculate the monetary value, including the disputed contingent valuation.[165]

According to the OPA and the assessment regulation, an assessment can either be made by the trustees and the trustees can claim against the responsible parties

156 33 U.S.C. §2706(d)(1).
157 Lee 2002, p. 16.
158 33 U.S.C. §2706(e)(1).
159 33 U.S.C. §2706(e)(2).
160 60 Fed. Reg. 39804.
161 15 C.F.R. §990.30.
162 *Ibid.*
163 15 C.F.R. §990.53(d)(2). 'Resource-to-resource restoration actions are designed to restore the damaged natural resource to an equivalent quantity of the same or comparable resource.' 'A service-to-service restoration action is one designed to restore lost services to an equivalent quantity.' See Letourneau & Welmaker 2000, p. 188.
164 15 C.F.R. §990.53(d)(3)(ii).
165 Letourneau & Welmaker 2000, p. 188. Contingent valuation method is a method to value natural resources by doing surveys and asking people what monetary value they would like to place on certain resources. It is not based on actual market transactions but is hypothetical by assuming people will respond to the surveys as in an actual transaction. See Cross 1989, p. 315-319.

for the costs afterwards, or responsible parties can conduct the assessment themselves with the approval of the government. A most common approach would be a cooperative assessment involving both the trustees and responsible parties in the assessment procedures.[166] A cooperative approach can be desirable since it can promote a fast restoration and avoid costly and time-consuming litigation. An early involvement of the responsible parties and even insurers can also make the risks more predictable for the insurers and can thus enhance the insurability of natural resources damage. Although it has been subject to many challenges since its promulgation, the regulation provides a practicable guidance in tackling the debatable and difficult task of natural resources damage assessment.[167]

Responsible Parties

One significant difference between the OPA and the international regime is that liability is not channeled to ship owners. The OPA identifies responsible parties based on the types of facilities involved. For vessels, any person owning, operating or demise chartering the vessel can be the responsible party. The OPA also defines the responsible party for onshore facilities, offshore facilities, deepwater ports and pipelines respectively.[168] Thus, unlike in the international regime, as far as vessels are concerned, not only ship owners, but also operators and charters, who may actually contribute to the oil pollution risks, can be held liable.

However, remarkably, cargo owners can escape liability under the OPA though they benefit from and contribute to the risks of oil transportation as well. During the promulgation of the OPA, there were proposals to establish a secondary liability on oil cargo owners. However, those proposals were not adopted.[169] This arrangement led to some criticism since the OPA provides the cargo owners no incentive to choose safer ships. There were also some recommendations to impose liability on cargo owners since the promulgation of the OPA.[170] Triggered by the Deep Horizon accident in 2010, liability and financial responsibility under the OPA became a hotly debated topic again. The Coast Guard Authorization Act (CGAA) was enacted in 2010. Although focusing on authorizations to the Coast Guard, it contains some general provisions influencing the OPA liability. One of those provisions extends liability to the cargo owner of single hull tankers.[171] Aware of the large risk of oil spill posed by single hull tankers, the OPA 1990 requires the phasing out of single hull tankers by 2015.[172] To further give cargo owners

166 Allen 2011.

167 For a more detailed analysis of the NOAA assessment regulation, see Letourneau & Welmaker 2000; Anderson 1997; Seevers 1996.

168 33 U.S.C. §2701(32).

169 H.R. CONF. REP. NO. 101-653, at 102(1990). See Kiern 2000, p. 489-490.

170 For example, Comments of the International Association of Independent Tanker Owners (INTERTANKO) on Preliminary Regulatory Impact Analysis for Financial Responsibility for Water Pollution (Vessels), 1993. See De La Rue & Anderson 2009, p. 61.

171 124 Stat. 2905 §713.

172 46 U.S.C. §3703a.

incentives to choose safer tankers, the CGAA extends liability to cargo owners if they choose to use the single tanker for transportation.[173]

Strict Liability and the Limitation of Liability

The same as in the international regime, strict liability is imposed on the responsible parties. A defense is available to responsible parties, if the damages or removal costs were caused solely by an act of God, an act of war, or by an act or omission of a third party with no contractual or employment relationship to the responsible party.[174]

The OPA also establishes limits on oil pollution liability according to different types of facilities. With the exception of an offshore facility (non-deepwater port), a cap is established for the total sum of removal costs and damages. The cap is the greater of a per incident cap and a per gross ton cap.[175] The cap was increased in 2006 by the Coast Guard and Maritime Transportation Act (CGMTA). Furthermore, the CGMTA also established different caps for single hull and double hull tankers. A further increase was made by the Coast Guard according to Consumer-price Index in 2009.[176] The existing caps for different facilities are as follows:

Table 2: Liability Caps under the OPA

Facilities	Existing liability limits
Single hull tanker> 3000GT	$3,200 GT or $23,496,000
Single hull tanker≤ 3000GT	$3,200 GT or $6,408,000
Double hull tanker>3000GT	$2,100 GT or $17,088,000
Double hull tanker≤ 3000GT	$2,100 GT or $4,272,000
Any vessel other than a tanker	$1,000 GT or $854,400
Offshore Facility except a deepwater port	All removal costs plus $75,000,000
Onshore facilities and a deepwater port	$350,000,000

In spite of those caps, a responsible party can lose his right to limitation if the incident was proximately caused by 'gross negligence or willful misconduct of, or the violation of an applicable Federal safety, construction, or operating regulation.'[177] A responsible party may also face unlimited liability if it fails to report an incident, or fails to provide requested cooperation in connection with removal activities or fails to comply with an order of the President.[178] Even when a responsible party can revoke the limitation under the OPA, it may still face unlimited liability if the applicable state statutes require so.

173 Papavizas & Kiern 2011.

174 33 U.S.C. §2703(a).

175 33 U.S.C. §2704(a).

176 US Coast Guard, Consumer Price Index Adjustments of Oil Pollution Act of 1990 Limits of Liability, Vessels and Deepwater Ports, Federal Register Vol. 74, No. 125, 1 July 2009, 31357-31369.

177 33 U.S.C. §2704(c)(1).

178 33 U.S.C. §2704(c)(2).

2.2.1.2. *State Statutes*

As discussed earlier, the OPA does not preempt state legislation. States can still have their own liability legislation concerning oil pollution. In the U.S., 30 of the 50 States have a coastline. All but six of the coast States have legislation on vessels liability.[179]

States may have diverging liability rules from the OPA by allowing a broader definition of the responsible parties, higher liability limits or even unlimited liability.[180]

For example, cargo owners are considered responsible parties explicitly in some States, such as Alaska and California, Maryland and New Jersey. In some other States, cargo owners are also potentially liable. Although they are not explicitly listed as responsible parties, the broad definition of responsible parties does not exclude cargo owners either.[181]

Some States impose strict and unlimited civil liability for cleanup costs, natural resources damages and private losses caused by oil pollution, including pure economic losses, such as Alaska, California, North Carolina and Rhode Island. In some other states, unlimited liability is only established for certain categories of damage, such as in Washington (for cleanup costs and damages to persons or property), Maryland (for cleanup costs, damage to real and personal property and natural resources damages), Massachusetts (for natural resources damages) and Florida (for natural resource damages, damage to real and personal property, and losses consequential upon property damage).[182]

A comprehensive discussion of the state oil pollution statutes is outside the scope here. However, it is worth noting that the states Statutes in addition to the OPA make vessels coming into American ports face potential unlimited liability. This is quite diverging from the international regime. Moreover, the stricter States' statutes often require a higher financial responsibility. Whether and to what extent the combination of unlimited liability and financial responsibility in States can contribute to better costs internalization and safer records also deserve more attention.

2.2.2. Financial Responsibility and Trust Fund

Liability itself cannot guarantee costs internalization and compensation. To ensure the availability of compensation in case of damage, the OPA requires potential responsible parties to establish and maintain evidence of financial responsibility up to a certain level.[183] Furthermore, to ensure the capacity of public authorities to

179 Force, Davies & Force 2011, p. 978.

180 For a short summary of the States' 'legislations, see De La Rue & Anderson 2009, Appendix 3, p. 1163-1181; Force, Davies & Force 2011, p. 978.

181 See De La Rue & Anderson 2009, Appendix 3, p. 1163-1181.

182 Force, Davies & Force 2011, p. 978.

183 33 U.S.C. §2716.

remove pollution in case of damage and to provide further protection to victims, the Oil Spill Liability Trust Fund was established.

2.2.2.1. *Financial Responsibility: General*

The OPA requires some vessels and offshore facilities to provide evidence of financial responsibility. The responsible party for

> (1) 'any vessel over 300 gross tons ... using any place subject to the jurisdiction of the U.S. or
> (2) any vessel using the waters of the exclusive economic zone to transship or lighter oil destined for a place subject to the jurisdiction of the U.S.'

is required to provided evidence of financial responsibility sufficient to meet the maximum amount of liability set in the OPA.[184]

The parties who are required to provide evidence of financial responsibility were expanded to the responsible parties of 'any vessel over 100 gross tons using any place subject to the jurisdiction of the United States' under the CGAA in 2010.[185]

For the identified offshore facilities, different levels of financial responsibilities are required based on the type of facilities. For an offshore facility located seaward of the seaward boundary of a State, the level is set at $35 million. For an offshore facility located landward of the seaward boundary of a State, the level is set as $10 million.[186] The responsible party of a deepwater port is required to provide financial responsibility sufficient to meet its maximum liability established under the OPA.[187]

Various methods can be used by potential responsible parties to meet their financial responsibility, including: insurance, surety bond, guarantee, letters of credit, qualification as a self-insurer or other evidence.[188] As a guarantor under the OPA, one has to accept the direct action by claimants. Furthermore, 'the guarantor may not invoke any other defense that might be available in proceedings brought by the responsible parties against the guarantor.'[189]

2.2.2.2. *P&I Clubs and Insurance*

If potential responsible parties can demonstrate the required financial responsibility, a certificate of financial responsibility (COFR) is issued. Traditionally, insurance has been the most commonly used method to meet financial responsibility.[190] Well before the promulgation of the OPA, P&I Clubs have been the primary providers of maritime liability insurance. However, although they

184 33 U.S.C. §2716(a).
185 124 Stat. 2905 §713.
186 33 U.S.C. §2716(c)(1).
187 33 U.S.C. §2716(c)(2).
188 33 U.S.C. §2716(e).
189 33 U.S.C. §2716(f).
190 Kim 2002, p. 571.

remain insurance providers, they refused to act as guarantors under the OPA.[191] Not only P &I Clubs, also many other insurers and shipping industries have shown their concerns or criticisms towards the OPA and the financial responsibility regulations issued by the Coast Guard. Major issues include: potential unlimited liability under the OPA and states' statutes; the capacity of the reinsurance market and the possibility of a direct action against guarantors.[192] The critics even predicted that the OPA of 1990 would result in the so-called 'train wreck' scenario, which refers to the withdrawal of most vessels from U.S. trade and the possibility of a related disruption in the U.S. economy.[193]

In spite of the doubts, the 'train wreck' never materialized. Although refusing to act as guarantors under the OPA, the P&I Clubs still provide oil pollution coverage in the U.S. at the price of additional premiums. Moreover, several alternatives started to emerge as the 'guarantors' after the issue of the Coast Guard's Interim Final Rule on Financial Responsibility, such as the Shoreline Mutual Insurance Association and the First Line and Arvak Ltd. Some larger tanker operators can also satisfy their responsibility through surety bond or self-insurance.[194] Those instruments have been developed for the financial responsibility purpose under the invisible hand of the U.S. oil market. It is worth noting that those instruments are not designed to replace but to complement the P&I policies. Usually, their coverage requires the membership in a P&I Club. The P&I Clubs still assume oil pollution costs in most cases, and only when the P&I Clubs fail to provide coverage, the risks fall within the scope of the new instruments.[195]

In addition to the financial responsibility required by the OPA, some State statutes also have provisions on financial responsibility. Some of them require financial responsibility from parties other than the ones under the OPA, such as owners of oil (California). Some States require a higher level of financial responsibility than the federal level, such as Alaska (for tank vessels and oil barges carrying crude oil: $400.20 each bbl. of storage capacity or $133,400,000, whichever is greater), California ($1 billion owners and operators of tankers, large barges and owners of oil) and Washington ($500 million for a tank vessels that carries oil as cargo).[196] To make it possible for potential responsible parties to meet their higher financial responsibility, some States allow the use of membership in a P&I Club as evidence of financial responsibility, such as Virginia and California. The California statute does not provide a right of direct action. Virginia lists the instruments that are subject to direct action, such as insurance, guaranty or surety. For P&I Clubs, Virginia does not require it to be subject to direct action.[197]

191 De La Rue & Anderson 2009, p. 57.

192 De La Rue & Anderson 2009, p. 56-58; Kiern 2000, p. 561-562.

193 See U.S. Coast Guard, Preliminary Regulatory Impact Analysis: Financial Responsibility for Water Pollution Vessels, 1993, p. 4-5.

194 De La Rue & Anderson 2009, p. 61-63.

195 Kim 2002, p. 581-582.

196 De La Rue & Anderson 2009, Appendix 3.

197 De La Rue & Anderson 2009, p. 60.

2.2.2.3. *Trust Fund*

Although the COFR can ensure a certain extent of compensation, damage may remain uncompensated if the limit of the COFR or liability is exceeded. In response to an accident, the public authorities may need to take removals and make natural resources assessment, and then claim the costs from liable parties. The prolonged claim procedure may influence the capacity of public authorities to give a prompt response. To fill in those gaps, the OSLTF is established.

The Fund is available for the removal costs of the trustees consistent with the National Contingency Plan, the costs incurred by the trustees in assessing the natural resources damage and developing and implementing the restoration plans, removal costs incurred by other parties determined to be consistent with the National Contingency Plan, uncompensated removal costs and damages and related administrative costs.[198] Unless under some limited situations, the claims should be presented to the responsible parties or their guarantors first.[199] The OPA limits payments of the OSLTF for removal costs and damages in each incident to $1 billion. A sublimit for natural resources damages payments is set at $500 million per incident.[200] The per incident limit for removal costs and damages was increased to $2.7 billion under the Energy Policy Act of 2005.[201]

The OSLTF is financed from several sources: the tax on crude oil transported to or produced in the U.S., the transfers from the previously existing pollution funds, interest on the Fund principal from U.S. Treasury investments, recovery of costs from responsible parties or their guarantors and penalties.[202] The tax on oil is the major part of the fund, which was set as 5 cent per barrel initially for a five-year period. In 1994 the balance of the OSLTF reached $1 billion and the tax expired. The tax was reinstated by the Energy Policy Act of 2005 and increased by the Energy Improvement and Extension Act of 2008 to 8 cents per barrel until 2016 and 9 cents per barrel in 2017. After paying removal costs and damages, the OSLTF can recover from responsible parties. According to an implementing report by the OSLTF, the annual recovery from the financial year 2004 to 2008 fluctuates from $7 million to $16 million, which means that 16 per cent of the OSLTF removal and claims expenditures were recovered.[203]

2.3. *Compensation System for Damage Caused by Hazardous Substances*

A compensation system for hazardous substances is established under the CERCLA 1980. In 1970s and 1980s, a few seriously polluted sites have drawn the public's

198 33 U.S.C. §2712(a).

199 33 U.S.C. §2713.

200 26 U.S.C. §9509(c)(2)(A).

201 119 Stat. 1058-1059 & 1361.

202 26 U.S.C. §9509(b).

203 Oil Spill Liability Trust Fund Annual Report FY2004—FY2008, 3-4 available at: <www.uscg.mil/npfc/docs/PDFs/Reports/OSLTF_Report_FY04-FY08.pdf>, last accessed on 8 May 2013.

attention, especially the 'Valley of the Drums' and 'Love Canal'.[204] The existing environmental legislation at that time usually dealt with the pollution at operating sites. For example, the Resources Conservation and Recovery Act (RCRA) has a special subtitle on hazardous waste, which regulates the disposal of hazardous waste to prevent future pollution. It also establishes a Corrective Action program to remediate existing pollution, but only on operating sites. The CERCLA tries to fill the gaps.[205] It was passed at the last moment of the 96th Congress. The CERCLA aims at identifying contaminated sites and assuring cleanup at those sites by regulatory instruments and liability rules. One significant difference between the CERCLA and earlier statutes is that it applies not only to the current active sites but also to sites that are no longer active.[206]

The CERCLA addresses hazardous substances dumped, spilled or otherwise released into the environment. The term 'hazardous substances' is defined broadly to include many substances defined in the previous legislations, but excluding petroleum.[207] Therefore different compensation systems are established for hazardous waste and oil in the U.S. The CERCLA has two goals: encouraging a quick response to the contamination and ensuring that the costs are borne by responsible parties.[208] To realize those goals, the CERCLA establishes a response action regime and liability framework. The same as the regime for oil pollution, the CERCLA distinguishes response action for a release and liability for natural resources damage.

2.3.1. Liability under the CERCLA

2.3.1.1. Response Action and Natural Resources Damage

According to the CERCLA, if there is a release of hazardous substances or a threat of such a release into the environment, the President is authorized to take response action according to the national contingency plan.[209] The President delegates most of his authority under the CERCLA to the EPA.[210] EPA can arrange two kinds of actions: removal action and remedial actions. Removal action is a short-time reaction to cleanup or removal of the released hazardous substances from the environment; while the remedial actions are long-term actions, which are consistent

204 The Valley of the Drums is a toxic waste site in Kentucky. It became a collection site for toxic wastes since 1960s and a fire in 1966 at the site made it catch the public attention. But at that time, there were no laws addressing the storage of toxic wastes and the site remained unregulated for another decade. Love Canal was a neighborhood in New York. It became the subject of attention after the release that it had formerly been used to bury tons of toxic waste in the mid 1970s. Those events have been regarded as triggering the passing of the CERCLA of 1980.

205 Grad 1982, p. 35-36.

206 *Ibid.*; Judy & Probst 2009, p. 193.

207 42 U.S.C. §9601(14).

208 *Control Data Corp. v S.C.S.C. Corp.*, 53 F. 3d 930, 936 (8th Cir. 1995).

209 42 U.S.C. §9604(a).

210 Exec. Order No. 12,580, 52 Fed. Reg. 2923 (23 January 1987).

with a permanent remedy taken instead of or in addition to removal actions.[211] EPA can either take those measures by itself and recover it from the responsible parties or ask the responsible parties to take the actions by themselves.[212] It is reported that the response actions can be taken in the form of either consensual agreements between the EPA and responsible parties, such as administrative orders on consent and consent decrees, or non-consensual instruments such as unilateral administrative orders and judgments.[213] In the recent years, there is a tendency of a decreased use of litigation and greater use of consensual measures.[214]

The CERCLA also created a trust fund to finance the response actions, the Hazardous Substances Trust Fund. This is the fund commonly known as superfund. The Fund can provide payment to the governmental response costs or the response costs caused by the other parties according to the national contingency plan.[215] The fund makes it possible for the EPA to take response actions before recovering the costs from the responsible parties. Before the legislation, there had been already many contaminated sites in the U.S. To clean up these sites, a national priority list (NPL) was established, under which the EPA lists the highest priority sites. The Trust Fund only finances remedial actions in the NPL sites.[216] In addition to the response actions taken by EPA, the States can also contribute to the cleanup: they are required to pay for 10 per cent of remediation projects funded by the federal superfund.[217] Moreover, the CERCLA does not preempt but encourages state administered remediation projects for non-NPL sites.[218]

Sometimes after the response actions, the damaged environment may still not be fully restored. Under this situation, the federal or state natural resources trustees or Indian tribes can take restoration measures or assess the damage.[219] Response actions, together with natural resources damage assessment try to provide for a complete compensation for the damage to the environment itself.

The first natural resource damage assessment regulation was promulgated under the authorization of the CERCLA. The assessment made according to this regulation has an effect of a rebuttable presumption.[220] This effect provides the trustees with a favorable status, since they only need to comply with the NRDA regulation, then the burden of proof is shifted to the responsible parties. The assessment regulation was published in 1986 by the Department of Interior (DOI). According to the DOI assessment rule, a trustee can recover the following costs: damages assessed according to this rule; costs of emergency restoration; assessment

211 42 U.S.C. §9601(23)(24).

212 42 U.S.C. §9604(a)(1), 9606(a).

213 US Government Accountability Office, Superfund: Litigation has Decreased and EPA Needs Better Information on Site Cleanup and Cost Issues to Estimate Future Program Funding Requirements, GAO-09-656, 2009, p. 21-22.

214 *Ibid.*, p. 6-7.

215 42 U.S.C. §9611(a).

216 40 C.F.R. 300.425 (2008).

217 42 U.S.C. §9604(c).

218 Schmid 2008, p. 504-505.

219 42 U.S.C. §9607(f).

220 42 U.S.C. §9607(f)(2)(C).

costs and interest loss.[221] To assess 'damages', the DOI rule establishes a four-stage procedure: pre-assessment phase, assessment plan phase, assessment phase and post-assessment phase. In the assessment phase, two types of procedures are established: Type A for small release of oil and hazardous waste which can be assessed by a simplified procedure based on computer modeling and Type B for a large and complex release for which individual assessment is necessary.[222] In type B procedure, two types of costs/losses need to be evaluated: the costs of 'restoration, rehabilitation, replacement, and/or acquisition of equivalent resources' and interim losses ('the compensable value of the services lost to the public through the completion of the restoration, rehabilitation, replacement, and/or acquisition of the equivalent of the injured resources and their services to baseline').[223] When determining restorations, the original regulation established a hierarchy of assessment methodologies and favored a market-oriented approach. This was reflected in the lesser of rule: damage was awarded to the lesser of the resource's replacement costs or the diminution of the property value.[224] That rule was annulled in the case *Ohio v DOI*, which favored the use of restoration as the primary assessment approach and allowed compensation for the non-use value.[225] Those revisions were adopted in the revised assessment regulation in 1994. When determining interim losses (called 'compensable value' in the DOI rule), the DOI rule allows compensation for both the use value and non-use value of natural resources. Lots of methods can be used to evaluate interim losses, including more traditional ones like market price methodology and the ones to estimate willingness to pay in imitated market conditions, like the contingent valuation methodology.[226] The assessment of non-use value is still one highly controversial aspect of the NRDA law.[227]

2.3.1.2. *Responsible Parties and the Scope of Liability*

The CERCLA imposes liability on quite a broad group of responsible parties, including: the owner and operator of a vessel or a facility; the owner and operator when the hazardous substances are disposed; those who arrange for disposal (generators) and certain transporters.[228] Though the CERCLA aims at applying the polluter pays principle, the liability provision means that, in fact, it is not only the polluters, but also the parties who may benefit from the cleaned sites pay for the cleanups and restorations. For example, many of the current owners and operators of polluted sites contribute little to the pollution.[229] Therefore, the liability is

221 43 C.F.R. §11.15(a).
222 42 U.S.C. §9601(c)(2)(A-B).
223 43 C.F.R. §11.83.
224 43 C.F.R. §11.35(b)(2-3) (1987).
225 *Ohio v Department of Interior* (880 F.2d 432, 442 D.C. Cir. 1989).
226 43 CFR §11.84(c).
227 Boyd 2003, p. 271.
228 42 U.S.C. §9607(a).
229 Judy & Probst 2009, p. 214.

imposed on many parties who were not asked to burden the cleanup costs by the previous legislation. The CERCLA stated that it became effective on the date of enactment.[230] Many courts have held the liability under the CERLCA to be retroactive.[231] The retroactive liability, together with some other characteristic of the superfund liability, such as a broad definition of responsible parties, joint and several liability has lead to a chilling effect on brownfield redevelopment[232] and the lack of liability insurance capacity in 1980s.[233] Those issues are discussed in the following analytical section.

The CERCLA requires the responsible party to be liable for both response costs and natural resources damage. There are four specific headings established under the liability provision: the costs of removal or remedial action incurred by federal government, state or Indian tribes; such costs incurred by other parties according to the national contingency plan; natural resources damage and certain assessment and study costs.[234] One difference between the liability under the CERCLA and the OPA is that, under the former, damage to private parties is excluded from the statute liability framework while, under the latter, a comprehensive liability framework is established for both NRD and private damage.

2.3.1.3. *Stringent Liability and the Relief of Liability*

In line with the liability for oil pollution, strict liability is also established for damage caused by hazardous substances under the CERCLA. There are limited defenses available for the responsible parties. If they can prove by a preponderance of the evidence that the (threat of) release is caused solely by an act of God, an act of war, an act or omission of a third party without contractual relationship or the combination of the above factors, they can be exonerated from liability.[235] This is quite limited, since only when those factors are the sole cause, the liability can be exonerated. If the potential responsible parties also contribute to the (threat of) release, those factors cannot be used to reduce their liability.

During the development of the legislation for the CERCLA, in the original proposal S. 1480, there was a specific provision which would impose joint and several liability when several potential responsible parties would be involved. This provision was later removed and the senate was in favor of general references to the preexisting common law and statutory framework.[236]According to the common law, the court usually held the liability under the CERCLA to be joint and several unless the harm is divisible. Considering the complex characteristics of the hazardous substances contamination, liability usually cannot be distinguished according to a

230 42 U.S.C. §9652(a).

231 *United States v Olin Corp.*, 107 F.3d 1506, 1511-12 (11th Cir. 1997); *United States v Mottolo*, 695 F. Supp. 615, 622 (D.N.H.1988).

232 McMorrow 2004, p. 1094-1095.

233 See Abraham 2001, p. 94-98.

234 42 U.S.C. §9607(a).

235 42 U.S.C. §9607(b).

236 Grad 1982, p. 10, 19.

potential responsible party's role at the site. Thus few parties can avoid joint and several liability.[237] However, according to a recent case Burlington N. & Santa Fe Ry. Co. v. United States, liability can be divisible considering a few factors, such as how long the property was leased and how much of the contaminated property was owned by a certain owner and so on. Therefore the liability under the CERCLA is not always joint and several. Instead, it can be apportioned when there is 'a reasonable basis for determining the contribution of each cause to a single harm.'[238]

The stringent liability established under the CERCLA 1980 had a chilling effect on the brownfield redevelopment. For the sites not qualified on the NPL list or of which the extent of contamination is only suspected, the extensive liability discouraged the owners and prospective purchasers to cleanup and redevelop such sites.[239] After the promulgation of the CERCLA, a few amendments have been passed to relieve the onerous liability and to encourage the redevelopment of brownfields. For example, the Congress introduced an innocent landowner defense in 1986. The EPA was authorized to enter into settlements with *de minimis* landowners and generators and to provide the latter with a covenant not to sue.[240] A significant relief was introduced by The Small Business Liability Relief and Brownfields Revitalization Act in 2001, under which the innocent landowner defense was updated, and a 'contiguous property owner defense' and 'bona fide prospective purchaser' was added.[241] Those provisions reduced the number of potential responsible parties and prompted quick settlements under the CERCLA.

In line with the compensation system for oil pollution, the liability established under the CERCLA is also capped.[242]

Table 3: Liability Caps under the CERCLA

Types of Facility	Cap
Vessels carrying hazardous waste (except incineration vessels)	Greater of $300 per gross ton, or $5,000,000
Other vessels	Greater of $300 per gross ton, or $500,000
Motor vehicle, aircraft, hazardous liquid pipeline facility or rolling stock	$50,000,000 or a less amount decided by the President, but no less than $5,000,000
Incineration vessels or any other facilities	All costs of response plus $50,000,000

The responsible party will lose his right to limit the liability, if (1) the (threat of) release of a hazardous substance was the result of his willful misconduct or willful negligence, or the primary cause was a violation of applicable safety, construction or operating standards or regulations, and (2) he fails or refuses to provide all reasonable cooperation and requested assistance about response actions.[243] To

237 Judy & Probst 2009, p. 245.
238 Randle, Sands, McKillop & Levy 2009, p. 137-138.
239 McMorrow 2004, p. 1095-1096.
240 McMorrow 2004, p. 1096-1099.
241 Reisch 2002, p. 100-102.
242 42 U.S.C. §9607(c)(1).
243 42 U.S.C. §9607(c)(2).

ensure that the potential responsible parties have the capacity to compensate in case of damage, the CERCLA also imposes financial requirement on certain parties.

2.3.2. Financial Responsibility and the Trust Funds under the CERCLA

2.3.2.1. Financial Responsibility: General

The CERCLA imposes financial responsibility on certain parties. The owner or operator of each vessel over 300 gross tons or any offshore facility is required to provide evidence of financial responsibility up to the greater of $300 per gross ton and $5,000,000 (a vessel carrying hazardous substances as cargo)/$500,000 (others). One of, or the combination of, the following instruments can be used to realize the financial responsibility: insurance, guarantee, surety bond, or qualification as a self-insurer.[244] The CERCLA also asks the EPA to promulgate the financial requirements for other facilities according to the 'degree and duration of risk associated with the production, transportation, treatment, storage, or disposal of hazardous substances' within five years of the enactment of the CERCLA.[245]

In spite of the requirement to establish new financial responsibility, CERCLA has never resulted in the implementation of this provision. This lack of implementation can be explained by the pragmatic attitude taken by the EPA. Financial assurance rules have been established for many onshore facilities under the RCRA, such as facilities that treat, store, and dispose of hazardous waste and underground petroleum storage tanks.[246] This may to some extent explain the failure of the introduction of new financial responsibility, since highest priority facilities have already fallen into the RCRA regime.[247] However, there are still some gaps left. For example, some types of waste are excluded from the RCRA coverage, such as wastes associated with mining activities. Another gap is that there is usually a lack of financial requirements from hazardous waste generators.[248] After more than 20 years delaying in promulgation, four environmental groups filed a suit against the EPA to push the latter to promulgate financial assurance regulation in 2008.[249] Research also shows that without financial responsibility, the mining sector proposes a high risk of insolvency. It is reported that the 63 hard rock mining sites on the Superfund's NPL may cost up to $7.8 billion to clean up. Given the long-term nature of the cleanup liability at mines and the financial statutes, it is estimated that $2.4 billion of them will have to be borne by taxpayers.[250] Hence the EPA has designated the hard rock mining industry as its priority for the development of

244 42 U.S.C. §9608(a)(1).

245 42 U.S.C. §9608(b)(1).

246 RCRA, Subtitle C and D.

247 Boyd 2003, p. 283.

248 GAO, Environmental Liabilities, EPA Should Do More to Ensure that Liable Parties Meet Their Cleanup Obligations, GAO-05-658, 2005, p. 33-34.

249 See *Sierra Club v Johnson, et al.*, No. C 08-01409, 2009 US Dist. LEXIS 14819 (N.D. Cal. 25 February 2009).

250 GAO 2005, p. 36.

financial responsibility requirements in 2009 and planned to propose a financial responsibility rule in 2013. Furthermore, the EPA also tries to indentify additional classes of facilities, such as those in the chemical manufacturing, petroleum and coal products manufacturing, and the electric power generation, transmission, and distribution industries.[251]

2.3.2.2. *Insurance*

Environmental liability insurance is an important instrument to realize the obligation to provide financial responsibility. It came to be developed since the 1970s in the U.S. with the appearance of mass tort suits. Historically, environmental liability was covered under the General Public Liability Policies (GPL). Two categories of pollution were covered: the sudden and accidental pollution and gradual pollution. The early policies were occurrence policies, which provide coverage for liability arising out of damage that occurs during the policy period. Later, with the increasing claims for historical pollution, a gradual pollution exclusion provision was introduced in the GPL policies.[252]

The pollution liability insurance market became more pressing with the promulgation of the CERCLA. As discussed in the theoretical part, insurance functions well when there is some extent of predictability concerning the frequency and magnitude of loss.[253] However, the joint effect of strict liability, joint and several liability and retroactive liability under the CERCLA introduced unanticipated liability for both the potential liable parties and insurers. In addition to the statutory characteristics, the judicial tendency to resort to the deep pocket and to invalidate the exclusion provisions in the insurance contract also alleviates the unpredictability problem. Those issues led to the tightening of the environmental liability insurance market in the mid of the 1980s.[254] An 'absolute pollution exclusion' clause was introduced to preclude all pollution coverage regardless whether it was gradual or accidental.[255]

Later, the insurance market began to recover. To avoid the unpredictability, the insurers shifted their policies from occurrence based to claims made. Thus the insurance policies cover only liability arising out of claims made during the policy period. Some specific environmental liability insurance policies (EIL) also began to appear, providing more extensive coverage.[256] Those policies not only provide potential responsible parties to meet their financial responsibility under the CERCLA, but also contribute to the redevelopment of brownfield redevelopment. Such policies include professional liability coverage, owner/operator liability

251 EPA, Superfund Financial Responsibility, <www.epa.gov/superfund/policy/-financialresponsibility/>, last accessed on 8 May 2013.

252 Richardson 2001, p. 297-298.

253 Chapter 4, Section 2.1.

254 Abraham 1988.

255 Pendygraft, Plews, Clark & Wright 1988, p. 152.

256 Richardson 2001, p. 298-299.

coverage, cleanup cost-cap or stop loss coverage, legal defense coverage and reopener or regulatory action coverage.[257]

2.3.2.3. *Trust Funds*

In addition to the financial responsibility instruments, the CERCLA also established two trust funds to complement the compensation capacity. The Superfund was established to finance the cleanups and uncompensated NRD. In addition, the CERCLA 1980 also required the creation of a post-closure liability trust fund. If a facility has (1) complied with regulatory requirements and (2) has been closed according to the regulations, the facility and surrounding area has been monitored as required and there is no substantial release likelihood, the liability can be transferred to such a fund. The fund was financed by a tax on disposed hazardous waste.[258] The unlimited liability that would be transferred to the trust fund raised concerns during the reauthorization of the CERCLA in 1985. The Congress and the EPA found that the fund would not be sufficient to pay for liability and decided to suspend the transfer of liability to the fund and to repeal the tax.[259]

The Superfund can be used to cover: governmental response costs, any claim for necessary response costs incurred by other parties under the national contingency plan, unsatisfied claims under the Clean Water Act and NRD claims under the CERLCA, related assessment and restoration costs, grants for technical assistance and lead contaminated soil.[260] The superfund has four major financial sources: taxes on crude oil and certain chemicals; an environmental tax assessed on corporations based upon their taxable income; appropriations from the general fund; fines, penalties and recoveries from responsible parties; and interest. Taxes used to be the primary source of superfund (68 per cent of the trust fund revenues from 1981 through 1995). However, the taxes expired in 1995, and since then the major source has been appropriation from the general fund (59 per cent from 1996 through 2007). The annual average appropriation was $1.3 billion from 1981 through 2007. If adjusted for inflation, the appropriation has declined in recent years.[261] The expiration has led to a sharp reduction in the balance of the Superfund, which peaked at $4.7 billion at the start of fiscal year 1997 and declined to $173 million at the start of fiscal year 2007.[262]

The dwindling of the Superfund does not only reduce the available compensation but also has a serious influence on the cleanup of contaminated sites

257 K. Chilton & P. Meyer, *Environmental Insurance for Brownfields Redevelopment: A Feasibility Study*, 1998, p. 11-15, available at: <www.huduser.org/portal/publications/-econdev/envins.html>, last accessed on 8 May 2013.

258 42 U.S.C. §9607(k).

259 GAO, Hazardous Waste: Funding of Postclosure Liabilities Remains Uncertain, GAO/RCED-90-64, 1990, p. 11.

260 42 U.S.C. §9611.

261 GAO, Superfund: Funding and Reported Costs of Enforcement and Administration Activities, GAO-08-841R, 2008, p. 3.

262 *Ibid.*, p. 8.

and restoration of the natural resource damage. Firstly, after the expiration of the superfund tax, fewer sites have been added to the NPL. This has disproportionally affected orphan and complex sites. Furthermore, since the NRD assessment often follows the response actions, the slow remediation process also influences the claims for NRD. Thirdly, the CERCLA only leaves limited room for States to settle with the responsible parties and to encourage the cleanup of polluted sites. Many States also do not have sufficient funding to address non NPL sites.[263]

2.4. *Empirical Data about Natural Resources Damage in the US*

A comparatively comprehensive compensation regime for natural resources damage has been established in the U.S. for more than 20 years. Many debates have taken place concerning the liability rules and assessment regulations. To what extent the NRD claims matter in practice can be shown by some empirical data. Before the Deep Horizon accident in 2010, the largest NRD case was the Exxon Valdez recovery, which caused approximately $2.1 billion in cleanup costs and $1 billion in natural resources damage.[264] The total cost resulting from the Deep Horizon accident is still unknown, but it is expected much higher than the Exxon Valdez accident.[265]

In the early days after the promulgation of the CERCLA and OPA, the scale of the NRD cases was small. A U.S. General Accounting Office (GAO) report in 1995 showed that as of April 1996, federal trustees had made 50 NRD settlements for a total of $106 million under the CERCLA, with recoveries ranging from $4,000 to $24 million for each case.[266] As for claims under the OPA, the major federal trustee, NOAA had collected $45 million to repair natural resource damage from 20 oil spills as of March 1999. Fewer than 1 per cent of the oil spills in the U.S. coastal zone involved an NRD assessment.[267]

The claims for NRD have been increased over the years. There is no official data about the frequency and scale of NRD claims. Research carried out by Environmental Law Institute (ELI) in 2007 tried to provide a comprehensive picture of the NRD settlements, judgments and claims through collecting information from the major federal trustees and the OSLTF.[268] This research studied 273 cases involving one or more federal trustees until 2006. The indentified cases and the expenditures from OSLTF on natural resources claims from 1989 until 2006 reached a total of $1.8 billion which equals an annual cost of $87.65 million.[269] Among the

263 Schmid 2008, p. 513-525.

264 Boyd 2003, p. 294.

265 Hollaender & Kaminisky 2000, p. 206.

266 US General Accounting Office, Outlook for and Experience with Natural Resource Damage Settlements, GAO/RCED-96-71, 1996.

267 Boyd 2003, p. 295.

268 Environmental Law Institute, Mitigation of Impacts to Fish and Wildlife Habitat: Estimating Costs and Identifying Opportunities 2007, available at: <www.landscope.org/-connect/find/partners/eli/fish_wildlife_habitat_mitigation/>, last accessed on 8 May 2013.

269 *Ibid.*, p. 77.

273 cases discussed, 132 cases were CERCLA related, leading to settlement and judgment values of $722,433,600 and 48 cases were OPA related, with the value of $126,290,530.[270] The ELI report also takes into account the NRD payment made by the OSLTF: from 1992 until 2004, the total NRD assessment and claims expenditures amounted to $16.6 million.[271] Another $2.2 million has been expended on NRD related issues by the OSLTF from 2004-2008.[272]

In addition to the expenditures on NRD, removals (response) costs are also an important part to compensate for the damage to the environment. As discussed earlier, the removal costs are treated differently from damages (NRD) in the U.S. However, the removal is the first step to clean up the pollution and to compensate for the damage to the environment. Hence, when discussing the costs caused by ecological damage, removal costs constitute an import part of it. There is no authoritative data on the removal (response) costs arising under the CERCLA and the OPA. The payment from the OSLTF on removal costs can provide a perspective on it. The annual expenditure of the OSLTF on removal costs varies from approximately $30 million to $110 million from 1995 until 2008.[273]

However, it has to be admitted that the data above may still be far from the actual costs to the environment. Not all damage is qualified as natural resource damage under the CERCLA and the OPA; some damage may not be detected and assessed. Even considering only the claims and expenditures actually made on NRD and removal related issues, the above estimation may still be under inclusive. Firstly, most of the data is based on the claims made by trustees; however the trustees may rely on additional appropriations to undertake assessment and restorations. Not all the costs by additional appropriations were considered in the statistics. Secondly, some States also allow claims for natural resources damage under the State statutes, and comprehensive data about the claims under the State law are still lacking. Thirdly, sometimes the responsible parties undertake restoration actions themselves rather than paying for the monetary damages. Those costs are not considered in the above data, either.[274] Considering those limits, the removal costs and NRD expenditures in practice may be substantially higher than estimated above.

2.5. *Economic Analysis*

After the brief introduction to the compensation regime under the federal statutes, as well as the scale of compensation for ecological damage in practice, this section tries to give a critical analysis of compensation regime under the CERCLA and the OPA respectively.

270 *Ibid.*, p. 80.

271 *Ibid.*, p. 81.

272 Oil Spill Liability Trust Fund Annual Report FY2004-FY2008, 7.

273 Oil Spill Liability Trust Fund Annual Report FY2004-FY2008, 6; US Coast Guard, Report on Implementation of the Oil Pollution Act of 1990, p. 10.

274 ELI 2007, p. 77-78.

2.5.1. Liability Rules

2.5.1.1. Liability Rules under the OPA

As under the international regime, strict liability is established under the OPA. Although oil pollution is a bilateral accident, the potential tortfeasors have much more influence on the risks than victims. They also have better control of the vessels and offshore facilities, and thus strict liability can give them better incentives to adopt an efficient care and activity level.

One difference between the OPA and the international regime is that liability under the former is not channeled to ship owners. The OPA has defined liable parties broadly. For the vessels, the parties who own, operate or demise charter the vessels are the responsible parties. For other facilities, the owners, operators, or licensees may be liable.[275] In spite of the broad definition, cargo owners are not defined as responsible parties under the OPA. Hence, the oil companies may have less incentives to choose safer vessels for transportation. But there is no provision precluding liability for other parties as under the international regime. The OPA does not preempt State law, which means oil companies may still be potentially liable under some State statutes, such as in Alaska and California. Moreover, a recent federal legislation also imposed liability on cargo owners for single hull tankers. This also gives oil companies incentives to choose safer double hull tankers. Moreover, a balance of interest was achieved in the negotiation of the OPA. Although oil companies are not 'potentially responsible parties', they have to contribute to the trust fund to cover some uncompensated claims.

The OPA also allows responsible parties to limit liability, except that the responsible parties for a non-deepwater port offshore facility have to pay the full removal costs. Admittedly, the limit is not high, and sometimes it has proved to be insufficient to cover bigger accidents. According to a report by U.S. Coast Guard, as of 2010, there have been 51 (substantial threats of) oil discharges by vessels which have lead to or are likely to result in excess of liability limits. As for the offshore facilities, the damage caused by Deep Horizon accident is predicted to be much higher than the limit of $75 million.[276] With the combination of strict liability and an insolvency risk, the responsible parties will have insufficient incentives to take efficient care. However, different from the international regimes, the responsible parties, although they may have the right to limit their liability, have a greater chance of being exposed to uncapped liability. Under the international regimes, liable parties only lose their right to limitation if the damage is caused by their act or omission intentionally or recklessly.[277] Under the OPA, if the responsible parties violate a federal safety, construction or operating regulation, or if they failed to meet some reports and cooperation obligation concerning the accident, they will also be deprived of the right to limit their liability. Moreover, many States also

275 33 U.S.C. §2701(32).

276 US Coast Guard, Oil Pollution Act Liability Limits: Fiscal Year 2010 Report to Congress, 2011.

277 Art. V, 2 of the1992 CLC.

impose unlimited liability on responsible parties for oil pollution. To this extent, the OPA can lead to better costs internalization than the international regime.

In addition to the higher chance of exposure to unlimited liability, another difference in determining the extent of liability is that the OPA allows more risk differentiation. The CLC sets the cap according to the tonnage of the vessels. While after the revision of 2006, the U.S. system allows different caps considering not only the tonnage, but whether the vessels are tankers or not and whether the tankers are single hull or double hull. Although, generally speaking, the combination of caps and strict liability can lead to under-deterrence, better incentives can be achieved by tailoring the caps to the different risks.

Another crucial characteristic of the U.S. system is that it allows a broader scope of compensable damage, especially when it comes to the compensation for the damage to the environment itself. The CLC limits the compensation for 'impairment of the environment other than loss of profit' to 'costs of reasonable measure of reinstatement actually undertaken or to be undertaken'. Another related heading is the costs of preventive measures.[278] In the U.S., the compensable scope is much wider: it allows compensation for removal costs and natural resource damage. The NRD incorporates costs of restoration and alternative measures, interim losses and assessment costs. Rather than only abstract liability provisions, a concrete and practicable assessment regulation is established in the U.S., which gives guidance on the procedures and methods used in the assessment. According to the OPA and assessment regulation, not only the restoration measures actually undertaken or to be undertaken are compensable. The trustees or responsible parties can also choose some other alternative measures, some as replacement, or even natural recovery. The losses pending the restoration are also compensable.[279] Although some methods allowed under the regulation have been subject to intense debates,[280] the attempt to allow compensation for the full value of the natural resources makes a further step towards efficient costs internalization. Moreover, the concrete guidance on natural resources damage assessment also helps to make risks, which are difficult to quantify, more predictable.

This, however, does not mean that the assessment procedure established in the U.S. is perfect. It has been criticized that, in practice, the NRD assessment costs, as opposed to restoration costs, make up a large portion of total NRD settlements and judgments costs.[281]

In addition to the theoretical analysis, some empirical data also shows the effect of the OPA on the oil spills in the U.S. According to a Congressional Research Service (CRS) report, the U.S. oil imports and consumption have increased steadily

278 Art. I, 6 of the 1992 CLC.

279 33 U.S.C. §2706(d)(1).

280 Debates have taken place on the methods to evaluate the non-market value of the natural resources, especially the contingent valuation methods. For example, Dobbins 1994; Bohm 1994; Kanner & Nagy 2005.

281 ELI 2007, p. 78.

through the years.[282] The volume and number of oil spills for incidents have decreased steadily from 1973 till 2004 (from approximately 15 million gallons in 1973 to less than 2 million gallons in 2004).[283] Specifically, the volume of oil spills from vessels in U.S. waters in the 1990s reduced dramatically compared to that in the 1980s. It is believed that the OPA is one of the key contributing factors: it expanded the federal response authority and made oil pollution liability stricter. Moreover, some States also enacted additional oil spill liability legislation after the Exxon Valdez spill in 1989. Those legislations also contributed to the declines of oil spills in the U.S. Another empirical research by Hendrickx shows that the OPA had a desirable effect on decreasing oil spills not only by its stricter liability framework and also the certification system and safety regulation concerning the ships.[284]

2.5.1.2. *Liability Rules under the CERCLA*

The liability under the CERCLA has some similarities to that under the OPA. Responsible parties are liable for the cleanup costs and NRD regardless whether they are negligent or not, which gives them incentives to take efficient care. The liability for damage caused by hazardous substances is also capped under the CERCLA. This, when combined with strict liability may lead to under-deterrence.

Some differences exist between the two systems as well. For example, the liability under the CERCLA covers only cleanup costs and NRD, while under the OPA, a comprehensive liability regime also extends to damage suffered by private parties. This may be explained partially by the different functions the two statutes play. The OPA provides compensation for damage resulting from oil pollution. The purpose of the CERCLA is to promote the cleanup of hazardous waste sites and place the costs on those responsible for the contamination.[285] Therefore, rather than a compensation instrument for all the damage caused by the (threat of) hazardous substances release, the liability under the CERCLA functions more like a funding mechanism for response. The damage suffered by individuals is still subject to common law and other statutes.

Liability under the CERCLA is not channeled to specific parties. The superfund liability is quite extensive; it even extends to the ones who are beneficiaries of the activities but did not contribute to the risks. Together with the retroactive liability and joint and several liability, the public authorities can shift the cleanup costs of the existing contaminated sites to private parties. This practice, on the one hand, broadened the funding source of the cleanup; on the other hand, had a chilling effect on the redevelopment of brownfield and tightened the liability insurance market in 1980s.

282 Jonathan L. Ramseur, Oil Spills in US Coastal Waters: Background, Governance and Issues for Congress, CRS 7-5700, RL 33705, p. 25.

283 *Ibid.*, p. 2-3.

284 Hendrickx 2007, p. 256-257.

285 *Burlington N. & Santa F Ry.Co. v United States*, 129 S.Ct. 1870, 1872 (2009); *Control Data Corp. v S.C.S.C. Cop.*, 53 F.3D 930, 936 (8th Cir. 1995).

As an immediate response to the Love Canal disaster, the CERCLA has been subject to many debates since its issuing. For example, Viscusi and Hamilton have criticized that the cleanup undertaken by the EPA is too costly compared to its benefits. After a re-evaluation of the risk assessments done by the EPA for 145 superfund sites in the 1990s, they concluded that the cleanup undertaken by the EPA was not efficient: 95 per cent of the superfund expenditure was spent to eliminate only 0.5 per cent of cancer risk. They identified a few elements in the Superfund program contributing to the inefficient calculation of individual lifetime risk, without considering population exposure, not using a measure of central tendency, assumption of future residential land use in areas that are not currently inhabited and the lack of consideration of cost-benefits in remediation decisions.[286] Some recent literature also shows the inefficiencies of superfund programs, such as the research by Greenstone & Gallagher (2008) and Gamper-Rabindran & Timmins (2010).[287] Those critical papers show the difficulties in determining the criteria for remediation. For example, more research is needed to answer the questions 'how clean is clean; (which target should be achieved, permanent remedy or only restrict exposure) and which remedy should be chosen.

Another criticism on the CERCLA is the high transaction costs it brought. An early research by RAND shows that transaction costs account for 21 per cent of the private hazardous waste expenditure during 1984-89.[288] The public expenditure on transaction costs is high as well. A 2008 GAO report shows that from 1999 to 2007, only 77 per cent of the superfund expenditure was dedicated to remedial and removal activities, with the rest on enforcement (indentify liable parties and require or claim costs from them) (13-15 per cent) and administrative activities (8-10 per cent).[289] The high transaction costs, together with some other inefficiencies even lead to some scholars favoring a funding from diffuse sources, such as tax rather than liability to clean up contaminated land.[290] The concerns about high transaction costs, however, become less serious in recent years.[291] One explanation may be that transaction cost (percentage) declines significantly once cleanup begins. The RAND research found that transaction costs share in the cleanup-ongoing stage and the construction-complete stage is much lower than their share in the site-study stage.[292] Therefore, with the more NPL Sites came into the stage of cleanup and construction being completed 30 years after the coming into force of the CERLCA, the transaction costs issues may become less serious. Another tendency contributing to the phenomena is the litigation under the CERCLA has decreased significantly in recent years. It is reported that from 1994 to 2007, the superfund cases filed annually in

286 See Hamilton & Viscusi 1999; their other criticisms about the cost-benefit of superfund program can be found at: Viscusi & Hamilton 1999; Hamilton & Viscusi 1997.

287 Greenstone & Gallagher 2008; Gampe-Rabindran, Mastromonaco & Timmins 2010, Valuing the benefits of superfund site remediation: Three approaches to measuring localized externalities, available at: <www.nber.org/papers/w16655>, last accessed on 8 May 2013.

288 Dixon 1992, p. 39.

289 GAO 2008, p. 3-4.

290 Sigman 2000.

291 Judy & Probst 2009, p. 237.

292 Dixon 1992, p. 51, 59.

U.S. district courts decreased by 50 per cent.[293] This is because, first, fewer sites were listed on the NPL in recent years; second, the EPA promotes settlements with responsible parties and third, the early cases have created some legal uncertainties clarified so there is less need to go to the court.[294] Since the litigation costs compose a substantial part of the transaction costs, this tendency may decrease the transaction costs share as well. Furthermore, even though the transaction costs brought by the CERCLA liability system are high, study shows that public financing is costly as well since it requires increases in distortionary taxes.[295]

In spite of the high transaction costs, liability rules can play a positive role as well. For example, liability, together with financial responsibility creates incentives for the polluters to take precautionary measures. Empirical research shows liability for cleanup, especially strict liability, has desirable effects on toxic waste management.[296] Besides, liability also allows the achievement of settlement between the EPA and the responsible parties and allows the EPA to force recalcitrant parties to conduct response actions.[297] Without liability, to rely solely on public revenue or environmental taxes to fund the cleanup of contaminated land is also not an easy task.

Furthermore, there are some criticisms not on the liability itself but on the form of liability which contributes to the inefficient incentives. For example, research shows that joint and several liability discourages the redevelopment of brownfields.[298] Several proposals have been made to limit the application of joint and several liability. As discussed earlier, in a recent superfund case, the Supreme Court also limited the scope.[299] Furthermore, retroactive liability cannot create preventive incentives for current decisions. Some scholars even argues that retroactive liability may even weaken the incentives since the insolvency risk is hence increased.[300]

The above analysis shows the implications of using liability to clean up pollution and to create precautionary incentives. In the U.S., a combined system of administrative regulation and liability rules is used. In case of contaminated lands, the EPA needs to clean up or require cleanup from the liable parties. When applicable, NRD assessments also need to be done by the trustees. In taking those cleanup or assessment, it is essential to determine the target to be achieved after cleanup/restoration and to choose the applicable remedy. When liability is used to both fund the program and to create preventive incentives, its design deserve more attention, such as whether joint and several liability and retroactive liability can be used. Careful design is also necessary to reduce the potential high transaction costs a liability system can create.

293 GAO 2009.

294 *Ibid.*

295 Dixon 1995.

296 See Alberini & Austin 2002; Alberini & Frost 2007; Stafford 2003.

297 Judy & Probst 2009, p. 227-229.

298 Sigman 2010; Change & Sigman 2010.

299 *Burlington N. & Santa Fe Ry. Co. v United States*, 129 S. Ct. 1870 (2009).

300 Boyd & Kunreuther 1997.

2.5.2. Financial Responsibility

2.5.2.1. *Financial Responsibility under the OPA*

The OPA required responsible parties for certain vessels and offshore facilities to provide evidence of financial responsibility up to a certain level. The compulsory requirement can relieve the problem of insolvency and ensure at least a certain level of compensation and costs internalization. However, the capacity of the market to provide guarantees or the responsible party to provide self-insurance is not unlimited. Hence a cap of financial responsibility needs to be set. The question then is which level is efficient. A tradeoff needs to be made here. A higher amount can ensure better victim protection and costs internalization; but a high financial responsibility may create a challenge for the limited capacity of the market and for smaller ship owners. A lower requirement is more insurable, but may then create less than efficient incentives.

The OPA of 1990 introduced stricter liability rules and financial responsibility requirements than under the preexisting legislation.[301] The stricter regimes led to much criticism in the beginning and there was a prediction that this would lead to the 'train wreck' scenario. This reflected the limited capacity of the market to some extent. But the scenario never materialized. The modest financial requirement imposed by the OPA has been proved to be insufficient compared to the catastrophic losses oil spills can cause. A GAO report shows that among the 51 major oil spills (with removal costs and damages exceeding $1 million) it reviewed, 10 have exceeded the liability limit, resulting in claims of about $252 million in total. The liability limits (financial responsibility) for certain types of vessels are disproportionally low compared to the risks they create, this is for example the case for tank barges. It is reported that the potential damage in excess of liability limit, together with additional NRD claims from the already cleaned-up spills, costs from previously sunken vessels that will discharge oil in the future, spills from unidentifiable source and potential catastrophic spills may create a challenge for the viability of the OSLTF.[302] The Deep Horizon accident in 2010 has led to discussions on the sufficiency of the liability limits again, especially for offshore facilities.

Admittedly, the increase of the level of financial responsibility has to take into account the capacity of the market and other adverse effects it may produce. For example, after the promulgation of the OPA, many ship owners tried to externalize the costs by subdividing their fleet into single-vessel companies. Although this provides no guarantee that they can hide true vessel ownership in the U.S. judicial system, the adverse effect of reorganization on safety needs to be considered.[303]

301 For example, some early legislation covers only clean-up costs claimed by the government but not damage claimed by private parties. The earlier legislation also does not cover losses pending the restoration. See Kiern 2000, p. 502-507; see also Verheij 2007a, p. 177.

302 GAO report: Oil Spills: Costs may impact viability of oil spill liability trust fund, GAO-10-795T, 2010.

303 Kim 2002, p. 567-569.

The OPA allows responsible parties to use different instruments to fulfill their financial responsibility, including insurance, surety bond, self-insurance and financial guaranty. This is in line with the theoretical model: the responsible parties can choose the instrument they prefer to meet their financial responsibility, thus the problem of high concentration of the market can be alleviated. However, in the field of marine oil pollution, most vessels chose to cover their liability by the P&I Clubs before the OPA. After the promulgation of the OPA, the P&I Clubs refused to be guarantors under the act because of the direct action requirement and the stricter liability. This can be explained by the special characteristic of the P&I Clubs. Instead of being pure insurers, the clubs are mutuals among the ship owners. This requires that each member of the mutual possesses a comparable level of risk. The P&I Clubs are organized on a global basis. The stricter liability in the U.S. creates higher risks in the U.S. rather than somewhere else.[304] Nevertheless, the P&I Clubs still managed to provide coverage in the U.S. by adding the U.S. Oil Pollution Clause. This requires vessels coming into the U.S. to provide additional premiums to obtain the coverage.[305] Furthermore, the market for the financial responsibility certificate also led to the emergence of new guarantors. The guarantors usually only provide coverage when the P&I Clubs fail to do so. In practice, the cases when the claims against those guarantors are triggered are still rare. The P&I Clubs remain the major providers of compensation. Moreover, the P&I Clubs can use some defenses to coverage, which give the insured incentives to control the risks. However, those defenses are not available for guarantors due to the direct action requirement. Thus, to what extent the requirement of those guarantors can contribute to a better risk reduction than is merely an additional ticket to entry into the U.S. market remains unclear.[306]

When the discussion concerning an increase in the liability limit and the financial responsibility appeared on the agenda again after the Deep Horizon accident, it led again to strong opposition from insurers.[307] The practice in some States with higher limits may provide an interesting perspective. As mentioned earlier, some States impose unlimited liability on responsible parties. The potential responsible parties are also required to provide a higher level of COFR. One example is California, where the owners and operators of tankers, large barges and owners of oil are required to provide financial responsibility of $1 billion. This is substantially higher than the requirement in the federal level. To make the coverage possible, the states allow the use of the membership of the P&I Clubs as COFR. The existing capacity of the P&I Clubs to cover oil pollution is $1 billion.[308]

304 Hearing before the Committee on Transportation and Infrastructure House of Representatives, One Hundred Eleventh Congress, Second Session, 9 June 2010, p. 60.

305 Wang & Faure 2010, p. 22.

306 Kim 2002, p. 588.

307 Hearing before the Committee on Transportation and Infrastructure House of Representatives, p. 60-62.

308 <www.igpandi.org/Group+Agreements/Pool+reinsurance+programme+2013-14>, last accessed on 8 May 2013.

2.5.2.2. *Financial Responsibility under the CERCLA*

A compulsory financial responsibility is also imposed on the responsible parties under the CERCLA. The owner or operator of certain vessels and offshore facilities are required to provide such evidence of financial responsibility. The modest cap required by the statute results in it having the same effect as the OPA: a certain level of guarantee is achieved, but the risk of a major release which may lead to insolvency problems remains high. Leaving aside the limited amount required by the CERCLA, several other problems also prevent the financial responsibility to create efficient incentives. Firstly, companies can organize and restructure themselves in ways that may allow them to limit their liability.[309] In addition, there are also some implementation problems. The EPA allows companies to choose between different instruments which have varying degrees of financial risk. This, on the one hand, alleviates the problem of market concentration that may emerge when only one instrument (insurance) is used; on the other hand, it creates a high insolvency risk when the choice is allowed without considering information on the extent of default risk that the company may pose. For example, the financial test is criticized broadly as an insufficient guarantee instrument but it is still broadly used. The EPA also increased the financial risk by not providing adequate oversight and enforcement of the financial responsibility provisions.[310]

The CERCLA also required the EPA to establish financial responsibility for other facilities. However, the EPA failed to implement this provision. Though many onshore facilities are subject to financial responsibility requirements under the RCRA, there are still many gaps left. They are not less likely to create serious release and insolvency problems. The failure of implementation also increases the risk of unpaid cleanups and reduces the preventive incentives.

Liability insurance is an important instrument for the industry to meet their financial responsibility. The environmental liability insurance market began to develop in the 1970s. However, the retroactive liability under the CERCLA and extensive explanation of liability and insurance provisions create many unanticipated burdens for the insurers, and created serious problems in the 1980s. This phenomenon shows that some extent of predictability is essential for the insurability of liability insurance.

It is worth noting that, though financial responsibility is established under the CERCLA, such a provision cannot guarantee the implementation itself. A GAO report shows that the EPA has not implemented the financial responsibility provision sufficiently. For example, the EPA allows companies to choose between financial assurance mechanisms freely without considering their default risk. The use of a financial test is criticized of possessing a high default risk and needs more oversight. The EPA is also criticized of providing inadequate oversight over the enforcement of financial assurance requirements and even of maintaining no

309 GAO 2005, p. 21.

310 *Ibid.*, p. 39-53.

relevant enforcement data.[311] The lack of implementation creates a risk for the recovery of cleanup costs by the government. This example shows the importance of maintaining efficient implementation of financial requirements.

2.5.3. Trust Funds

2.5.3.1. Trust Fund under the OPA

A compensation fund, the OSLTF, was established under the OPA. It provides finance for both the removals and restoration undertaken by trustees and for uncompensated claims. According to the law and economics criteria, a compensation fund is only efficient if it is financed by the one who actually contributed to the risks and the contribution reflects the extent of risks they created. The OSLTF is financed in several ways: by a tax on crude oil, transfers from preexisting funds, cost recoveries, interest and penalties. The tax is the major source of finance. In the other words, the fund is mainly financed by the oil industry. Since the OPA only imposes liability on the shipping industry, the fund strikes a balance between the interests of the industries. Since the oil industry also contributes to the risks, the requirement of their contributions can give them incentives to adopt a more efficient activity level. However, the tax is based on the volume of oil imported or produced, and there is no differentiation based on the safety conditions and records. Thus the financing structure is also criticized by some scholars as sub-optimal, since it gives no incentives on the care level the oil industry may choose.[312]

2.5.3.2. Trust Fund under the CERCLA

The Superfund is established under the CERCLA. The fund has its revenue mainly from environmental taxes before 1995 and general appropriation afterwards. The expiration of superfund taxes results in the fund being primarily financed by the public fund. Therefore, it is the general public rather than polluters who pay for the damage. Even under the former tax regime, one may still argue that it is less desirable than other individual financial responsibility instruments, such as insurance. This is because, generally speaking, public funds allow less risk differentiation. However, this does not mean that the Superfund is a pure subsidy: the Superfund is not only a compensation instrument; it also makes the cleanup and NRD assessment possible. As far as damage to the environment is concerned, removal of the pollutant and restoration of the environment is more important than providing monetary compensation. Thus the public authorities play an important role in providing removals/remedials and NRDA. The Superfund is also an important instrument to pre-finance such activities before the responsible parties are identified and held liable. The decreasing of the Superfund has shown a slow down of the cleanup of contaminated sites and the decrease of NRD litigations.

311 GAO 2005, p. 39-53.
312 Wang & Faure 2010, p. 24-26.

Since the compensation for ecological damage requires the combination of the administrative system and the tort system, an effective financing instrument is significant to make the administrative activities and later the claims possible.

Chapter 8

COMPENSATION SYSTEMS IN THE EU AND SOME EU MEMBER STATES

In the EU, efforts have been made to establish a comprehensive compensation framework for environmental damage. The Directive on Environmental Liability with regard to the prevention and remedying of environmental damage (the ELD) was issued in 2004 to create an EU-wide compensation and response regime for certain types of environmental damage. However its application is limited. It covers only damage to protected species and natural habitats, water damage and land damage. For environmental damage falling out of this scope, a patchwork regime is still operating in the EU. This section first tries to give a short introduction about the regime established under the ELD and its implementation in the Member States. As far as the damage beyond the scope of the ELD is concerned, a cherry-picking approach is used to examine a few compensation instruments used in EU Member States. Those examinations try to give an idea about how compensation instruments are used in the EU to compensate for ecological damage.

1. The Compensation System under the ELD

For a long time, the EU has made an effort to establish a comprehensive compensation framework for environmental damage. Such an attempt started from a sector-based approach, specifically in the waste sector. A proposal was made in 1989 to introduce a directive on civil liability for damage caused by waste.[1] This sector-based approach was soon replaced by a broader regime covering environmental liability generally. A Green Paper[2] and White Paper[3] on environmental liability were published in 1993 and 2000 respectively. In the early proposal on civil liability in the waste sector and the Green Paper, the individual and collective compensation mechanisms under a civil law system were discussed. The White Paper, however, introduced a differentiated system: a civil law system

1 Proposal for a Directive on Civil Liability for Damage Caused by Waste, *OJ* C251/3 (1989); as amended COM(1991) 219 final, *OJ* 1991 C192/6.

2 EC, Green Paper on Remedying Environmental Damage, COM(1993) 47 final, 14 May 1993.

3 EC, White Paper on Environmental Liability, COM(2000) 66 final, 9 February 2000.

for traditional damage and an administrative law system for damage to biodiversity and the contamination of sites. The civil law system led to fierce criticism since it touched upon the tort law which had been built up in Member States over decades. This resulted in that the final Directive focused only on an administrative system for the prevention and remediation of environmental damage. A shift from a civil law approach to a administrative law approach happened during the legislative drafting process of the ELD.[4]

The ELD aims at establishing 'a framework of environmental liability based on the polluter pays principle, to prevent and remedy environmental damage.'[5] In the considerations preceding the Directive, the aims are also clarified. For example in Consideration 2, it holds that the polluters shall be held financially liable, 'in order to induce operators to minimise the risks of environmental damage so that their exposure to financial liabilities is reduced.' Consideration 3 also discusses the objective explicitly: 'to establish a common framework for the prevention and remedying of environmental damage at a reasonable cost to society.' Hence the ELD has the objectives of prevention and restoration of damage, as well as establishing a 'common framework'.

The ELD introduced a preventive and remedial system for environmental damage, which is defined to include: damage to protected species and natural habitats, water damage and land damage. The term 'damage' is defined as 'a measurable adverse change in a natural resource or measurable impairment of a natural resource service which may occur directly or indirectly.'[6] The ELD gives the Member States some discretion in defining the protected species and natural habitats, not only the ones listed in the Directive 92/43/EC and the Directive 79/409/EEC, but also the ones designated by the Member States for equivalent purpose are covered. The water damage includes water under the Water Framework Directive. In defining those two categories, a 'significant adverse effects' threshold is used.[7] For damage to protected species and natural habitats, there should be 'significant adverse effects on reaching or maintaining the favourable conservation status of such habitats or species.' To assess the 'significance', one needs to refer to the baseline, taking account of the criteria set in Annex I.[8] Hence to determine whether the damage poses 'damage to protected species and habitats', several factors need to be considered: the baseline condition, favorable conservation status and significance. These actually impose a high threshold. For water damage, it should also significantly adversely affect 'the ecological, chemical and/or quantitative status and/or ecological potential'.[9] While as for the land damage, only

4 For a detailed discussion on this shift, see Winter, Jans, Macrory & Krämer 2008, p. 165.

5 Art. 1 of the ELD.

6 Art. 2(2) of the ELD.

7 For the damage to protected species and natural habitats, the damage should have significant adverse effects on reaching or maintaining the favorable conservation status of such habitats or species. For water damage, it should have significant adverse affects on the ecological, chemical and/or quantitative status and/or ecological potential of the waters. Art. 2(1)(a, b) of the ELD.

8 Art. 2(1)(a) of the ELD.

9 Art. 2(1)(b) of the ELD.

when the land damage creates a significant risk for human health, it is covered under the ELD.[10] In other words, if the land damage has no direct concerns for human health, the administrative procedure under the ELD will not be triggered. This definition is, on the one hand, quite limited compared to the natural resources damage defined in the U.S. system since it applies only to a few categories of environmental components. On the other hand, it also covers some types of damage not covered in the U.S. system, such as the damage caused by non-contamination events.[11] For example, in neither the definition of 'damage to protected species and natural habitats' nor 'water damage', there is a requirement of 'contamination'.[12] In the U.S., under both the CERCLA and the OPA, there needs to be a (threat of) release of oil or hazardous waste into the environment.[13] Therefore, natural resources damage and response costs in the U.S. are related to contamination.

The ELD covers only occupational activities but not individual behavior in daily life. It establishes different liability frameworks according to the types of activities. For the damage caused by activities listed in the Annex III as hazardous activities, a strict liability is established. A negligence rule is still applicable if the non hazardous activities lead to damage to protected species and natural habitats.[14] Two defenses are introduced in the ELD: damage caused by armed conflicted, hostilities, civil war or insurrection or a natural disaster.[15] The ELD also allows Member States to decide whether to allow two other defenses: the compliance defense and state of arts defense.[16] The ELD imposes liability on 'operators' which are defined as anyone operating or controlling the occupational activities or according to the national law, having decisive economic power over the technical functioning of the activity. The operator includes but is not limited to the holder of a permit/authorization or the person registering/notifying such an activity.[17] Therefore the liability under the ELD is not channeled to specific parties. Anyone who has a substantive contribution to the risks may be held liable. The ELD is vague on who should bear the burden of proof of the causal link between damage and the operator. The common practice would be the plaintiff, in the case of the ELD, the public authorities have the burden of proof. When multiple parties are involved, the ELD allows the Member State to decide whether joint and several liability or several liability is applicable.[18] Different from the CERCLA, the liability under the ELD is not retroactive: the Member States have the obligation to transpose the directive into their domestic law by 30 April 2007 and only activities that happened after the coming into force of national law are covered. Another

10 Art. 2(1)(3) of the ELD.
11 Bio Intelligence Serv. 2009, p. 49.
12 Art. 2 of the ELD.
13 15 C.F.R. §990.11; 43 C.F.R. §11.10.
14 Art. 3(1) of the ELD.
15 Art. 4(1) of the ELD.
16 Art. 8(4) of the ELD.
17 Art. 2(6) of the ELD.
18 Art. 9 of the ELD.

significant characteristic of the ELD is that no cap is introduced for the liability of the operators.

As mentioned earlier, the ELD adopts a system of a more administrative nature rather than a pure civil liability approach. In case of damage or an imminent threat of such damage, the operators are required to take preventive or remedial action and to report to the competent authorities. The competent authorities have the authority but not an obligation to require the operators to provide related information or take the necessary actions themselves.[19] However, if the operators have indeed taken such measures, they shall recover the costs from the liable operators.[20] Under such a system, whether liability will be triggered is, to a large extent, determined by whether the competent authorities will take actions in case of (an imminent threat of) damage.

To ensure that operators have compensatory capacity in case of damage, the ELD considers the potential use of financial security. There was fierce debate on whether to introduce a mandatory financial security requirement. Although many NGOs and a few Member States supported such a mandatory system, this was opposed by most Member States.[21] This reluctance is reflected in the final text. The ELD only requires the Member State to promote the development of financial security instruments and the Commission to present a report on the availability of such instruments by 2010 (Article 14).[22] An article leading to less attention is Article 8 (2), which requires the authorities to recover preventive/remedial costs '*inter alia*, via security over property or other appropriate guarantees' from liable parties. This provision has led to some confusion in its implication. Bocken argues that it establishes an *ex post* obligation on the operators to provide an appropriate guarantee to cover their liability under the Directive if an incident giving rise to liability has taken place.[23]

The ELD required Member States to transpose it into domestic law by 30 April 2007. However, only three Member States met this deadline; Italy, Lithuania and Latvia. Only on 1 July 2010, had all Member States transposed it completely.[24] Much liberty is given to the Member States to determine some critical elements of the liability, such as the scope of the regime, available defenses, whether joint and several liability is used and so on. The Member States can also adopt more stringent provisions in relation to the prevention and remedying of environmental damage.[25] It is reported that the Member States vary in transposing the ELD into domestic laws.[26] Some countries have gone further than the ELD by giving a broader definition of environmental damage, such as Poland, Austria, England and

19 Arts. 5, 6 of the ELD.
20 Art. 8(2) of the ELD.
21 See De Smedt 2009, p. 11.
22 Art. 14 of the ELD.
23 Bocken 2006, p. 13-32.
24 De Smedt & Faure 2010, p. 784.
25 Art. 16(1) of the ELD.
26 European Commission 2010.

Sweden.[27] They include species and habitats protected under national/regional protection regimes. For the scope of activities covered by strict liability, a number of Member States also include further activities not listed in Annex III.[28]As for more than one tortfeasor, most Member States apply joint and several liability, with the exception of Denmark, Finland, France, Slovakia and Slovenia, which choose proportionate liability.[29] Member States also vary in the adoption of defenses: fewer than half of the Member States allow both permit and state-of-the-art defenses; some allow neither of them, while others choose only one of the defenses.[30]

It is reported that the general awareness of the ELD among stakeholders is low, especially with SMEs.[31] The implementation is also limited. There have been still very limited ELD cases so far.[32]As of the end of 2011, experience with the ELD is very limited, with the exception of Poland. In Germany, only three cases reached the court and no cases in Belgium and the Netherlands.[33] In addition to the lower awareness among stakeholders, many other issues also account for the slow implementation, such as lack of political will, difficulties in determining the 'significant' threshold for environmental damage, baseline data and evaluation.[34]

As far as the financial security is concerned, only a minority of Member States have a compulsory financial guarantee requirement: The Czech Republic, Slovakia, Spain and Hungary. Most Member States are reluctant to do so, considering the lack of experience. Even for the Member States which decided to introduce a compulsory system, a gradual approach is adopted, either in terms of time, covered industrial sectors (for example starting from IPPC industries) or covered liabilities (for example starting from primary remediation).[35] Furthermore, there are some other Member States that require an *ex post* financial security mechanisms after an incident giving rise to liability has taken place, such as the Flemish and Walloon Region in Belgium.[36] Many insurers also oppose a compulsory financial security system, considering limited market demands, inhomogeneous risks across countries

27 CEA, Navigating the Environmental Liability Directive, A Practical Guide for Insurance Underwriters and Claims Handlers, April 2009, p. 10, available at: <www.insuranceeurope.eu/uploads/Modules/Publications/1240585425_eld-best-practice-guide-update.pdf>, last accessed on 8 May 2013.

28 European Commission 2010, p. 3.

29 *Ibid.*, p. 3-4.

30 *Ibid.*

31 Stakeholder & Practitioner Workshop Implementation of the ELD in the EU Report, 8 November 2012, 9-11, available at: <http://ec.europa.eu/environment/legal/liability/pdf/-workshop/report.pdf>, last accessed on 8 May2013.

32 European Commission, Environmental Liability, available at: <http://ec.europa.eu/-environment/legal/liability/index.htm>, last accessed on 8 May 2013.

33 European Commission 2010. Presentation of Dr. Ulrich Klein on the Transposition and Application of the Environmental Liability Directive in Germany on the 10th ELD government experts meeting of 7th November 2011 in Brussels, and <http://ec.europa.eu/environment/-legal/liability/index.htm>, last accessed on 8 May 2013, as well as mail from Mr. Hans Lopatta, desk officer for the ELD with the EU Commission.

34 Stakeholder & Practitioner Workshop Implementation of the ELD in the EU Report.

35 Bio Intelligence Serv. 2009, p. 28-33.

36 De Smedt & Faure 2010, p. 805.

and industrials, moral hazards and so on.[37] As required under the ELD,[38] a report was published in 2010 on the financial security. In this report, the Commission concludes that 'there is not sufficient justification at the present time for introducing a harmonized system of mandatory financial security.' It requires a further review of the option of mandatory financial security before 2014.[39]

A survey of major industrial sectors in 2010 showed that most companies (75 per cent of the respondents) use insurance to cover their liability risks.[40] In the existing insurance market, there are already some products available to cover the general environmental liability, on the one hand, and the ELD liability, on the other hand. For example, some General Third Party Liability policies (GTPL) cover third-party damage caused via the environment. It is the primary insurance policy used by companies to cover their environmental liability.[41] Some Environmental Impairment Liability Policies (EIL) also exist, covering both third-party damage and some public liability (cleanup requirements). The EIL coverage is often taken during a land transaction, merger or acquisition. However, it is reported that the EIL market is still quite limited.[42] Experts hold that there is still little awareness of the influence of the ELD among many insurers and polluters.[43] Some stand-alone policies to cover ELD liability also began to emerge in recent years, such as the policy provided by Allianz AGF, AXA and XL Insurance.[44] However, those products also have some limitations, such as the exclusion of gradual environmental damage, sub-limits or exclusions for compensatory remediation and damage from non-pollution events.[45] During the development of insurance products, the insurers also express their concerns about the insurability of such risks, risk assessments and potential compulsory requirements.[46]

In addition, a proposal is also made to establish a compensation fund or risk-sharing pool to cover environmental liability in Europe.[47] A major accident hit the MAL alumina factory near Kolontar, Hungary in 2010, which led to 10 deaths,

37 Stakeholder & Practitioner Workshop Implementation of the ELD in the EU Report, 8-9; Nils Hellberg, ELD Stakeholder Workshop, Financial Security, available at: <http://ec.europa.eu/environment/legal/liability/pdf/workshop/pres_Hellberg.pdf>, last accessed on 8 May 2013.

38 Art. 14(2) of the ELD.

39 European Commission 2010, p. 10.

40 Ad-Hoc Industry, Natural Resource Damage Group, Report: Survey of Industrial Companies, Insurance and Other Financial Security Instruments and Remediation of Environmental Damage under the EU Environmental Liability Directive, p. 3, 2010, Brussels, available at: <www.endseurope.com/docs/100219b.pdf>, last accessed on 8 May 2013.

41 *Ibid.*

42 Bio Intelligence Serv. 2008, p. 40-41.

43 Interview with Mr. Kremers.

44 For a summary of the available insurance products in the EU market, see Bio Intelligence Serv. 2009, p. 41-43.

45 Bio Intelligence Serv. 2009, p. 57-62.

46 CEA 2009; Swiss Re, Insuring Environmental Damage in the European Union, 2007.

47 A project is commissioned by the European Commission, DG Environment to study the feasibility of creating a fund to cover environmental liability and losses occurring from industrial accidents. See <http://eldfund.biois.com/>, last accessed on 8 May 2013.

several hundred people injured and 1,000 hectares of land contaminated. The estimated cost is around 115 million Euro, beyond the capacity of the polluter. In response to this accident, the Hungarian government made a proposal to establish 'a European Industrial Disaster Risk-Sharing Facility (the facility) to be funded by an annual levy on targeted industries and companies.'[48] This facility can provide coverage if the sum of traditional and environmental damage from a major industrial accident may exceed 100 million Euro. It assumes that the companies can provide coverage up to 100 million Euro by insurance or other guarantees.[49] This proposal aims at providing a quick response to a major accident or limit the financial exposure of each company to 100 million Euro.[50] However, some details of this facility structure are unclear and many challenges remain. For example, a compulsory financial security up to 100 million Euro is assumed. However, whether a compulsory financial security should be established under the ELD is still under dispute. The SMEs may have neither capacity not incentives to get financial coverage up to 100 million Euro. Environmental risks across countries and sectors are inhomogeneous, which may lead to cross-subsidy via pooling. The project is still ongoing in exploring the feasibility of such a fund or pooling system. Whether an environmental fund can be established to cover environmental risks in Europe still remains to be seen.

2. Compensation Mechanisms Used in Member States

As discussed above, the application of the ELD is quite limited. For the other damage to the environment falling outside of its coverage, the compensation still relies on domestic laws in Member States. Such a regime is shaped by regulatory instruments, liability rules and also other specific compensation mechanisms. A full discussion of all those instruments in different Member States is outside the scope of this research. Here a few compensation mechanisms used by different Member States are chosen to show how those mechanisms can be used in the EU Member States to contribute to the prevention and compensation of ecological damage. From the mechanisms discussed in the theoretical part: liability insurance, direct insurance, risk sharing agreement and environmental funds, this research chooses one example for each.

In Germany, the Environmental Liability Act of 1990[51] requires mandatory liability insurance from some industries. This, although it has never been fully

48 European Commission, DG ENV, Study to Explore the Feasibility of Creating a Fund to Cover Environmental Liability and Losses Occurring from Industrial Accidents, 7, available at: <https://docs.google.com/viewer?a=v&pid=sites&srcid=YmlvaXMuY29tfGVsZC1md-W5kfGd4OjYxYWYyYmM0YTNkNWIwNg>, last accessed on 8 May 2013.

49 *Ibid.*

50 *Ibid.*

51 Umwelthaftungsgesetz (UmweltHG), Environmental Liability Act of 10 December 1990 BGBl I 1990, 2634 (as amended), an unofficial English translation can be found at: <www.utexas.edu/law/academics/centers/transnational/work_new/german/case.php?id=1396>, last accessed on 8 May 2013.

implemented, promoted the development of stand-alone environmental liability insurance. The transposition of the ELD into its domestic law may lead to new developments in the field of environmental liability insurance as well. Another interesting example in Germany is the solidarity agreement contracted among nuclear operators. This risk-sharing agreement helps to make the high amount of compensation possible. In the Netherlands, a shift from liability insurance to direct insurance happened in the late 1990s to avoid the unpredictability caused by the judicial system. Another example in the Netherlands, the SUBAT,[52] was established to clean up the historic pollution at fuel sites. It shows how environmental funds can be used to compensate for ecological damage. These examples of specific compensation mechanisms will now be briefly discussed.

2.1. *Liability Insurance in Germany*

In Germany, a comprehensive environmental liability statute was published in 1990, the Environmental Liability Act. However, this Act covers only damage caused via the environment, such as death, personal injury or property damage due to an environmental issue. It does not apply to the damage to the environment itself.[53] The Act is restricted to damage caused by stationary installations but does not apply to damage caused by ships and other means of transport.[54] The Act created a list of facilities. The operator is strictly liability for the damage caused by those facilities.[55] To further relieve the burden from plaintiffs, the Act adopts a presumption of causation: the causation is presumed if a facility is inherently suited to cause the damage. However, an exception of the presumption is also allowed: the presumption does not apply if the facility has been properly operated (the special operation duties have been complied with and no disruption of operations has occurred).[56] This exception has resulted in many criticisms, which argue that it allows the operator to rebut the presumption easily.[57] The Act introduces two separate caps for death/personal injury and property damage.[58] The compensation for restoration costs is allowed to a limited extent: if property damage also impairs nature or scenery, and the injured person restores the environment, then these costs are recoverable.[59] One interesting characteristic of this Act is that it requires the operators of certain facilities to provide financial security, which can be in the form of liability insurance, an indemnity agreement or a guarantee made by the government or a licensed credit institution.[60]

52 *Stichting Uitvoering Bodemsanering Amovering Tankstations* (Foundation for the Remediation of Petrol Stations).
53 Art. 1 of the UmweltHG.
54 Brinkmann 1998.
55 Art. 1 of the UmweltHG.
56 Art. 6 of the UmweltHG.
57 For a discussion of those criticisms, see Schwarze 2001, p. 317.
58 Art. 15 of the UmweltHG.
59 Art. 16 of the UmweltHG.
60 Art. 19 of the UmweltHG.

After the legislation, the legal literature cast doubt on whether the Act could actually improve the status of the environmental victims. Those criticisms arose as a result of some internal characteristics of the Act, such as the rebuttal of the presumption.[61] In legal practice, the Environmental Liability Act also created little changes: only six court decisions were delivered on the basis of this act from its enactment until 2005. Of those cases, only one was for the plaintiff.[62] In spite of little changes in legal practice, however, the empirical studies show that in a thriving legal and insurance context, the Act has lead to the decrease of environmental accidents.[63] This phenomenon can be explained by the active role played by environmental liability insurance in improving environmental safety and the influence of the Act on the insurance industry.

Since the 1970s, the environmental risks had usually been added to the general third-party insurance as a free extra. However, the situation began to change after the coming into force of the Environmental Liability Act. The German Insurance Association (GDV) published an Environmental Impairment Liability Insurance model (the UHV model) in 1993[64] and specific environmental liability insurance products began to develop thereafter. Although the Act established a list of facilities, of which the operators are required to maintain compulsory financial security, this has never been implemented. It was reported that there were 130,000 installations which had taken out insurance policies by 2003. But only less than one per cent of them were compulsory.[65] It is the widespread risk aversion among industries rather than the compulsory legislative requirement that created the incentives to seek environmental liability insurance coverage. In the 1970s, environmental risk cover was provided without further risk investigation as a 'free supplement'. After the introduction of the Environmental Liability Act, the environmental liability insurance became more risk-based. Before the issuing of coverage, safety audits and risk evaluations have to be made. A risk analysis branch was established by all major German Industry Insurers. Now, many Germany industries seek environmental insurance coverage as part of their overall framework of commercial liability rather than regulatory requirement under the Environmental Liability Act. This has contributed to lower accident rates since 1993.[66]

After the coming into force of the Environmental Liability Act, an important change in the German system is the transposition of the ELD. To transpose the Directive, the German Parliament adopted the Environmental Damage Prevention and Remediation Act (EDA)[67] and amendments to the Federal Protection of Nature

61 Schwarze 2001, p. 316-320.

62 Schwarze & Hoffmeister 2010, p. 419.

63 Schwarze 2001; Schwarze & Hoffmeister 2010.

64 Schwarze & Hoffmeister 2010, p. 419.

65 *Ibid.*

66 Schwarze 2001, p. 318-320.

67 *Gesetz über die Vermeidung und Sanierung von Umweltschäden* (USchadG)(Act serving to implement the Directive of the European Parliament and of the Council on Environmental Liability with Regard to the Prevention and Remedying of Environmental Damage), Part I p. →

Act and the Water Resources Management Act. The EDA addressed some issues left open by the Directive. The ELD gives the Member States the option to broaden the definition of protected species and natural habitats by adding areas protected under the national law.[68] This option is not used in Germany to broaden the application of the Directive. The EDA dedicates the choice to use the permit defense and the 'state of art' defense to the states.[69] As far as multiple liable parties are involved, joint and several liability is adopted in the EDA. The EDA does not oblige a financial guarantee.[70] From the transposition until 2011, there are three pending court cases involving the EDA. Most cases are settled by agreement or administrative decision before going to the court and the focus is on damages to protected species and natural habitats.[71]

The transposition of the ELD, although not mandating environmental insurance, has started to bring some further changes to the environmental liability insurance market in Germany. In 2007, the GDV published a non-committal model wording to provide the coverage for ELD damage (the 'USV model wording').[72] The USV model tries to establish a stand-alone policy specifically for the ELD liability. A manifestation trigger is used in the policy wording. All three types of environmental damage under the ELD are covered, except the damage occurring on the insured's premises. In relation to damage to protected species and natural habitat and water damage, the insurance covers primary, complementary and compensatory remediation, with a sublimit for compensatory remediation. A sudden and accidental disruption of the insured's normal operation is required for coverage.[73] A variety of products have begun to develop after the publishing of the USV model wording.[74]

2.2. *Direct Insurance in the Netherlands*

In addition to liability insurance, other insurance policies can also be used to cover ecological damage. For example, under a direct insurance policy, the polluters can seek insurance coverage for damage suffered by third parties, if such damage

666, an English copy can be found at: <www.bmu.de/files/pdfs/allgemein/-application/pdf/u_schad_g_eng.pdf>, last accessed on 8 May 2013.

68 Art. 2(1)(a) of the ELD.

69 Art. 1, section 9(1) of EDA.

70 For the details of the transposition, see De Smedt & Faure 2010; Rehbinder 2007.

71 U. Klein, *Transposition and Application of the Environmental Liability Directive in Germany, presented on 10th National Experts Meeting on ELD*, 2011, available at: <http://ec.europa.eu/environment/legal/liability/pdf/ELD_germany.pdf>, last accessed on 8 May 2013.

72 *Allgemeine Versicherungsbedingungen für die Umweltschadensversicherung*, see M. Schubert, *The Environmental Liability Directive and the German Insurance Association's Model Wordings*, June 2007, available at: <www.genre.com/sharedfile/pdf/InsuranceIssues200706-en.pdf>, last accessed on 8 May 2013.

73 *Ibid.*

74 N. Hellberg, *Update on German ELD Market, presentation at the Stakeholder and Practitioner Workshop on the implementation of the ELD in the EU*, 2011, available at: <http://ec.europa.eu/-environment/legal/liability/pdf/CEA_german_update.pdf>, last accessed on 8 May 2013.

originates from the insured site. Such a direct insurance policy emerged in the Netherlands since 1998.

In the Netherlands, before 1998 there were mainly three types of insurance cover for environmental damage, the AVB, the MAS and fire insurance. This system, however, was subject to much criticism. For example, the division of the cover under the AVB and the MAS was whether the risks are sudden or gradual. However, sometimes, there is a very thin line between these two. In addition, the insured's own site was neither covered under the AVB nor the MAS. Most importantly, liability coverage relates to the uncertainties of the tort system. This is especially true with environmental liability, where more uncertainties exist in proving causation and quantifying the damage. Hence, for the insurers to base the coverage on liability means higher unpredictability from the judicial system.[75]

The drawbacks mentioned above triggered the shift towards a direct insurance policy in 1998: the environmental damage insurance (MSV). The MSV policies were provided by an environmental insurance pool managed by the Verenigde Assurantiebedrijven Nederland (VAN).[76] At that moment, the insurers had limited capacity and were inexperienced in covering environmental damage. Hence, pooling together made it possible for the smaller insurers to be able to provide such coverage and gain experience.[77] The MSV is a combination of first-party insurance and direct insurance: it covers all the environmental damage occurring on or from the insured site, provided it concerns pollution of the soil or of the water.[78] It is worth noting that the VAN provides different types of environmental damage insurance. There were nine types of policies from which the insured could choose by the end of 2004.[79] Thus the insured could choose to cover only damage on his own land or also cover the damage suffered by third parties.[80] The MSV had some characteristics which distinguished it from environmental liability insurance. First, it also covered the damage on the insured's own sites, which is usually excluded by liability insurance. Secondly, the trigger of coverage was no longer liability, but the satisfaction of the requirements concluded in the contract: the insured risks on or from the insured sites. Hence the uncertainties caused by the involvement of the judge were avoided.[81] This insurance also benefited the victims, since they had the right of direct action against the insurance, thus it was no longer necessary for them to use liability against the insured.[82] The MSV policies had limitations as well. For example, in addition to the MSV, the liability system still applied. The insured could choose only to buy coverage for his own site but not extend it to third parties; or

75 Faure 2002b, p. 299-300.

76 VAN manages or provides knowledge support for several different insurance pools in the Netherlands; see the website of VAN, <www.verenigdeassurantiebedrijven.nl/-index.php/onze_disciplines/de_passies_van_van/nederlandse_milieupool/>, last accessed on 8 May 2013.

77 Interview with Mr. Kremers.

78 Faure 2002b, p. 299-300.

79 Interview with Mr. Kremers.

80 Faure 2007a, p. 92.

81 Faure 2007a, p. 93-95.

82 Faure 2007a, p. 92.

since the trigger of coverage is no longer liability, liability may not fall into the scope of MSV contract. Under this situation, since the MSV was meant to replace liability insurance, this means that the polluters cannot get coverage from the liability insurance market anymore and may not be able to cover his liability.[83]

After the introduction of the MSV, a few important shifts have happened. The first one is that the environmental pool, which used to provide MSV policies, was dissolved in 2007. This can be explained by two reasons: firstly, the strict competition law in the EU and the Netherlands made such a pool more difficult; besides, some members of the pool have gained sufficient experience and capacities to provide insurance by themselves.[84] Thus, rather than providing insurance itself, the VAN has become a knowledge center, which provides expertise support to insurers. This allows more room for competition in the market. The VAN still provides standard insurance policy drafts and has a tool to help insurers calculate premiums. But the individual insurers have the choice to determine their own insurance clauses and premiums.[85] Furthermore, the environmental damage insurance itself has also undergone some important changes. Since there were nine types of policies by the end of 2004 so that even the insurance intermediaries had difficulties to understand the differences, the VAN changed its policies into two types in 2005, the Environmental Impairment Insurance (EII) and the Underground Storage Tank Insurance.[86] The EII provides both coverage to the insured's own sites and the third parties' sites. It covers 'costs of remediation of the insured location and the third-party locations; damage and costs resulting from the remediation and damage to property caused by the pollution.'[87] The Clause has also an article guiding the split of the insured amount between the insured and the third parties. If the insured amount is insufficient to cover the whole damage, it shall be divided in the ratio of 50:50 between the insured location and the third-party location.[88] This is a complaint of the insured since, in case of a big accident, he has to wait until the insurer makes sure there will be no more claims which make the total damage in excess of the insured amount.[89] Another important shift is that the insurance no longer provides a right of direct action to the third parties. Thus if a third party wants to claim for the costs suffered on his site, he has to go through the insured.[90] This means that the third party's (victim's) status has been weakened. In spite of this change, the EII is still first-party and direct insurance but not a liability insurance. The trigger of coverage is not the liability but the pollution itself. It intervenes when the concentration of a pollutant in or on the soil/surface water exceeds the relevant government standards.[91] The administrative standards here,

83 Faure 2007a, p. 95.
84 Interview with Mr. Kremers.
85 *Ibid.*
86 The Clauses provided by Mr. Kremers, on file with the author.
87 Art. II.1 of the 2005 EII.
88 Art. II.3.2 of the 2005 EII.
89 Interview with Mr. Kremers.
90 *Ibid.*
91 Art. I.9 of the 2005 EII.

however, do not necessarily implicate liability. In the Netherlands, the government has the discretion to require remediation or not in case of soil pollution. Thus when pollution exceeds government standards, even if the government does not order remediation, the damage can be covered.[92]

The environmental damage insurance provides an alternative to liability insurance. It gives the polluters opportunities to cover damage on its own sites and also to cover the damage to other without resorting to the sophisticated tort system. However, admittedly, it has limitations as well. It targets SMEs most. While the bigger companies have more choices: they can use captives or buy liability insurance from U.S. insurers or their subsidiaries in Europe.[93] The expert mentioned that from 1998 until now, there were only approximately 700 claims against the VAN based on the environmental damage insurance, which led to a total amount of compensation of 8 million Euro.[94]

2.3. *Risk-sharing Agreements in Germany*

As discussed in the theoretical part, risk-sharing agreements can be used as an alternative to traditional insurance for less predictable risks, especially for the highly technical sectors. The practice in the nuclear industry provides such an example. In the U.S., the Price-Anderson Act establishes an obligation for the operators of nuclear power plants to pool together to cover their liability through a retrospective premium system.[95] A pooling system for the nuclear industry is also established in Germany, but through a contractual arrangement rather than as a result of legislative obligations. Before the description of the pooling system, this section briefly introduces the nuclear liability rules in Germany.

2.3.1. Nuclear Liability

In Germany, the nuclear liability clauses are mainly stipulated in the Act on the Peaceful Utilization of Atomic Energy and the Protection against its Hazards (Atomic Energy Act).[96] The Atomic Energy Act was passed in 1959 and was revised in 1985, 2002 and 2011. In addition to this act, Germany is also a party to the Paris Convention and to the Brussels Supplementary Convention, as well as to the Joint Protocol. According to the Atomic Energy Act, 'the Paris Convention and the Joint Protocol shall apply as national law in the Federal Republic of Germany, unless its provisions depend on reciprocity as effected by the entry into force of the Convention.'[97]

92 Interview with Mr. Kremers.

93 *Ibid.*

94 *Ibid.*

95 See *supra* Chapter 7.1.

96 *Atomgesetz (AtG) – Gesetz über die friedliche Verwendung der Kernenergie und den Schutz gegen ihre Gefahren*, an English copy can be found at: <www.oecd-nea.org/law/nlb/nlb-70/supplement.pdf>, last accessed on 8 May 2013.

97 Section 25 of the Atomic Energy Act.

The 1959 Atomic Energy Act aims both at promoting the use of nuclear energy and preventing damages.[98] The pro-nuclear policies continued until the 1990s when the Government started to make an effort to phase out nuclear energy. An agreement was reached on the gradual phasing out of nuclear energy in 2000 between the government and the industry[99] and the Act on the Structural Phase-Out of Nuclear Power for the Commercial Production of Electricity was passed in 2002. According to the act, there will be no further licenses issued for new nuclear power plants or reprocessing facilities, and the existing nuclear power plants are supposed to be closed before 2020.[100] In 2010, the German government decided to extend the operation of the existing nuclear reactors by an average of 12 years. However, the catastrophic Fukushima accident in 2011 in Japan triggered the change of nuclear policy in Germany again. The Government stopped the extension plan and came back to the phase-out policy again. The nuclear power will only serve as a bridging tool before the replacement with renewable energy. An amendment to the Atomic Energy Act was passed in the same year.[101]

According to existing act, the nuclear liability in Germany has the following characteristics. As in the international regime, the liability is channeled to the nuclear operators and the operators are strictly liable for the damage caused by a nuclear incident.[102] The liability is even stricter in Germany, since the defenses under the international regimes are no longer available, such as an armed conflict, hostilities, civil war, insurrection or a grave natural disaster of an exceptional character. But if the damage occurs abroad, the defenses are not applicable only if that country provides the same benefits to Germany.[103] One significant difference between the German and the international system is that an unlimited liability system is established in Germany. Only if the damage is caused by an armed conflict, hostilities, civil war, insurrection or a grave natural disaster of an exceptional character, the liability is limited to the maximum amount of the government indemnification.[104] To provide coverage for the potential liability, the operators are required to seek financial security. The administrative authority shall determine the type, terms and amount of the financial security. 'The amount of the financial security shall be fixed subject to a maximum of EUR 2.5 billion.'[105] Financial security can either be the third-party liability insurance or be in other forms. If the liability is not covered by or cannot be satisfied by the financial security, the federation shall indemnify the operator. The maximum amount of indemnification is set at 2.5 billion Euro, and the obligation of payment is the maximum amount minus the amount that is covered by financial security.[106] After

98 Carroll 2008, p. 90.

99 Mez & Piening 2002, p. 161.

100 For the phase out of nuclear power in Germany, see Vorwerk 2002.

101 <www.bmu.de/english/nuclear_safety/information/doc/4300.php>, last accessed on 8 May 2013.

102 Section 25(1) of the Atomic Energy Act.

103 Section 25(3) of the Atomic Energy Act.

104 Section 31(1) of the Atomic Energy Act.

105 Section 13 of the Atomic Energy Act.

106 Section 34(1) of the Atomic Energy Act.

the payment of the indemnification, recourse is possible if the operator disobeys specific obligations, or the operator caused the damage willfully or by gross negligence, or if the operator did not seek financial security to the required extent.[107]

2.3.2. Pooling System

In the beginning of the 1970s, an increase of the financial security up to DEM 1 billion was on the legislative agenda. The first DEM 500 million should be covered by private means while the government should indemnify the remaining half. The insurers and nuclear operators negotiated to cover liability up to DEM 500 million fully by insurance. The first DEM 200 million was covered by insurers while for the remaining DEM 300 million, the insurer only fronted the contract. The remaining DEM 300 million was re-insured by the operators of nuclear power plants as a whole. This arrangement remained valid until 2002.[108]

In 2002, the amendment to the Atomic Energy Act increased the amount of financial security up to 2.5 billion Euro and allowed financial security in other forms rather than through liability insurance.[109] At that moment, insurers only covered 256 million Euro. To face the dramatically increased requirement about financial security, although they were prepared to negotiate increased coverage, the insurers could not provide full coverage.[110] The operators started to find alternatives: in 2001, the four parent companies of the 19 nuclear power plants negotiated and concluded a 'Solidarity Agreement' to provide the remaining coverage mutually.[111]

Under this agreement, the companies pool together to cover the second tier (in addition to the first 246 million Euro by the insurers) up to 2.24 billion Euro. Each party has an obligation to contribute a percentage of the total amount in case damage is attributed to one of the parties. The percentage for each nuclear power station is calculated according to the square root of the thermal reactor output. The percentage of power plants is then attributed to the parent companies on the basis of their participation (Clause 1(3)). This allocation of liability is different from that in the U.S., where each operator bears the same quota. In Germany, the allocation of the contribution is based on the generating capacity. As in the U.S., the obligation to make the contribution only comes due after a damage in excess of the insurance capacity happens. However, the possibility for the operators to contribute is even smaller in Germany: the partners only have to pay if neither the operator nor the parent company are in a position to pay up to 2.5 billion Euro (Clause 1(5)). In this case, the solidarity agreement is only a guarantee for the payment of liable parties. After the payment, the partners have a right of claim for repayment on the liable parties. But the liability for third parties prevails over the claims for recourse (Clause 1(8)). In addition to mutually guaranteeing the coverage of liability, the partners also need to provide help in handling the claims such as to make available

107 Section 37 of the Atomic Energy Act.

108 Pelzer 2007, p. 43.

109 Sections 13, 14 of the Atomic Energy Act; Carroll 2008, p. 91.

110 Carroll 2008, p. 91.

111 Pelzer 2007, p. 44, note 24.

legal and commercial staff capacity and infrastructure. For these kind of aids, the partners cannot ask for repayment. The partners also provide help for the use of independent contractors, up to the amount of 122, 218 million Euro (Clause 2). To ensure the availability of assets in case of damage, the partners need to submit an auditor's certification each year (Clause 3).

Both the U.S. and Germany establish a retrospective pooling system to cover the nuclear liability. Both pooling systems provide a second tier of compensation: the first tier is made available through commercial insurance. However, some differences do exist: the U.S. system is established on the legislative obligations arising from the PAA, while the German arrangement is achieved through contract. The content is hence subject to changes in the contract. In addition, each parties' contribution is the same in the U.S. and is pre-determined in the statute, thus the total amount of compensation is dependent on the number of operators. In Germany, the maximum amount of compensation is set in the contract and the percentage of each partner depends on their involvement in the power plants and the generating capacity. In the U.S., the individual liability of the operator is limited to the capacity of financial security, while in Germany, the operators are still liable if the capacity of the pooling is depleted. Another major difference lies in the conditions of payment. In the U.S., the obligation of retrospective premiums comes due when damage in excess of the insurance capacity happens and the operators cannot get recourse from the liable parties. In Germany, the pooling is only a guarantee: it pays in case the liable party fails to pay and the partners who have paid can claim for the repayment.

Research has shown that a risk-sharing agreement is desirable when there is not sufficient actuarial information *ex ante* to make the risk insurable and mutual monitoring is possible.[112] In the field of the nuclear industry, the operators have better access to risk information than the insurers. Furthermore, for a risk-sharing agreement to work, the *ex ante* information about the probability of damage is sufficient, while the insurers also need the information on the magnitude of potential damage for insurance to work. Another important issue is that the German nuclear pooling system is established within a country where the same economic conditions and legal framework apply. Mutual monitoring is also possible through the nuclear regulatory authorities, thus the pool members have a comparable level of nuclear risks. By confining the pooling scope within one country, the difficulties perplexing the establishment of an international/European pool are avoided.[113] Another advantage of the pooling system is that it is less costly. The *ex ante* payment of a premium is necessary for insurance and the premium is not refundable if no accident happens. In the German risk-sharing pool, the obligation to pay for contributions only comes due in case of damage passing the triggering level and the liable parties fail to provide sufficient compensation.

112 Skogh 1998, p. 247-264 and Skogh 2000, p. 521-537.

113 Pelzer 2007, p. 54-55; Carroll 2008, p. 94-96.

2.4. *Environmental Funds in the Netherlands*

The theoretical part has shown that environmental funds can intervene in the compensation for ecological damage in different situations: for example, they can provide a higher level of compensation; only intervene when liable parties and other guarantors/insurers failed to do so, advance the cleanup procedures and so on. A full discussion of all the environmental compensation funds used in the EU is outside the scope here. This part discusses one example, the SUBAT,[114] a compensation fund for historical pollution in fuel outlet sites in the Netherlands, to show how environmental funds can play a role to compensate for ecological damage.

The SUBAT foundation is a fund established to promote the cleanup of historic soil pollution in fuel outlets. To understand why this fund was created and how it works, the soil pollution liability rules in the Netherlands are addressed briefly here. The soil pollution in the Netherlands is mainly shaped by two acts: the *Interimwet Bodemsanering* (Ibs)[115] the *Wet Bodembescherming* (Wbb)[116] and their amendments. The Ibs came into force in 1983 and remained valid until 1994. Under the Ibs, the central government, represented by the Minster of the Environment (MOE), plays an important role in determining the decontamination of polluted sites. The MOE can recover the costs of research and cleanup on two grounds: the tort of negligence and unjust enrichment. The Wbb was published in 1986 and in 1994 a revision merged the Ibs and the Wbb. Under the Wbb, a more decentralized and privatized approach is adopted. Furthermore, in addition to taking cleanup measures and claiming for the costs, the government can also give decontamination orders to the polluters, owners of polluted soil or holders of a ground lease. Although the Wbb tried to apply the polluter pays principle to historical soil pollution, this attempt has been proved to be unsuccessful. The decontamination orders for historical pollution were hardly given.[117] Thus, a great difference between the liability rules for soil pollution in the Netherlands and under the CERCLA is that under the former system, a retroactive liability for historic damage was not established.

Since the attempt to find liability for historic pollution is not successful, the cleanup for such sites falls on either the government or some other instruments. The SUBAT foundation is such an instrument. In 1991 a regulation was announced to tighten the requirement for gas stations to prevent soil pollution, and the regulation was finally adopted in 1994.[118] For many stations, the requirements were not easy to fulfill. But to terminate the activities there were costs as well. The SUBAT was created to make it possible for (smaller) gas stations to terminate their activities without paying for the historic pollution. The SUBAT came into force in 1991. The owners of gas stations who wanted to terminate their commercial activities within

114 *Stichting Uitvoering Bodemsanering Amovering Tankstations.*

115 The Interim Act on the Decontamination of the Soil.

116 The Soil Protection Act.

117 Verheij 2007b, p. 39-49.

118 Faure 2007a, p. 81.

three months after signing the agreement with the SUBAT and would not open a new gas station at the same site could apply to it before 31 August 1993. The cleanup of those sites would then be executed by the SUBAT and financed by the oil companies. The Companies can pass it on in the oil price. Therefore, it is the consumers who in the end pay the price for the cleanups of soil pollution.[119] However, this cannot be immediately judged as inefficient. The short deadline to apply for the fund ensures that only historic pollution is covered. This makes sure that no distortive incentives are given to the future actors.

The SUBAT is just one example showing how environmental funds are used to help cleanup historic pollutions. There are some other similar funds available in other Member States as well. For example, in Belgium, an environmental fund called *'Bofas'* (*Bodemsaneringsfonds voor tankstations*) also exists to help the decontamination of petrol station soil.[120]

3. Critical Analysis

The ELD introduced a response and compensation system for ecological damage in Europe. However, its application is limited to damage to protected species and habitats, protected waters and land. For the other types of damage, the domestic law of the Member State applies. Hence, a full analysis of the compensation system in Europe is not feasible here. This section tries to briefly analyze the compensation system introduced by the ELD and to discuss a few examples of compensation mechanisms mentioned above.

The ELD adopts two types of liability: strict liability for the dangerous activities listed in the Annex and negligence liability for the others. This is in line with the economic criteria. Strict liability gives the parties engaging in dangerous activities incentives to adopt both efficient care levels and efficient activity levels. While for a normal activities, negligence rules can (when the due care levels are set properly) on the one hand create efficient care incentives, and on the other hand deter the actors from engaging too much in socially unbeneficial activities. The ELD does not introduce a cap on liability. Thus a solvent liable party will face the whole damage it causes. Although the ELD imposes liability on 'operators', it defines the term 'operator' broadly to include anyone 'who operates or controls the occupational activity or, where this is provided for in national legislation, to whom decisive economic power over the technical functioning of such an activity has been delegated.'[121] Furthermore, it does not bar the liability of others who may contribute to the risk or prevent the recourse towards other parties. Hence, the ELD does not channel liability to specific parties, which corresponds with the economic criteria as well. In spite of these desirable characteristics, whether the ELD will create an optimal deterrent effect depends on many practical issues as well. The application of the ELD is quite limited in terms of the types of damage. Thus many activities

119 Faure 2007a, p. 81-82.

120 Verheij 2007b, p. 63.

121 Art. 2(6) of the ELD.

still fall outside of its scope. Moreover, the ELD system is not a pure civil liability system; it has many public law elements as well.[122] Whether the liability procedure will be triggered to a large extent depends on whether the government takes or orders preventive or remedial measures.

Earlier discussion shows that the ELD aim at establishing a common framework of the prevention and remediation of environmental damage. However, it leaves plenty of room for implementation by the Member States. There are many differences among Member States in transposing the ELD, including whether to expand the scope of application, types of activities covered under the strict liability, defenses and liability in the case of multiple tortfeasors. As for the prevention function of the ELD, the above analysis shows that the adoption of strict liability/negligence rules and non-channeling of liability is in line with economic criteria. However, there are some other features which may jeopardize the preventive effect as well. For example, the application of the ELD is very limited. Although some Member States have chosen to broaden the scope of protected species and natural habitats, the overall application of the ELD is still narrow. Moreover, the threshold of significant damage is not always easy to satisfy. The above analysis shows that for damage to the protected species and habitats, there should be significant effects on the favorable conservation status and a measurable adverse change in relation to the baseline condition. These criteria create a high threshold. For example, research shows that this threshold is not likely to be exceeded in many cases in the Netherlands.[123] Moreover, to determine the baseline condition, much information is needed, which poses another challenge in many situations.[124] The limited environmental components covered by the ELD, together with the high threshold, make the coverage of the Directive very narrow.

In addition, the available defenses also influence the preventive effect of liability rules. The permit defense allows the operators not to bear the remedial costs if 'he demonstrates that he was not at fault or negligent and that the environmental damage was caused by an emission or event expressly authorised by, and fully in accordance with the conditions of, an authorisation conferred by or given under applicable national laws and regulations which implement the legislative measures adopted by the Community specified in Annex III, as applied at the date of the emission or event.'[125] Under a state-of-the-art defence, the operator is not liable if his activity 'was not considered likely to cause environmental damage according to the state of scientific and technical knowledge at the time when the emission was released or the activity took place.'[126] With a permit defense, the potential liable annex III Operators may not have strong incentives to invest in risk-minimizing measures hence they can escape liability as long as they comply with

122 See Section 1.

123 Backes *et al.* 2005, p. 11-22, cited in Van den Broek 2009, note 23.

124 Van den Broek 2009, p. 123-124.

125 Art. 8(4) of the ELD.

126 Art. 4(b) of the ELD.

the permit. The limited scope of the ELD, together with the availability of permit and state of art defenses may limit the preventive effect of the Directive.[127]

The ELD encourages the development of the financial security market, but does not itself introduce a mandatory financial security system. This is also criticized by economists as causing under-deterrence. When strict liability applies, if the liable operator is insolvent, he will only have incentives to internalize the costs up to its own assets, but not the whole costs it has created. Under this situation, compulsory financial security can relieve the under-deterrence problem.[128] However, during the promulgation of the ELD, many states opposed the introduction of a compulsory financial security system. Although a few insurance products began to develop after the adoption of the ELD, most insurers still argue against a compulsory scheme.[129] In 2010, six years after the issuing of the ELD, the report reviewing the financial security practice still held that a mandatory system could not be justified.[130] This is understandable since until now, there have been very limited ELD cases which do not create a high demand of the insurance products and, on the other side of the coin, the insurance products are also limited and insurers are reluctant to accept a compulsory system. Under this situation, a mandatory system does not necessarily lead to better practice. A progressive mandatory system is under way. For example, in Spain, a mandatory system was introduced. However, it is reported that many operators still are not aware of this or are waiting for more information about the exact products they need to buy.[131]

In addition to the ELD, this section also discussed some other instruments used by the Member States to cover environmental damage. These limited examples are not meant to sketch a full picture of the compensation system in Europe but just to show how different compensation instruments can contribute to better prevention and compensation. In addition, those instruments are not designed specifically for ecological damage, but are used also to cover traditional damage caused via the environment.

In Germany, the Environmental Liability Act of 1990 established the basic rules for death/personal injury and property damage due to environmental issues. This Act also requires mandatory financial security from some facilities. However, in practice, such a provision is rarely used. It is reported that most environmental liability insurance policies are taken voluntarily but not because of the mandatory requirement.[132] In spite of this, the interaction between liability and insurance contributed substantially to the decrease of environmental accidents.

In addition to liability insurance, other alternatives are also available on the European market. In the Netherlands, direct insurance was introduced in 1998 in

127 De Smedt & Faure 2013.

128 Faure & Grimaud 2003, p. 39.

129 De Smedt & Faure 2013.

130 European Commission 2010.

131 ACE, A Guide to Environmental Liability in Europe, at 10-11, available at: <www.acegreen.com/files/documents/newsfiles/ELD%20Guide%202011.pdf>, last accessed on 8 May 2013.

132 See Section 2.1.

place of liability insurance. Under such a policy, the trigger of coverage is damage itself but not liability. Hence it can avoid the uncertainties resulting from decisions by the judges and the insured's own sites are also covered. This policy was originally provided by an environmental pool, which allows the individual insurers to accumulate experience and capacity. Now the pool has dissolved and different insurers can provide their own environmental damage insurance products. Hence more competition is possible.

The German nuclear operators face a high financial security requirement: 2.5 billion Euro, which is difficult to obtain from the insurance market. Hence the nuclear operators pool together to provide guarantees for themselves. If a disaster leads to higher damage than the amount available from insurance, and the liable operator's capacity, all the nuclear operators in Germany will make contributions to cover such damage. Such a risk-sharing arrangement is possible and desirable in some highly technical sectors, since the operators themselves have better information than the regulators or judges and can monitor each other for achieving a better safety record.

The theoretical discussion shows that compensation funds usually have some disadvantages when compared to more market-based instruments in terms of higher administrative costs and less preventive incentives. The SUBAT established in the Netherlands is a compensation fund to help the gas stations who want to close their operation to cover their historical pollution. Although it is the oil companies but not the gas stations themselves who contribute to the fund, it cannot be judged as inefficient immediately. Since it covers only historical pollution and has a short notice period, it will not create distortive incentives for future behavior. Moreover, such a fund also makes the cleanup of the gas station sites and the closure of some gas stations possible.

Chapter 9

CONCLUSION OF PART II

The compensation systems in the international regimes, in the U.S. and in the EU are analyzed respectively in Part II. This section first gives a short summary of those different systems, and compares them briefly. It is worth noting that this part does not intend to give a tentative checklist of all the similarities and differences between the systems. However, the examination of how compensation systems for similar types of damage work differently among jurisdictions can shed light on how to design compensation systems in various scenarios. After the comparison, this part tries to analyze whether the practice is in line with the theoretical standing points elaborated in Part I. For example, do the compensation models, with the combination of regulation, liability rules and compensation instruments work in practice? Does the application of a compulsory and voluntary financial security system comply with the theoretical criteria? How are different compensation instruments used in various situations? If a conflict between theory and practice exists, the question also arises what the reasons are for this divergence.

1. A Comparison of the Systems

In the international regime, two areas have been chosen, oil pollution and nuclear damage. Two sets of conventions sketch the compensation framework for vessel-induced oil pollution, the CLC (Fund convention Supplementary Convention) and the Bunker Convention. Under the CLC, liability is channeled to ship owners. The liability is capped, and guaranteed by a compulsory financial requirement. Risk-sharing pools, the P&I Clubs are used broadly to cover oil pollution liability. In addition, international compensation funds (the 1971 Fund and the 1992 Fund) have been established to complement the compensation. The international compensation system for nuclear damage has also some similar characteristics: the liability is channeled, strict and capped. Insurance is used to guarantee the first layer of compensation. State funds and collective funds can be used to provide a second and third layer of compensation.

A comprehensive compensation system for ecological damage has been developed in the U.S. under federal statutes, especially the CERCLA and the OPA. In case of a release (threat) of hazardous waste/oil pollution, these acts establish an obligation to respond and restore. Assessment standards and procedures are also developed according to these acts. Considering their similarities, those two systems are discussed together in the same section. They have direct relevance for ecological damage, under the title of responses costs (removal costs) and natural resources damage. The liability is strict and mostly capped.[1] Unlike in the international regime, liability is not channeled to specific parties. Under the OPA, not only ship owners but also other risk contributors can be held liable. Only the oil companies can escape liability and contribute in the form of compensation funds. The CERCLA defines the liable parties so broadly that it is often criticized for causing a chilling effect on the redevelopment of brownfields. In addition, one unique feature of the CERLCA liability is that it applies retroactively. Both systems establish a financial requirement for certain types of potentially liable parties. Different instruments can be used to realize such an obligation. Funds are created under both systems, which do not only compensate for the claims unsatisfied under the liability framework, but also play an important role in advancing the costs of response/removal measures.

Another interesting system in the U.S. is the compensation system for nuclear damage. The PAA establishes a nuclear liability framework, which is strict, economically channeled to the operators and capped. A multi-layer compensation system is established: the first layer is guaranteed by liability insurance (ANI Insurance). In case of damage in excess of this amount, all the nuclear power plant operators are asked to contribute retrospectively up to $111.9 million. The liability insurance, together with risk-sharing pools, makes the possible compensation for a nuclear incident amount to up to $12.2 billion.

The ELD is also briefly discussed in this part. The ELD only applies to specific categories of damage to the environment. Two forms of liability are created (strict/negligence rules). The development of financial security mechanisms are encouraged under this system. However, the Directive itself is still cautious in requiring a compulsory financial security system. Much discretion has been left to the Member States. Since the transposition of the Directive has only been completed recently, it is too early to judge how it works in practice. A comprehensive study on how different compensation instruments are used in all the EU Member States is outside the scope here. This research chooses a few examples to show how the individual instruments are used in the Member States, such as the liability insurance and risk-sharing pools established in Germany, and the direct insurance and environmental fund (SUBAT) in the Netherlands.

A short comparison of the systems is provided as follows. The covered risks, liability rules and financial security systems are the main factors to be compared.

1 Noticing one exception under the OPA: for the offshore facility except a deepwater port, the removal costs are not capped. 33 U.S.C. §2704 (a).3.

Table 4: Comparison of Different Compensation Systems

Systems	Covered Risk	Liability		
		Liability Standards	Channeling	Standing (for ecological damage)
International Oil Pollution(CLC, Fund Convention and Supplementary Convention)	Traditional damage, Preventive costs, Impairment of the environment limited to reasonable restoration costs	Strict liability	Channeled to ship owners; Recourse towards other parties possible	
International Nuclear Damage (second generation of NEA system as an example)	Personal and property damage; the economic loss arising from personal injury and property damage; the costs of reinstatement of the impaired environment; the loss of income deriving from a direct economic interest in any use or enjoyment of the environment, and the costs of preventive measures	Strict liability, less defense available	Channeled to nuclear operators; recourse not possible	
The U.S. PAA system	Public liability, mainly bodily injury and property damage	Strict liability for 'extraordinary nuclear occurrence'; others depend on state law	No legal channeling	
The OPA	Traditional damage; removal costs, natural resources damage (including restoration costs, interim losses, assessment costs)	strict	Not channeled; broad definition of liable parties, to include ship owners, operators and bareboat charters	Federal, state and local natural resources trustees and Indian tribes have standing according to federal legislation; common law also allows claims for natural resources damage; citizen suits make it possible for individuals to push public authorities to act.
The CERCLA	Response costs and natural resources damage	Strict and retroactive	Not channeled, very broad definition of liable parties, including: the owner and operator of a vessel or a facility; the owner and operator when the hazardous substances are disposed of; those who arrange for disposal (generators) and certain transporters	

Systems	Covered Risk	Liability		
		Liability Standards	Channeling	Standing (for ecological damage)
ELD	Three types of environmental damage: damage to protected species and habitats; water damage and land damage	Strict liability for listed dangerous activities and negligence when other activities lead to damage to protected species and habitats.	Not channeled. Operators are liable parties. But the definition of operators is broad.	The competent authority in Member States

Systems	Liability	Financial Security and Compensation Instruments			
	Cap of Liability	Compulsory or voluntary financial security	Insurance	Risk-sharing Pool	Compensation Funds
International Oil Pollution (CLC, Fund Convention and Supplementary Convention)	Depends on size of vessels, varies from 4.51 million SDRs to 89.77 million SDRs Loss the right of limit when causing damage intentionally or recklessly	Yes, up to liability cap; can use insurance or other financial security such as bank guarantees.		Most commonly used instrument. Covers both general liability and oil pollution liability	Two layers of compensation funds, makes total amount of compensation up to 750 million SDRs. Covered risk: no liability established; liability caps exceeded; liable parties and guarantors unable to pay; Contribution: oil companies pay; based on amount of oil transported
International Nuclear Damage (second generation of NEA system as an example)	No less than 700 million Euro; Member States may allow unlimited liability	Yes, up to liability cap	Usually provided by national pools		Two tiers of compensation funds from Installation State or Collective States; up to 1500 million Euro; How to contribute depends on Member States

Systems	Liability	Financial Security and Compensation Instruments			
	Cap of Liability	Compulsory or voluntary financial security	Insurance	Risk-sharing Pool	Compensation Funds
The U.S. PAA system	Individual financial security plus retrospective premiums	Yes, financial security for liability up to the maximum amount from private sources; Insurance, indemnity agreement with government can be used. Financial security for property damage also required, priority given to decontamination	Liability insurance provided by a national insurance pool ANI; the available amount now is $375 million	When damage exceeds the insurance capacity, a retrospective premium scheme applies; each nuclear operator shall pay up to $111.9 million; total amount reaches $12.2 billion. Property insurance provided by a mutual pool NEIL	No fund established
The OPA	Capped with the exception of the removal costs of offshore facility except a deepwater port; caps for vessels depends on both types and size; lose right of limitation if gross negligence, intention or violating administrative regulation; state law may establish higher cap or even unlimited liability	Mandatory; for vessels, required amount is liability cap; for other facilities, depends on types of facilities. Different instruments can be used: insurance, surety bond, guarantee, letter of credit, qualification as a self-insurer or other evidence		P&I Clubs also play a primary role; insurance and other mutual pools often fill the gaps left by P&I Coverage.	Fund established; covers uncompensated claims and pre-finances removals, natural resources damage assessment costs; Contributions: tax on oil companies; transfers; interest of investments, recovery from liable parties and their guarantors.

Systems	Liability	Financial Security and Compensation Instruments			
	Cap of Liability	Compulsory or voluntary financial security	Insurance	Risk-sharing Pool	Compensation Funds
The CERCLA	Capped, amount depends on type of facilities	Mandatory for certain vessels and offshore facilities; EPA is considering also establishing a mandatory requirement for the mining industry	Liability insurance crisis in 1980s; now EIL policies broadly available		Superfund established; Cover uncompensated claims, response costs, and restoration costs. Taxes on polluters used to be the major financial source; after tax expired, the major source is appropriation from general fund
ELD	Not capped	Voluntary; a few Member States introduced a mandatory system	Products started to be available; but the penetration rate is low.		Not yet. The proposal to establish an environmental fund is under research.

This part discusses compensation systems for similar types of damage in different jurisdictions, such as compensation systems for vessels-induced marine oil pollution and nuclear damage in the international regime and in the U.S. The compensation system for natural resources damage in the U.S. and the compensation systems under the ELD are also comparable. Those examples provide interesting examples showing the possible designs for one type of damage and their pros and cons. For the compensation systems for oil pollution, the U.S. system allows much broader compensable ecological damage (removal costs plus natural resources damage) compared to the international system (preventive and reasonable restoration costs). Strict liability and mandatory financial security is adopted in both systems. But instead of channeling liability to ship owners, as in the international system, the U.S. system allows other parties, such as operators or bareboat charters, to be held liable. Futhermore, the liability cap in the U.S. is higher than the international one, with the potential of uncapped liability under State law. The U.S. cap is also more tailored to the risks of different vessels, with more possibility for the liable parties to lose the chance of limitation. Such a design can create more preventive incentives for the potential liable parties.[2]

Another example is compensation for nuclear damage. Some similar principles are adopted in both the international and U.S. systems, such as absolute liability and capped liability with mandatory financial security. However, there are significant differences as well. Liability is channeled to nuclear operators in international conventions. Hence other parties who may also contribute to risks are exonerated from liability and have no incentives to take sufficient precaution. In the U.S., liability is only economically but not legally channeled to nuclear operators: all parties causing damage can be sued in the court when a cause of action can be established; but the heaviest burden to seek financial coverage is imposed on the nuclear operators. A nuclear accident may turn out to be catastrophic and insurers alone cannot provide sufficient compensation. To provide a higher level of compensation, in addition to the commercial insurance market, the international conventions established a two-layer system financed by funds from the installation state or all states. This is criticized by economists as being subsidies to the nuclear industry and hence creating distortive preventive incentives. The U.S. system establishes a retrospective premium system, sharing the risks among operators and it also makes possible the coverage of more than $12 billion. The risk-sharing system is supported by economic literature as creating mutual monitoring incentives for operators.

The more relevant systems for ecological damage are the compensation systems for natural resource damage in the U.S. (under the CERCLA and the OPA) and the compensation system under the ELD. The coverage is different under two systems: the former covers various types of natural resources, while the latter covers only three types: protected species and habitats, water damage and land damage. There is a differentiation of removals and natural resources damage while such a differentiation does not exist for the ELD, which establishes one system for both preventive and remedial measures. A comprehensive assessment procedural has been in place in the U.S. for a long time and much experience with compensation has been accumulated over the years. In the EU, however, very few ELD cases have been reported and the evaluation of such damage is still a challenge. A compulsory financial security system is established under the CERCLA and the OPA, but in the EU, several years after the coming into force of the ELD, it is still reported that a compulsory system is not feasible. The lack of ELD cases and the reluctance of insurers to accept a compulsory system might be explanations for this voluntary system.

2 For a more detailed comparison between the compensation systems for vessel-induced marine oil pollution in the U.S. and the international conventions, see Liu, Faure & Wang, 2013.

2. The Compensation Models

The empirical part focused on the liability rules and compensation instruments in different jurisdictions. However, this does not suggest that regulation in those areas is absent or unimportant. On the contrary, regulation has been used as the primary instrument to prevent environmental damage in many areas discussed above. For example, a series of international conventions are in place regarding nuclear safety.[3] There is also an international comprehensive regulatory framework to prevent marine oil pollution.[4] The regulatory systems for ecological damage are complicated issues on which literature has addressed extensively. This research focuses on the compensation systems for ecological damage, hence I focused on liability rules and compensation mechanisms. However, in spite of the limitation of the research scope, one has to notice that in the areas discussed above, regulation, liability rules and compensation instruments are used in combination. Regulation is used broadly in the oil sector, nuclear sector and the sector producing hazardous waste to prevent environmental damage. Liability rules are also established in diverse forms. Among those systems, there are both examples of compulsory financial security and voluntary financial security systems. Most examples fall in the former category, and usually different compensation instruments can be used to satisfy the obligation to provide financial security. For example, under the CERLCA and the OPA, the potential liable parties can choose between insurance, surety bonds, guarantees, letters of credit, qualifications as a self-insurer or other evidence.[5] Under the ELD, the establishment of a financial requirement is also considered. However, considering the lack of experience of the new types of liability and insufficiency of products covering such risks in the market, many Member States oppose the establishment of a compulsory system at this moment. Hence, a voluntary system is adopted under the ELD.

As regards the choice between a compulsory and a voluntary financial security system, there are two interesting cases. The German Environmental Act of 1990 requires the operators of certain facilities to provide financial security. However, at that moment, the environmental impairment insurance policies were not generally available. This provision has not been broadly implemented. The insurance market began to develop in the subsequent years. However, with the interaction between liability and insurance, many industrials started to have the incentives to seek insurance coverage. It is reported that only 1 per cent of environmental insurance policies taken out by the industrials are compulsory. This interactive framework also helps to reduce the environmental accidents.[6] In the U.S., CERCLA requires certain vessels and offshore facilities to provide financial security. It also requires the EPA to promulgate financial requirements for other risky facilities. However,

3 Examples are the Convention on Nuclear Safety, Convention on Physical Protection of Nuclear Material, the Convention on Assistance in Case of a Nuclear Accident or Radiological Emergency and the Convention on Early Notification of a Nuclear Accident.

4 For the details, see Wang 2011, p. 19-52.

5 33 U.S.C. §2716(e); 42 U.S.C. §9608(a)(1).

6 Schwarze & Hoffmeister 2010, p. 419.

this provision remains unimplemented until now. There are indeed, some sectors which possess high environmental risks and may become insolvent. For example, mining is such a sector. It is reported that the 63 hard rock mining sites on the Superfund's NPL may cost up to $7.8 billion to clean up. Given the long-term nature of the cleanup liability at mines and the financial statutes, it is estimated that $2.4 billion of them will have to be borne by taxpayers.[7] In response to this problem, the EPA started a procedure to develop a financial responsibility requirement for mines.[8] Those two examples show an interesting interaction between liability rules and financial security mechanisms. In a voluntary system like Germany, liability rules give risk-averse industries incentives to seek environmental insurance coverage. The insurance, with risk differentiation and risk control policies, in the other way around, create preventive incentives for the industries. Therefore such an interactive, voluntary system also helps to reduce environmental accidents. In the U.S., a voluntary system is also applicable for mines. However, the cleanup for mining has a long-term nature. A mine has a high risk of insolvency even after the cleanup starts. Thus they have little incentives to internalize the costs by seeking insurance coverage. Different from the case in Germany, the voluntary system in the U.S. failed to create sufficient preventive incentives.

3. The Choice between Different Compensation Instruments

In the compensation system discussed in Part II, different combinations of compensation instruments are used. Liability insurance is the most commonly used instrument, which is usually used to guarantee the first layer of compensation.

The liability insurance market depends, to a large extent, on the predictability of liability. A retroactive and expanding explanation of liability and coverage provisions has contributed to the liability insurance crisis in the U.S. in the 1980s. To avoid the judicial uncertainties, a direct insurance was adopted in the Netherlands in late 1990s. Direct insurance covers certain types or risks decided in the insurance contract, regardless whether liability can be established or not.

The use of risk-sharing agreements can be found in the shipping and nuclear sector. In those sectors, ship owners/nuclear operators have much better information about their risks and precaution measures than insurers. They can also control their risks at a comparable level through some self-regulation or regulating bodies. The development of P&I Clubs and nuclear pools are in line with the theoretical criteria.

Environmental funds can also be found in some systems. They can either provide an upper layer of compensation or intervene when no liability can be established. Environmental funds can also cover cleanup and restoration costs before liability is established, hence to accelerate cleanup and restoration procedures.

7 GAO 2005, p. 36.

8 See Chapter 7, Section 3.2.

4. The Divergence of Practice from Theoretical Models and Indicators

4.1. *Inefficiency of Liability Rules*

The above analysis identifies some parallels between the theory and practice. However, the practice can in some cases also divert from the theory sometimes. One example is that liability rules are not always formulated in an efficient manner.

The theory has clarified that liability rules for ecological damage should be uncapped, un-channeled and un-retroactive. However, the contrary features can often be found in liability rules in practice. For example, the international liability systems for oil pollution and nuclear damage are both channeled (to ship owners or nuclear operators) and capped. This can sometimes be explained from a private interest perspective.[9] The legislative or negotiating procedure is usually filled with bargaining by different interest parties. For example, the international nuclear liability conventions are shaped to a large extent by the U.S. delegates. To protect the nuclear suppliers from liability for damage abroad, the channeled liability was adopted. The investor was afraid of the potential catastrophic liability, the pro-nuclear government also wanted to encourage the development of nuclear industry. Thus the nuclear liability compensation established a rather low liability limit.

4.2. *Broad Use of Environmental Funds*

The theoretical analysis shows that environmental funds have some disadvantages compared to other individual compensation instruments. For example, the financing structure makes the funds usually less risk related. The operators of funds have less incentives to control costs and so on. Thus the funds should play only a complementary role. However, in practice, the funds are used very broadly. Thus, these are sometimes criticized as a subsidy to the liable parties, such as the fund for nuclear damage under the international regimes. A private interest approach can again provide an explanation here (governments do not want to overburden their nuclear sector). In addition, sometimes, funds are established to balance the interests among different parties. For example, the CLC channeled liability to ship owners. Such an agreement was reached on the condition of the establishment of a fund financed by the oil industry. Besides, funds can also be used only to finance the cleanup of historical pollution (SUBAT). Therefore, it will not create distortive incentives in the future.

4.3. *Implementation of Financial Requirement*

Theoretical research shows that the combination of regulation, liability rules and financial requirements can be used to prevent and compensate for ecological

9 For the discussion of liability for oil pollution in the international system, see Wang 2007, p. 197-241. For the international compensation system for nuclear damage, see Faure & Vanden Borre 2008.

damage. However, a simple financial requirement cannot guarantee the implementation itself. For example, the CERCLA establishes financial responsibility for certain vessels and offshore facilities. A report shows that the EPA has not implemented the financial responsibility provision sufficiently. For example, the EPA allows companies to choose among financial assurance mechanisms freely without considering their default risk. The use of a financial test is criticized as possessing a high default risk and needs more oversight. The EPA is also criticized as providing inadequate oversight over the enforcement of financial assurance requirements and even as maintaining no related enforcement data.[10] The lack of implementation creates a risk for the recovery of cleanup costs by the government. This example shows the importance of maintaining an efficient implementation of financial requirements.

The various systems discussed above can provide experiences for China in developing its own compensation system. For example, a clear definition of compensable damage and an evaluation procedure like the case in the U.S. is very important to establish liability for ecological damage. The insufficient data on the baseline situation of the damaged environment, which has puzzled the EU in implementing the ELD, also poses a big challenge in China. To overcome the hurdle of standing, especially for public natural resources, public authorities can be granted *locus standi* explicitly in statutes. To allow NGOs or individuals to claim in case the public authorities fail to use its authority (properly) can create additional deterrence. Furthermore, a compulsory financial security, such as is the case under the CERCLA and the OPA, can help to alleviate the insolvency problem. However, the feasibility depends on the demand for compensation instruments (such as insurance) and the willingness and capability of the market to provide such policies. A case on point is the voluntary financial security under the ELD.

[10] GAO 2005, p. 39-53.

PART III

TOWARDS AN EFFICIENT COMPENSATION SYSTEM IN CHINA

Chapter 10

COMPENSATION SYSTEM FOR ECOLOGICAL DAMAGE IN CHINA AND POLICY RECOMMENDATIONS

1. Introduction

The above parts have proposed theoretical models, with the combined use of regulation, liability rules and compensation instruments to prevent and compensate for ecological damage. Then a few examples in the international regime and different jurisdictions were provided to show how the theoretical models work in practice. In this part, how ecological damage is compensated in China will be sketched.

Many have pointed to the fact that the spectacular economic growth in China has come at a high price, including the serious damage to the natural resources.[1] There is increasing literature available now, both in and outside of China, on the available governance instruments that China is increasingly using to prevent natural resources damage.[2] An issue which received relatively less attention in China so far was the *ex post* compensation of victims and restoration of the environment for such a type of damage, especially when no individual damage is involved. This section will try to address the question whether an obligation to restore the environment exists and to what extent natural resources damage, apart from personal injury and property damage, can be compensated in China. In addition, how compensation instruments can and are used in China to promote compensation for ecological damage is also discussed here. As in other compensation systems discussed in Part II, the compensation scheme for ecological damage cannot always be separated from the scheme for traditional damage. Hence, when relevant, liability rules and compensation instruments for traditional damage are discussed as well.

The general environmental liability rules have been established under the General Principles of Civil Law, the Environmental Protection Act and the newly

1 See Chapter 1, Section 2.

2 See for example, Ma & Ortolano 2000; Mol & Carter 2006; Carter & Mol 2007; Lan, Simonis & Dudek 2007; Chen 2009.

adopted Tort Liability Law. However, the general rules make no explicit reference to natural resources damage. General hurdles exist in *locus standi*, causation and the assessment of natural resources damage.[3] The only domain in which compensation seems to work better, not only on paper but also in practice, is the domain of marine oil pollution in which international conventions have played an important role.[4] In this research, we use marine oil pollution as an example to show how natural resources damage can be compensated in China. Not only liability rules are discussed, this research also tries to find out how a financial security system can contribute to better prevention and compensation. Therefore, the insurance products, especially the coverage provided by P&I Clubs and environmental compensation funds are also addressed. It is also worth noting that a compensation system for traditional damage caused via the environment and ecological damage is not always separable: the liability rules may apply to both kinds of damage; an insurance policy may cover property damage, personal injury and preventive/cleanup costs (which are usually regarded as one part of natural resources damage). Thus when relevant, the compensation system for traditional damage is introduced briefly as well.

This chapter will be structured as follows: after an introduction (1) the role of liability rules in compensating for ecological damage is sketched (2), followed by a discussion on insurance (3) and the specific case of marine oil pollution (4). For every topic first the theoretical possibilities to provide compensation are sketched; next the question is addressed what role these mechanisms play in practice. A short economic analysis (5) is provided, followed by some policy recommendations (6).

1.1. *Theory: Environmental Liability in Past and Present*

In this section, the legal framework on whether and if yes, how liability for natural resources damage can be established are sketched. To address this topic, two questions need to be answered: firstly, whether an obligation exists to clean up the pollution and restore the damaged environment in case of ecological damage; and secondly, whether the liability provisions under the civil law and environmental statutes also cover ecological damage. It is worth noting that liability for ecological damage and general environmental damage (personal injury and property damage caused via the environment) share many similarities. The general liability rules under civil law and environmental statutes may apply in both cases. Hence, when relevant, how general environmental damage can and is compensated are also examined here.

1.1.1. Basis of Liability

Traditionally, both civil law and environmental statutes provide remedies to environmental damage in China. Environmental liability rules can be found in both

[3] See Chapter 10, Section 1.1.
[4] See Chapter 10, Section 3.

General Principles of Civil Law of 1986 (GPCL)[5] and the Environmental Protection Law of 1989.[6] However, there were some differences remaining between the provisions in these two acts.

Article 124 of the GPCL stipulates that

> Any person who pollutes the environment and causes damage to others in violation of State provisions for environmental protection and the prevention of pollution shall bear civil liability in accordance with the law.[7]

The violation of a relevant regulation is a prerequisite to establish liability according to this provision. However, this requirement is contradictory to Article 41 of EPA:

> A unit that has caused an environmental pollution hazard shall have the obligation to eliminate it and make compensation to the unit or individual that suffered direct losses.[8]

The latter introduces strict liability, without requiring the violation of a regulation. Debates have been held on which provision of them prevails and how to interpret the requirement of violation under the CPCL in both Chinese legal scholarship and cases law.[9] The introduction of a New Tort Liability Law (TLL)[10] in 2009 has brought a stop to these debates. The TLL has a specific chapter (Chapter VIII) on environmental liability, which reiterates the strict liability for environmental damage without a requirement to violate a relevant regulation:

> Where any harm is caused by environmental pollution, the polluter shall assume the tort liability.[11]

Although the environmental liability provisions under the CPCL and the EPA have not been formally abrogated, it is likely that they may play only a minor role in practice in the future since the victims are expected to rely on the New TLL.[12]

5 The General Principles of Civil Law (promulgated by the Nat'l People's Cong., 12 April 1986, effective 1 January 1987) [hereinafter GPCL]. An unofficial English copy can be found at: <www.china.org.cn/china/LegislationsForm2001-2010/2011-02/11/content_21898337.htm>, last accessed on 8 May 2013.

6 Environmental Protection Act (promulgated by the Standing Comm. Nat'l People's Cong., promulgated and effective on 26 December 1989) [hereinafter EPA]. An unofficial English copy can be found at: <www.china.org.cn/english/environment/34356.htm>, last accessed on 8 May 2013.

7 Art. 124 of the GPCL.

8 Art. 41 of the EPA.

9 See on these debates, Faure & Hu 2011, p. 231-233.

10 Tort Liability Act of the People's Republic of China (promulgated by the Standing Comm. Nat'l People's Cong., 1 January 2010, effective 1 July 2010) Art. 70 (China) [hereinafter 2009 Tort Liability Law]. An unofficial English copy can be found at: <www.procedurallaw.cn/english/law/201001/t20100110_300173.html>, last accessed on 8 May 2013.

11 Art. 65 of the TLL.

1.1.2. Scope of Liability: Is Natural Resources Damage Covered?

The provisions mentioned above provide the basis for environmental liability in China. However, it is not clear about which type of damage is covered. For example Article 124 of the GPCL does not clarify what constitutes 'damage to others'. To apply this provision in practice, one has to consider the general provisions under the GPCL, such as Article 106:

> Citizens and legal persons who through their fault encroach upon state or collective property or the property or person of other people shall bear civil liability. Civil liability shall still be borne even in the absence of fault, if the law so stipulates.[13]

This provision requires encroachment of property or person to establish liability, without using the term 'property right' or 'personal right'.[14] The expression is clearer under Article 2 of the TLL:

> Those who infringe upon civil rights and interests shall be subject to the tort liability according to this Law.
> The term 'civil rights and interests' used in this act includes the right to life, the right to health, the right to name, the right to reputation, the right to honor, right to self image, right of privacy, marital autonomy, guardianship, ownership, usufruct, security interest, copyright, patent right, exclusive right to use a trademark, right of discovery, equities, right of succession, and other personal and property rights and interests.[15]

This provisions use infringement of 'civil rights and interest' instead of 'civil right' as a requirement to establish liability. The term 'civil rights and interest' is further defined by listing specific rights and interests.[16] The catch-all expression also enables an interest to be protected under the TLL even without being established as a 'civil right' and not being explicitly included in the list.[17] However, it is still unclear whether the damage to the environment itself without personal injury and property damage can be compensated under the TLL.

As examined in Part II of this research, in both the U.S. and Europe, this type of damage has been explicitly admitted as compensable. For example, the Comprehensive Environmental Response, Compensation and Liability Act (CERCLA) and the Oil Pollution Act have introduced liability for natural resources

12 According to Art. 83, the Law on Legislation, 'In the case of national law, administrative regulations, local decrees, autonomous decrees and special decrees, and administrative or local rules enacted by the same body, if a special provision differs from a general provision, the special provision shall prevail; if a new provision differs from an old provision, the new provision shall prevail.' Law on Legislation (promulgated by the Standing Comm. Nat'l People's Cong., 15 March 2000, effective 1 July 2000), Art. 2 (China).

13 Art. 106 of the GPCL.

14 *Ibid.*

15 Art. 2 of the TLL.

16 *Ibid.*

17 The language of 'and other personal and property rights and interests' in Art. 2 implies that the rights and interests listed before are not exclusive. See Palmer 2011, p. 115, note 45.

damage explicitly.[18] In U.S. legislation, nature resources means land, fish, wildlife, biota, air, water, ground water, drinking water supplies, and other such resources 'belonging to, managed by, held in trust by, appertaining to, or otherwise controlled' by the United States, any state or local government or Indian tribe.[19] The scope of natural resources is defined by their relationship to public authority, but not the ownership. Thus not only the public natural resources, but also the privately-owned ones are encompassed under the heading of 'natural resources damage'. In Europe, the Environmental Liability Directive establishes a liability framework for three types of environmental damage.[20] In this research, the term natural resource damage is used, which refers to the damage to the environment itself without involving personal injury and property damage. However, in China, nether the GPCL, the EPA or the newly adopted TLL clarifies the issue for ecological damage. It is not clear, on the one hand, whether an obligation exists to clean up and to restore the damaged environment; and, on the other hand, whether this type of damage, such as the enjoyment of the environment, costs of preventive measures and restoration measures are compensable.

In addition to those general legislations, there are some specific environmental statutes in China which also contain an environmental liability provision, such as the Water Pollution Prevention Act,[21] the Air Pollution Prevention Act[22] and the Solid Waster Pollution Prevention Act.[23] However, either the WPPA or the APPA has specific relevance to natural resources damage. Although Article 85 of the SWPPA obliges polluters to restore the damaged environment, there is no practical guidance on how to determine the existence of such damage and how to restore the environment.[24]

As far as soil pollution is concerned, there have been no specific liability rules in China yet. Applying the general environmental liability rules to soil pollution creates some practical difficulties: given the long-term characteristics of soil pollution, the polluters are even more difficult to identify; or the polluters may cease to exist before the manifestation of the pollution. However, the cleanup and compensation for soil pollution are now subject to a variety of different, rather piecemeal types of stipulations. For example the SWPPA provides in Article 35 that

18 See 42 U.S.C. §9607(a); 33 U.S.C. §2702(b)(2). For a more detailed discussion about natural resources damage liability in the U.S., see Boyd 2003.

19 15 C.F.R. §990.30.

20 Instead of 'natural resources damage', the term 'environmental damage' is used in the ELD, which refers to damage to protected species and habitats, protected waters and land. Art. 2(1) of the ELD.

21 Water Pollution Prevention Act (promulgated by the Standing Comm. Nat'l People's Cong., 11 May 1984, revised 28 February 2008, effective 1 June 2008) [hereinafter WPPA], Art. 85.

22 Air Pollution Prevention Act (promulgated by the Standing Comm. Nat'l People's Cong., 5 September 1987, effective 1 June 1988, revised 29 April 2000) [hereinafter APPA], Art. 62.

23 Solid Waste Pollution Prevention Act (promulgated by the Standing Comm. Nat'l People's Cong., 30 October 1995, revised 29 December 2004, effective 1 April 2005) [hereinafter SWPPA], Art. 85.

24 Art. 85 of the SWPPA: 'It is necessary to get rid of dangers, compensate losses according to law and take measures to restitute to the previous environmental condition if any environmental pollution by solid wastes is caused.'

the entities discharging industrial solid wastes need to take preventive measures at sites before the termination of the activities.[25] In response to this situation, the Ministry of Environmental Protection (former Environmental Protection Agency) has issued a few normative documents, guiding on the prevention of and liability for soil pollution. The notice on the strengthening of pollution prevention during the relocation of industrials[26] and the opinion on strengthening the soil pollution prevention[27] are two such examples. Those two documents provide such basic guidance on the allocation of the soil pollution liability. The Huanban 2004 document stipulates that producers of dangerous waste must access the risk before relocation to a new industrial site takes place and that the soil of the old site should be cleaned up by former polluters. The Huanfa 2008 document has some similar relevance. It is based on the polluter pays principle and holds the former operators as primary liable parties. When the polluters have already ceased their existence, or cannot be identified, the government or the transferee of the land-use right shall take prevention measures. However, these documents are still too abstract to solve many problems in practice. There are also several drafts that intended to provide more detailed technical guidance on the assessment, monitoring and management of polluted sites.[28] These drafts have not entered into force yet. In addition to those

25 Art. 35 of the SWPPA: 'Where it is necessary for the entities discharging industrial solid wastes to be terminated, measures for preventing and controlling pollution shall be taken in advance to the facilities and sites for storing and treating industrial solid wastes, and the untreated industrial solid wastes shall be disposed properly to prevent environmental pollution.'

26 The notice on the strengthening of pollution prevention during the relocation of industrials (issued by the National Environmental Protection Agency Office, 1 June 2004, Huanban (2004) No. 47) (hereinafter Huanban 2004 document), available at: <www.mep.gov.cn/gkml/zj/bgt/200910/t20091022_173879.htm> (Chinese), last accessed on 8 May 2013.

27 The opinion on strengthening the soil pollution prevention (issued by Ministry of Environmental Protection, Department of Nature and Ecology Conservation, 6 June 2008, Huanfa (2008) No. 48) (hereinafter Huanfa 2008 document), available at: <www.mep.gov.cn/gkml/hbb/bwj/200910/t20091022_174598.htm> (Chinese), last accessed on 8 May 2013.

28 The guidelines for risk assessment of contaminated sites; the guidelines for soil remediation of contaminated sites; the technical guidelines for environmental monitoring of sites, the technical Specification for Environmental Site Investigation and the temporary regulation on management of soil pollution on polluted sites. The guidelines for risk assessment of contaminated sites (drafted by Ministry of Environmental Protection, Department of Science, Technology and Standards, 29 September 2009, Huanbanhan [2009] No. 1020), available at: <www.sepa.gov.cn/info/bgw/bbgth/200910/t20091009_162122.htm> (in Chinese), last accessed on 8 May 2013; [the guidelines for soil remediation of contaminated sites] (drafted by Ministry of Environmental Protection, Department of Science, Technology and Standards, 10 December 2009, Huanbanhan [2009] No. 1312), available at: <www.mep.-gov.cn/gkml/hbb/bgth/200912/t20091223_183351.htm> (in Chinese), last accessed on 8 May 2013; The technical guidelines for environmental monitoring of sites (drafted by Ministry of Environmental Protection, Department of Science, Technology and Standards, 1 February 2010, Huanbanhan [2010], No. 99), available at: <www.mep.gov.cn/gkml/hbb/bgth/201002/t20100208_185580.htm> (in Chinese), last accessed on 8 May 2013; The technical Specification for Environmental Site Investigation →

national documents and drafts, some local authorities also made efforts to issue their own documents on prevention and restoration of soil pollution, such as Shengyang, Beijing and Chongqing.[29]Those local documents have many similarities to the national normative documents mentioned above.

The only field where damage to the environment is explicitly admitted as compensable and an operable procedure exists is marine oil pollution. The Marine Environment Protection Act[30] authorizes public authorities to claim for such damage and such cases do exist in practice.[31]

1.1.3. Standing

As far as compensation for ecological damage is concerned, one major obstacle is who has the *locus standi* to claim for such costs, since usually no individual interest is involved.

The Civil Procedure Law in China prescribes the conditions for a cased to be accepted by the court. According to Article 108:

> The following requirements must be met when an action is initiated:
> (1) the plaintiff must be an individual, legal person or any other organization that has a direct interest in the case;
> (2) there must be a specific defendant;
> (3) there must be a concrete claim, facts and cause of action; and

(drafted by Ministry of Environmental Protection, Department of Science, Technology and Standards, 5 August 2009, Huanbanhan [2009] No. 793), available at: <www.sepa.gov.cn/info/bgw/bbgth/200908/t20090812_157381.htm> (in Chinese), last accessed on 8 May 2013; The temporary regulation on management of soil pollution on polluted sites (drafted by Ministry of Environmental Protection, Department of Policies, Laws and Regulations, 15 December 2009, Huanbanhan [2009] No. 1321), available at: <www.mep.gov.cn/gkml/hbb/bgth/200912/t20091223_183364.htm> (in Chinese), last accessed on 8 May 2013.

29 For example, in Shenyang, a regulation on contaminated sites remediation and restoration was promulgated in 2007. See The Temporary Regulation on Environmental Remediation and Restoration at Contaminated Sites in Shenyang (issued by Shenyang Environmental Protection Agency and Shenyang Plan and Land Resources Bureau, effective as of 1 June 2007, Shenhuanbao [2007] No. 87). In Beijing, some local standards have been issued to guide the assessment of soil pollution, such as the Environmental Site Assessment Guideline (issued by Beijing Bureau of Quality and Technical Supervision, effective as of 1 January 2010, DB11/T 656-2009). The Chongqing government also issued a document on soil pollution restoration. Notice on Promoting the Restoration at Contaminated Industrial Sites in Chongqing (issued by Government Office of Chongqing, 26 June 2008, Yubanfa [2008] No. 208).

30 Marine Environmental Protection Act (promulgated by the Standing Comm. Nat'l People's Cong., 23 August 1882, revised 25 December 1999, effective 1 April 2004) [Herinafter MEPA], an English copy can be found at: <www.mlr.gov.cn/mlrenglish/laws/200710/-t20071012_656329.htm>, last accessed on 8 May 2013.

31 For details about compensation for marine oil pollution, see *infra* Section 4.

> (4) the action must be within the scope of acceptance for civil lawsuits of the people's courts and within the jurisdiction of the people's court where it is filed.[32]

This provision defines the plaintiff narrowly as the one who has a direct interest in the case. This constrained standing makes it possible that a plaintiff is denied access if he cannot show direct physical or economic damage.[33] Thus an important hurdle exists when there is no individual damage involved in pollution incidents. This problem may be relieved in the new revised Civil Procedure Law which allows for public interest litigation.[34] It holds that: if environmental pollution and activities infringing on many consumer's legal rights harm public interests, the authorities and organizations prescribed by law can bring a suit in people's court.[35] In this case, public authorities and NGOs may claim for ecological damage which concerns public interests. However, the provision limited the standing of the parties which are prescribed by law. This means, only if the statutes authorize explicitly, one party can file a litigation in the court. For example, under the MEPL, the public authorities in charge of marine environmental supervision and management can claim for losses, if there is damage to the marine ecosystem, to marine fishery resources and to the marine protected areas.[36] However, in other areas, when there is no such specific legislation, standing still poses a significant hurdle.

When the standing or other requirement for accepting a case is not satisfied, the court can issue a verdict to reject the case according to the Civil Procedure Law:

> Article 112 When a people' s court receives a written or oral complaint and finds after review that it meets the requirements for acceptance, it shall file the case within seven days and notify the parties; if the complaint does not meet the requirements for acceptance, the court shall, within seven days, give a verdict that the complaint be rejected. If the complainant has an objection against the order, he or she may file an appeal.[37]

This provision, in theory, requires a verdict if the court decides not to accept the case. In the other words, the rejection of a case is supposed to be accompanied by a written rationale, which gives the plaintiff the possibility to appeal such a decision. However, in practice, such a step is often skipped by the judges.[38]

32 Civil Procedure Law (promulgated by the Nat'l People's Cong., promulgated 9 April 1991, effective 9 April 1991, revised 29 October 2007 and 31 August 2012) [hereinafter CPL]. An unofficial English copy (till the 2007 revision) can be found at: <http://china.findlaw.cn/jingjifa/shewaifalv/swflfg/20110414/91492.html>, last accessed on 8 May 2013.

33 Alford & Shen 1997, p. 147; Briggs 2006, p. 327.

34 The Civil Procedure Law is revised in 2012, which adds a new provision: Art. 55 on public interest litigation. Standing Committee of National People's Congress Decision to Revise Civil Procedure Law in People's Republic of China (promulgated by Standing Committee of National People's Congress, 31 August 2012, effective as of 1 January 2013) (Hereinafter Civil Procedure Law Revision).

35 Point 9, Civil Procedure Law Revision.

36 Art. 90 of the MEPA.

37 Art. 112 of the CPL.

38 Stern 2011, p. 297; for details, see Section 1.2.2.

The limited standing under civil procedure law, the conservative attitude towards public interest litigation and the arbitrariness existing in accepting a case have led to hot discussion in the literature.[39] The trend to broaden the standing has been reflected in the recent revision procedure of the Civil Procedure Law. Recently, China started to revise its Civil Procedure Law. Two Drafts have been submitted to the Standing Committee of National People's Congress for discussion in November 2011[40] and April 2012.[41] The revision was finally promulgated in August 2012. The first draft added a public interest litigation provision holding 'if an activity which pollutes the environment or violates many consumers' rights poses an infringement on the public interest, related public authorities or social organizations can file litigation.' Such a provision opens a possibility for public authorities and NGOs to file a lawsuit when there is no individual damage involved. However, this provision in the first draft was criticized as too obscure and was changed to 'public authorities and social organizations prescribed by law' in the second draft. This narrow approach was adopted in the final revision. The new law holds that: if environmental pollution and activities infringing on many consumers' legal rights harm public interests, the authorities and organizations prescribed by law can bring a suit in the people's court.[42] In this case, public authorities and NGOs may claim for ecological damage which concerns the public interest. However, the provision limits the standing to the parties which are prescribed by law. This means, only if the statutes explicitly authorize, one party can file a case in court. For example, under the MEPL, the public authorities in charge of marine environmental supervision and management can claim for losses, if there is damage to the marine ecosystem, to marine fishery resources and to the marine protected areas.[43] However, in other areas, when there is no such specific legislation, standing still constitutes a significant hurdle.

Another provision in the revision aims at protecting the plaintiffs' right to sue. It revises Article 112 of the Civil Procedure Law by adding one sentence that 'the Court should protect the parties' right to sue according to law.' It further clarifies that 'if the requirements for accepting a case are not satisfied the court shall make a written verdict to reject the case.' Compared to the existing provisions, the revision explicitly requires that the verdict to reject a case should be written, so that the parties' right to appeal can be better protected.

39 For example, see Stern 2011; Briggs 2006; Wang 2006, p. 220-223; Lee 2008, p. 814-822.

40 The First Draft of Amendment to Civil Procedure Law (Hereinafter first draft). The draft can be found at: <www.npc.gov.cn/npc/xinwen/syxw/2011-10/29/content_1678367.htm>, last accessed on 8 May 2013 (Chinese).

41 An introduction to the second draft can be found at: <www.npc.gov.cn/huiyi/cwh/1126/201204/25/content_1718870.htm>, last accessed on 8 May 2013 (Chinese).

42 Point 9 of the Civil Procedure Law Revision.

43 Art. 90 of the MEPA.

1.1.4. Causation, Multiple Tortfeasors and Burden of Proof

With respect to these issues that can have a crucial bearing on the effectiveness of environmental liability as a compensation mechanism, the Tort Liability Act of 2009 also produced interesting innovations. Article 66 of the Tort Liability Act holds that:

> Where any dispute arises over an environmental pollution, the polluter shall assume the burden to prove that it should not be liable or its liability could be mitigated under certain circumstances as provided for by law or to prove that there is no causation between its conduct and the harm.[44]

This means that the burden of proof has been completely shifted to the polluter. As a consequence of the strict liability introduced in Article 65 it is the polluter who now bears the burden of proving that one of the situations of exemption or mitigation of liability applies. The polluter also has the burden to prove the disconnection between his activity and the harm which could potentially be a dangerous situation for defendants to the extent that they may not be able to show that their activity was not the source of a particular damage suffered by the plaintiffs.[45] However, this is not revolutionary either, since a rule of the reversal of proof burden had already been put in place in a judicial explanation for Civil Procedure Law in 1992.[46] Moreover, the concern of over-deterrence can be balanced through the implementation of this reversal of burden provision in practice, which is often criticized as problematic.[47]

As far as the multiple-polluter case is concerned the New Tort Liability Act holds in Article 67 that:

> Where the environmental pollution is caused by two or more polluters, the seriousness of liability of each polluter shall be determined according to the type of pollutant, volume of emission and other factors.[48]

This means that in case of multiple polluters a causal apportionment of liability will take place according to the type of pollutants and the volume of emissions. It is striking that Chinese law does follow a proportional approach in case of multiple tortfeasors, but chooses for a shifting of the burden of proof (in Article 66) in case of uncertainty over causation.

44 Art. 66 of the TLL.

45 Economic analysis shows that this rule may lead to over-deterence to the potential polluters, See Faure & Hu 2011, p. 241. Literature also shows concern that the Tort Liability Law allows the reversal of burden without requiring even an indication of a casual relationship. See Koziol & Zhu 2010, p. 357-358.

46 The Opinion of Supreme Court on the Application of Civil Procedure Law (issued by the Judge Committee of the Supreme Court on 14 July 1992, faban (92) No. 22), available at: <www.law-lib.com/law/law_view.asp?id=54915>, last accessed on 8 May 2013.

47 See Moser &Yang 2011, p. 10897; for more detailed information about the practice, see the following Section 1.2.2.

48 Art. 67 of the TLL.

The new rules dealing with multiple injurers were also mentioned as an innovation by Chinese experts we interviewed.[49] Before the promulgation of the Tort Liability Act, in most cases a joint and several liability rule applied. The GPCL stipulates that:

> If two or more persons jointly infringe upon another person's rights and cause him damage, they shall bear joint liability.[50]

The GPCL and Environmental Protection Act have no specific provisions on multiple tortfeasors in case of environmental liability; thus the general rule under GPCL applies. However, with the introduction of the Tort Liability Act, liability needs to be decided according to the contributions of each polluter and a several liability rule applies.

1.1.5. Quantification

Even if ecological damage is admitted as compensable, a practical problem that will still arise is how to quantify such type of damage. The damaged environmental elements may not always have a market value. Moreover, even if they do, the natural resources does not only have use value which can be quantified easily, but also equally, if not more important non-use value, such as cultural, esthetic and ecological values.[51] Therefore, how this type of damage can be quantified is a critical question to be addressed. In the U.S., concrete natural resources damage assessment rules have been established under the CERCLA and the OPA.[52] After a long debate, a restoration-based approach has been established to assess the natural resources damage. This means, when natural resources damage happens, it should be restored to the initial status and the restoration costs and lost natural resources service during the restoration period are compensable.[53] When restoration is not

49 Interview with Prof. Wang Canfa, Professor of environmental law at the China University of Political Science and Law (CUPL) and director of the Centre for Legal Aid to Pollution Victims (CLAPV) in Beijing, China (23 August 2011) (interview transcript on file with the author).

50 Art. 130 of the GPCL.

51 Use value is dereived from the actual use of the environment and can be evaluated by market values. Non-use value, however, cannot be measured by the market value. Usually, existence value and instrinsic value are regarded as non-use value. For discussion about different types of values natural resources have, see Dobbins 1994, p. 898-908; Kanner & Nagy 2005, p. 421-424.

52 The CERCLA authorized the Department of Interior to develop assessments rules for natural resources damage (hereinafter DOI rules). See 43 C.F.R. §§11.10-11.93. The OPA directed the National Oceanic and Atmospheric Administration to develop natural resources damage assessment rules. See 15 C.F.R. §§990.10-990.66 (hereinafter NOAA rules).

53 Initially, a lesser of rule was adopted under the DOI rules: the damages are limited to the lesser of the restoration costs and the diminution of use-value. A shift towards a restoration-based approach was triggered by two cases: *Ohio v Department of Interiors* (880F.2d 432, 442 (D.C.Cir.1989) and *Colorado v Department of Interior* (880 F.2d 481 (1st Cir. 1989).

possible, other monetary valuation methods can be used.[54] Although this method is far from perfect and has caused hot debates, the assessment rules themselves have provided an operational procedure to quantify the natural resources damage and have lead to many restoration efforts in practice.[55] In Europe, although compensation for environmental damage under the ELD is still in its early development stage, the Directive itself has provided a short guidance on the quantification issues. For the biodiversity damage and damage to the protected waters, a restoration-based approach is also adopted.[56] However, in China, how natural resources damage can be assessed remains unclear, with the exception of the fishery losses caused by water pollution.[57] For example, quantifying soil pollution is still difficult: not only prior information on background levels is missing, appropriate standards to evaluate to what extent a polluted site should be restored are also lacking.[58] In response to this situation, the government began its effort to develop methodologies to assess natural resources damage and started trials in some areas recently.[59] The Ministry of the Environmental Protection published a Recommendation Methods on Assessing Environmental Damage in 2011.[60] The Recommendation gives general guidance on how to assess pure environmental damage, but is not a binding standard to be applied in the court. Hence it certainly does not solve all problems of ecological damage assessment. Recently, a Temporary Assessment Rule for Pollution Damage Caused by Environmental Accidents was published by the MEP and declared open for consultation from related public authorities and environmental research institutes.[61] It establishes a

54 See Boyd 2003, p. 271-275.

55 Although there is no official data on how many natural resources claims have been made after the promulgation of the CERCLA and the OPA, an ELI research shows that till 2006, there are at least 273 cases involving one or more federal trustees claiming for natural resources damage. Among those, 132 cases were CERCLA related, leading to settlement and judgment values of $722,433,600 and 48 cases were OPA related, with the value of $126,290,530. See Environmental Law Institute, Mitigation of Impacts to Fish and Wildlife Habitat: Estimating Costs and Identifying Opportunities (2007).

56 Annex II of the ELD.

57 See Section 3.1.1.

58 Interview with Dr. Cai, a researcher on the restoration of soil pollution from South China Institute of Environmental Sciences (11 August 2011), interview transcript on file with the author.

59 Several Opinions on Evaluation the Environmental Pollution Damage (promulgated by the Ministry of the Environmental Protection, 25 May 2011, Huanfa [2011] no. 60), available at: <www.mep.gov.cn/gkml/hbb/bwj/201105/t20110530_211357.htm>, last accessed on 8 May 2013 (Chinese).

60 Recommendation on Methods on Assessing Environmental Damage is developed by Chinese Academy for Environment Planning, designated by the Ministry of Environmental Protection. It is not legally binding, available at: <www.mep.gov.cn/-gkml/hbb/bwj/201105/W020110530352486511962.pdf>, last accessed on 8 May 2013 (Chinese).

61 Temporary Assessment Rule for Pollution Damage Caused by Environmental Accidents (Consultation Draft), published by the Ministry of Environmental Protection, available at: <www.zhb.gov.cn/gkml/hbb/bgth/201301/t20130128_245592.htm>, last accessed on 8 May 2013.

procedure for responding to, assessing and restoring pollution damage caused by sudden accidents. How such a rule will be finalized and implemented still remains to be seen.

1.2. *Practice*

The above section sketches the legal framework of liability for ecological damage in China. However, the obstacles towards sufficient compensation do not only exist on paper but also in practice. This section shows to what extent the damaged environment is restored and how compensation is provided in practice.

1.2.1. Who Acts after an Accident?

After an environmental incident happens or pollution is detected, the first issue that arises is whether an investigation and cleanup is undertaken, rather than compensation. Especially, when there is only damage to the environment but not individual loss, cleanup or restoration is an important step towards damage assessment and compensation.

It may not be surprising that the interviews we held all confirmed that the most important player in demanding restoration after environmental pollution is the government. Historically many enterprises involved in heavy industry were (and to a large extent still are) state-owned enterprises. This may, on the one hand, obviously have disadvantages when it comes to the incentives of public authorities to 'be harsh' on polluters; on the other hand, it also leads to an acceptance that government may be responsible for historic pollution and therefore the primary party who should take action especially to clean up historically polluted sites.[62] When referring to the government in China, this can, depending upon the division of competences, either be the central government or the local authorities. However, usually they only take cleanup action in response to emergency situations; longer term restoration does not take place.[63]

There may, however, be situations in which the government will attempt to shift costs to polluters. The Ministry of the Environment made clear that a policy referred to as 'from two to three' has been initiated. This basically refers to the desire of policy makers to change industry in China from the heavy secondary industry to the (less polluting) tertiary industry. Under this policy identifying soil pollution after the industry is relocated becomes important as well as restoration given the changing use of the site.[64] In some cases local governments undertake the

62 Interview with a representative of the Ministry of Environmental Protection (MEP), in Beijing, China (23 August 2011) (interview transcript on file with the author).

63 Interview with Professor Wang Jin, professor of environmental law, interview in Beijing, China (24 August 2011) (interview transcript on file with the author).

64 For more detailed information about the relocation of industries in some urban cities in China, see Word Bank, Overview of the Current Situation on Brownfield Remediation and Redevelopment in China, Report No. 57953, September 2010, available at: →

restoration themselves and pass on the restoration costs in the land transfer fees or require the redevelopers (to whom the land use rights have been transferred) to undertake the restoration under the supervision of the government.[65] The restoration at the Beijing Hongshi Paint Plant site provides an example. That site once housed a pesticide plant which was later transformed into a paint plant. Site assessment showed that the contaminated soil in this site amounted up to 140,000 m3. After the plant was relocated, the government called for bids for its redevelopment. According to the bidding document, the winning bidder is required to prepare and implement a restoration plan in accordance with the contaminated soil disposal plan, formulated by Beijing Environmental Protection Bureau. This leads to the developers spending tens of millions RMB on soil remediation.[66]

This also shows that in China, different than for example in Europe, there is not always a clear distinction between public authorities acting under private law (hence using tort law) or under administrative law. In fact many cases, for example of soil pollution, may be handled rather via administrative law rather than via civil law. This allows the government to use administrative law for example to require the polluter to do more risk assessment, clean up the sediment et cetera.[67] This also means that the nature of the remedies that are used in practice, more particularly in soil pollution cases are often more of an administrative nature[68] or even of an economic nature[69] than the classic remedies in tort law.[70] The remedies applied in practice will still be further discussed below.[71]

1.2.2. Barriers to Access to Justice

Many barriers to access to justice exist in practice, not only for ecological damage but also more traditional damage. The most important barrier also mentioned during our interviews is that the court has to accept the case. In a Chinese court, the filing division (*li'an ting*) determines whether to accept a case or not. As discussed earlier, according to the Civil Procedure Law, the judges have to give a verdict if a case is rejected. However, in practice, such a step is often skipped and the plaintiffs are left without the record of refusal.[72] In practice, a case may be rejected because an administrative solution is forthcoming or because law is not regarded as a proper solution for that dispute.[73] It is even more problematic when it comes to the

<http://documents.worldbank.org/curated/en/2010/09/13132932/overview-current-situation-brownfield-remediation-redevelopment-china>, last accessed on 8 May 2013.

65 Interview with a MEP representative.

66 World Bank 2010, p. 23.

67 Interview with a MEP representative.

68 Hence forcing polluters to restore the polluted environment.

69 For example passing on clean-up costs to the developer of the site.

70 Such as they are *inter alia* enumerated in Art. 15 of the TLL.

71 See Section 1.2.4.

72 Stern 2011, p. 297.

73 *Ibid.*

(politically) sensitive cases.[74] One example where the court refused to accept the environmental case without a written verdict is the petro-chemical case. An explosion occurred at a petro-chemical plant owned by PetroChina Cooperation due to an operational fault in 2005. This explosion and consequent emergency measures led to large amount of toxic substances being spilled into the Songhua River.[75] This led to a temporary stop of water supply in Harbin city and direct economic loss up to RMB 1.5 billion in this city alone.[76] Apart from direct economic loss, this incident led to significant ecological loss as well. However, the constrained standing provision in the Civil Procedure Law created a challenge for claims for such loss. After this pollution, some experts filed a civil public interest litigation with the nature as joint-plaintiff to the High People's Court of Heilongjiang. However, nature has no standing according to Chinese law and the experts suffered no direct loss. Hence the case was not accepted by the court.[77]

Another hurdle for victims to go to court is the acceptance fee system. According to Civil Procedure Law, when parties file a civil litigation, they should pay an acceptance fee according to law.[78] The acceptance fee paid by plaintiffs is usually 0.5 per cent to 4 per cent of the compensation requested. This can be costly for the victims who have already sustained serious harm. Although the law allows an application for a reduction, waivers or a postponement of the payment of the fee,[79] the reliance on such fees for the operational budget creates disincentives to grant waivers in some courts.[80] Sometimes the lawyers of the plaintiffs do not even want to apply for waivers because they fear that this will bias judges against the plaintiffs.[81]

Even in the situation where the court accepts to hear the case plaintiffs still face substantial problems.

A major problem reported is that the combination of the economic and political power of the industry actors that act as defendants and the fact that many judges suffer a lack of expertise in handling pollution cases may seriously jeopardize the plaintiff's chances of winning an environmental case in court.[82]

A pollution incident may lead to damage to a large number of victims. The Civil Procedure Law allows class action (know as representative action in China) if one party of the litigation engages more than two persons and the object of the action is the same.[83] The class actions are supposed to make litigation more efficient and to create a wide influence of the suit and thus make it easier for the victims to get a remedy. However, in practice, there is a trend to restrict the use of the class

74 Interview with Kathinka Fürst, researcher in Amsterdam University and affiliated to CLAPV. (23 August 2011) (Interview transcript on file with the author).

75 Wang, Huang & Yan 2008, p. 273-275.

76 Wang, Yu, Li & Li 2008, p. 291.

77 Wang, Yu, Li & Li 2008, p. 301.

78 Art. 107 of the CPL.

79 *Ibid.*

80 Moser & Yang 2011, p. 10897.

81 Briggs 2006, p. 327.

82 This was especially reported in the interview with Prof. Wang Canfa.

83 Arts. 53, 54 of the CPL.

action. This restriction is reflected in the 'Notice regarding Problems with the Acceptance of Class Action Lawsuits by the People's Courts' issued by the Supreme People's Court in 2005.[84] This notice stipulates that the courts can split up class action suits if it is difficult to handle as class action suit; it also redirects jurisdiction over the class action to courts at lower level. The court has the tendency to split up class actions because this can increase the court fee charged and the number of cases, which is an important index of the court's workload.[85] Since class actions involving many victims are more likely to draw wide media coverage and interest of higher-level authorities, the courts tend to solve it individually and to avoid possible turbulence that might be brought by a large group of losing plaintiffs.[86] However, keeping the cases solved at local level may strengthen the effects of local protectionism.[87]

Of course there may be genuine problems in proving the environmental case of the victim as well. Classic problems that victims of environmental damage face (like providing proof of the damage and of a causal relationship with the behavior of the defendant) obviously pose a serious problem in China as well.[88] This of course stands in some contrast to the possibilities victims of environmental pollution have on paper in China. Remember that according to Article 66 of the Tort Liability Act the burden of proof of exemptions, as well as of the absence of causation, has been shifted to the polluter.[89] However, experts reported that in practice proving causation remains a problem. Actually, the reversal of proof existed in Chinese law before the promulgation of the Tort Liability Law.[90] However, it is not fully implemented in practice. In practice, before the judges decide to shift the burden of proof to the polluters, sometimes they require a different degree of preliminary proof from the plaintiffs. For example, sometimes the court requires the plaintiff to prove that the defendant is polluting the environment in order the shift the burden. In some other cases, the victims are required to produce preliminary evidence on that 'it is more likely than not that the defendant polluted the environment and caused the victim harm'. A lower burden scenario requires three criteria to be met: 'the plaintiff has suffered a quantifiable loss; this harm has been proven to be caused by pollution and in the relevant

84 Notice regarding Problems with the Acceptance of Class Action Lawsuits by the People's Courts (promulgated by the Sup. People's Ct. 30 December 2005, effective 1 January 2006).

85 Zhao 2004, p. 170.

86 *Ibid.*

87 Wang 2007, p. 215.

88 Interview with Prof. Wang Canfa. See also Van Rooij 2010, p. 69.

89 See Section 1.1.4.

90 See Supreme People's Court Opinion on Several Issues concerning the Application of Civil Procedure Law (adopted by the Judicial Committee of the Supreme Court on 14 July 1992, effective as of 14 July 1992), Art. 74, 'Parties in a litigation should provide proofs for his claims. However, in the following tort cases, if the defendant denies the facts asserted by the plaintiff, the burden of proof lies with the defendant:... claims for damages caused by environmental pollution....' An English copy can be found at: <www.lawinfochina.com/display.aspx?lib=law&id=6690&CGid=>, last accessed on 8 May 2013.

temporal and physical space there is a possible source of this environmental pollution'. In some extreme cases, the victims are even asked to provide direct evidence that the harm was caused by pollution.[91] This variance in practice shows that without clear criteria in determining causation and practical guidance on applying the shift of burden of proof, a simple reversal of burden of proof provision cannot guarantee its implementation in practice.

1.2.3. NGOs

An interesting question is obviously to what extent civil society in China also plays a role in aiming to achieve a restoration of environmental harm in China. The environmental NGOs (eNGOs), as an important part of civil society, are still in their early stage of development in China. The 'China Society for Environmental Sciences', established in 1978 is the first eNGO in China. It was only by the 1990s, that eNGOs started to develop fast.[92] It is reported that there were , eNGOs in China in 2005 and the number grew to 3,539 in 2008.[93] Scholars attribute the rapid rise of eNGOs to the changes in political opportunities, mobilized organizational resources, the influence of international communications, internet, media and so on.[94] However, it is also broadly recognized that substantial obstacles still exist for the full prosperity of eNGOs. For example, NGOs face strict legal and administrative barriers, which make their legitimacy a serious concern. In China, the NGOs can have different forms, such as social organizations (*shehuituanti*), foundations (*jijinhui*), private not-for-profit organizations (*minban zuzhi*), grassroots NGOs (*caogen zuzhi*) and so on.[95] When talking about NGOs, the first form one may think of is social organizations, which are 'organizations formed by citizens voluntarily which carry out activities aimed at realizing the common aspirations of their members in accordance with organizational articles of association'.[96] To be qualified as a social organization, an organization needs to satisfy a series of substantive and procedural requirements, including finding a government institution as a sponsor and registering with the Bureau (Ministry) of Civil Affairs.[97] These hurdles mean that many organizations cannot register as social organizations.[98] According to the regulations, an organization conducting activities as a social organization without being registered is an illegal social organization.[99] Hence some NGOs try to find shortcuts to gain a legal status, such as being

91 See McMullin 2009, p. 168-171.

92 Bao 2009, p. 2-3.

93 Zhan & Tang 2011, p. 2.

94 See Yang 2005, p. 46; Zhan & Tang 2011, p. 3-6; Ru & Ortolano 2009.

95 Li 2011, p. 1.

96 The Regulation on Registration and Administration of Social Organizations (promulgated by State Council, 25 September 1998), Art. 2.

97 Li 2011, p. 2-5.

98 Lau 2009, p. 9.

99 See The Regulation on Registration and Administration of Social Organizations, Art. 2 and the Provisional Measures for Suppression of Illegal Civic Organization (promulgated by Ministry of Civil Affairs, 10 April 2000), Art. 2. See also Li 2011, p. 3.

affiliated with the universities, or registering with the Bureau of Industry and Commerce as a business entity. However, not all organizations are able to find a shortcut and still many remain unregistered.[100] It is estimated that only 23.3 per cent of the eNGOs are registered with the Bureau (Ministry) of Civil Affairs and the rest are illegal.[101] Of those which are indeed registered, only a small fraction of them are registered as 'social organizations' with tax-exempt status; others are registered as Private non-profit organizations, corporations or student-led social organizations.[102] One important characteristic of those eNGOs in China is that there is a strong influence from the government, since many of the NGOs are actually government-organized NGOs (GONGOs). It is reported that until 2005, among the 2,768 eNGOs in China, 49.9 per cent are GONGOs, 40.3 per cent are student-led organizations, 2.5 per cent are branches of international NGOs and only 7.3 per cent are grassroots eNGOs (citizen-organized NGOs).[103] This characteristic, together with the political and institutional background, makes the eNGOs reluctant to take some 'confrontational' actions. Most of their efforts are made to promote environmental consciousness, sustainable development and public participation.[104] Very few of them try to help pollution victims through lawsuits and challenge local firms, since 95 per cent of them practice under the principle of 'help, but not make trouble; participate, but not intervene; supervise, but not replace; act, but not violate.'[105] But recently, the eNGOs start to play a larger role in the domain of policy advocacy, such as helping the victims to file a lawsuit, challenging local firms and even influencing the functioning of the State. For example, some NGOs provide legal aid to pollution victims and support them in lawsuits against polluters. The Centre for Legal Assistance to Pollution Victims (CLAPV) is such an organization. It is reported that CLAPV received over 10,000 complaints from citizens during its 8 years of operation and got directly involved in 104 of them.[106]

In addition to supporting the individuals to file a suit against polluters, the eNGOs recently also started to file public interest litigations on their own behalf. For example, it is reported that the All China Environment Federation, a big GONGO in China, filed four public environmental litigations in 2011, two of which have been successful and the others are still in the judicial procedure.[107] The case *All China Environment Federation v Jiangsu Jiangyin Container Inc.* was reported as the

100 Lau 2009, p. 9.

101 Bao 2009, p. 7.

102 Zhan & Tang 2011, p. 5, 8-9.

103 Bao 2009, p. 7. But in recent years, there is increasing role being played by the grassroots NGOs. See Ru & Ortolano 2009.

104 Yang 2005, p. 52.

105 Bao 2009, p. 8.

106 Van Rooij 2010, p. 70.

107 The Environmental Rights Supporting Service by All China Environment Federation in 2011, available at: <http://gongyi.163.com/12/0213/11/7Q504CLB00933KC8.html>, last accessed on 8 May 2013 (in Chinese).

first environmental civil public litigation filed by the NGO and accepted by the court.[108]

The increasing role played by eNGOs in environmental litigation is in line with the introduction of environmental court in some local and intermediate courts in recent years. It is reported that there are 86 environmental courts at different levels in China as of October 2011,[109] of which, the environmental courts in the intermediate courts of Guiyang, Wuxi and Kunming are the most reported ones. Those courts have issued some documents to give guidance to the scope of cases the environmental courts shall accept. Some have mentioned specifically that eNGOs are eligible plaintiffs to file public interest litigations.[110] These documents alleviate the legal obstacles for eNGOs to bring public litigation in some local courts. However, a written rule alone cannot guarantee the sufficient involvement of eNGOs.[111] In spite of the rapid introduction of environmental courts nationwide, the caseload for these courts is still low, especially as far as pubic interest litigations brought by eNGOs are concerned.[112] Moreover, it is pointed out that the courts are still very cautious to accept some controversial cases against powerful defendants.[113]

1.2.4. Remedies

Before the discussion on remedies for ecological damage, this part examines briefly how general environmental damage is remedied in China. The Chinese legal framework allows different types of remedy to the pollution victims, which can be divided into two large categories: compensation and elimination of harm.[114]

108 Behind the First Environmental Public Interest Litigation: Fight for 'Environmental Right', available at: <www.chinanews.com/gn/news/2009/11-28/1989095.shtml>, last accessed on 8 May 2013 (in Chinese).

109 Zhang & Zhang 2012.

110 See J. Gao, *Development of Environmental Courts in China: the Promises and Challenges and Implementation for Environmental Public Interest Litigation*, 2009, available at: <www.iucnael.org/zh/component/search/?searchword=environmental+court&ordering=&searchphrase=all>, last accessed on 8 May 2013.

111 In spite of the rapid establishment of environmental courts in recent years, many scholars are cautious about the potential achievement of such instruments. See Wang & Gao 2010; Goelz 2009.

112 See Zhao & Jiang 2012, p. 16-17 (in Chinese).

113 Thus although both legislation and practice start to open up space for public interest litigation, literature holds the actual effect of this new type of litigation is still moderate. See Liu 2012.

114 According to the TLL, the 'forms of tort liability' includes:
1. cessation of infringement;
2. removal of obstruction;
3. elimination of danger;
4. return of property;
5. restoration to the original status;
6. compensation for losses;
7. apology; and
8. elimination of consequences and restoration of reputation.
See Art. 15 of the TLL.

Cessation of infringement, elimination of danger and restoration to the original status fall into the category of 'elimination of harm'. Cessation of infringement means an injunction to stop an ongoing infringing action; elimination of danger means to stop an action which has a substantial threat to the environment and restoration means to restore the polluted environment to its initial status.[115] However, in practice, the judges are reluctant to order such an injunction. This can be explained by several reasons: firstly, some injunction orders, such as suspension or closing down of a factory are usually regarded as political decisions so that the court rarely uses them without political commitment initiated from the government.[116] Moreover, even if the court chooses to order an injunction, such as 'cessation of infringement', this is still a too broad concept to use in practice. Without further clarifying how to accomplish such an injunction, it may still remain unenforced.[117]

Another related issue is that instead of these legal remedies, sometimes alternatives which are more of an economic nature are used. For example in the early stages of environmental awareness, precisely given the difficulties of environmental litigation, the goals of plaintiffs was often not directly to obtain compensation in terms of monetary damages, but rather job opportunities. Hence, the idea was that the injury of the victims should not necessarily be compensated with monetary damages, but rather with job opportunities, hence providing some social security. Obviously in those cases the question of quantification of damage just does not arise. However, it was indicated that with the development of the market economy and increasing environmental awareness this kind of socio-economic compensation to victims may no longer work and hence quantification of damages becomes an important issue.[118]

It is even more problematic as far as pure environmental damage is concerned. As was mentioned above, environmental damage is not specifically addressed in legislation with the exception of marine pollution[119] so that it is not clear if an obligation to clean up/restore the environment exists and how this type of damage can be compensated. As discussed earlier, quantification of such damage is a big problem. But the Ministry of Environmental Protection has launched an experiment in some provinces and cities with respect to environmental damage assessment in order to collect experience which could later be transformed into a national

115 Zhao 2004, p. 177.

116 *Ibid.*

117 One example is the case *Zhang Changjian et al. v Pingan Rongping Chemical Plant*. The plaintiffs claimed that the defendant caused serious damage to the biodiversity in the neighborhood and, especially, fishery losses to the villagers. The court ordered the defendant to compensate the victims and to stop the infringement immediately without further clarifying how to achieve this. Several years after the judgment, however, it was still unclear whether the defendant has taken any action in response to the order. See Wang 2006, p. 212-217.

118 Interview with Ms. Ma, a researcher on environmental health from South China Institute of Environmental Sciences (11 August 2011), interview transcript on file with the author.

119 See Section 1.1.5.

system.[120] According to the Recommendation on Methods on Assessing Environmental Damage, a restoration-based approach is adopted to evaluate the pure environmental damage.

The old habit of using a more socio-economic based approach to remedying environmental damage can still be noticed, especially in cases where the government seeks recovery of environmental harm. The case of soil pollution can illustrate this. The background for a restoration claim is usually the relocation of old industries. For example the relocation of hundreds of old industrial facilities from Beijing to the outskirts led to 8,000,000 m^3 brown fields that needed to be redeveloped.[121] According to the Beijing EPB document, before the industrial land was changed to other use, an environmental impact assessment should be undertaken to determine the extent of the soil pollution. The polluters should be held responsible for the cleanup.[122] However, in practice, recovery from polluters according to the polluter pays principle is not always possible. The former industry (polluters) may cease to exist or cannot afford the costs. Furthermore, polluters often are state-owned enterprises, and government has no strong incentive to pursue these SOEs. Instead of holding the SOEs liable, the government prefers to seek socio-economic remedies from the polluters such as asking them to make more investments and to promote the local economy.[123] Hence the government seeks institutional arrangements whereby on the one hand polluted soils can be redeveloped and on the other hand new developers have to pay in terms of a higher price for the development project or new developers are required to undertake restoration themselves. One of such example is the restoration at Hongshi paint plant site.[124] The legal basis for these transfers is that all land is state-owned or collectively owned and that the industry only has a use right on the land. Hence the government can require in the bidding document that the new developers who get the land-use right should take restoration measures.

1.3. *Summary*

The analysis above shows that difficulties towards compensating for ecological damage exist both on paper and in practice in China. A legal framework of environmental liability has been established under the GCPL, the EPA and especially, the newly adopted TLL. However, in those provisions, there is no explicit reference to ecological damage. It is still not clear whether the environmental liability provisions also apply if no individual damage is concerned. Furthermore, the constricted standing requirements in the Civil Procedure Law

120 Several Opinions on Evaluation the Environmental Pollution Damage (promulgated by Ministry of Environmental Protection, 25 May 2011, Huanfa [2011] No. 60) (in Chinese).

121 World Bank 2010, p. 4.

122 The Notice of Beijing Environmental Protection Agency on Evaluation of the Soil Environment at the Former Industrial Sites (Promulgated by Beijing Environmental Protection Agency, 6 July 2007, Jinghuafa [2007] No. 151) (in Chinese).

123 Interview with Prof. Wang Jin.

124 See Section 1.2.1.

limit the possibility to bring public interest litigations. Moreover, the lack of standards also makes assessment of ecological damage difficult.

In practice, the compensation seems even more difficult. It is very often the government that acts to obtain recovery of environmental harm. They do not have a strong incentive to pursue the SOEs. NGOs so far also play a limited role and litigation in general is limited, given the huge barriers concerning access to justice. These not only have to do with formal statutory barriers (like the court having to allow a case), but also problems concerning expertise and knowledge of the judiciary.

Most remedies are of a socio-economic nature, forcing, for example, a polluting industry to reinvest in the local economy. There seems to be less focus on long-term restoration of the environment or on providing individual restorative justice to victims. That seems to be a more general conclusion: the (economic) idea that environmental liability provides incentives for polluters to invest in efficient abatement technologies is of course based on the idea that potential polluters are exposed to the full social costs of their activity and will hence be sufficiently deterred by a finding of liability. The overview of practice shows that not only the probability of being held liable to pay damages is quite low (in fact it is often not the polluter but rather the new developer who is invited to compensate for the harm) and moreover the amounts of compensation paid only seem to be a fraction of true social losses caused by environmental harm. It is therefore, to say the least, doubtful that, given the current practice, environmental liability can play its preventive role in China.

2. Environmental Insurance

As discussed in the theoretical part, liability rules alone cannot guarantee sufficient compensation and efficient prevention of ecological damage. Compensation instruments can be used together to promote the realization of the goals. In China, however, those instruments are still in their early stage of development. A recent innovation is the development of the insurance market for environmental liability. It is worth noting, however, this instrument is not specifically developed for ecological damage but, more broadly, the general environment. Some of the insurance policies may also cover parts of ecological damage, such as the preventive costs. This part sketches the possibilities of environmental insurance in China, both in theory and in practice.

2.1. *Theory*

This section focuses generally on possibilities for polluters to obtain environmental insurance. Although some other insurance is available in specific fields, such as marine oil pollution[125] or for nuclear liability,[126] a full discussion of them is outside the scope here.

2.1.1. Statutory Background

An obvious question that can be asked is to what extent a statutory duty exists to purchase liability insurance. As we just mentioned, when the extent of the damage can exceed the individual wealth of the injurer, an insolvency problem may arise which may justify the introduction of mandatory insurance.[127] The literature has largely argued in favor of the introduction of environmental liability insurance in order both to guarantee effective compensation to victims and to avoid under-deterrence following the judgment proof problem.[128] As far as a general environmental insurance in China is concerned, we can be short: the environmental legislation generally in China is silent on compulsory insurance; exceptions only exist as far as marine oil pollution is concerned.[129] Also the New Tort Liability Act of 2009 is silent in this respect.[130] Actually, environmental insurance is quite a new product in China. Only until recently, the government started a policy to promote the development of the environmental insurance market. The Ministry of Environmental Protection and the China Insurance Regulatory Commission issued a document 'Opinion on the Development of Environmental Pollution Liability Insurance' in 2007, which requires local authorities to conduct research and experiments on environmental liability insurance.[131] There are a few voluntary environmental liability insurance programs promoted by some local governments. For example, Shenyang has promulgated a local regulation 'Shenyang Regulation on Preventing Pollution from Dangerous Waste'[132] in 2008, which is the first local regulation touching upon environmental liability insurance in China. It stipulates that the insurers are encouraged to establish products to cover environmental

125 To be discussed in the next Section 3.

126 Which we do not discuss at all in this contribution; for nuclear liability and its insurance in China see, Liu & Faure 2012.

127 When a serious insolvency risk exists, the insured only has incentives to buy insurance up to the amount of his assets, rather than for the whole damage he may cause. Under this situation, compulsory insurance will make the insured internalize the whole costs they create. See Wagner 2007, p. 110; Faure & Grimeaud 2003, p. 181-185.

128 See for example Faure & Hartlief 2003, p. 211-220.

129 To be discussed below in Section 3.

130 See in this respect also Faure & Hu 2011, p. 237.

131 Opinion on the Development of Environmental Pollution Liability Insurance (promulgated by Ministry of Environmental Protection and China Insurance Regulatory Commission, 4 December 2007, Huanfa [2007] No. 189), available at: <www.zhb.gov.cn/info/gw/-huanfa/200802/t20080220_118389.htm>, last accessed on 8 May 2013 (in Chinese).

132 Shenyang Regulation on Preventing Pollution from Dangerous Waste (promulgated by Shenyang Standing Committee of People's Congress, 28 November 2008), available at: <www.ln.gov.cn/zfxx/fggz/gwyfg_3/sy/200902/t20090204_330763.html>, last accessed on 8 May 2013 (in Chinese).

liability from dangerous waste; the potential polluters are encouraged to seek such insurance coverage.[133]

2.1.2. Theoretical Insurance Options[134]

Insurance experts[135] report that the insurance cover for environmental harm in China can theoretically take place under three different headings.

A first possibility is that there would be a general liability insurance policy to which enterprises can subscribe. This type of general liability insurance covers industrial accidents. To the extent that an industrial accident leads to environmental damage which can be related to bodily injury, property damage or even pure ecological loss, this would be covered under the general liability insurance policy. The broad definition of environmental liability makes it sometimes difficult to be differentiated from liability caused by other industrial activities. Hence the expert holds that a general liability insurance policy should not and cannot exclude pollution damage from its coverage.[136] An important feature of a general liability insurance policy is, however, that it covers accidents and hence excludes gradual pollution.

A second possibility is that there would be cover under the product liability insurance coverage. At first sight this seems like a different type of insurance policy. However, the product liability policy usually does not discriminate between environmental damage and other damage. One could imagine harmful products (for example related to food or feed or agricultural products) that cause environmental harm as well. Theoretically these liabilities are far reaching. Even gradual pollution can be covered under this policy. The example is provided of the case of a chemical manufacturer who supplies the food industry with substances that could lead to serious health effects. This would be covered under a product liability insurance policy.

The third type of policies would (different than the first two that are not exclusively focused on environmental harm) be specific environmental insurance policies. Usually this would be an environmental liability policy, hence covering third-party liability (rather than harm to the insured site itself). An environmental liability policy would hence cover (of course depending upon the formulation in the policy conditions) liability for damage from the premise itself and operations with

133 Art. 8 of Shenyang Regulation on Preventing Pollution from Dangerous Waste.

134 We do realize that this already touches upon practice. However, in this section we describe the insurance policies that could theoretically cover environmental harm; in the next Section 2.2., we will discuss to what extent these policies are used in practice and which problems arise in that respect.

135 We are in this respect especially grateful to Mr. Christian Lahnstein of Munich Reinsurance Company for providing us valuable information in an interview on 14 September 2011.

136 *Ibid.*

defined extensions, also beyond the narrow cover for industrial accidents under more traditional general liability insurance policies.[137]

2.2. *Practice*

2.2.1. Insolvency Risk – Compulsory Insurance?

A first important issue to mention is that experts report that defendants in some cases go bankrupt as a result of which victims cannot collect compensation.[138] In that case, government will pick up the bill and moreover the polluter can *ex ante* even count on government intervention for pollution costs, as a result of which he will have no incentive to increase their own costs by buying insurance coverage.[139] However, this may not be a problem in all pollution cases. For example an insolvency risk may not arise in the case of large SOEs. It is also reported that because a stringent liability rule was lacking, industry would have no incentives to demand liability insurance.[140] To the extent that a judgment proof problem arises, compulsory insurance, or at least a requirement of financial securities for selected industries that pose high pollution risks could solve this problem.[141]

There is, however, some debate concerning the introduction of compulsory insurance. The debate at the policy level concerns the fact that the policy maker is considering on the one hand to introduce compulsory insurance and on the other hand to make high polluting industries pay pollution fees. Some propose to assess environmental risks before the operation starts via the environmental impact assessment and subsequently to require via permitting polluters to provide financial guarantees before starting the operation.[142] In practice, some industry with high environmental risks is also required to seek insurance coverage in some local

137 First-party insurance and liability insurance are two important types of insurance. In the first-party insurance system, compensation is awarded directly by the insurer to the victim. It is the victim who buys the insurance and the insurer pays as soon as damage occurs and the damage is the insured risk. See Faure & Bruggeman 2008, p. 11-14. Liability insurance (third-party insurance) covers 'claims of victims against injurers who are liable in damages.' So the covered risk in liability insurance is the insured's liability for damage caused to other parties. See Wagner 2009, p. 377. Direct insurance has some similarity to first-party insurance and liability insurance. In a direct insurance policy, the potential injurer who possesses a particular site seeks insurance coverage also for the benefits of third parties who could suffer damage resulting from that particular site. Unlike the pure first-party insurance funded by victims, in a direct insurance policy, it is still the polluters that pay for the premium. The difference between direct insurance and liability insurance is that the trigger of coverage under direct insurance is the materialization of the insured risk rather than liability. See Faure 2002b, p. 294.

138 Interview with Prof. Wang Canfa.

139 *Ibid.*

140 Interview with a MEP representative.

141 Interview with Prof. Wang Canfa, Prof. Wang Jin.

142 Interview with Prof. Wang Jin.

areas.[143] However, some argue that charging a pollution fee from industry (which apparently needs to be paid by permitted installations)[144] and at the same time requiring the purchase of insurance would make polluters pay twice. Hence, it is suggested that part of the pollution fees paid by industry should be used to purchase environmental liability insurance.[145] This suggestion can be found in some local areas, where government provides some subsidies to pay environmental insurance premiums financed from the pollution fee charged from the enterprises.[146] In some other areas, the government links environmental insurance with other environmental subsidies, green credit policies and so on.[147]

2.2.2. Environmental Insurance in Practice

As we already mentioned it is generally held that environmental insurance in China is not very developed.[148] The reason has already been mentioned repeatedly: since polluters largely (can) count on government to intervene in the restoration of polluted sites, since liability risks are low and since (with the exception of marine oil pollution) the purchase of insurance is not compulsory, industry has little incentive

143 For example, in Wuxi (a city in Jiangsu Province), some enterprises are obliged to buy environmental insurance since 2011, depending on its location and type. See Opinions on Implementation of Environmental Liability Insurance in Wuxi (issued by the Government Office of Wuxi, 22 February 2011, Xizhengbanfa [2011] No. 48), available at: <www.wuxi.gov.cn/web101/BA01/C/02/875461.shtml>, last accessed on 8 May 2013 (in Chinese).
A compulsory system is also adopted in Changsha and Guangxi. See Rules on Management of Environmental Risky Enterprises (issued by Changsha Environmental Protection Bureau, August 10, 2009, Changhuanfa (2009) no.44) (hereinafter Changsha Rules), available at: <www.changsha.gov.cn/xxgk/gfxwj/szfgzbm/shbj/201007/t20100701_82840.html>, last accessed on 8 May 2013 (in Chinese); Opinions on Implementation of Environmental Liability Insurance (issued by Guangxi Environmental Protection Bureau, 4 August 2011, Guihuanfa [2011] No. 47), available at: <www.gxepb.gov.cn/xxgkml/ztfl/-hjglywxx/wrfz/201108/t20110823_6007.html>, last accessed on 8 May 2013 (in Chinese).

144 Interview with Prof. Wang Jin.

145 *Ibid.*

146 In China, the polluters need to pay a pollution fee. Sometimes it is argued that asking the polluters on the one hand to pay for pollution fee and on the other hand to buy the environmental insurance in fact makes them pay twice. Thus in some local areas, parts of the pollution fee is used to provide a subsidy to the enterprises which buy environmental liability insurance. Wuxi is such an example; see Opinions on Implementation of Environmental Liability Insurance in Wuxi.

147 Sichuan is such an example. According to a Sichuan EPB document, the government should give priority to companies who bought environmental liability insurance when deciding to offer subsidies on pollution control. In addition, the attendance of environmental liability insurance is also a criterion to examine the green credit performance. See the Opinion on the Implementation of Environmental Pollution Liability Insurance Policy (issued by the Environmental Protection Bureau of Sichuan and by the Sichuan Branch of the Chinese Insurance Supervisory and Management Commission, 30 November 2010, Chuanhuanfa [2010] No. 85).

148 Interview with Prof. Wang Jin.

to purchase environmental insurance. Logically an insurance market to provide products covering environmental risks has therefore barely developed.

In practice a variety of the insurance products that could cover environmental risks can indeed also be observed. A brief overview may illustrate this:[149]

1. The main insurance product covering environmental risks was the general liability insurance policy with an extension to pollution risks 3-4 years ago. Pollution is, however, not clearly defined in the policy and the scope may be limited. There is usually a sublimit on the coverage for pollution damage and, moreover, the general liability insurance only covers accidents and hence excludes gradual pollution.
2. Chinese insurers also started offering stand-alone environmental liability insurance after 2007. This mainly covers personal injury and property damage caused by pollution but does not cover specifically pure ecological damage. The cleanup costs (excluding cleanup at polluters' own sites) can be covered either directly under this policy[150] or via an added clause.[151] However, usually cleanup costs are covered to the extent that the cleanup may prevent further personal injury or property damage. Hence, cleanup costs are only covered as insured prevention costs. The cleanup costs to protect the environment itself (hence not related to the prevention of further personal injury or property damage) are not covered.[152] The restoration costs are usually also excluded.[153]
   The insurance policy uses claims-made clauses, which is customary in environmental liability coverage. It means that the claim for damages must have been received by the insured or his insurer during the period of insurance cover. Under a claims-made policy the claim has to be received within the period of insurance cover. Furthermore, the policy may also require that the incident which leads to pollution happens with in a certain retroactive

149 The information was provided in an interview with Ms. Zhang Jing and Ms. Jean Wu, representatives of the Munich Reinsurance Company, Beijing branch (15 September 2011) (transcript on file with the author).

150 See for example, Art. 4, the Clause of Environmental Pollution Liability Insurance (China Continent Property & Casualty Insurance Company Ltd.), available at: <www.iachina.cn:8080/iaclause/clause/html/20091207041357078.html>, last accessed on 8 May 2013.

151 See for example, The added clause on Cleanup costs to Environmental Liability Insurance provided by China Pacific Insurance Co. Ltd. (Hereinafter CPIC added clause) available at: <https://www.cpic.com.cn/cx/upload/Attach/infordisclosure/50867389.pdf>, last accessed on 8 May 2013.

152 Interview with Ms. Zhang Jing and Ms. Jean Wu, representatives of the Munich Reinsurance Company, Beijing branch (15 September 2011) (transcript on file with the author).

153 For example, CPIC added clause. A distinction is made between cleanup costs and restoration costs. The costs of measures to clean up the pollutants in insured sites are covered, while the costs of measures taken to restore the environment to its initial status are not.

period.[154] This increases the predictability to the insurers. Some authors are critical of claims-made policies, arguing that they could dilute the deterrent function of liability law.[155] As far as premiums are concerned insurers make a distinction between companies that constitute high environmental risks and companies that do not. For high risk companies the insurer will usually appoint an expert to do a risk assessment. The insured normally cooperates in this risk assessment. For lower-risk companies the premium will be based on a fixed premium rating table.[156] The compensation rate would still be low but experts hold it could increase in the future.[157]

3. A pollution site liability insurance policy. This is a third-party insurance policy, providing coverage for damage to third parties and remediation costs coming from polluted sites. There are still very few such products in China. Huatai Insurance Company is one of the insurers who provide such a product.[158] An analysis of their policy conditions makes clear how the coverage is structured:

   Two types of risk can be covered under this policy: new pollution and pre-existing pollution. The covered risk is defined as loss that 'the insured is legally liable to pay as a result of claims, remediation costs, and associated legal defense expenses' arising out of a 'pollution condition on, at, under or migrating from the covered locations'.[159] This loss should be claimed or first discovered during the policy period and reported to the insurer during the policy period/or the extended reporting period. The difference between new pollution and pre-existing pollution is that under the title of 'new pollution', the pollution conditions should first commence during the policy period; while under the title of 'pre-existing pollution', the pollution conditions should be first commenced prior to the inception date of the policy period.[160]

   This provision uses the term 'the insured is legally liable to pay'. Thus it is formally still liability insurance. The term 'claim' is defined broadly to include 'government actions, suits or other actions alleging responsibility or liability on the part of the insured for bodily injury, property damage or remediation costs arising out of pollution conditions'.[161] In other words, this policy covers both traditional third-party liability and remediation costs on premises for which the insured are legal liable to pay. It is worth noting that pure environmental damage is also covered under this policy. The term 'property

154 For Example, Art. 3, CPIC Environmental Pollution Liability Insurance, available at: <https://www.cpic.com.cn/cx/upload/Attach/infordisclosure/50885015.pdf>, last accessed on 8 May 2013.

155 See more particularly Abraham 1982, p. 123-131.

156 Interview with Ms. Zhang Jing and Ms. Jean Wu.

157 *Ibid.*

158 Huatai Insurance Company of China, Premises Pollution Liability Insurance, (hereinafter Huatai Premise Insurance), available at: <www.ehuataisz.com/uploadfile/200806/-Premises%20Pollution%20Liability%20Insurance.pdf>, last accessed on 8 May 2013.

159 Provision 1, Huatai Premise Insurance.

160 *Ibid.*

161 Provision V(C), Huatai Premise Insurance.

damage' is defined to include 'natural resources damages',[162] which mean 'damages for injury to or damage sustained by or destruction or loss of fish, wildlife, biota, land, air, water, groundwater, drinking water supplies and other similar resources belonging to managed by held in trust, by appertaining to or otherwise controlled by any government or local government authority'. 'Remediation costs' is defined as 'reasonable expenses incurred to investigate, quantify, monitor, mitigate, abate, remove, dispose, treat, neutralize, or immobilize pollution conditions to the extent required by environmental law'.[163] Thus unlike the environmental liability policies, the coverage under premise pollution liability insurance is by definition much broader. However, since the clause requires all the costs being what the insured is legally liable to pay, to what extent the broad provision under this policy will lead to broad compensation still depends on the liability provisions in law and on their explanations.

Huatai Insurance Company started to provide such a product since 2008. However, three years after such an attempt, it was reported that the progress was still slow, and the insured were mainly enterprises with foreign-related issues.[164]

In addition to the Premise Pollution Liability Insurance, there are also some other similar products with a less further going coverage. These products cover cleanup costs at polluters' own premises as added clauses to environmental liability insurance policies. The added clauses provided by Changan insurance[165] and Pingan insurance are such examples.[166] Under those policies, third-party damage and cleanup costs at the third parties' land are covered under the environmental liability insurance. The added clause only provides coverage to the cleanup costs on the insured's land.

4. Product liability insurance. As mentioned above,[167] the traditional product liability policies do not exclude pollution and hence in theory provide broad coverage. Of course the condition is that the environmental damage must be linked to a product for which the insured is liable. In that case (different than for example the General Liability Insurance Policy) also gradual damage would be covered. However, in practice, so far there has not yet been one case

162 Provision V(U), Huatai Premise Insurance.

163 Provision V(V), Huatai Premise Insurance.

164 L. Xie, *How to Promote Environmental Liability Insurance?*, available at: <www.sinoins.com/101288/101475/59505.html>, last accessed on 8 May 2013.

165 Environmental Pollution Liability Insurance, Added Clause on the Cleanup costs on Premise Sites, Changan Property and Liability Insurance Ltd., available at: <www.iachina.cn:8080/iaclause/clause/html/20091207035605187.html>, last accessed on 8 May 2013.

166 Environmental Pollution Liability Insurance, Added Clause on the Cleanup costs on Premise Sites, Pingan Insurance Ltd., available at: <http://property.pingan.com/upload/-20100701094634255.pdf>, last accessed on 8 May 2013.

167 Section 2.1.2.

of a claim for environmental damage under the product liability coverage. This therefore remains a largely theoretical possibility.[168]

5. Property insurance. There is also a general property damage insurance. This covers first-party damage to the site of the insured. In principle pollution risks are also covered by such an all-risk property policy, unless particular damage would be explicitly excluded. This could be the case if the property damage to the site of the insured is caused by a gradual erosion or pollution (for example not sudden pollution and considered not foreseeable by the insurer).

This shows that there are hence quite a few possibilities to purchase environmental insurance in the market, either explicitly or implicitly via general (liability or property) insurance policies. However, it also shows that the number of companies active in the environmental insurance market is limited. For example, China started experiments in some local areas to develop the environmental insurance market after 2007. It is reported that the revenue from environmental liability insurance only accounted 0.015 per cent of the total liability insurance revenue in the experiment areas in 2009. In Shenzhen, one of the experiment areas, only eight enterprises bought such insurance products in 2009.[169] This is also confirmed by Professor Wang Jin who held that only some larger insurance companies provide explicit coverage for environmental damage, whereby the type of coverage provided by the different companies is quite similar. There would only be a few differences for example as far as exclusions of liability, scope of coverage or premiums are concerned.[170]

2.2.3. Difficulties and Limits

Both experts and (re)insurance companies mention particular difficulties with the provision of environmental insurance in China. This should not come as a surprise given the relatively small number of insurance companies offering those policies and the relatively small number of insured. (Re)Insurers mention adverse selection as an important problem. Adverse selection is the phenomenon caused by information asymmetry.[171] Limited information on the side of insurers would make insurance particularly attractive for high-risk companies which could eventually endanger the insurability of risks.[172] In China the problem would arise that mainly high-risk companies are interested in purchasing environmental insurance.[173] The insured on their side also report that the amount of coverage would be too low. Often insurance coverage is only provided for one to two million RMB and only in

168 Interview with Ms. Zhang Jing and Ms. Jean Wu.

169 J. Li & J. Liang, *The Ministry of Environmental Protection Tries to Promote Compulsory Environmental Liability Insurance,* available at: <http://green.sina.com.cn/2012-07-09/103024739500.shtml>, last accessed on 8 May 2013 (in Chinese).

170 Interview with Prof. Wang Jin.

171 Akerlof 1970, p. 488-500.

172 See Faure 2007b, p. 81-97 and Thomas 2007, p. 105-132.

173 Interview with Ms. Zhang Jing and Ms. Jean Wu.

exceptional cases for 10 million to 300 million RMB (1.61 million to 48.17 million USD).[174] Those amounts may indeed be rather low by international standards. Moreover, not only are there complaints of low coverage; premiums are also considered to be high.[175] The premium would be around 6-8 per cent of the insured amount. Compared to on average 0.3 per cent in case of traditional liability insurance this would be quite high.[176] Given the difficulty to predict the character of environmental risks, insurers would, like the literature predicts,[177] ask an additional risk premium to cope with their insurer ambiguity. But these high premiums may not generate any willingness to pay on the side of firms.

2.3. *Summary*

As far as the ability of environmental insurance to provide adequate compensation for environmental harm in China is concerned, we can be short: the first and major problem is the lack of an adequate institutional and statutory background for creating a market for environmental insurance. In this respect we do not only refer to the absence[178] of an obligation to purchase liability insurance; we equally refer to the fact that most pollution risks are covered by government and that the probability for polluters to be faced with environmental liability is in general low. That obviously explains a low demand for environmental insurance.

Not surprisingly one can on the one hand notice quite a few theoretical possibilities of environmental insurance coverage in China, but on the other hand that there are few insurance companies offering environmental insurance and also few companies interested in purchasing it. Moreover, those who are interested are probably the high risk ones, thus creating a serious adverse selection problem. Furthermore, in spite that a few specific environmental liability policies have been developed to cover environmental risks, experts have emphasized the possibility to use general liability insurance and product liability policies to cover environmental liability.[179] Those, however, remain theoretical possibilities and are not usually used in practice. General liability policies often exclude pollution coverage; environment-related claims are rare in practice under product liability policies. However, as discussed above, it is not always easy to differentiate environmental liability from other liability covered under general liability policies or product liability policies. Hence, the expert holds that those policies should be developed to cover environmental liability in China as well.[180] The only positive element one can mention is that apparently the Chinese insurance and reinsurance markets have developed a variety of environmental products that are in principle able to cover

174 *Ibid.*

175 Interview with Prof. Wang Canfa.

176 Interview with Prof. Wang Jin.

177 See Kunreuther, Hogarth & Meszaros 1993.

178 With the exception of the case of marine environmental pollution to be discussed below in Section 3.

179 E-mail contacts with Mr. Lahnstein on 1 February 2013.

180 *Ibid.*

environmental risks. Premiums today are still relatively high, but increased possibilities of risk assessment may reduce uncertainties and hence premiums. The crucial issue is that an institutional environment is created in which a demand of environmental insurance can emerge. In that case China apparently has sufficiently possibilities to offer the necessary coverage.

3. Compensation for Vessel-induced Marine Oil Pollution

As mentioned earlier, marine oil pollution deserves a separate discussion since it is one of the few cases where the liability and compensation instruments for natural resources damage seem to be working relatively adequately. This may be explained by the fact that some international conventions China has joined oblige Member States to introduce a financial security (like compulsory insurance) for sea-going vessels to cover the risks of marine pollution. Moreover, a long tradition of coverage via the so-called protection and indemnity clubs for ships, to cover environmental pollution risks exists in the field of marine pollution. As in the above sections, the discussion on marine oil pollution in China will also follow the same structure: firstly, the legal framework is sketched; this is followed by the practice of compensation for oil pollution in China.

3.1. *Theory*

3.1.1. Scope of Compensable Damage and Quantification of Damage

The Marine Environmental Protection Act is the basic law in the field of marine environmental protection and pollution prevention. Article 90 of MEPA stipulates liability of marine pollution:

> Whoever causes pollution damage to the marine environment shall remove the pollution and compensate the losses; in case of pollution damage to the marine environment resulting entirely from the intentional act or fault of a third party, that third party shall remove the pollution and be liable for the compensation.[181]

In line with EPA and the new TLL, strict liability is established under the MEPA. However, it does not further explain what constitutes 'pollution damage'. This should be understood considering the Conventions China has joined and the related judicial explanations. China is a Member State of the CLC, which means that the definition of pollution damage under the CLC also applies to China. However, in practice, there are still debates on the application scope of the CLC. As mentioned in Section 2, the CLC applies to 'any sea-going vessel and seaborne craft of any type whatsoever constructed or adapted for the carriage of oil in bulk as cargo'.[182] The term 'oil' is explained as 'any persistent hydrocarbon mineral oil such as crude oil,

181 Art. 90 of the MEPA.
182 Art. I(1) of the 1992 CLC.

fuel oil, heavy diesel oil and lubricating oil'.[183] Thus when the pollution involves other types of vessels or crafts, or the damage is caused by non-persistent oil, the domestic Chinese law applies. However, even when damage is caused by the CLC ships and oil, there are still debates on whether the CLC applies only to ships with foreign related issues or to all types of sea-going vessels and seaborne crafts.[184] How to interpret 'foreign related issues' is also important in determining the application of the CLC.[185]

As far as domestic law is concerned, a judicial explanation issued by the Supreme Court guides the judgment on vessel-induced oil pollution in 2011.[186] This explanation applies to 'vessel-induced oil pollution damage as involved in oil pollution incidents of vessels that cause oil pollution damage or pose dangers of oil pollution damage in the territory or any other territorial sea of the People's Republic of China'.[187] The term 'oil pollution damage' is explained in a similar way to the CLC, which includes:

> 1. Costs of preventive measures to prevent or minimize vessel-induced oil pollution damage, and further loss or damage caused by preventive measures;
> 2. Property damage caused outside the vessel carrying oil by the vessel-induced oil pollution incident, and loss of earnings caused therefrom;
> 3. Loss of earnings caused by environmental damage resulting from oil pollution; and
> 4. Costs of reasonable measures which have been taken or are about to be taken to restore the contaminated environment.[188]

Under such a definition, two titles are related to natural resources damage: prevention costs and restoration costs. To further clarify the scope of compensable pure environmental damage (ecological damage), the explanation stipulates that:

> If a vessel-induced oil pollution incident causes environmental damage, the compensation for environmental damage shall be limited to expenses for reasonable measures which have been taken or are about to be taken to restore the environment. Such expenses include reasonable expenses of monitoring, assessment and research.[189]

Similar to the CLC, a cautious attitude is adopted to explain restoration costs: only the 'reasonable' measures that 'have been taken or are about to be taken' are considered compensable.

183 Art. I(5) of the 1992 CLC.

184 See Hu & Bo 2006, p. 193-205; Faure & Wang 2006b, p. 329-330.

185 For example, a foreign element may be: one party involved is foreigner; the cause of the cases happens abroad; the subject matter is located abroad. See Hu & Bo 2006, p. 198-199.

186 Provisions of the Supreme People's Court on Several Issues Concerning the Trial of Cases of Disputes over Compensation for Vessel-induced Oil Pollution Damage (promulgated by the Judicial Committee of the Supreme People's Court, effective on 1 July 2011) (hereinafter the 2011 Explanation).

187 Art. 1 of the 2011 Explanation.

188 Art. 9 of the 2011 Explanation.

189 Art. 17 of the 2011 Explanation.

When compensation for natural resources damage is concerned, an unavoidable question arises how to quantify such damage. As discussed earlier, there are general rules on assessing natural resources damage in China. An non-binding recommendation on assessment methods is published, which gives guidance on quantification of five types of damage: personal injury, property damage, emergency response costs, assessment costs and restoration costs.[190] When restoration is possible, the restoration costs mean the actual costs which have taken place. If the restoration is unavailable, the recommendation allows assessment with a stimulated restoration method.[191] In other words, the recommendation goes further than the 2011 explanation. The latter allows compensation for restoration costs only when restoration has been taken or are about to be taken. However, under the recommendation, compensation is still possible even if restoration is not possible. It is worth noting that the Recommendation is not a binding document. In the field of water pollution, there are indeed two standards guiding the quantification of fishery losses: the Rules on Calculating Fishery Losses Caused by Water Pollution Accidents[192] and the Calculation Methods for Economic Losses Caused by Fishery Pollution Accidents.[193] The 1996 rules apply both to the calculation of direct economic losses suffered by individuals and to natural fishery resources which are not owned by private parties. The 2008 standards further clarify the methods to assess natural fishery losses. These two documents together provide a practical guidance in assessing one type of natural resources damage/natural fishery losses. As for other types of natural resources damage, the Technical Guidelines for Ecological Damage Assessment on Marine Oil Spill provide more detailed guidance.[194] They allow compensation for direct marine ecological losses; restoration costs for the habitats and species as well as assessment costs.[195] However, this is not a legally binding compensation standard but a sectoral standard providing guidance on the assessment of marine pollution damage.

As far as marine environmental liability is concerned, there are several other issues worth mentioning here. The above discussion has shown that strict liability is established. The MEPA allows three types of defenses: the damage caused by war; irresistible natural calamities or negligence or other wrongful acts in exercise of functions of competent departments responsible for the maintenance of beacons or other navigation aids.[196] The MEPA is silent on how to determine liability if damage

190 Recommendation, Section 3.2.

191 Recommendation, Section 4.5.

192 The Rules on Calculating Fishery Losses Caused by Water Pollution Accidents (Promulgated by the Ministry of Fishery, effective on 8 October 1996, Nongyufa [1996] No. 14) (hereinafter 1996 Rules).

193 Calculation Methods for Economic Losses Caused by Fishery Pollution Accidents (Promulgated by the General Administration of Quality Supervision, Inspection and Quarantine and Standardization Administration, effective 1 June 2008, GB/T 21678-2008). (hereinafter 2008 Standard).

194 Technical Guidelines for Ecological Damage Assessment on Marine Oil Spill) issued by the Oceanic Agency, on 9 April 2007, as effective from 1 May 2007, HY/T 095-2007.

195 Section 8 of the Technical Guidelines for Ecological Damage Assessment on Marine Oil Spill.

196 Art. 92 of the MEPA.

is caused by multiple tortfeasors. One new characteristic of the TLL of 2009 is that the multiple tortfeasors are severally liable for the environmental damage they caused.[197] In line with this provision, the 2011 explanation also introduces several liability only as the primary form of liability to deal with the multiple tortfeasor issue:

> Article 3 When oil has escaped from two or more vessels, and pollution damage results therefrom, if the party who suffers the damage requests that the owners of all vessels involved undertake the liability for compensation, the owners of all vessels involved shall undertake their respective liability for compensation if the damage is reasonably separable according to the quantity of oil leaked, the harm caused by their oil and other relevant factors; if the damage is not reasonably separable, the owners of all vessels involved shall be jointly and severally liable, unless exonerated by law.[198]

Both in the U.S. as well as in the international regime, liability for oil pollution is capped (with the exception that for the offshore facility, except a deepwater port under the OPA, liability for removal costs is unlimited).[199] In China, neither the TLL nor the MEPA touch upon the cap of liability. However, the Commercial Maritime Code allows the liable party to limit its maritime liability.[200] It is worth noting here that the categories of claims which are subject to the limit under the CMC are much broader than types of oil pollution under the MEPA.[201] Since China is a Member State of the CLC, which established a separate limit of oil pollution liability, the limits set in the CMC do not apply to claims for oil pollution under the CLC.[202] As

197 Art. 67 of the TLL: 'Where the environmental pollution is caused by two or more polluters, the seriousness of liability of each polluter shall be determined according to the type of pollutant, volume of emission and other factors.'

198 Art. 3 of the 2011 Explanation.

199 33 U.S.C. §2704(a).

200 Commercial Maritime Code (Promulgated by the Standing Committee of the National People's Congress, effective as of 1 July 1993) (hereinafter CMC), Art. 207.

201 The limit under the CMC is established for:
'(1) Claims in respect of loss of life or personal injury or loss of or damage to property including damage to harbour works, basins and waterways and aids to navigation occurring on board or in direct connection with the operation of the ship or with salvage operations, as well as consequential damages resulting therefrom;
(2) Claims in respect of loss resulting from delay in delivery in the carriage of goods by sea or from delay in the arrival of passengers or their luggage;
(3) Claims in respect of other loss resulting from infringement of rights other than contractual rights occurring in direct connection with the operation of the ship or salvage operations;
(4) Claims of a person other than the person liable in respect of measures taken to avert or minimize loss for which the person liable may limit his liability in accordance with the provisions of this Chapter, and further loss caused by such measures.
All the claims set out in the preceding paragraph, whatever the way they are lodged, may be entitled to limitation of liability. However, with respect to the remuneration set out in sub-paragraph (4) for which the person liable pays as agreed upon in the contract, in relation to the obligation for payment, the person liable may not invoke the provisions on limitation of liability of this Article.'
Art. 207 of the CMC.

202 Art. 208(2) of the CMC.

mentioned earlier, there are debates on the application scope of the CLC in both academia and in case law.[203] This debate also considers the determination of the limit for oil pollution damage. To clarify this issue, the Regulation on the Prevention and Control of Vessel-Induced Marine Environment Pollution of 2009 stipulates:

> With regard to the limitation of liability for pollution damage caused by vessels, the provisions of Maritime Code of the People's Republic of China in respect of the limitation of liability for maritime claims shall apply. However, with regard to the limitation of liability for pollution damage caused by vessels carrying persistent oil in bulk to sea areas under the jurisdiction of the People's Republic of China, the provisions of the international treaties concluded or acceded to by the People's Republic of China shall apply.[204]

According to this provision, the CLC will apply as long as the damage is caused by vessels carrying persistent oil. It seems that the foreign-related issue is no longer necessary for the application of the CLC. If the damage is caused by an accident which does not fall into the scope of the CLC, such as damage caused by non-persistent fuel oil or fuel oil carried by vessels rather than by tankers, the limits under the CMC will apply. This idea is also confirmed in the 2011 Explanation.[205] In addition, the explanation clarifies that the costs of preventive measures are not subject to the CMC limitation, if the damage is caused by non-persistent fuel oil or fuel oil carried by vessels rather than oil tankers.[206]

3.1.2. Standing

To claim for natural resources damage, one major obstacle in the Chinese legal system relates to the question of who has the *locus standi*. According to the Civil Procedure Law, only the party who has 'a direct interest in the case' can bring a lawsuit to the court. However, when there is only damage to the natural resources, especially public natural resources, which are not privately owned, who has the standing to claim for the damage remains difficult. This obstacle has excluded much public interest litigations in China.[207] However, this is less of a problem for marine pollution. The MEPA authorizes public authorities explicitly to claim for marine pollution damage:

> For damages to marine ecosystems, marine fishery resources and marine protected areas which cause heavy losses to the State, the department invested with power by the provisions of this law to conduct marine environment supervision and

203 For a different interpretation of the application scope in cases law, see Zhang 2010, p. 360-369.

204 Regulation of the People's Republic of China on the Prevention and Control of Vessel-Induced Marine Environment Pollution (Promulgated by the Executive Meeting of the State Council, effective on 1 March 2010) (Hereinafter Regulation on Vessel-Induced Pollution), Art. 52.

205 Art. 19 of the 2011 Explanation.

206 Art. 20 of the 2011 Explanation.

207 The claims for pollution of Songhua jiang by PetroChina in 2005 provide an example. See Section 1.2.2.

> administration shall, on behalf of the State, put forward a compensation demand to those held responsible for the damages.[208]

In China, many natural resources are owned by the States. This provision authorizes the competent public authorities to represent States to claim for compensation. The competent public authorities are limited to 'the department invested with power by the provisions of this law to conduct marine environmental supervision and administration.' According to the MEPA, there are mainly four types of public authorities involved: environmental protection agencies, ocean agencies, marine safety agencies and fishery administrations.[209] The environmental protection agencies are responsible for protecting the ocean from land-based pollutants and coastal construction projects; ocean agencies are responsible for the supervision and administration of the marine environment, for preventing pollution caused by marine construction projects and dumping of wastes in the sea; the marine safety agencies are in charge of marine environmental protection in the port waters and the investigation and handing of pollution accidents; the fishery administrations are responsible for pollution inside the fishing port waters and protecting the ecological environment in fishing zones.[210] Hence the latter three parties play a major role in claiming for marine natural resources damage. When a vessel accident leads to marine pollution, the marine safety agency 'shall have the right to adopt forcible measures to avoid or reduce pollution damage.'[211] In other words, the marine safety agency is responsible for the prevention measures and cleanup in case of an accident and can claim such costs later. In addition to such measures, if the accident leads to other environmental losses, such as lost ecological capacity, the ocean agency can claim for the damage. The fishery administration is the party who can claim for the lost natural fishery resources.

3.1.3. Mandatory Financial Security

As already mentioned, China is a Member State of the CLC, which introduces an obligation for ship owners to seek insurance coverage for the potential liability under the convention. Influenced by the CLC, the MEPA 1999 requires the establishment of vessels oil pollution liability insurance and oil funds and authorizes the State Council to promulgate concrete rules on those issues.[212] However, such concrete rules were only issued in 2009, through the Regulation on Vessel-Induced Pollution. The Regulation obliges the vessels navigating in the Chinese sea area (with the exception of vessels of less than 1,000 tons gross tonnage carrying cargoes other than oil) to buy insurance or seek other financial security coverage.[213] The amount of financial security they seek shall be no less than the

208 Art. 90 of the MEPA.
209 Art. 5 of the MEPA.
210 *Ibid.*
211 Art. 71 of the MEPA.
212 Art. 66 of the MEPA.
213 Art. 53 of the Regulation on Vessel-Induced Pollution.

amount required under the CMC and the Conventions China accedes to.[214] To implement this provision, another document was published in 2010 to further clarify the types of vessels which are subject to compulsory financial security requirement and the amount they need to take.[215]

Table 5: The Types of Vessels and Required Financial Security in China[216]

<table>
<tr><td colspan="2">Vessels with persistent oil as cargo</td><td colspan="2">Vessels with non-persistent oil as cargo and non oil tankers larger than 1,000 tons gross tonnage</td></tr>
<tr><td>Types of vessels (gross tonnage)</td><td>Amount of financial security</td><td>Types of vessels (gross tonnage)</td><td>Amount of financial security</td></tr>
<tr><td rowspan="2">(1) Lower than 5,000 tons</td><td rowspan="2">4.51million SDR</td><td>(1) 20-21 tons (not include 21)</td><td>27,500 SDR</td></tr>
<tr><td>(2) 21-300 tons (not include 300)</td><td>(1) + 500 SDR per ton</td></tr>
<tr><td rowspan="4">(2) Higher than 5,000 tons</td><td rowspan="4">(1)+ 631 SDR per ton; but the total amount is no more than 89.77 million SDR</td><td>(3) 300-500 tons</td><td>167,000 SDR</td></tr>
<tr><td>(4) 501-30,000 tons</td><td>(3) + 167 SDR per ton</td></tr>
<tr><td>(5) 30001-70,000 tons</td><td>(4) + 125 SDR per ton</td></tr>
<tr><td>(6) Above 70,001 tons</td><td>(5) + 83 SDR per ton</td></tr>
</table>

The Implementation Rules require that Chinese vessels should buy insurance from the insurers determined by the Marine Safety Agency or acquire other financial security, such as a letter of guarantee and letter of credit from insurers or other financial institutions determined by the Marine Safety Agency.[217] The requirements for commercial insurance companies and Protection & Indemnity Clubs to be determined as qualified are also clarified in the Implementation Rules.[218] In 2012, 23 insurance companies and P&I Clubs have been acknowledged by the Marine Safety Agency, including the China Shipowners Mutual Assurance Association (CSMAA), some commercial insurers and some members of the International Group of Protection & Indemnity Clubs (IG Group).[219]

3.1.4. Compensation Funds

Although China acceded to the CLC in 1980, it is not a Member State of the IOPCF of 1971 and 1992. Hence it is not obligatory for the Chinese oil industry to contribute

214 *Ibid.*

215 Implementation Rules on the Civil Liability Insurance for Vessel-Induced Oil Pollution Damage (promulgated by the Ministry of Transport, effective as of 1 October 2010) (hereinafter Implementation Rules).

216 Arts. 5, 6 of the Implementation Rules.

217 Art. 8 of the Implementation Rules.

218 Arts. 9, 10 of the Implementation Rules.

219 Notice on the Lists of Insurance Companies Providing Oil Pollution Damage Liability Insurance for Chinese Vessels of 2012, issued by China Maritime Safety Administration, on 19 December 2011, Haichuanbo [2011] No. 862. The International Group of Protection & Indemnity Clubs (IG Group) composed of thirteen principal Protection and Indemnity Clubs, which provide liability cover for approximately 90 per cent of the world's ocean-going tonnage.

to the Oil Pollution Fund. However, the Regulation on Vessel-Induced Pollution requires the establishment of a domestic Vessel-induced Oil Pollution Damage Compensation Fund (Hereinafter Compensation Fund).[220] According to the regulation, all the cargo owners or their agents who receive persistent oil cargo carried by sea within sea areas of china shall contribute to the Compensation Fund.[221] The concrete rule to manage the fund was published recently.[222]

The Regulation on the Compensation Fund sets the contribution at RMB 0.3 per ton of persistent oil. The fund can be used to compensate or indemnify when: (1) the total amount of compensation exceeds the ship owner's limitation of liability; (2) the legal defenses are available; (3) the ship owner and its insurer/guarantor cannot provide full compensation and (4) the liable ships cannot be identified.[223] Three exceptions are clarified when the Compensation Fund does not apply: damage is caused by wars, insurrections or non-commercial vessels/military ships held by the government; claimants cannot prove oil pollution is caused by ships or the damage is fully or partially caused by the victims' fault.[224] One major difference between the Chinese Compensation Fund and the IOPCF is that the former establishes a priority list to provide compensation in case of insufficient capacity of the fund. On the one hand, for the claims caused by different accidents, the Compensation Fund shall deal with the compensation according to the time of application; on the other hand, if the claims are caused by the same accident, the compensation shall be provided according to the following order: emergency response costs; cleanup costs; direct economic losses suffered by fishery and the tourism industry; the costs of measures to restore the marine ecosystem and natural fishery resources; monitoring costs incurred by the management committee of the Compensation Funds; and lastly, other costs approved by the State Council.[225] The upper limit of compensation for one accident is set at RMB 30 million.[226]

3.2. *Practice*

3.2.1. Filing Claims

As sketched above, there is an important role for administrative agencies and more particularly the Maritime Safety Agency (MSA) in filing and handling claims. The MSA is in charge of the response and cleanup of pollution in the case of an accident. The State Oceanic Agency (SOA) is on the other hand responsible for the restoration

220 Art. 56 of the Regulation on Vessel-Induced Pollution.

221 *Ibid.*

222 Management Regulation of Collection and Use of the Vessel-induced Pollution Damage Compensation Fund (promulgated by Ministry of Finance and Ministry of Transport, effective as of 1 July 2012, Caizong [2012] No. 33) (hereinafter Regulation on Compensation Fund).

223 Art. 15 of the Regulation on Compensation Fund.

224 Art. 16 of the Regulation on Compensation Fund.

225 Art. 17 of the Regulation on Compensation Fund.

226 Art. 18 of the Regulation on Compensation Fund.

of the environment. Most claims for ecological damage are hence brought either by the MSAs or by SOAs. Fishermen are usually the victims who brought claims for their individual losses. Most disputes are settled rather than brought to court.[227]

A problem in the claims management is that different authorities have competence for claiming compensation in case of marine oil pollution and that the division of power especially between local authorities and the central government in this respect is not always very clear.[228] Usually in maritime environmental litigation it is the State Oceanic Agency that claims for pure environmental damage whereas the Maritime Safety Agency will claim for the cleanup costs and impose fines as well.[229] The compensation for damage caused by the ship 'Tasman Sea' is an example of a case that was brought by the oceanic and fishery agencies. In 2002, 'Tasman Sea' collided near Tianjin and led to the leaking of oil. The release led to serious damage to the fishery industry and to the marine environment. The Tianjin Oceanic Agency and Fishery Agency brought claims against the ship shortly after the accident. The Oceanic Agency claimed for the loss of oceanic environmental capacity, losses of marine biodiversity, restoration costs and assessment costs. The Fishery Agency claimed for natural fishery losses. In the first judgment in 2004, the defendants were required to pay the Oceanic Agency losses of environmental capacity and assessment costs of more than RMB 10 million, and the Fishery Agency more than RMB 15 million as natural fishery losses. This case was reported as the first case in China where the natural resources damage is compensated.[230] However, the division of authority is not always clear. In practice, sometimes the prosecutors, environmental agencies or local government also claim for ecological damage.[231]

3.2.2. Compensation via P&I Clubs

As mentioned earlier, only certified insurers and P&I Clubs can provide insurance coverage for oil pollution in China. By far the CSMAA is one of the largest parties that provides coverage for oil pollution liability. This is a so-called Protection and Indemnity Club (P&I Club), composed of ship owners that conclude a risk sharing agreement and in that way mutually cover each other's losses.[232] The CSMAA as P&I Club formally rather functions as a risk-sharing agreement and not as an insurer (in the sense that risks are mutually shared and not shifted to a third party).[233] However, from the victim's perspective the crucial point is that P&I Clubs

227 *Ibid.* and interview with CSMAA.

228 Interview with Prof. Wang Jin.

229 Interview with representatives of CSMAA.

230 See Ma & Du 2006, p. 224, 231. Noting that this case has been appealed by the defendant to the High Court of Tianjin, and the final judgment is not made public yet. See also Zhu 2007b, p. 28-30.

231 See Section 4.2.4.

232 See Coghlin 1984, p. 403-416.

233 See on the differences between risk sharing and insurance also Faure & Hartlief 2003, p. 167-168.

compensate the losses for which the members (usually ship owners) are covered. It is worth noting that CSMAA is not a member of the IG Group.

Clause 12 of the Rules of CSMAA clearly provides that pollution risks are covered. According to this clause the following risks are included:

A. Liability for loss, damage or contamination;
B. Any loss, damage or expense which the member incurs, or for which he is liable, as a party to any agreement approved by the directors, including the costs and expenses incurred by the member in performing his obligations under such agreements;
C. The costs of any measures reasonably taken for the purpose of avoiding or minimizing pollution or any resulting loss or damage together with any liability for loss of or damage to property caused by measures so taken;
D. The costs of any measures reasonably taken to prevent an imminent danger of the discharge or escape from the entered ship of oil or any substance which may cause pollution;
E. The costs of liabilities incurred as a result of compliance with any order or direction given by any government or authority, for the purpose of preventing or reducing pollution or the risk of pollution, provided always that:
   a. such compliance is not a requirement for the normal operation or salvage or repair of the entered ship; and
   b. such costs or liabilities are not recoverable under the Hull Policies or the Hull Certificates of the entered ship.[234]

There is usually in the rules of the P&I Club no specific title concerning restoration costs. These are, however, covered under the title of cleanup costs. Cleanup costs need to be reasonable and real. Interim losses are according to the CSMAA difficult to be evaluated and are usually not compensated.[235]

The CSMAA has an acceptance policy whereby a ship before it will be covered is first inspected. A classification society is designated to undertake the inspection. When the CSMAA is of the opinion that the ship is not qualified it can either ask for improvement of safety measures or otherwise decide to refuse cover. Refusal of coverage will have as a consequence that the ship owner will have to try to seek coverage from another P&I Club or insurer since insurance coverage is mandatory. After the ship has been accepted insurers will do random inspections on the basis of the presumed quality properties of the ship and the age.[236] There is little difference as far as the premium that is charged for on the one hand sea-going ships and on the other hand inland ships that remain in China. A risk differentiation is applied.[237] The differences in premiums are usually based on technical differences between the

234 Rule of China Shipowner Mutual Assurance Association, Rule 12, available at: <www.cpiweb.org/en_baoxiantiaokuan/3.8.jsp>, last accessed on 8 May 2013.

235 Interview with representatives of CSMAA.

236 *Ibid.*

237 See on the importance of risk differentiation as remedy to cure the moral hazard risk in insurance Shavell 1979a.

ships and also on the past loss experience. In case of a heavy claim record the contribution of the ship will be increased. Evaluation of risks takes rather place on the basis of the entire fleet of a ship owner than on the basis of one individual ship.[238] CSMAA only provides coverage but above the amount of the so-called retention it will purchase reinsurance from the international group of P&I Clubs.[239] Reinsurance is purchased on a yearly basis.

3.2.3. Remedies

Usually compensation is claimed in marine pollution cases for loss of fishery resources (a distinction is made between the direct losses, involving the fish that immediately died as a result of the pollution and middle and long-term losses).[240] Cleanup costs to remove the oil are claimed as well. Cleanup costs and losses by cleanup measures can usually be calculated relatively easily because these costs are known.[241] Such costs are easier to be claimed back; also because the public authorities can apply the arrest of the vessels for the costs of cleanup and preventive measures.[242] Difficulties often arise on the one hand concerning the acceptability of restoration measures that are proposed or executed by government and concerning the evaluation of the damage.[243]

To ensure the timely cleanup of pollution, a regime of vessel pollution cleanup agreements is established. Some types of vessels are required to sign cleanup agreements with qualified institutions before they arrive at Chinese harbors.[244] With such an agreement, a vessel pollution cleanup institution needs to provide response and cleanup services in case an accident happened to the vessels. Such cleanup institutions have to be certified by the Maritime Safety Agencies.[245]

As mentioned earlier, there is increasing practice in claiming for marine ecological damage. Sometimes public authorities will claim for this damage. However, at the national level, there has been no regulation yet on how to use such damages claimed from the polluters. Recently, some local areas started to publish some regulations on this issue, such as the Guangdong temporary regulatory document on the use of money compensated for the oceanic and fishery

238 Shavell 1979a.

239 See on the functioning of this international group of P&I Clubs, Faure & Van den Bergh 2002.

240 In The Rules on Calculating Fishery Losses Caused by Water Pollution Accidents, a distinction is made between direct economic losses and natural fishery losses. Those are usually called direct losses and middle and long-term losses in practice. For details about these concepts, see Li 2006, p. 7-8 (in Chinese).

241 Interview with Prof. Wang Jin.

242 See Maritime Procedure Law (promulgated by Standing Committee of National People's Congress, revised on 25 December 1999, effective as of 1 July 2000), Art. 21.

243 *Ibid.*

244 Regulation on Emergency Prevention and Response to Vessels-Induced Marine Environmental Pollution (Promulgated by the Ministry of Transport, effective as of 1 June 2011), Arts. 25-27.

245 *Ibid.*, Arts. 17-19.

environment.[246] The document applies to damages awarded to oceanic and fisheries agencies for the marine environmental damage and fishery losses in the Guangdong province.[247] It holds that the money should be put in a specific financial account of the provincial of local treasury and should be used to monitor the environment and to protect and eventually remediate the marine environment.[248]

3.2.4. Cases

In recent years, there are increasingly cases involving claims for marine (inland waters) ecological damage in China. This research uses 34 marine environmental pollution cases adjudicated by Guangzhou Maritime Court from 1991 until 2009 as an example to show how ecological damage can be compensated in China.[249] In many of those cases, fishery associations or local communities claimed for their fishery losses. In addition to those claims for individual losses, there are also 14 cases where compensation is awarded for ecological damage and which can be classified as public interest litigation. There are generally two types of plaintiffs in those cases: public authorities and procuratorates (two cases). Maritime safety agencies and oceanic and fishery agencies are the most common plaintiffs falling into the category of public authorities. Eight cases involve oceanic and fishery agencies as plaintiffs and two cases involve maritime safety agencies. Other authorities also filed a few cases, such as the environmental protection bureaus (one case), the environmental and health management departments (one case) and local governments (one case).[250] Those cases show how the judges are dealing with standing issues for public interest litigation in the marine pollution area in China. However, it has to be remembered that the actual number of pollution cases may be

246 Guangdong temporary regulatory document on the use of money compensated for the oceanic and fishery environment (promulgated by Guangdong Department of Finance, effective as of 25 October 2006, Yuecaizong [2006] No. 205).

247 *Ibid.*, Art. 2.

248 *Ibid.*, Arts. 3-4.

249 The information is based on the Project 'Water Pollution Public Interest Litigation' held jointly by Sun Yat-sen University Faculty of Law and the Guangzhou Maritime Court. The author is grateful to Prof. Li Zhiping of Sun Yat-sen University and to Ms. Yang of Guangzhou Maritime Court for providing the helpful information. It is worth noting that not all the tort cases adjudicated in the Maritime Court are related to vessels. According to a Supreme Court judicial explanation, the maritime court accepted not only cases concerning damage and pollution caused by vessels, but also 'cases on disputes over claims for damages arising out of shipping, production, operations on the sea or on water areas leading to the sea'. See [Some Provisions of the Supreme People's Court on the Scope of Cases to be entertained by Maritime Courts] (Adopted by the Judicial Committee of the Supreme People's Court, promulgated on and effective as of 9 August 2001), Art. 1(5). An English copy can be found at: <www.gzhsfy.org/english/shownews.php?id=9772>, last accessed on 8 May 2013. Hence, in addition to vessels related tort cases, Guangzhou Maritime Court also accepts 'cases on disputes over claims for damage arising out of pollution on the sea or water areas leading to the sea.' See <www.gzhsfy.org/english/shownews.php?id=9772>, last accessed on 8 May 2013.

250 See Li & Zhan 2011, p. 240-241.

substantially larger given that many cases are settled rather than being brought to court.[251] Moreover, this sample only relates to one maritime court in China, being the one of Guangzhou in the South of China.

In the examined cases, compensation is awarded for different types of pollutions. For the claims made by fishery associations and local communities, the awarded damages are usually fishery losses suffered collectively by them. The other cases consider public interest litigation, where public authorities/procuratorates claim for cleanup costs, natural fishery losses, assessment costs and, at times, restoration costs.[252] Several cases could even be found that involved compensation for environmental damage in addition to compensation for the above-mentioned types of losses. There are two cases involving claims for ecological damage in addition to cleanup costs and fishery losses: *Guanzhou Haizhu District Procuratorate v Chenzhongming*[253] and *Guanzhou Panyu District Procuratorate v Lu Pingzhang*.[254] However, it is worth noting that these two cases do not involve vessel-induced pollution. In both cases, it is the defendants' enterprise that caused water pollution. The environmental damage is assessed broadly according to the assessment report in these cases. In the first case direct economic losses were assessed, those included monitoring costs, the water resources fee and cleanup costs. In addition also indirect economic losses resulting from the environmental damage were assessed. However, it was held that these were in fact too difficult to evaluate and hence could not be compensated. The same was the case for water treatment costs. Those were equally considered to be too difficult to evaluate and hence were not compensated. In the second case three kinds of losses were included under 'environmental damage', being direct damage (damage to agriculture, fishery and ecology) and indirect damage (economic losses of other productive and consumption systems which were caused via water damage and costs of preventive measures) and losses of enjoyment. The indirect losses were calculated on the basis of abstract mathematical models; the losses of enjoyment were neglected and hence not compensated.[255]

The representatives of the P&I Club held that in practice oil pollution is an important risk and leads to quite a few cases involving the discharge of bulk oil and cargo oil; however, so far China has not suffered from major vessel-induced pollution incidents of the type of Erika or Prestige. There have hence been more

251 Interview with representatives of CSMAA and interview with a representative of the Oceanic and Fisheries Administration of the Guangdong province, (15 August 2011), interview transcript on file with the author.

252 For example, in the case of the *Zhuhai Environmental Protection Bureau v Taizhou Donghai Marine Transport Limited and China Vessels Fuel Supplying Fujian Company* (hereinafter Fujian company), one of the defendant's vessel had a collision accident leading to oil release and serious marine pollution in Zhuhai. The EPB adopted both clean-up measures and restoration measures, and it claimed for both costs and assessment costs in the court. In the judgment, the Fujian company was required to pay for all those costs. *Zhuhai Environmental Protection Bureau v Taizhou Donghai Marine Transport Limited and China Vessels Fuel Supplying Fujian Company* (1999) Guanghaifashizi No. 88) on file with Guangzhou Maritime Court.

253 (2008) Guanghaifachuzi No. 382, on file with Guangzhou Maritime Court.

254 (2009) Guanghaifachuzi No. 247, on file with Guangzhou Maritime Court.

255 Assessment reports on file with Guangzhou Maritime Court.

many smaller cases, none of which ever exceeded the limit of the insurance coverage.[256]

3.2.5. Challenges

Although the evaluation and compensation of marine pollution damage seems considerably better than the evaluation and compensation in land-based pollution cases, there are still some problems in this domain as well. These were already mentioned and most of them relate to the evaluation of environmental damage. The two cases we examined in more detail that were dealt with by the Guangzhou Maritime Court also show the practical difficulties. Because experts are not able to evaluate indirect economic losses or losses of enjoyment these are simply neglected. These difficulties all have to do with the fact that a standard for evaluating marine environmental damage is lacking and that there are no consistent methods in the evaluation of the damage.[257] This absence of evaluation methods obviously leads to uncertainty and potentially to under-compensation.[258]

4. Economic Analysis of the Chinese Compensation System for Natural Resources Damage

In this chapter, an assessment of the way in which environmental damage is compensated in China is provided by looking both at the theoretical compensation possibilities based on the statutes and rules and at their application in practice. Of course one should be careful with drawing too strong normative conclusions on the basis of this analysis given the limits of our empirical method. I do realize that although some interviews were conducted with representatives from government agencies, academia, NGOs and insurers, the number of interviews was limited and most of them were concentrated in the Beijing and Guangzhou areas. The main results can be summarized as follows.

4.1. Liability Rules

Economic analysis shows that regulation and liability can be used together to create efficient preventive incentives to potential polluters. Subject to some characteristics, such as the presence of information asymmetry or an insolvency problem, the availability of legal action and administrative costs, either regulation or liability rules alone can provide optimal deterrence.[259] This is also true in China. Although China has made many improvements in establishing its environmental legal framework[260] and strengthening the enforcement,[261] it is still subject to formidable

256 Interview with representatives of CSMAA.

257 Interview with Prof. Wang Jin.

258 Although the representatives of the P&I Club held that rather the opposite (over-compensation) may be the case: Interview with representatives of CSMAA.

259 Shavell 1984a, p. 359-364.

260 Wang 2006, p. 202-203.

enforcement challenges as other industrializing countries, such as weak agency capacity and a capture-prone governance text.[262] In this case, liability rules can be used to fill the regulation and enforcement gaps.

The general civil laws and specific environmental statues work together to create the legal basis for environmental liability. The newly adopted Tort Liability Law in China reiterates a few principles and clarified some issues concerning environmental liability. A comparatively comprehensive and sound liability system for personal injury and property damage caused via the environment has been established on paper. A strict liability rule is confirmed, which in theory can lead to efficient care and activity levels. A reversal of the burden of proof can relieve the victims from the heavy burden to prove the causation between the pollution and their damage.[263] There is no general provision channelling liability to specific parties and there are no general caps introduced on the liability. All those seem in line with economic theory. However, the review of practical experiences showed that to an important extent, for a variety of reasons, possibilities that the formal legislation offers are simply not sufficiently used.

As far as ecological damage is concerned, one can find serious obstacles not only in practice but even on paper. Firstly, the law does not impose an obligation to restore the damaged environment on specific parties (polluters or specific public authorities). Secondly, the law is also unclear on whether ecological damage is compensable with the exception of the case of marine pollution. Although general environmental liability provisions can be found in the new Tort Liability Law, it does not have an explicit reference to pure ecological damage. Moreover, hurdles do not only exist in substantive law but also in procedural law. Ecological damage may be widespread and hence not involve individual victims. In China the law only allows very limited public interest litigation, which hence poses a serious challenge of the *locus standi*. Last but not least, lack of assessment standards also makes compensation for ecological damage quite difficult in practice.

Indeed, those legislative gaps, naturally, have lead to the insufficient restoration and compensation in practice. One could for example notice a strikingly important role for government in the cleanup of polluted sites. On the one hand one could consider this to be positive as government apparently takes responsibility for cleanup actions. However, there are in practice quite a few difficulties with this. One point is that actions are usually not directed against polluters. Especially since historic pollution is often caused by state-owned enterprises, incentives of government to act against polluters may be limited. Moreover, often government will provide restoration itself and simply charge a higher price for the new development of the land.

This shows that precisely because of this government intervention there is, in China, probably a different type of role for government in these restoration cases, whereby public and private law instruments are mixed, as well as legal and

261 Van Rooij 2006.

262 McAllister, Van Rooij & Kagan 2010.

263 Section 1.1.4.

economic instruments. Since actions are usually addressed against new developers rather than against polluters it is very doubtful that the actions provide efficient incentives for prevention to potential polluters. Compensation should in fact not be a goal in itself but rather provide incentives to potential polluters to abate pollution to efficient levels. Given the low probability of a liability suits and low amounts of damages awarded it is very doubtful that in China environmental liability can today exercise its deterrent effect, notwithstanding the recent changes, *inter alia* as a result of the Tort Liability Law of 2009.

4.2. *Insurance*

Theoretical analysis shows that compensation instruments can complement the liability rules in both the compensation and prevention of ecological damage.[264] However, in China, when liability rules fail to provide efficient deterrent effects, one can easily question when a comprehensive compensation system, including a variety of compensation instruments such as insurance, environmental funds and capital markets, works. Although insolvency risks may arise from time to time, especially for SMEs, there is no general requirement of financial security with the exception of marine oil pollution caused by sea-going vessels. The judgment-proof problem may on the one hand leave the victims uncompensated and the damaged environment unrestored, and on the other hand lead to under-deterrence. Unsurprisingly, the low liability deterrence together with a lack of a mandatory financial security system makes the various compensation instruments less developed in China. Only in recent years, the environmental insurance market started to develop. Now the insurance possibilities are widely available, including general liability insurance policies, specific environmental liability pollution and so on. Personal injury and property damage caused via environmental incidents are generally covered. Cleanup costs may be also covered to the extent that it can prevent further personal injury and property damage. Some insurance policies[265] even cover remediation costs on and from the insured sites. However, there are gaps as well: the restoration costs in addition to cleanups are usually uncovered. The premiums today are still high and adverse selection is a serious challenge facing the insurers. The above analysis shows that the insurance is not broadly used. That should as such not come as a surprise since environmental liability does not constitute a serious threat, thus removing incentives of polluters to seek insurance coverage. However, if environmental liability were to develop further, in practice the insurance market is apparently increasingly able to provide at least basic coverage for this environmental liability risk.

264 Faure & Liu 2012.

265 Huaitai Insurance.

4.3. *Marine Oil Pollution*

There is more positive news as far as the compensation function in marine oil pollution cases is concerned. It has a more comprehensive legal framework for ecological damage. The MEPL allows competent public authorities to claim for marine environmental damage. Legislation also explicitly includes some parts of ecological damage are compensable, such as prevention costs and reasonable restoration costs. This can be explained by the influence of the international conventions (the CLC) China has joined. It, nevertheless, has inefficient features as well. The compensable scope of ecological damage is similar to that under the CLC, but it is much more limited than the definition of natural resources damage in the OPA in the U.S. The U.S. system allows also compensation for loss of ecological service pending restoration and compensation even when restoration is impossible. Admittedly, one has to realize this broader compensation is only feasible when clear assessment rules are in place, which counts as another weakness in the Chinese system. The legislation itself does not explicitly channel liability to ship owners. However, in practice, it is usually the ship owners that are held liable, which fails to create incentives for any parties who may equally contribute to the risks.[266] Since China is one party to the CLC, the CLC limitation of liability also applies in China. When an accident falls out of the CLC scope, the limitation established in the CMC applies, which is even lower than the CLC limits. The capped liability, combined with strict liability can provide insufficient preventive incentives to ship owners.

Another positive part of the compensation system for marine oil pollution is its more-developed insurance market. A Chinese P&I Club (CSMAA) has a long history of covering pollution risks of vessels, including preventive and restoration costs. A mandatory insurance system was adopted in the Regulation on Vessel-Induced Pollution in 2009, which can relieve the judgment-proof problem. The recent legislation also led to the establishment of an oil compensation fund. All the cargo owners or their agents who receive persistent oil cargo carried by sea within sea areas of china shall make contributions to the fund based on the amount of oil they received (RMB 0.3 per ton). The fund complements the compensation capacity of the ship owners and their insurers. To base the contribution only on the amount of oil but not on the actual oil pollution risks, however, has been criticized in literature since it cannot create sufficient preventive incentives for the oil industry. From an economic perspective not only should the duty of compensation be placed on the ones actually contributing to the risks, but the contributions should also be in proportion to the amount of risks they create.[267] To base the contributions to the fund only on the amount of oil received merely gives the oil industry incentives to change their activity level but not their care level (for example choosing safer vessels to transfer the oil cargo). In the U.S., when proposing an oil fund, Wood has proposed a system with better risk differentiation: to base the initial contribution on the historical oil spill created by each vessel and to allow a decrease of the

266 Wang 2011, p. 249.

267 Wang 2011, p. 338-344.

contribution if better safety measures are adopted.[268] However, such a system incurs high administrative costs. To what extent risk differentiation is desirable depends on the tradeoff between the benefits in saving primary costs and the increasing of tertiary costs.

5. Recommendations to Improve the Compensation System

The above analysis has shown that, especially from the historical perspective, important steps have been taken in recent years, both at the legislative level (implementation of strict liability and the attempt to introduce public interest litigation) and to some extent at the practical level as well (developing of expertise in maritime courts with marine oil pollution and development of an environmental insurance market). These developments are undoubtedly important, but claims for ecological damage in China realistically still face substantial hurdles in obtaining fair compensation, as a result of which today environmental liability in China is neither fulfilling its compensatory nor its preventive role. In that respect there still is a long way to go. This section tries to provide some recommendations for the future development of the compensation system.

5.1. Legal Framework

The above analysis has shown some legislative gaps that prevent effective compensation for ecological damage in China. To this end, the economic analysis and international experience can provide some hints on the improvement of the Chinese legal framework.

Although a comparative comprehensive tort system for traditional damage caused via the environment has been established at least on paper, the legislation for ecological damage is not so promising in China. There is still much left to be done at the legislative level. First of all, a clear obligation in response to an environmental accident needs to be established. Either the polluters or the public authorities shall take cleanup/restoration measures in case of ecological damage. A clear division of authority is necessary to ensure an effective and timely response. For example, in the U.S., various government authorities are regarded as trustees of specific natural resources and need to respond to pollution and claim for the damage.[269] Authorizing the public agencies can also help to solve the standing problem, which is an important hurdle facing the natural resources damage claim cases. The tendency to open up space for public interest litigation will give the public authorities or NGOs the possibility to claim for ecological damage. In addition, an explicit definition of the compensable ecological damage is also important, which needs to be accompanied by feasible evaluation standards. Although there are already quite some cases where compensation for ecological damage was awarded in case of marine oil pollution, how such a new title can be

268 Wood 1975, p. 47-50.

269 See Chapter 3, Section 2.2.

quantified is still haunting the judges. Without a clear evaluation standard, compensation for ecological damage will probably only remains a solution on paper.

The development of compensation instruments is still in its early stage in China. Some possibilities started to emerge in the insurance market from 2007, but it is reported that the coverage is still low and the premiums are comparatively high. The development of environmental liability insurance is possible only when there is a serious liability threat facing the potential polluters. This of course needs an improvement of the liability system. Furthermore, more policy supports, including subsidies in early stages, linking insurance coverage with pollution fees, and a green credit policy can also contribute to the prosperity of the insurance market. Although China is not a member of the IOPCF, it started to establish its own oil compensation fund in 2012. This opened up the attempt to use other compensation instruments to cover ecological damage in addition to insurance. Making the contributions to the fund more risk related rather than just based on the amount of oil imported will have better deterrent effect.[270] Theoretical models show that in case of insolvency risks, compulsory financial security can create more efficient incentives.[271] However, when insurance and other compensation instruments are still in their infancy in China, a general compulsory financial security system may not be feasible. With the further development of the insurance market and alternative instruments, a progressive introduction of a compulsory system can be of help.[272]

5.2. *Practical Issues*

The traditional environmental liability rules in China are as such not bad. However the practical obstacles prevent them from creating either sufficient compensation or efficient deterrence. This experience reveals that the practical problems deserve at least equal attention, if not more, as the legislative ones.

Theoretical models show that liability rules and compensation instruments can complement regulation in providing efficient deterrence. However, rather than separate parts, regulation, liability and compensation instruments for ecological damage compose an interlinked system. Admittedly, environmental regulation is subject to weak capacity and capture problems. However, to expect the court system to act as an independent power, and thus another pathway to fill this gap, may not work out as well as predicted by the theories in China.[273] To some extent, the courts in China 'are better conceived of as cost-benefit-weighing government

270 For the discussion of risk differentiation in compensation funds, see Wang 2011, p. 338-343.

271 See Chapter 4, Section 2.4.

272 The progressive introduction of a compulsory system can either start from specific areas or certain industries. Actually, in practice, in some local areas such as Changsha, the government already requires certain high-risk industry to seek insurance coverage. See Art. 4 of the Changsha Rules.

273 McAllister, Van Rooij & Kagan 2010.

regulators rather than neutral arbiters.'[274] Sometimes, the judges still need to make policy considerations and rely on support from the government. For example, the judges may be reluctant to accept sensitive cases, including collective cases; to prove the existence of harm, pollution action and sometimes, even causal links, the victims need the support from local environmental agencies (in forms of daily monitoring data or on field evaluation reports). As far as ecological damage is concerned, the link between the public and tort system is more obvious: it is the public authority who determines whether and how to restore the damaged environment and later, and whether to start a claim procedure from polluters. Hence, the improvement of the liability system also depends on the strengthening of regulation. The development of the insurance market is also not a separate issue. Without serious liability threats, the polluters will have no incentives to buy liability insurance. Policy supports, such as linking insurance with pollution fees or other encouraging measures can also promote the development of insurance. Therefore, one cannot expect the liability rules and compensation instruments alone to solve the problems created by ecological damage. On the contrary, regulation, liability rules and compensation instruments need to develop hand-in-hand.

Although an effective compensation system for ecological damage also needs a decent regulatory framework and the Chinese regulatory system is still subject to many weaknesses, one should not necessarily to be too pessimistic. In recent years, China is trying to enhance its environmental enforcement and the literature has noticed a convergence towards a more coercive and formal way of enforcement (although one has to admit many enforcement gaps still exist and the convergence is fragile).[275] Moreover, although judges have incentives to avoid political controversy, they do occasionally make innovations at the margins by providing new interpretations or validating new types of claims.[276] The attempt to establish environmental courts and allowing public interest litigation also constitutes a further step towards compensation for ecological damage. Although there is still a long way to go for China to come towards a sound compensation system for ecological damage, opportunities for efforts and improvements are available at both the legislative as well as at the practice level.

[274] McMullin 2009, p. 183.

[275] See Lo 2006; Van Rooij 2010.

[276] Stern 2010, p. 91-93.

Chapter 11

CONCLUSION AND POLICY RECOMMENDATIONS

In recent years we have witnessed several astonishing environmental catastrophes. The explosion at the drilling unit Deepwater Horizon in the Gulf of Mexico on 20 April 2010 led to the leakage of 4.9 million barrels of oil.[1] This means not only enormous economic losses but costly environmental restoration as well. One year later, a nuclear accident only comparable to the Chernobyl accident stroke Fukushima. Its impact on human health and the environment is still in evolution and has far from fully manifested yet. However, the radiation-related impact on wild species has started to be reported.[2] These disasters have shocked the international community and triggered the reflection on the sufficiency of our existing environmental regulatory and liability system in protecting the public and the environment.

Well before these accidents, people have realized the significant influence of environmental incidents on not only humans and economic interests but also on the environment itself. A series of accidents have led to the establishment of a compensation system for ecological damage in many jurisdictions.[3] In the U.S., the damage to the environment is defined as 'natural resource damage'. The federal acts OPA and CERCLA introduced a liability system and authorized the promulgation of detailed assessment procedures decades ago. Across the Atlantic, a liability system for damage to protected habitats and species, water damage and land damage started to be established by the ELD a few years ago. It tries to establish a common framework for the prevention and remediation of environmental damage. This development reflects people's awareness of the value of the environment itself

1 National Commission on the BP Deepwater Horizon Oil Spill and Offshore Drilling, Report to the President, Deep Water: The Gulf Oil Disaster and the Future of Offshore Drilling, January 2011, 167, available at: <www.oilspillcommission.gov/sites/default/-files/documents/DEEPWATER_ReporttothePresident_FINAL.pdf>, last accessed on 8 May 2013.

2 Hiyama, A., The Biological Impacts of the Fukushima Nuclear Accident on the Pale Grass Blue Butterfly, <www.nature.com/srep/2012/120809/srep00570/full/srep00570.html>, last accessed on 8 May 2013.

3 Such as the promulgation of the OPA in response to the Exxon accident and the promulgation of the CERCLA after the Love Canal pollution.

and of the insufficiency of replying solely with regulation to prevent ecological damage. Such legislation has moved forward in compensating for ecological damage, as a type independent from personal injury and property damage caused via the environment. However, the OPA faces the challenges created by the Deep Horizon accident. The comparatively new ELD system still has limited practical influence and it is criticized by law and economics literature as failing to create sufficient preventive incentives.[4]

In the context of, on the one hand, an increasing development of the legislative framework for ecological damage compensation and, on the other hand, costly ecological damage keeping happening from time to time, the question arises *how a compensation system should be designed to better prevent and compensate for ecological damage*. Before answering this question, two issues need to be clarified. Firstly, why is a comprehensive compensation system necessary? The second issue concerns the relationship between such systems for ecological damage specifically and for environmental damage (including personal injury and property damage caused via the environment) generally. Traditionally, regulation is used to prevent ecological damage, however, with inconsistent success. Accidents keep occurring, leading to huge damage and the damaged environment remains often unremedied. Hence, liability rules started to develop, hand-in-hand with compensation instruments, such as insurance, guarantees and environmental compensation funds. Liability rules and compensation instruments do not only promote the restoration/compensation of the ecological damage but can also enhance the deterrent effect as well. Therefore, the existing system to deal with ecological damage is a comprehensive one, including regulation, liability rules and compensation instruments. As for the second question, many unique characteristics of ecological damage make it more suitable to be solved with its own liability rules than under the general environmental liability framework. The term 'ecological damage' is determined to contain both damage to public natural resources and privately-owned ones which have an ecological value in excess of their economic value. As far as damage to public natural resources damage is concerned, no individual victim is concerned. Hence the standing requirement constituted an important hurdle under traditional tort law. Even when private natural resources are concerned, since it is more valuable ecologically than economically, restoration is a more suitable remedy than monetary compensation. The emphasis on restoration means more government involvement will be necessary. The government can initiate or order the restoration and control the restoration standards. All those characteristics make liability for ecological damage different from traditional environmental liability. Furthermore, when liable parties cannot be identified, or liability cannot be established, some compensation instruments can play a role, such as direct insurance and environmental funds. Even in these cases, one has to notice that the compensation instruments should also promote the restoration of the damaged environment and not only provide monetary compensation. When individual compensation instruments are concerned, both

4 De Smedt & Faure 2013.

stand-alone instruments for ecological damage and general ones for both ecological damage and traditional damage exist. A typical stand-alone instrument is the Premise Liability Insurance Policy which covers cleanup costs on the insured sites. Most compensation instruments, however, cover both types of damage. Those categories of instruments have been examined in this thesis as well.

1. Summary

In the theoretical part (Part I) of this research, three compensation models for ecological damage were proposed. Chapter 2 gives a brief justification for the choice of the term ecological damage and some unique characteristics of natural resources damage. Economic analysis was used to examine the legal rules, individual compensation instruments and compensation models (Chapters 3-5). Calabresi divides the social costs of accidents into three categories: the primary costs (precaution costs plus the actual damage), second costs (concerning loss spreading) and tertiary costs (administrative costs) and holds that a liability regime should minimize the total social costs.[5] This thesis uses Calabresi's theoretical framework to analyze liability rules for ecological damage (Chapter 3) and to compare the potential of different compensation instruments to minimize those three categories of costs (Chapter 4). Chapter 5 presents three compensation models with the combination of regulation, liability rules and compensation instruments, based on whether compulsory financial security is required or not. It also identifies some indicators to choose among different compensation instruments.

Part II tests the theoretical models in practice. This part chooses examples of compensation systems for ecological damage, or more generally for environmental damage, to show whether the practice is in line with the theoretical models presented in the first part. Firstly, the compensation system for natural resources damage in the U.S. and the compensation system established by the ELD in EU are chosen because they provide two well-known examples of systems specifically for ecological damage. The U.S. system establishes a comprehensive system under the CERCLA and the OPA, including a liability system, detailed assessment standards and mandatory financial security. Moreover, the comparatively longer history of such a system makes it not only a scheme on paper but also in practice. The ELD system is a much newer one, which presents many differences from the American one, such as the definition of ecological damage (natural resources damage/environmental damage), liability standards and voluntary financial security system. In addition to these two systems, this thesis introduced some other systems for general environmental liability as well. This is because, on the one hand, the instruments for traditional damage and ecological damage are not always separable (for example the international regime for oil pollution covers both preventive/restoration costs and personal injury and property damage); on the other hand, given the limited experience in the field of ecological damage, some instruments used for general liability provide helpful insight as well. For example,

5 Calabresi 1970.

the compensation systems for nuclear damage (international and the US) provide interesting examples in showing how a potential catastrophe can be covered. The ELD is still a new product in Europe, which means in many Member States, the specific system for ecological damage is still under development. How they deal with general environmental liability hence sheds light on the possibilities of the future compensation systems for ecological damage. Part II was structured as follows: the international compensation regime for oil pollution and nuclear damage was first sketched (Chapter 6). Chapter 7 discussed the compensation system for natural resources damage established under the OPA and the CERCLA, as well as the compensation system for nuclear damage. In the EU, the ELD provides an EU-wide scheme for three types of environmental damage. On the Member States level, a patchwork approach was followed to show the functioning of one example for each compensation instruments (Chapter 8). The empirical research showed that on the one hand many parts of theoretical models and indicators are followed in practice; on the other hand, divergences exist as well. Some explanations for such divergences have also been presented.

After presenting the theoretical models and examining their use in practice, this thesis tried to answer the question *whether, and if yes, how such a compensation system can be established in China*. Part III (Chapter 10) examines the existing compensation system in China by looking at not only legal framework but also practice. A short economic analysis and policy recommendations were also provided.

2. Proposed Models to Prevent and Compensate for Ecological Damage

This thesis tried to identify some models which can, on the one hand, create efficient incentives to prevent ecological damage and, on the other hand, promote restoration and compensation. Economic analysis was used to examine the potential combined use of regulation, liability rules and individual compensation instruments.

Regulation is the primary instrument used to prevent ecological damage. Traditionally, ecological damage, apart from the economic value of privately-owned natural resources, was not compensable under tort law. This thesis used Steven Shavell's indicators to test whether regulation or liability rules are better suited to prevent ecological damage. The four indicators are: the availability of information about risk between private parties and administrative parties and the judges, the solvency of potential liable parties; the probability of legal action and the administrative costs.[6] Applying those indicators to ecological damage reveals that neither regulation nor liability rules dominate each other, and the joint use of two instruments can help to prevent ecological damage.

Economics analysis also helps to design efficient liability rules. Although the tort system also plays an important role in compensation and social justice, the

6 Shavell 1984a, p. 359-364.

economic literature holds that its primary role is deterrence.[7] As far as ecological damage is concerned, this means: strict liability should apply to damage created by environmental hazardous activities while negligence rules to others. Liability should not be retroactive or channeled to specific parties. Caps on liability will create insufficient care level, especially when combined with strict liability. Those above-mentioned principles can basically also apply to more general environmental liability.[8] The compensation scheme for ecological damage has its own characteristics as well. One important component of ecological damage is damage to public natural resources, which concerns no individual victims. This creates an important hurdle of *locus standi* in many jurisdictions. Economic criteria need the party who has most incentives and capability to restore or recover the damage to have standing, usually being the public authorities. To give a public interest group secondary *locus standi* can mobilize the social resources to initiate ecological damage claims. Furthermore, clear assessment procedures (standards) are especially important to make ecological damage claims feasible.

Even the combination of regulation and liability rules cannot guarantee efficient prevention and compensation. On the one hand, both instruments focus on prevention and the role of compensation is partially neglected. On the other hand, problems such as insolvency of liable parties, low probability of claims and asymmetry information result in inefficient deterrence.[9] To overcome those inefficiencies, compensation instruments, such as liability insurance, first-party and direct insurance, risk-sharing agreement, environmental compensation funds and other guarantees can play a role.

The above analysis shows that the joint use of regulation, liability rules and compensation instruments is desirable to prevent and compensate for ecological damage. However, such a conclusion will still be too vague to apply in practice: what is the interaction between those instruments? Shall a mandatory financial coverage via the compensation instruments be imposed on the potential liable party? How can a choice be made between various compensation instruments? Although this thesis also tries to provide some thoughts in answering those questions, the author is also aware of the complexities of different jurisdictions, including the regulatory capacities, independence of the judicial system, development of the market, legal cultures and so on. Hence this research does not aim at creating one universal model which will fit into all situations. Instead, it proposes three models, based on whether a compulsory financial security system is required or not, from which one can choose the most suitable one according to the situation of a certain society. Moreover, this thesis also identifies a few indicators to help the choice of compensation instruments. This is still a simplification of the reality. One has to admit that there is no priority among those indicators and to apply them in practice needs a delicate balance among them. Another issue one cannot neglect is that the compensation system for ecological damage is a

7 Faure & Grimeaud 2003, p. 19-20.
8 Faure & Grimeaud 2003, p. 18-66.
9 See *infra* Chapter 3, Section 3.

comprehensive one and regulation, liability rules and compensation instruments are interrelated. They should go hand in hand rather than develop independently. For example, although liability rules can complement regulation in creating deterrent incentives, government intervention is of special importance in initiating an ecological damage claim. Liability for ecological damage is not a pure civil liability framework but has many public law elements. The public authorities should be able to detect the damage. It is also they that determine to order restoration or initiate restoration themselves and later recover the costs. Hence, with low capacity or incentives of the public authorities, to make liability for ecological damage into practice is also less likely. Furthermore, many compensation instruments are also based on liability, such as liability insurance and risk-sharing agreements. This, in other words, means that those compensation instruments also need the government's behavior (such as the detection of ecological damage, require, approve or take response and restoration measures) to trigger the coverage.

Three compensation models are presented in this thesis. The first one is a financial responsibility system which imposes an obligation on some environmentally hazardous enterprises to seek financial coverage. Strict liability, combined with insolvency problems will lead to under-deterrence. Compulsory financial security systems can alleviate the judgment-proof problems. In addition, by establishing a principal-agent relationship between the assurance providers and the firm engaged in environmentally-risky activities, a compulsory system can incentivize the principal to monitor the environmental safety of the agents. A compulsory system can also help to promote timely, low-cost public access to compensation. However, these advantages of a compulsory system do not mean it is feasible for all situations. In an immature market, where the assurance providers are incapable or unwilling to provide coverage, a mandatory system will not lead to actual implementation. If there is a high concentration market, a mandatory system will lead to low quality-high price products. Hence, the first model will be like this: firstly, regulation can be used to prevent ecological damage. In case of damage, liability rules apply, which can be either strict of negligence (depending on whether the activity is hazardous), and should not be retroactive, channeled and capped. When the market is mature and competitive, a mandatory financial security system can be established. Since no market is able to provide unlimited coverage, the financial security needs to be capped. The cap can be set in the law; determined by government on a case-by-case basis or estimated by the operators themselves and approved later by the government. The question of who has such power depends on regulatory capacity, probability of the government to be captured and corresponding costs of even approach. The potential liable parties can choose among different compensation instruments to satisfy their financial obligations. If the damage exceeds the total assets of the liable parties and the ceiling of financial security, or if liability cannot be established, some collective compensation instruments can be used, such as environmental compensation funds.

When the market is immature or uncompetitive, a voluntary financial security promoting the development of compensation instruments with gradual introduction of a mandatory system is more feasible (the second model). Under such a system, the role of the government is to promote the predictability of environmental liability and incentivize the operators to seek financial coverage.

A third model is based on the second one. Under such a system, liability is still unlimited and the operators have the choice to seek financial coverage or not. However, an optional financial cap is introduced. If the operators seek financial coverage up to this level, they can limit their liability to such a level. Such a system give them incentives to get financial coverage up to a higher level which will otherwise not be achieved.

In all these three models, if liability cannot be established (either because liable parties can not be identified or the difficulties in proving liability), or liable parties and their financial guarantors become insolvent, some collective compensation instruments, such as environmental funds, can play a role. An important issue here is to make polluters finance instruments to create additional deterrent incentives.

When examining the empirical examples, one can find many parallels between the theory and the practice. Firstly, in all these systems discussed, there is combined use of regulation, liability rules and compensation instruments. Various regulatory instruments, including license and permits, taxes and emission trading and so on, have long been in place to prevent environmental/ecological damage. However, with the limited scope of this research, they are not discussed in detail. Liability rules are one key element of those systems. In many fields, oil pollution, nuclear damage and damage caused by hazardous substances under the CERLCA, strict liability is established. This is in line with the economic criteria, since the operators have better information about their own risks and strict liability gives them incentives to control activity levels as well. Under the ELD, both strict liability and negligence rules are applicable, depending on whether they are caused by dangerous activities or not. In both the U.S. and the ELD system, public authorities have standing to claim ecological damage from liable parties and detailed assessment standards are established in the U.S. In addition, both compulsory and voluntary financial security systems can be found in practice. In all the discussed international and U.S. systems, compulsory financial security is adopted. In nuclear, oil transportation and hazardous substances sectors, the possibility of catastrophic accidents creates high insolvency risks. A mandatory financial security system can relieve this problem. In the EU, since the ELD has just been transposed a few years ago, and the insurance and other guarantee markets for its financial security products are still in its infancy, a voluntary system is more feasible.

In spite of these parallels, sometimes divergences from the theory can be found in the practice as well. For example, when one examines the liability rules under the international regimes, liability is both capped and channeled. These characteristics, especially when combined, lead to under deterrence. These can be explained from the private interest perspective. The international conventions are the compromise of different interests. The U.S., although ratifying neither of these conventions, played an important role in shaping them. In the 1960s, the negotiation period for the international nuclear liability conventions, the U.S. was the biggest nuclear supplier and did not want to be exposed to liability for damage abroad.

Hence, it supported a system which channeled all liability to nuclear operators and exonerated other parties such as nuclear suppliers from liability. Such a channeling provision was later borrowed by the CLC without cautiously questioning its reasonableness. A cap on liability was established because the governments did not want to overburden the newborn nuclear industry. A similar logic also applied to the CLC. In addition, it was believed that a limitation on liability makes insurance more available. However, liability and a cap on financial security (insurance) do not necessarily need to converge. Moreover, the cap for oil pollution liability has proved to be insufficient after nearly every major accident. An unlimited system can already be found in some states of the U.S. Another divergence of the practice from theory is the retroactive liability under the CERCLA. This imposes an obligation on private parties to clean up historical pollution. The retroactive liability, together with some other characteristic of the superfund liability, such as a broad definition of the responsible parties, joint and several liability has lead to a chilling effect on brownfield redevelopment[10] and to the lack of liability insurance capacity in the 1980s.[11]

3. Indicators to Choose Compensation Instruments and a Multilayered Compensation Approach

In the models presented above, different compensation instruments can be used to cover ecological damage. However, this does not mean that every instrument fits for all situations and that all instruments shall be used together. This will be too costly an approach to follow. Individual analysis of each instrument shows that each has its own strengths and weakness and there is no hierarchy between them. This thesis identifies a few indicators which can aid operators in choosing suitable instruments, and inform lawmakers and insurance providers in developing the products and the market.

Before discussing the indicators, six types of compensation instruments are examined individually: liability insurance, first-party and direct insurance, risk-sharing agreements, environmental funds, guarantees provided by operators themselves or third parties and capital market. Calabresi's social costs theory is also used here to show each instrument's potential to reduce the three types of social costs. Liability insurance is the most commonly used instrument. It helps to increase the utility of the risk-averse potentially-liable parties by transferring the risks of higher damage with a fraction of it as premium. The deterrence effect remains as long as the premium corresponds with the actual risk. The moral hazard and adverse selection problems can be controlled through deductibles, monitoring and risk differentiation. Pooling, re-insurance and capital markets can be used to increase the capacity of insurers. However, a competitive market is essential for liability insurance to function efficiently, especially when financial security is compulsory. As alternatives, other types of insurance also play a role, such as first-

10 McMorrow 2004, p. 1094-1095.

11 See Abraham 2001, p. 94-98.

party insurance and direct insurance. First-party insurance is relevant in the sense that polluters can use it to cover the cleanup costs of their own sites. Besides, direct insurance also emerges in several markets to cover the damage to third parties' sites. Under such a policy, however, the trigger of coverage is no longer liability but damage itself. As long as damage, as prescribed in the insurance contract, happens and originates from the insured sites, it is covered by direct insurance. This policy can preclude the uncertainties from the judges in deciding liability. Under a risk-sharing agreement, the polluters pool together to provide coverage for themselves, hence they are both the insurer and the insured. A risk-sharing agreement can provide coverage to some unpredictable or high-magnitude risks which the traditional insurance market may be reluctant to cover. This is because it is no longer necessary to make contributions before accidents and information about one party's probability to cause damage is sufficient. A risk-sharing agreement is suitable for highly technical sectors where the operators themselves have better information than insurers about their risks and can control the risks through mutual monitoring. Many different types of environmental funds also exist which can either provide a higher layer of compensation or intervene when no liability is established. It can also finance the cleanups before a claim against the polluters is made. However, it has the disadvantages of poor risk differentiation and high administrative costs. A fund needs to be organized in a way not to dilute the deterrent incentives. The contributions need to be paid by potential injurers but not by the general revenues and should be more risk related. Security mechanisms can also be used, including the self-provided ones, such as mortgages, liens, deposits, trusts funds and escrow guarantees and the ones provided by third parties, such as bank guarantees, sureties and corporation guarantees. It is worth noting that most of those instruments are only available or affordable for big companies. Some of those instruments can save costs, such as self-insurance and captives while others are quite expensive, such as bank guarantees and sureties. Governmental oversight is also important to ensure money is actually available at the moment of damage. Furthermore, the possibility to use the capital market to cover ecological damage is also briefly sketched in the theoretical part. Capital markets can be used either to increase insurance capacity or are used directly by operators of environmental hazardous installations to cover their potential liability.

After discussing the compensation instruments individually, this research identifies some factors to choose among the various instruments. The first factor influencing the choice is whether a liable party is identifiable. When a liable party can be identified, liability rules can be used. Thus liability insurance can play an important role as well. Otherwise, alternatives can be used such as environmental funds. A second issue is the nature of damage. For a catastrophe with low frequency and high severity, risk-sharing pools can be used; the financial market can also enhance the capacity of insurers. Self-insurance is better suited for more predictable risks or, in other words, general damage. The size of enterprises also influences the choice. Self-provided guarantees and bank guarantees are more available to big enterprises. For SMEs, direct insurance and environmental funds can help them to cover the damage they caused. The sector in which potential injurers are operating composes the fourth indicator. In high technical sectors, a risk-sharing agreement is a desirable choice, while for more traditional sectors, insurers have better expertise.

Those indicators give some suggestions in choosing among various instruments. Admittedly, to use them in practice to establish a comprehensive system is still a sophisticated issue. Those indicators do not mean to choose one most suitable instrument in each specific situation, or to use several instruments equally when they all fit. This thesis tries to adopt a multilayered approach to establish a compensation system. Firstly, when a liable party can be identified, tort law provides the deterrent incentives. Under this situation, liability insurance can be used to provide the first layer of compensation. Insurers are specialized in risk spreading and differentiations. However, it does not fit into all scenarios. When the insurance market is highly concentrated, liable parties can resort to alternatives. For example, a risk-sharing agreement can be established for highly technical sectors; larger companies can also use captives and other guarantees. Even when insurance is in place, the capacity is not unlimited. In case of catastrophes, other instruments, such as risk-sharing agreements and environmental funds can play a role in providing an upper layer of compensation. Secondly, when liable parties can not be identified, there are more limited choices since many compensation instruments are financed by the liable parties. First-party insurance can cover on-site remediation costs. Direct insurance can also have a limited role to play. The trigger of direct insurance is damage but not liability. In other words, the insured is not necessarily the liable party. However, it still requires a causal links between the insured sites and the damage. If such a link cannot be established, the damage is not covered. In this case, environmental funds can come into play.

The empirical part provides examples of how compensation instruments work in different sectors and jurisdictions. One can find that liability insurance play a role in most situations. For example, in both the international and U.S. compensation systems for nuclear damage, liability insurance provides the first layer of compensation. Since a nuclear damage may turn out to be catastrophic, insurers pool together to provide the coverage. However, there is only one pool in each jurisdiction and nuclear operators are not free to seek insurance from pools in other jurisdictions. It hence leads to high concentration in the market and high prices for insurance products. Recently, some alternatives start to emerge, such as the risk-sharing pool established in Belgium: ELINI. It is a world-wide pooling system where nuclear operators provide coverage for their own liability. However, it does not mean to compete with national pools but to provide additional coverage for the new headings under the revised Paris Convention, which the traditional pools are reluctant to cover. In the U.S., the risk-sharing agreement plays another role: to provide a second layer of coverage. There are more than 100 nuclear reactors in the U.S., which makes the total available amount from the pool and insurance reach $12.2 billion. A similar nuclear pooling system can also be found in Germany. The CERCLA establishes a compensation system for damage caused by hazardous substances. Since the 1970s, environmental liability insurance began to develop. However, the market became more restricted with the promulgation of the

CERCLA. The joint effect of the stringent legislation (strict liability, joint and several liability and retroactive liability) and the judicial tendency (tendency to resort to the deep pocket and to invalidate the exclusion provisions in the insurance contract) makes the risk unpredictable for both potential liable parties and insurers. That hence led to the tightening of the environmental liability insurance market in the mid 1980s. The insurance market started to recover later and insurers shifted the policies from occurrence-based to claims-made policies to avoid the unpredictability. To way to avoid an unpredictable judicial expansion of environmental liability in the Netherlands is to shift from liability insurance to direct insurance. The direct insurance policy (MSV) was introduced in 1998. In the early stage, insurers provided coverage through a pool to gain experience. Such a pool dissolved in 2007 when insurers accumulated experience and capacity to provide their own products. Hence there is more competition in the market now. This direct insurance targets SMEs, since the big enterprises have more choice such as using captives, buying liability coverage from the U.S. insurers and so on. One exception where liability insurance plays only a minor role is oil pollution. Here the risk-sharing pools, Protection and Indemnity Clubs, provide liability cover for approximately 90 per cent of the world's ocean-going tonnage (of vessels). Those Clubs are non-profit-making mutual insurance associations which pool together ship owners and charters to cover their third-party liabilities. In mid-19th century, the increasing marine shipping risks made the traditional insurance market reluctant to provide coverage. Hence the P&I Clubs began to develop. The OPA adopted a more stringent system than the international conventions for oil pollution, which requires many potential liable parties to obtain a Certificate of Financial Responsibility (COFR). Afraid of facing high risks, the P&I Clubs refused to act as the guarantors under the OPA but still provide insurance coverage. To fill the gap of P&I coverage and the coverage required by the OPA, alternatives started to emerge such as the products provided by Shoreline Mutual Insurance Association and the First Line and Arvak Ltd. Some larger tanker operators also managed to get COFR through surety bond or self-insurance. Environmental funds are also used widely to cover environmental damage. It plays a positive role since it can intervene when no liable parties can be identified (such as the OSLTF and the IOPF for oil pollution) and promote cleanups by allowing payment from it before the public authorities claim from liable parties (such as the OSLTF). How to contribute to funds is an important criterion to assess the efficiency of compensation funds. In the international nuclear conventions, the government has the obligation to make funds available. To finance funds by public revenue is criticized as a subsidy and dilutes the deterrent incentives. In other cases, such as the OSLTF and the IOPF, it is mainly the polluters who make contributions. However, still those contributions are more related to the activity level (amount of oil) but not to actual risks. As far as SUBAT is concerned, it is not the fuel outlet operators that pay for their own cleanups. Since it covers only historical risks, it does not create distortive incentives for the future operators.

This thesis proposes some models to compensate for ecological damage and discussed several compensation instruments. However, simply having those instruments in place does not necessarily lead to implementation. For example, the CERCLA establishes financial responsibility for certain vessels and offshore facilities. Insufficient implementation has been reported, such as allowing companies to choose among financial assurance mechanisms freely without considering their default risk and providing inadequate oversight over the enforcement of financial assurance requirements.[12] The lack of implementation creates a risk for the recovery of cleanup costs by the government. In the EU, the ELD system is still new and many stakeholders have not realized its actual influence yet. This means at this stage, that the deterrent effect of such a system is still limited. Those examples show that to have a system in place is one thing and let it really work and achieve desirable preventive and compensation effect is another. When implementation in western countries has already not been an easy task, how such a system can be made to work in China is really a challenge.

4. The Compensation System in China and Policy Recommendations

This research also tries to answer the questions *how ecological damage is compensated in China and how such a system can be improved.* Although China has made many improvements in environmental regulation, it is still subject to many challenges such as weak agency capacity and a capture-prone governance. In this context, one can expect to use liability rules to overcome some of these inefficiencies. However, the examination of the Chinese tort system and environmental statutes shows that there is neither a general obligation for polluters to restore the damaged environment to its initial status, nor explicit liability for ecological damage (with the exception of marine oil pollution). Although general environmental liability provisions can be found in the Tort Liability Law, it does not have an explicit reference to pure ecological damage. Furthermore, hurdles also exist in procedural law: public interest litigation is not formally allowed, which makes claims for ecological damage difficult since usually no individual victims are involved. How ecological damage can be assessedf is another challenge and the assessment guidance is under development. Unsurprisingly, difficulties in practice accompany legislation gaps. Polluters usually have no incentives to restore the environment and government plays an important role in cleaning up polluted sites. Usually such cleanups are taken when re-development is involved. Government usually does not act directly against the polluters which can be explained by the fact that state-owned enterprises contribute highly to historical pollution. Rather than claiming from polluters, government often passes on the cleanup costs to new developers.

Considering the infancy stage of establishing liability for ecological damage, it is not surprising that there is an undeveloped market for corresponding compensation instruments. The former Environmental Protection Agency published a document requiring the development of environmental liability insurance in 2007.

12 GAO 2005, p. 39-53.

Afterwards, many insurers started to provide environmental insurance products. Those include stand-alone environmental liability insurance (covers both traditional environmental damage and cleanup costs) and pollution site liability insurance (covers damage to third-party and remediation costs coming from the polluted sites). Product liability insurance and property insurance may have limited relevance as well. Cleanup costs may be covered to the extent that they can prevent further personal injury and property damage. But the restoration costs in addition to cleanups are usually uncovered. The premiums today are still high and adverse selection is a serious challenge facing the insurers. In spite of the increasing availability of insurance products in the market, environmental liability insurance is not widely used. That should as such not come as a surprise since environmental liability does not constitute a serious threat, thus removing incentives of polluters to seek insurance coverage.

However, there is not only gloomy news on the Chinese system. A more comprehensive legal framework has been established for marine oil pollution. The MEPL allows competent public authorities to claim for marine environmental damage. Being a member of the CLC, China adopts many CLC elements in its domestic law/practice. For example, legislation explicitly states that some parts of ecological damage are compensable, such as prevention costs and reasonable restoration costs. Liability is channeled to ship owners and capped. In practice, there are already a few cases where ecological damage caused by marine oil pollution is partially compensated. Insurance products for marine oil pollution liability have long been available and a mandatory financial security system was introduced in 2009. A national compensation fund imitating the IOPF was also started to emerge in 2012. The compensation system for marine oil pollution, although far from perfect, sheds light on how a general compensation system could move forwards.

The young compensation system for ecological damage leaves much room for improvement. The Kuznets curve shows that there is a strong relationship between the wealth of a nation and its environmental quality and the best way to increase environmental quality is to decrease poverty.[13] However, this does not mean that we need to wait until the economy is more developed. Literature shows that developing countries need to take into account their own limits and situations to design more suitable regulatory instruments.[14] This also works in designing a compensation system, which should be tailored to China's own situation. The limited scope of this research cannot provide a tentative framework and schedule for the future development. However, by examining the theoretical models and existing situations in China, a few recommendations can be made to give an insight into the possible pathways.

First of all, a clear obligation in response to an environmental accident needs to be established. Either the polluters or the public authorities shall take

13 See on the original Kuznets curve Kuznets (1995) and for the environmental Kuznets curve, Esty & Porter 2005.

14 Faure, Goodwin & Weber 2010.

cleanup/restoration measures in case of ecological damage. A clear division of authorities is necessary to ensure an effective and timely response. Authorizing the public agencies can also help to solve the standing problem, which is an important hurdle facing the natural resources damage claim cases. The tendency to open up space for public interest litigation will give the public authorities or NGOs the possibility to claim for ecological damage.

Futhermore, an explicit definition of the compensable ecological damage is also important, which needs to be accompanied by feasible evaluation standards. Although there are already quite some cases where compensation for ecological damage was awarded in case of marine oil pollution, how such a new title can be quantified is still haunting the judges.[15] Without a clear evaluation standard, compensation for ecological damage will probably only exist on paper.

The development of compensation instruments should go hand-in-hand with environmental liability rules. Although liability rules for traditional damage are, generally speaking, not bad, many gaps and ambiguities exist with liability rules for ecological damage. A more clear liability system together with actual implementation can incentivize the polluters to seek insurance coverage. More policy supports, including subsidies in an early stage, linking insurance coverage with pollution fees, green credit policy can also contribute to the prosperity of the insurance market. As far as the compensation fund for marine oil pollution is concerned, to make the contributions to the fund more risk related rather than just based on an amount of oil imported will have better deterrent effect. Admittedly, the benefits in creating deterrence by risk differentiation should be balanced with the administrative costs incurred by such differentiation. A general compulsory financial security system is still not feasible at this stage where insurance and other compensation instruments are still in their infancy. With the further development of the insurance market and alternative instruments, a progressive introduction of a compulsory system can be of help.

Another issue worth noting is that regulation, liability rules and compensation instruments should be developed hand-in-hand, to enhance each other, but not separately. For ecological damage to be awarded in the court, public authorities need to detect the damage and take restoration measures. Without liability, one would not expect strong incentives from polluters to seek insurance coverage. Insurance development also needs policy support in the early stage.

5. Limits and Future Research

This research tries to propose some models to prevent and compensate for ecological damage. Economic analysis is used to design such systems, especially in defining liability rules and comparing individual compensation instruments. A few examples are given in the empirical part to show how the theoretical models work in practice. Of those examples, only the U.S. system and the ELD system are specifically meant for ecological damage. Other systems are sketched here either

15 See Chapter 8, Section 4.2.4.

because parts of ecological damage is also covered there, together with more general environmental damage, or because they have some interesting characteristics which can shed light on the system for ecological damage. For the different systems discussed in the empirical part, focus has been put on the legal framework: how liability rules are defined and how a financial security regime is established. The available compensation instruments and their functioning are also analyzed in some jurisdictions. This research then compared this regime design and available instruments with theoretical models. Whether a system is efficient is evaluated in a direct way: whether the elements - liability rules, compulsory/-voluntary financial security and individual instruments - used correspond with theoretical models. However, major difficulties remain in judging the system directly: to what extent the introduction of liability and compensation system helps to reduce ecological damage or promote the restoration. It is very difficult to gather sufficient data to make a sound conclusion on this point. The U.S. natural resources damage compensation system has been established for several decades. However, there is no authoritative information on how many claims have been made and how much damages have been awarded so far. This research tries to collect some information from government reports, OSLTF operations and some academic research to shed light on the possible scale of NRD claims. This information shows that NRD compensation is not that big in terms of both number and amount. However, the data collected is far from the actual costs to the environment: not all damage is qualified as natural resource damage under the CERCLA and the OPA; some damage may not be detected and assessed. Even considering only the actual claims and restorations made, the information is incomplete: the actions taken by responsible parties themselves are not counted, payments of public trustees from additional appropriations or trustee claims based on state laws are not considered. Hence a whole picture of NRD compensation/restorations is still not clear. As far as the ELD system is concerned, since it is quite new, and the awareness of its actual influence is new. Hence a few cases are available in practice which cannot lead to strong conclusions.

This research focuses on damage caused during the operation of a facility/activity. However, environmental risks exist also in the construction and post-closure phase. Regulation usually requires preventive measures to be taken before the start of operation. Sometimes it may go further to require compensatory measures to offset the negative impacts of the construction of a project. In addition, environmental damage may have long tail features, which mean that it may happen or manifest after the closure of a facility. How to compensate for the damage caused in these two stages is also an interesting question. To what extent the presented models for damage caused during the operation can fit the situation in construction and post closure deserves more attention.

Further research can also be undertaken to analyze compensation instruments developed in recent years for natural catastrophes, which could also be used as ecological damage compensation instruments. Although natural catastrophes and ecological damage are distinct, there may be some parallels, specifically concerning the identification of liable injurers. Models of government intervention in case of natural catastrophes (such as government providing insurance or acting as reinsurer of last resort) could also be usefully called upon to provide compensation for ecological damage. These and other issues undoubtedly merit further research into this fascinating area of ecological losses.

SUMMARY
COMPENSATING ECOLOGICAL DAMAGE: COMPARATIVE AND ECONOMIC OBSERVATIONS

Environmental pollution and devastation accompanying rapid economic development have created alarming losses to human beings, property and the environment itself. Many efforts have been undertaken to create institutions aiming at the prevention and compensation of environmental harm. An area starting to attract attention only recently is ecological damage: the damage to public natural resources and private owned natural resources which has an ecological value in excess of its market value. The difficulties in its assessment and the legal status of natural resources make traditional tort law an insufficient solution for the protection and compensation of ecological damage. This doctoral research tries to answer the question: *how to achieve better prevention and compensation of ecological damage by mobilizing different instruments: regulation, liability rules and compensation mechanisms (such as liability insurance, direct insurance, risk sharing agreement, environmental funds, other guarantees and capital market)?* This research shows that liability rules and compensation mechanisms can not only promote the restoration of ecological damage, but also create incentives for potential polluters to prevent damage *ex ante*.

This research contains three parts: the theoretical part, empirical part and the case study of China. In the theoretical part, economic analysis of law is employed to help choose between the instruments to prevent and compensate for ecological damage. To apply Shavell's model on the choice between regulation and liability in the context of ecological damage shows that the combined use of them is desirable. Calabresi's social costs theory is used to analyze the design of liability rules: standing, liable parties, liability standards, causation and damages. Even with the combined use of regulation and liability, the optimal prevention and compensation of ecological damage still cannot be acheived, which makes other compensation mechanisms beneficial complements. The social costs theory helps also to compare different compensation mechanisms.

Considering the limits of the instruments discussed above, a desirable compensation model which can create efficient preventive incentives and can promote restoration and compensation at the same time should use regulation, liability rules and compensation mechanisms together. Based on whether it is mandatory for potential liable parties to seek coverage for their potential liability by various compensation mechanisms (such as insurance, risk-sharing agreements, a deposit and so on), this research proposes three compensation models: the compulsory one, the voluntary one and the voluntary one with an optional cap. The choice among these three models should be tailed to the characteristics of the risks and the specific situation of a certain society, such as the insolvency risks, availability of compensation mechanisms and willingness of the insurers or other guarantors to provide such coverage.

In all these three models, different compensation mechanisms can be used. However, leaving the potential liable parties choices does not mean all mechanisms should be used in all scenarios, which might be too expensive. This research identifies a few indicators in helping choosing the suitable mechanisms, such as whether liable party is identifiable, the nature of damage, the size of enterprises and so on.

One can find examples in line with or divergent from the theoretical models and indicators in practice. This research discusses different compensation systems: the international regimes (compensation for oil pollution and nuclear damage), U.S. systems (compensation systems for natural resources damage under the CERCLA and the OPA, as well as the nuclear compensation regime under the PAA) and EU systems (the ELD and a few examples of compensation mechanisms used in some Member States).

Those examples show many parallels with the theoretical models. In all these systems discussed, there is combined use of regulation, liability rules and compensation instruments. The use of strict liability and negligence rules in different situations are also in line with the economic criteria. In both the U.S. and the ELD system, public authorities have standing to claim ecological damage from liable parties and detailed assessment standards are established in the U.S. In addition, both compulsory and voluntary financial security systems can be found in practice, depending on the risks a sector possesses and the willingness of insurers/other guarantors to provide coverage.

In spite of these parallels, sometimes divergences from the theory can be found in practice as well, such as capped and channeled liability under the international regimes, broad use of compensation funds with insufficient risk differentiation and some implementation problems in the U.S. and the EU. Some of the divergences can be explained from the private interest perspective.

With regard to China, the examination of the tort system and environmental statutes shows that there is neither a general obligation for polluters to restore the damaged environment to its initial status, nor explicit liability for ecological damage (with the exception of marine oil pollution). The legislation gaps are accompanied by difficulties in practice, such as the lack of incentives for polluters to restore the environment and the important role played by government especially when re-development is involved. Considering the infancy stage of establishing liability for ecological damage, it is not surprising that there is an undeveloped market for corresponding compensation instruments. The environmental insurance market only started to become available a few years ago, but still faces high premiums and adverse selection problems. However, there is not only gloomy news on the Chinese system. A more comprehensive legal framework has been established for marine oil pollution, with regard to both liability rules and compensation mechanisms. Being a member of the international conventions contributes to the more promising scenario in compensating marine oil pollution.

By examining the theoretical models and existing situations in China, a few recommendations can be made to give an insight into the possible pathways. First of all, a clear obligation in response to an environmental accident needs to be established. Furthermore, an explicit definition of the compensable ecological damage is also important, which needs to be accompanied by feasible evaluation standards. Last but not least, regulation, liability rules and compensation instruments should be developed hand-in-hand, to enhance each other, but not separately.

SAMENVATTING
COMPENSATIE VOOR ECOLOGISCHE SCHADE: RECHTSVERGELIJKENDE EN ECONOMISCHE BESCHOUWINGEN

Milieuvervuiling en de daarmee gepaard gaande snelle economische ontwikkeling hebben tot alarmerende verliezen geleid aan mensen, eigendom en aan het milieu zelf. Veel maatregelen zijn ondernomen om instituties in het leven te roepen die tot doel hebben milieuschade te voorkomen en te compenseren. Één gebied dat in toenemende mate in de belangstelling staat is dat van de zogenaamde ecologische schade. Het gaat daarbij om schade aan publieke goederen (natuur), maar ook om schade aan natuurlijke hulpbronnen die in privaat bezit zijn, maar die een ecologische waarde hebben die hoger is dan hun marktwaarde. Moeilijkheden bij het ramen van de ecologische schade en moeilijkheden met betrekking tot de juridische aard van de ecologische schade maken de toepassing van het klassieke aansprakelijkheidsrecht moeilijk. Het aansprakelijkheidsrecht wordt doorgaans niet als voldoende oplossing gezien voor de bescherming en compensatie van ecologische schade. Dit proefschrift tracht een antwoord te bieden op de vraag: hoe kan een betere preventie en compensatie van ecologische schade worden gerealiseerd door verschillende instrumenten in te zetten: regulering, aansprakelijkheidsregels en compensatiemechanismen (zoals aansprakelijkheidsverzekering, directe verzekering, risicospreidingovereenkomsten, schadefondsen en andere garanties en kapitaalmarkten)? Dit onderzoek laat zien dat aansprakelijkheidsregels en vergoedingsmechanismen niet alleen het herstel van ecologische schade kunnen bevorderen, maar ook prikkels kunnen creëren voor potentiële vervuilers om de schade *ex ante* te voorkomen.

Dit onderzoek bestaat uit drie delen: een theoretisch deel, een empirisch onderdeel en een case study over China. In het theoretisch deel is de economische analyse van het recht gebruikt om een keuze te maken tussen instrumenten die kunnen bijdragen aan de preventie en compensatie van ecologische schade. Het model van Shavell betreffende de keuze tussen regulering en aansprakelijkheid is toegepast op het terrein van de ecologische schade en laat zien dat een gecombineerd gebruik van beide instrumenten de voorkeur geniet. De sociale kostentheorie van Calabresi is ook gebruikt om de optimale vormgeving van aansprakelijkheidsregels te onderzoeken. Daarbij is aandacht gegeven aan de vraag die het recht heeft om een vordering uit te brengen, de aansprakelijke partijen, aansprakelijkheidsstandaards, oorzakelijk verband en schadevergoeding. Zelfs met een gecombineerd gebruik van regulering en aansprakelijkheid kan een optimale preventie en compensatie van ecologische schade niet worden gerealiseerd. Dit betekent dat andere compensatiemechanismen een nuttig complement zullen vormen. De sociale kostentheorie helpt ook om verschillende compensatiemechanismen te vergelijken.

Elk van de instrumenten die hierboven werd besproken heeft ook bepaalde limieten. Idealiter zou een optimaal compensatiemodel efficiënte prikkels tot preventie genereren en ook compensatie van ecologische schade bevorderen door een gecombineerd gebruik van regulering, aansprakelijkheidsregels en vergoedingsmechanismen. Afhankelijk van de vraag of de aansprakelijke partijen financiële garanties dienen te stellen om hun potentiële aansprakelijkheid te dekken door verschillende compensatiemechanismen (zoals verzekering, risicospreiding-overeenkomsten of een garantie) stelt dit onderzoek drie verschillende compensatiemodellen voor: een verplicht model, een vrijwillig model en een vrijwillig model met een optionele beperking van de aansprakelijkheid. De keuze tussen die drie modellen dient afhankelijk te worden gesteld van de specifieke risico's, maar ook van de specifieke situatie in een bepaalde maatschappij, zoals het insolventierisico, de beschikbaarheid van compensatiemechanismen en de bereidheid van verzekeraars of andere garantiegevers om dergelijke zekerheidsstellingen ook te verschaffen.

In alle drie modellen kunnen verschillende compensatiemechanismen worden gebruikt. Echter, het feit dat de potentieel aansprakelijke partijen de mechanismen kunnen kiezen betekent niet dat alle mechanismen in alle scenario's ook dienen te worden gebruikt; dit zou vermoedelijk te duur zijn. Dit onderzoek identificeert enkele indicatoren die helpen bij de keuze van een geschikt instrument. Daarbij speelt ondermeer een rol of de aansprakelijke partij kan worden geïdentificeerd, wat de aard van de schade is evenals de omvang van het bedrijf van de schadeveroorzaker.

Verschillende voorbeelden kunnen worden gevonden in de praktijk die ofwel een bevestiging vormen van de besproken theoretische modellen ofwel daarvan afwijken. Dit onderzoek bespreekt verschillende compensatiemechanismen: internationale vergoedingsmechanismen (meer bepaald voor schade veroorzaakt door olieverontreiniging op zee en schade veroorzaakt door nucleaire ongevallen), het vergoedingssysteem in de VS (meer bepaald de vergoeding van schade aan natuurlijke hulpbronnen volgens de CERCLA en de OPA evenals het nucleaire compensatieregime onder de PAA) en de Europese systemen (meer bepaald de milieuaansprakelijkheidsrichtlijn en enkele voorbeelden van vergoedingssystemen die worden gebruikt in enkele andere lidstaten).

Deze voorbeelden vertonen enkele parallellen met de theoretische modellen. In alle systemen die werden besproken is sprake van een gecombineerd gebruik van regulering, aansprakelijkheidsregels en compensatiemechanismen. Het gebruik van risicoaansprakelijkheid en schuldaansprakelijkheid in verschillende situaties correspondeert ook in belangrijke mate met de economische uitgangspunten. Zowel in de VS als in het systeem van de milieuaansprakelijkheidsrichtlijn in Europa hebben overheden het recht om vergoeding te vorderen voor ecologische schade van schadeveroorzakers. In de VS zijn er daarenboven ook gedetailleerde standaards ontwikkeld voor de schaderaming. Daarenboven kunnen zowel vrijwillige als verplichte systemen van financiële zekerheid in de praktijk worden aangetroffen, afhankelijk van de risico's die door een bepaalde sector worden gecreëerd en de bereidheid van verzekeraars of andere verstrekkers van financiële zekerheid om dekking te verschaffen.

Ondanks deze parallellen kunnen soms ook afwijkingen van de theorie in de praktijk worden aangetroffen. Dit betreft bijvoorbeeld het fenomeen van de financiële limitering van aansprakelijkheid en de kanalisering van aansprakelijkheid. Beide kunnen in de internationale regimes worden aangetroffen. Dit geldt ook voor het wijdverspreide gebruik van compensatiefondsen met onvoldoende risicodifferentiatie evenals voor bepaalde implementatieproblemen in de VS en de EU. Sommige van de afwijkingen die men in de praktijk kan aantreffen van het theoretisch model kunnen vanuit de private belangengroepentheorie worden verklaard.

Een onderzoek naar het aansprakelijkheidssysteem en het milieurecht in China laat zien dat er geen algemene verplichting bestaat voor vervuilers om het vervuilde milieu in zijn oorspronkelijke staat te herstellen. Evenmin is er een uitdrukkelijke aansprakelijkheid voor ecologische schade (met uitzondering voor de zeewaterverontreiniging door olie waar dit wel geregeld is). Er zijn derhalve in China behoorlijke gaten in de wetgeving. Deze worden overigens ook vergezeld van toepassingsmoeilijkheden in de praktijk, zoals het gebrek aan prikkels voor de verontreinigers om het milieu daadwerkelijk te herstellen; een belangrijk probleem bestaat ook in de rol die door de overheid wordt gespeeld, in het bijzonder wanneer vervuilde gronden opnieuw worden ontwikkeld en in gebruik genomen. Het ontwikkelen van een aansprakelijkheidssysteem voor ecologische schade in China staat eigenlijk nog in de kinderschoenen. Het is derhalve ook niet verwonderlijk dat er een onderontwikkelde markt is voor de corresponderende vergoedingssystemen. Milieuschadeverzekering kwam slechts enkele jaren geleden tot ontwikkeling in China en kent nog altijd vele problemen door hoge premies en antiselectie. Echter, er is niet alleen negatief nieuws wat betreft de compensatie voor ecologische schade in China. Een beter ontwikkeld systeem is uitgewerkt voor zeewaterverontreiniging door olie, zowel wat betreft de aansprakelijkheidsregels als wat betreft de compensatiemechanismen. China is ook aangesloten bij enkele internationale verdragen en dit leidt tot een beter scenario wat betreft de vergoeding van schade veroorzaakt door zeewaterverontreiniging door olie in China.

Op basis van de theoretische modellen en de bestaande vergoedingsmogelijkheden in China kunnen enkele aanbevelingen worden geformuleerd ter verbetering. In de eerste plaats dient een heldere rechtsplicht tot optreden na een milieu-incident te worden gecreëerd met het oog op een herstel van de schade. In de tweede plaats dient ook een definitie van ecologische schade die voor vergoeding in aanmerking komt te worden ingevoerd, hetgeen vergezeld dient te gaan van heldere standaards voor de berekening van de schadeloosstelling. Tenslotte dienen regulering, aansprakelijkheidsregels en compensatiemechanismen tegelijk te worden ontwikkeld zodat zij elkaar ook wederzijds kunnen versterken.

POLICY DOCUMENTS, GUIDELINES, REPORTS

ACE 2011
A Guide to Environmental Liability in Europe, 2011.

Ad-Hoc Industry, Natural Resource Damage Group 2010
Report: Survey of Industrial Companies, Insurance and Other Financial Security Instruments and Remediation of Environmental Damage under the EU Environmental Liability Directive, 2010.

ANI 2006
Insuring the Future of Nuclear Power Worldwide, 2006.

Berger 2009
Berger, M., *Managing Nuclear Risks in the United States*, a presentation at INLA Inter Jura Congress 2009

Bio Intelligence Serv. 2009
Study on the Implementation Effectiveness of the Environmental Liability Directive (ELD) and Related Financial Security Issues 47, 2009.

CEA 2009
Navigating the Environmental Liability Directive: a Practical Guide for Insurance Underwriters and Claims Handlers, 2009.

China Ministry of Environmental Protection

- The notice on the strengthening of pollution prevention during the relocation of industrials (issued by the National Environmental Protection Agency Office, 1 June 2004, Huanban (2004) No. 47)

- Opinion on the Development of Environmental Pollution Liability Insurance (promulgated by Ministry of Environmental Protection and China Insurance Regulatory Commission, 4 December 2007, Huanfa (2007) No. 189)

- The opinion on strengthening the soil pollution prevention (issued by Ministry of Environmental Protection, Department of Nature and Ecology Conservation, 6 June 2008, Huanfa (2008) No. 48)

- The guidelines for risk assessment of contaminated sites (drafted by Ministry of Environmental Protection, Department of Science, Technology and Standards, 29 September 2009, Huanbanhan (2009) No. 1020)

- The guidelines for soil remediation of contaminated sites (drafted by Ministry of Environmental Protection, Department of Science, Technology and Standards, 10 December 2009, Huanbanhan (2009) No. 1312)

- The technical guidelines for environmental monitoring of sites(drafted by Ministry of Environmental Protection, Department of Science, Technology and Standards, 1 February 2010, Huanbanhan (2010), No. 99)

- The technicial Specification for Environmental Site Investigation] (drafted by Ministry of Environmental Protection, Department of Science, Technology and Standards, 5 August 2009, Huanbanhan (2009) No. 793)

- The temporary regulation on management of soil pollution on polluted sites] (drafted by Ministry of Environmental Protection, Department of Policies, Laws and Regulations, 15 December 2009, Huanbanhan (2009) No. 1321)

- Several Opinions on Evaluation the Environmental Pollution Damage (promulgated by the Ministry of the Environmental Protection, 25 May 2011, Huanfa (2011) No. 60)

Department of Energy & Climate Change 2011
Implementation of Changes to the Paris and Brussels Conventions on Nuclear Third Party Liability: A Public Consultation 2011.

Environmental Law Institute 2007
Environmental Law Institute, Mitigation of Impacts to Fish and Wildlife Habitat: Estimating Costs and Identifying Opportunities, 2007.

European Commission

- Legal Study for the Accession of Euratom to the Paris Convention on Third Party Liability in the Field of Nuclear Energy, TREN/CC/01-2005, 2005.

- Commission Guidance Document on Article 6(4) of the 'Habitats Directive' 92/43/EEC: Clarification of the Concepts of Alternative Solutions, Imperative Reasons of Overriding Public Interest, Compensatory Measures, Overall Coherence, Opinion of the Commission, January 2007.

- Report from the Commission to the Council, the European Parliament, the European Economic and Social Committee and the Committee of the Regions: Under Article 14 (2) of Directive 2004/35./CE on the Environmental Liability with Regard to the Prevention and Remedying of Environmental Damage, COM (2010) 581 final, 12 October 2010.

- DG ENV, Study to Explore the Feasibility of Creating a Fund to Cover Environmental Liability and Losses Occurring from Industrial Accidents, 2012.

Gampe-Rabindran, Mastromonaco & Timmins 2010
Gampe-Rabindran, S., Mastromonaco, R. & Timmins, C., *Valuing the benefits of superfund site remediation: Three approaches to measuring localized externalities*, 2010.

Gao 2009
Gao, J., *Development of Environmental Courts in China: the Promises and Challenges and Implementation for Environmental Public Interest Litigation*, 2009.

Gittus 2006
Gittus, J., *Introducing Nuclear Power to Australia: An Economic Comparison, A Report Prepared for the Australian Nuclear Science and Technology Organization*, 2006.

Hellberg 2011
Hellberg, N., *Update on German ELD Market, presentation at the Stakeholder and Practitioner Workshop on the implementation of the ELD in the EU*, 2011.

International Oil Pollution Compensation Fund
- Claims Manual, December 2008 edition.

- Oil Spill Liability Trust Fund Annual Report FY2004—FY2008

Klein 2011
Klein, U., *Transposition and Application of the Environmental Liability Directive in Germany, presented on 10th National Experts Meeting on ELD*, 2011.

Marsh & McLennan Companies 2006
The Catastrophe Bond Market at Year-End 2005: Ripple Effects from Record Storms, 2006.

National Commission on the BP Deepwater Horizon Oil Spill and Offshore Drilling
- The Amount and the Fate of the Oil, Staff Working Paper No. 3, 6 October 2010.

- Report to the President, Deep Water: The Gulf Oil Disaster and the Future of Offshore Drilling, 2011.

NBER Working Paper
Gampe-Rabindran, S., Mastromonaco, R. & Timmins C., Valuing the Benefits of Superfund Site Remediation: Three Approaches to Measuring Localized Externalities. NBER Working Paper Series, Working Paper 16655, 2010.

Sigman, H., The Pace of Progress at Superfund Sites: Policy Goals and Interest Group Influence, NBER Working Paper Series, Working Paper 7704, 2000.

Change, H. & Sigman, H., An Empirical Analysis of Cost Recovery in Superfund Cases: implications for Brownfields and Joint and Several Liability, NBER Working papers 16209, 2010

Rand 1992
Dixon, L., Superfund and Transaction Costs: the Experiences of Insurers and Very Large Industrial Firms, CT-102, 1992.

Rothwell 2002
Rothwell, G., *Does the U.S. Subsidize Nuclear Power Insurance?, Stanford Institute for Economic Policy Research*, 2002.

Schubert 2007
Schubert, M., *The Environmental Liability Directive and the German Insurance Association's Model Wordings*, 2007.

Swiss Reinsurance
- Insuring Environmental Damage in the European Union, 2007.
- Insurance-linked Securities Market Update, 2010.

US Coast Guard
- Preliminary Regulatory Impact Analysis: Financial Responsibility for Water Pollution Vessels, 1993
- Report on Implementation of the Oil Pollution Act of 1990.
- Oil Pollution Act Liability Limits: Fiscal Year 2010 Report to Congress, 2011.

US Congressional Budget Office 2008
Nuclear Power's Role in Generating Electricity 2008.

US Department of Energy, Energy Information Administration 1992
Federal Energy Subsidies: Direct and Indirect Interventions in Energy Markets, SR/EMEU/92-02, 1992.

US Government Accountability Office
- Analysis of the Price-Anderson Act, EMD-80-80, 1980.

- Hazardous Waste: Funding of Postclosure Liabilities Remains Uncertain, GAO/RCED-90-64, 1990.

- Outlook for and Experience with Natural Resource Damage Settlements, GAO/RCED-96-71, 1996.

- Superfund Program: Updated Appropriation and Expenditure Data, GAO-04-475R Superfund Program, 2004. (2004a)

- Report to Congressional Requesters, Nuclear Regulation: NRC's Liability Insurance Requirements for Nuclear Power Plants Owned by Limited Liability Companies, GAO-04-654, 2004. (2004b)

- Environmental Liabilities, EPA Should Do More to Ensure that Liable Parties Meet Their Cleanup Obligations, GAO-05-658, 2005.

- Federal Electricity Subsidies: Information on Research Funding, Tax Expenditures, and Other Activities That Support Electricity Production, GAO-08-102, 2007.

- Superfund: Funding and Reported Costs of Enforcement and Administration Activities, GAO-08-841R, 2008.

- Superfund: Litigation has Decreased and EPA Needs Better Information on Site Cleanup and Cost Issues to Estimate Future Program Funding Requirements, GAO-09-656, 2009.

- Oil Spills: Costs may impact viability of oil spill liability trust fund, GAO-10-795T, 2010.

US Nuclear Regulatory Committee
- The Price-Anderson Act- The Third Decade: Report to Congress, NUREG-0957, 1983.

- NRC's 1998 Report to Congress on the Price-Anderson Act SECY-98-160,1998.

Versicherungsforen Leipzig 2011
Study: Calculating a Risk-appropriate Insurance Premium to Cover Third-Party Liability Risks that Result from Operation of Nuclear Power Plants, 2011.

Wilks 2009
Wilks, G., *Managing Nuclear Risks: Nuclear Property Insurance, a presentation at Managing Nuclear Risks, Win 2009 Global Conference*, 2009.

World Bank 2010
Overview of the Current Situation on Brownfield Remediation and Redevelopment in China, Report No. 57953, 2010.

LEGISLATION

International Conventions

Lugano Convention on Civil Liability for Damage Resulting from Activities Dangerous to the Environment, 32 International Legal Materials, 1228.

Convention on Biological Diversity, 1760 UNTS 79; 31 ILM 818.

Convention on Access to Information, Public Participation in Decision-Making and Access to Justice in Environmental Matters.

Convention on Long-Range Transboundary Air Pollution, 18 International Legal Materials 1442.

Convention on Civil Liability for Damage Caused during Carriage of Dangerous Goods by Road, Rail and Inland Navigation Vessels, UN Doc ECE/TRANS/79.

Convention on Liability and Compensation in Connection with Carriage of Hazardous and Noxious Substances by Sea, International Transport Treaties, Suppl. 20 (October 1996), I-573–600/Cm. 3580 RMC I, II.7.125.

Convention for the Prevention of Marine Pollution from Land-Based Sources 1974, 1546 UNTS 119; 13 ILM 352 (1974); UKTS 1978, No. 64.

Geneva Convention on Long-range Transboundary Air Pollution, 1302 UNTS 217.

Convention on the Protection and Use of Transboundary Watercourses and International Lakes, 31 International Legal Materials 1312.

Basel Protocol on Liability and Compensation for Damage Resulting from Transboundary Movements of Hazardous Waste and Their Disposal, UN Doc. UNEP/CHW.1/WG/1/9/2.

Convention on Civil Liability for Damage from Pollution by Oil resulting from Exploration for and Exploitation of Seabed Mineral Resources, 14 International Legal Materials 1450.

International Convention on Civil Liability for Oil Pollution Damage 1969, 973 UNTS 3, RMC I. 7.30, II. 7.30.

The Protocol of 1992 to Amend the International Convention on Civil Liability for Oil Pollution Damage 1969 (1992 CLC) , Misc 36 (1994), Cm 2657, RMC I, 7.51, II. 7.51. International Convention on the Establishment of an International Fund for Compensation for Oil Pollution Damage, 1110 UNTS 57, Cmnd 5061.

The Protocol of 1992 to Amend the International Convention on the Establishment of an International Fund for Compensation for Oil Pollution Damage, RMC I.7.111, II,1,7,111, Misc 37(1994), Cm 2658.

Protocol of 2003 to the International Convention on the Establishment of an International Fund for Compensation for Oil Pollution Damage, 1992. (The Supplementary Fund Protocol), RMC II. 7.115.

The International Convention on Civil Liability for Bunker Oil Pollution Damage, RMC I.7.130, II 7.130.

1971 Brussels Convention Relating to Civil liability in the Field of Maritime Carriage of Nuclear Material, UNE 944 UNTS 255.

Paris Convention on Third Party Liability in the Field of Nuclear Energy as amended by the Additional Protocol of 28 January 1964 and by the Protocol of 16 November 1982, and by the Protocol of 12 February 2004, 956 UNTS 251.

Convention Supplementary to the Paris Convention on Third Party Liability in the Field of Nuclear Energy (as Amended by the Additional Protocol of 28 January 1964 and by the Protocol of 16 November 1982, and by the Protocol of 12 February 2004), 1041 UNTS 358.

Vienna Convention as Amended by the Protocol of 12 September 1997 to Amend the Vienna Convention on Civil Liability for Nuclear Damage, 1063 UNTS 358.

Joint Protocol relating to the Application of the Vienna Convention and the Paris Convention.

Convention on Supplementary Compensation for Nuclear Damage (CSC).

Convention on Nuclear Safety.

Convention on Physical Protection of Nuclear Material.

Convention on Assistance in Case of a Nuclear Accident or Radiological Emergency and the Convention on Early Notification of a Nuclear Accident.

Convention for the Protection of Human Rights and Fundamental Freedoms.

EU Law and Directives

Proposal for a Directive on Civil Liability for Damage Caused by Waste, *OJ* C251/3 (1989); as amended COM(91) 219 final, *OJ* 1991 C192/6.

Green Paper on Remedying Environmental Damage, COM(1993) 47, final, 14 May 1993.

White Paper on Environmental Liability, COM(2000), 66 final, 9 February 2000.

Directive 2004/35/EC of the European Parliament and of the Council on Environmental Liability with regard to the Prevention and Remedying of Environmental Damage, *OJ* 2004 L143/59.

Council Directive 85/337/EEC, on the Assessment of the Effects of Certain Public and Private Projects on the Environment, *OJ* 1985 L175/40.

Council Directive 79/409/EEC of 2 April 1979 on the Conservation of Wild Birds, *OJ* 1979 L103/1.

The Directive 2009/147/EC of the European Parliament and the Council of 30 November 2009 on the Conservation of Wild Birds, *OJ* 2010 L20/7.

Council Directive 92/43/EEC of 21 May 1992 on the Conservation of Natural Habitats and of Wild Fauna and Flora, *OJ* 1992 L206/7.

Directive 2000/60/EC of the European Parliament and of the Council establishing a framework for the Community action in the field of water policy, *OJ* 2000 L327/1.

Directive 2003/4/EC of the Parliament and of the Council of 28 January 2003 on Public Access to Environental Information and Repealing Council Directive 90/313/EEC, *OJ* 2003 L41/26.

Directive 2003/35/EC of the European Parliament and of the Council of 26 May 2003 Providing for Public Particpation in respect of the Drawing up of Certain Plans and Programes relating to the Environment and Amending with regard to Public Participation and Access to Justice, *OJ* 2003 L156/17.

Directive 2011/92/EU of the European Parliament and of the Council of 13 December 2011 on the Assessment of the Effects of Certain Public and Private Projects on the Environment, *OJ* 2012 L26/1.

Directive 2008/1/EC of the European Parliament and of the Council of 15 January 2008 concernig Intergrated Pollution Prevention and Control, *OJ* 2008 L24/8.

Council Regulation No. 1367/2006 on the application of the provisions of the Convention on Access to information, Public Participation in Decision-making and Access to Justice in Environmental Matters to Community institutions and bodies, *OJ* 2006 L264/13.

Commission Regulation No. 267/2010 of 24 March 2010 on the Application of Article 101(3) of the Treaty on the Functioning of the European Union to Certain Categories of Agreements, Decisions and Concerted Practices in the Insurance Sector, *OJ* 2010 L83/1.

Council Decision 2005/370/EC of 17 February 2005 on the Conclusion, on Behalf of the European Community, of the Convention on Access to Information, Public Participation in Decision-making and Access to Justice in Environmental Matters, *OJ* 2005 L124/1.

Commission Proposal for a Regulation on the application of the Provisions of the Convention on Access to information, Public Participation in Decision-making and Access to Justice in Environmental Matters to EC institutions and bodies, COM(2003) 622 final, 24 October 2003.

The Proposal for a Council Decision on the Conclusion, on Behalf of the European Community, of the Convention on Access to Information, Public Participation in Decision Making and Access to Justice regarding Eenvironmental Matters, COM(2003) 625 final, 24 October 2003.

Commission Proposal for a Directive on Access to Justice in Environmental Matters, COM(2003) 624 final, 24 October 2003.

European Parliament Legislative Resolution on the Proposal for a European Parliament and Council Directive on Access to Justice in Environmental Matters, *OJ* 2004 C103 E/626.

Proposal for a Directive on Civil Liability for Damage Caused by Waste, *OJ* 1989 C251/3; as amended COM(1991) 219 final, *OJ* 1991 C192/6.

Opinion of the European Economic and Social Committee on the Proposal for a Directive of the European Parliament and of the Council on Access to Justice in Environmental Matters, *OJ* 2004 C117/14.

US law

Act to Amend the Atomic Energy Act of 1954, Pub. L. No. 85-256, 71 Stat. 576 (1957).

Comprehensive Environmental Response, Compensation, and Liability Act, 42 U.S.C. §§9601-9674.

Clean Water Act of 1977, 33 U.S.C. §§1251-1387.

Energy Independence and Security Act of 2007, Pub. L. No. 110-140, § 934 (a) (1) (D), 121 Stat. 1492, 1744.

Energy Reorganization Act of 1974, Pub. L. No. 92-438, 42 U.S.C. § 5801.

Federal Liability Risk Retention Act, 15 U.S.C. §§3901-3906.

Office of the Secretary of the Interior, Natural Resource Damage Assessments, 43 C.F.R. §§11.10-11.93.

Oil Pollution Act of 1990, 33 U.S.C. §§2701-2761.

Oil Pollution Act Regulations, Natural Resource Damage Assessment, 15 C.F.R. §§990.10-990.66.

Resource Conservation and Recovery Act, 42 U.S.C. §6901

Germany
Bürgerliches Gesetzbuch (BGB)

Gesetz über die friedliche Verwendung der Kernenergie und den Schutz gegen ihre Gefahren (Atomgesetz – AtG) (Atomic Energy Act)

Gesetz über die Vermeidung und Sanierung von Umweltschäden (USchadG) (Act serving to implement the Directive of the European Parliament and of the Council on Environmental Liability with Regard to the Prevention and Remedying of Environmental Damage).

Umwelthaftungsgesetz (UmweltHG), Environmental Liability Act.

UK
English Environmental Protection Act 1990 (c. 16).

Finland
Act on Compensation for Environmental Damage of Finland, 737/1994.

New Zealand
New Zealand Environment Act of 1986, 1986 No. 127.

The Netherlands
Burgerlijk Wetboek (BW), Dutch Civil Code.

Interimwet Bodemsanering (Ibs), The Interim Act on the Decontamination of the Soil.

Wet Bodembescherming (Wbb), The Soil Protection Act.

France

Code de l'environnement.

China

Air Pollution Prevention Act](promulgated by the Standing Comm. Nat'l People's Cong., 5 September 1987, effective 1 June 1988, revised 29 April 2000)

Civil Procedure Law (promulgated by the Nat'l People's Cong., 9 April 1991, effective 9 April 1991, revised 29 October 2007 and 31 August 2012)

Commercial Maritime Code (Promulgated by the Standing Committee of the National People's Congress, effective as of 1 July 1993)

Environmental Protection Act (promulgated by the Standing Comm. Nat'l People's Cong., 26 December 1989, effective 26 December 1989)

General Principles of Civil Law (promulgated by the Nat'l People's Cong., 12 April 1986, effective 1 January 1987)

Law on Legislation (promulgated by the Standing Comm. Nat'l People's Cong., 15 March 2000, effective 1 July 2000)

Marine Environmental Protection Act (promulgated by the Standing Comm. Nat'l People's Cong., 23 August 1882, revised 25 December 1999, effective 1 April 2004)

Solid Waste Pollution Prevention Act (promulgated by the Standing Comm. Nat'l People's Cong., 30 October 1995, revised 29 December 2004, effective 1 April 2005)

Tort Liability Law of the People's Republic of China (promulgated by the Standing Comm. Nat'l People's Cong., 1 January 2010, effective 1 July 2010)

Water Pollution Prevention Act (promulgated by the Standing Comm. Nat'l People's Cong., 11 May 1984, revised 28 February 2008, effective 1 June 2008)

The Opinion of Supreme Court on the Application of Civil Procedure Law (issued by the Judge Committee of the Supreme Court on 14 July 1992, faban (92) No. 22)

Supreme People's Court Opinion on Several Issues concerning the Application of Civil Procedure Law (adopted by the Judicial Committee of the Supreme Court on 14 July 1992, effective as of 14 July 1992)

Some Provisions of the Supreme People's Court on the Scope of Cases to be entertained by Maritime Courts (Adopted by the Judicial Committee of the Supreme People's Court, promulgated on and effective as of 9 August 2001)

Notice regarding Problems with the Acceptance of Class Action Lawsuits by the People's Courts (promulgated by the Sup. People's Ct. 30 December 2005, effective 1 January 2006)

Provisions of the Supreme People's Court on Several Issues Concerning the Trial of Cases of Disputes over Compensation for Vessel-induced Oil Pollution Damage (promulgated by the Judicial Committee of the Supreme People's Court, effective on 1 July 2011)

Regulation of the People's Republic of China on the Prevention and Control of Vessel-Induced Marine Environment Pollution (Promulgated by the Executive Meeting of the State Council, effective on 1 March 2010)

Regulation on Emergency Prevention and Response to Vessels-Induced Marine Environmental Pollution (Promulgated by the Ministry of Transport, effective as of 1 June 2011)

Management Regulation of Collection and Use of the Vessel-induced Pollution Damage Compensation Fund (promulgated by Ministry of Finance and Ministry of Transport, effective as of 1 July 2012, Caizong [2012] No. 33)

The Rules on Calculating Fishery Losses Caused by Water Pollution Accidents (Promulgated by the Ministry of Fishery, effective on 8 October 1996, Nongyufa [1996] No. 14)

Calculation Methods for Economic Losses Caused by Fishery Pollution Accidents (Promulgated by the General Administration of Quality Supervision, Inspection and Quarantine and Standardization Administration, effective 1 June 2008, GB/T 21678-2008)

Implementation Rules on the Civil Liability Insurance for Vessel-Induced Oil Pollution Damage (promulgated by the Ministry of Transport, effective as of 1 October 2010)

Shenyang Regulation on Preventing Pollution from Dangerous Waste (promulgated by Shenyang Standing Committee of People's Congress, 28 November 2008)

The Regulation on Registration and Administration of Social Organizations (promulgated by State Council, 25 September 1998)

The Provisional Measures for Suppression of Illegal Civic Organization (promulgated by Ministry of Civil Affairs, 10 April 2000)

BIBLIOGRAPHY

Abraham 1982
Abraham, K.S., 'Cost Internalization, Insurance and Toxic Tort Compensation Funds', *Virginia Journal of Natural Resources*, 1982, 2, p. 123-148.

Abraham 1988
Abraham, K.S., 'Environmental Liability and the Limits of Insurance', *Columbia Law Review*, 1988, 88, p. 942-988.

Abraham 1993
Abraham, K.S., 'Cleaning up the Environmental Liability Insurance Mess', *Valparaiso University Law Review*, 1993, 27, p. 601-636.

Abraham 2001
Abraham, K.S., 'The Rise and Fall of Commercial Liability Insurance', *Virginia Law Review*, 2001, 87, p. 85-109.

Akerlof 1970
Akerlof, G., 'The Market for "Lemons": Quality, Uncertainty and the Market Mechanism', *Quarterly Journal of Economics*, 1970, p. 488-500.

Alberini & Austin 1999
Alberini, A. & Austin, D.H., 'Strict Liability as a Deterrent in Toxic Waste Management: Empirical Evidence from Accident and Spill Data', *Journal of Environmental Economics and Management*, 1999, p. 20-48.

Alberini & Austin 2002
Alberini, A. & Austin, D., 'Accidents Waiting to Happen: Liability Policy and Toxic Pollution Releases', *The Review of Economics and Statistics*, 2002, 84, p. 729-741.

Alberini & Frost 2007
Alberini, A. & Frost, S., 'Forcing Firms to Think about the Future: Economic Incentives and the Fate of Hazardous Waste', *Environmental Resource Economics*, 2007, 36, p. 451-474.

Alberton 2007
Alberton., M., 'Saint George and the Dragon: Transposing the Environmental Liability Directive in Italy', *Environmental Liability*, 2007, p. 235-242.

Alford & Shen 1997
Alford, W. & Shen, Y., 'Limits of the Law in Addressing China's Environmental Dilemma', *Stanford Environmental Law Journal*, 1997, 16, p. 125-150.

Allen 2011
Allen, C., 'Proving Natural Resource Damage under OPA 90: Out with the Rebuttable Presumption, in with APA-Style Judicial Review?', *Tulane Law Review*, 2011, 85, p. 1039-1074.

Ameye 2009
Ameye, E., 'Legal Study on Nuclear Third Party Liability for DG TREN of the European Commission', in: Pelzer, N. (ed.), *European Nuclear Liability Law in a Process of Change*, Baden-Baden: Nomos, 2009, p. 147-156.

Ameye 2010
Ameye, E., 'Channeling of Nuclear Third Party Liability towards the Operator: Is it Sustainable in a Developing Nuclear World or is there a Need for Liability of Nuclear Architects and Engineers?', *European Energy and Environmental Law Review*, 2010, p. 33-58.

Anderson 1997
Anderson, C.B., 'Damage to Natural Resources and the Costs of Restoration', *Tulane Law Review*, 1997, 72, p. 417-492.

Anderson 1998
Anderson, D., 'Development of Environmental Liability Risk Management and Insurance in the United States. Lessons and Opportunities', *Risk Management and Insurance Review*, 1998, p. 1-23.

Arlen 1993
Arlen, J., 'Compensation Systems and Efficient Deterrence', *Maryland Law Review*, 1993, 52, p. 1093-1136.

Arlen 2000
Arlen, J., 'Tort Damages', in: Bouckaert, B. & De Geest, G. (eds.), *Encyclopedia of Law and Economics, Volume II, Civil Law and Economics*, Cheltenham: Edward Elgar, 2000, p. 682-734.

Aronson 1988
Aronson, B., 'Compensation of Pollution-related Health Damage in Japan', *Social Science and Medicine*, 1988, p. 1043-1052.

Backes 2002
Backes, Ch., 'Netherlands', in: Ebbesson, J. (ed.), *Access to Justice in Environmental Matters in the EU,* The Hague: Kluwer Law International, 2002, p. 379-398.

Balkenborg 2001
Balkenborg, D., 'How Liable Should a Lender Be? The Case of Judgment-Proof Firms and Environmental Risk: Comment', *The American Economic Review*, 2001, 91, p. 731-738.

Banks 2004
Banks, E., *Alternative Risk Transfer: Integrated Risk Management through Insurance, Reinsurance and the Capital Markets,* Chichester: John Wiley & Sons Ltd, 2004.

Banks 2005
Banks, E., *Catastrophic Risk: Analysis and Management*, Chichester: John Wiley & Sons Ltd, 2004.

Bao 2009
Bao, M., 'Environmental NGOs in Transforming China', *Nature and Culture*, 2009, 4, p. 1-16.

Barboza 1997
Barboza, J., 'The ILC and Environment Damage', in: Wetterstein, P. (ed.), *Harm to the Environment: The Right to Compensation and the Assessment of Damages,* Oxford: Clarendon Press, 1997, p. 73-82.

Bator 1958
Bator, F.M., 'The Anatomy of Market Failure', *The Quarterly Journal of Economics*, 1958, p. 35-379.

Bawcutt 1991
Bawcutt, P., *Captive Insurance Companies: Establishment, Operation and Management*, Woodhead: Faulkerner, 1991.

Benedek 1998
Benedek, D., 'Non-pecuniary Damages: Defined, Assessed and Capped', *Revue Juridique Themis*, 1998, 32, p. 607-662.

Bennet 2001
Bennet, P., 'Mutual Risk, P&I Insurance Clubs and Maritime Safety and Environmental Performance', *Marine Policy*, 2001, p. 13-21.

Ben-Shahar 2009
Ben-Shahar, O., 'Causation and Forseeability', in: Faure, M. (ed.), *Tort Law and Economics*, Cheltenham: Edward Elgar, 2009, p. 83-108.

Bergkamp 2001
Bergkamp, L., *Liability and Environment: Private and Public Law Aspects of Civil Liability for Environmental Harm in an International Context*, The Hague: Kluwer Law International, 2001.

Berkovitz 1989
Berkovitz, D.M., 'Price-Anderson Act: Model Compensation Legislation? - The Sixty-Three Million Dollar Question', *Harvard Environmental Law Review*, 1989, 13, p. 1-68.

Betlem 2006
Betlem, G., 'Transnational Operator Liability', in: Betlem, G. & Brans, E. (eds.), *Environmental Liability in the EU: The 2004 Directive Compared with U.S. and Member State Law*, London: Cameron May, 2006, p. 149-188.

Bishop 1983
Bishop, W., 'The Contract-Tort Boundary and the Economics of Insurance', *Journal of Legal Studies*, 1983, 12, p. 241-266.

Blanchard 2000
Blanchard, P., 'Responsibility for Environmental Damage under Nuclear and Environmental Instruments: A Legal Benchmarking', *Journal of Energy & Natural Resources Law*, 2000, 18, p. 233-253.

Bocken 1987
Bocken H., 'Alternatives to Liability and Liability Insurance for the Compensation of Pollution Damages (Part 1)', *Tijdschrift voor Milieuaansprakelijkheid*, 1987, p. 83-87.

Bocken 1988
Bocken, H., 'Alternatives to Liability and Liability Insurance for the Compensation of Pollution Damages (Part 2)', *Tijdschrift voor Milieuaansprakelijkheid*,1988, p. 3-10.

Bocken 1997
Bocken, H., 'The Compensation of Ecological Damage in Belgium', in: Wetterstein, P. (ed.), *Harm to the Environment: The Right to Compensation and the Assessment of Damages*, Oxford: Clarendon Press, 1997, p. 143-158.

Bocken 2002
Bocken, H., 'Alternative Compensation Systems for Environmental Liabilities, General Report', Theme IB, AIDA XIth World Congress, 2002, available at: <www.aida.org.uk/worldcongress02/theme1b_generalreport.doc>, last accessed on 25th January, 2013.

Bocken 2006
Bocken, H., 'Financial Guarantees in the Environmental Liability Directive: Next Time Better', *European Environmental Law Review*, 2006, p. 13-32.

Bocken 2009
Bocken, H., 'Alternative Financial Guarantees for Environmental Liabilities under the ELD', *European Energy and Environmental Law Review*, 2009, p. 146-170.

Bocken, Lambrechts, Boes, De Nauw, Faure & Lavrysen 1996
Bocken, H., Lambrechts, W., Boes, M., De Nauw, A., Faure, M. & Lavrysen, L., 'The Flemish Draft Decree on Environmental Policy: An Outline', in: Bocken, H. & Ryckbost, D. (eds.), *The Codification of Environmental Law: Proceedings of the International Conference*, The Hague: Kluwer Law International, 1996, p. 11-40.

Bohm 1994
Bohm, P., 'CVM Spells Responses to Hypothetical Questions', *Natural Resources Journal*, 1994, 34, p. 37-50.

Bongaerts & Debièvre 1987
Bongaerts, J. & Debièvre, A., 'Insurance for Civil Liability for Marine Oil Pollution Damages', *Geneva Papers on Risk and Insurance*, 1987, p. 145-157.

Borch 1961
Borch, K., 'The Utility Concept Applied to the Theory of Insurance', *The Astin Bulletin*, 1961, p. 245-255.

Borch, Aase & Sandmo 1990
Borch, K.H., Aase, K.K. & Sandmo, A., *Economics of Insurance*, Amsterdam: Elsevier Science Publishers, 1990.

Boyd 2002
Boyd, J., 'Financial Responsibility for Environmental Obligations: Are Bonding and Assurance Rules Fulfilling Their Promise?', in: Swanson, T. (ed.), *An Introduction to the Law and Economics of Environmental Policy: Issues in Institutional Design*, Oxford: Elsevier Science, 2002, p. 417-485.

Boyd 2003
Boyd, J., 'A Market-Based Analysis of Financial Insurance Issues Associated with U.S. Natural Resource Damage Liability', in: Faure, M. (ed.), *Deterrence, Insurability, and Compensation in Environment Liability: Future Development in the European Union*, Vienna: Springer/Wien/New York, 2003, p. 258-302.

Boyd & Hunreuther 1997
Boyd, J. & Kunreuther, H., 'Retroactive Liability or the Public Purse?', *Journal of Regulatory Economics*, 1997, p. 79-90.

Boyer & Laffont 1997
Boyer, M. & Laffont, J.J., 'Environmental Risks and Bank Liability', *European Economic Review*, 1997, p. 1427-1459.

Boyer & Porrini 2002
Boyer, M. & Porrini, D., 'The Choice of Instruments for Environmental Policy: Liability or Regulation?', in: Swanson, T. (ed.), *An Introduction to the Law and Economics of Environmental Policy: Issues in Institutional Design*, Oxford: Elsevier Science, 2002, p. 245-268.

Boyer & Porrini 2008
Boyer, M. & Porrini, D., 'The Efficient Liability Sharing Factor for Environmental Disasters: Lessons for Optimal Insurance Regulation', *The Geneva Papers on Risk and Insurance-Issues and Practices*, 2008, p. 337-362.

Brans 2001
Brans, E.H.P., *Liability for Damage to Public Natural Resources*, The Hague, Kluwer Law International, 2001.

Briggs 2006
Briggs, A., 'China's Pollution Victims: Still Seeking a Dependable Remedy', *Georgetown International Environmental Law Review*, 2006, 18, p. 305-334.

Brighton & Askman 1997
Brighton, W. & Askman, D.V., 'The Role of Government Trustees in Recovering Compensation for Injury to Natural Resourcs', in: Wetterstein, P. (ed.), *Harm to the Environment: The Right to Compensation and the Assessment of Damages*, Oxford: Clarendon Press, 1997, p. 143-158.

Brinker 1998
Brinker, S., 'Opening the Door to the Indeterminate Plaintiff: An Analysis of the Causation Barriers Facing Environmental Toxic Tort Plaintiffs', *UCLA Law Review*, 1998, 46, p. 1289-1331.

Brinkmann 1998
Brinkmann, T., 'German Law: Damage to Persons, Property, and the Environment', *Tulane Maritime Law Journal*, 1998, 22, p. 545-568.

Bruggeman 2010a
Bruggeman, V., 'Capital Market Instruments for Natural Catastrophe and Terrorism Risks: A Bright Future?', *Environmental Law Reporter*, 2010, 40, p. 10136-10153.

Bruggeman 2010b
Bruggeman, V., *Compensating Catastrophe Victims: A Comparative Law and Economic Approach*, The Hague: Kluwer Law International, 2010.

Burrow 1999
Burrow, P., 'Combining Regulation and Legal Liability for the Control of External Costs', *International Review of Law and Economics*, 1999, p. 227-244.

Calabresi 1970
Calabresi, G., *The Costs of Accidents: A Legal and Economic Analysis*, New Haven/London: Yale University Press, 1970.

Carroll 2008
Carroll, S., 'Perspective on the Pros and Cons of a Pooling-type Approach to Nuclear Third Party Liability', *Nuclear Law Bulletin*, 2008, p. 75-97.

Carroll 2009
Carroll, S., 'European Challenges to Promoting International Pooling and Compensation for Nuclear Reactor Accidents', Nonproliferation Policy Education Center Working Paper Series, available at SSRN.

Carter & Mol 2007
Carter, N. & Mol, T. (eds.), *Environmental Governance in China*, London/New York: Routledge, 2007.

Cartwright 2005
Cartwright, M.E., 'Superfund: It's No Longer Super and It Isn't Much of a Fund', *Tulane Environmental Law Journal*, 2005, 18, p. 299-322.

Change & Sigman 2010
Change, H. & Sigman, H., 'An Empirical Analysis of Cost Recovery in Superfund Cases: implications for Brownfields and Joint and Several Liability', NBER Working Papers 16209, 2010.

Chen 2009
Chen, G., *Politics of China's Environmental Protection: Problems and Progress*, Singapore: World Scientific Publishing, 2009.

Chilton & Meyer 1998
Chilton, K. & Meyer, P., *Environmental Insurance for Brownfields Redevelopment: A Feasibility Study*, 1998.

Coghlin 1984
Coghlin, T.G., 'Protection and Indemnity Clubs', *Lloydd's Maritime and Commercial Law Quarterly*, 1984, p. 403-416.

Coleman 1992
Coleman, J., *Risks and Wrongs*, New York: Cambridge University Press, 1992.

Cooter & Ulen 2012
Cooter, R. & Ulen, T., *Law and Economics*, Massachusetts: Pearson Education, 2012.

Coroner 2006
Coroner, F., 'Environmental Liability Directive: How Well Are Member States Handling Transposition?', *Environmental Liability*, 2006, p. 211-214.

Crocker & Snow 1986
Crocker, K. & Snow, A., 'The Efficient Effects of Categorical Discrimination in the Insurance Industry', *Journal of Political Economy*, 1986, p. 321-344.

Cropw & Knott 1994
Cropw, P. & Knott, D., 'Deadline Near for Compliance with U.S. Oil Spill Liability Rules', *Oil and Gas Journal*, 1994, p. 14-39.

Cross 1989
Cross, F.B., 'Natural Resource Damage Valuation', *Vanderbilt Law Review*, 1989, 42, p. 269-342.

Crossen & Niessen 2007
Crossen, T. & Niessen, V., 'NGO Standing in the European Court of Jurstice – Does the Aarhus Regulation Open the Door?', *Review of European Community and International Environmental Law*, 2007, p. 332-340.

Cummins 2008
Cummins, J.D., 'Cat Bonds and Other Risk-Linked Securities: State of the Market and Recent Developments', *Risk Management and Insurance Review*, 2008, p. 23-47.

Cummins & Tennyson 1996
Cummins, J.D. & Tennyson, S., 'Moral Hazard in Insurance Claiming: Evidence from Automobile Insurance', *Journal of Risk and Uncertainty*, 1996, 12, p. 29-50.

Cummins & Weiss 2009
Cummins, J.D. & Weiss, M.A., 'Convergence of Insurance and Financial Market: Hybrid and Securitized Risk-Transfer Solutions', *The Journal of Risk and Insurance*, 2009, 76, p. 493-545.

Currie 2008
Currie, D.E.J., 'The Problems and Gaps in the Nuclear Liability Conventions and an Analysis of How an Actual Claim Would Be Brought under the Current Existing Treaty Regime in the Event of a Nuclear Accident', *Denver Journal of International Law and Policy*, 2008, 35, p. 85-128.

Danzi 2009
Danzi, E., 'Some Reflections on the Exclusion of Nuclear Damage from the Scope of Application of the Environmental Liability Directive', in: Pelzer, N. (ed.), *European Nuclear Liability Law in a Process of Change*, Baden-Baden: Nomos, 2009, p. 191-212.

Danzon 1984
Danzon, P.M., 'Tort Reform and the Role of Government in Private Insurance Markets', *Journal of Legal Studies*, 1984, 13, p. 517-550.

De La Rue & Anderson 2009
De La Rue, C. & Anderson, C.B., *Shipping and the Environment: Law and Practice*, London: Informa, 2009.

De Sadeleer 2006
De Sadeleer, N., 'Polluter-Pays, Precautionary Principles and Liability', in: Betlem, G. & Brans, E. (eds.), *Environmental Liability in the EU: The 2004 Directive Compared with U.S. and Member State Law*, London: Cameron May, 2006, p. 89-147.

De Smedt 2009
De Smedt, K., 'Is Harmonisation Always Effective? The Implementation of the Environmental Liability Directive', *European Energy and Environmental Law Review*, 2009, p. 2-18.

De Smedt & Faure 2010
De Smedt, K. & Faure, M., 'The Implementation of the Environmental Liability Directive: A Law and Economics Analysis of the Transposition of the ELD in Belgium, the Netherlands and Germany', *Zeitschrift für Europäisches Privatrecht*, 2010, p. 783-808.

De Smedt & Faure 2013
De Smedt, K. & Faure, M., 'The ELD's Effects in Practice', in: Bergkamp, L. & Goldsmith, B. (eds.), *The EU Environmental Liability Directive: A Commentary*, Oxford: Oxford University Press, 2013, p. 299-314.

Deakin, Johnston, & Markesinis 2003
Deakin, S., Johnston, A. & Markesinis, B., *Markesinis and Deakin's Tort Law*, Oxford: Oxford University Press, 2003.

Dionne & Spaeter 2003
Dionne, G. & Spaeter, S., 'Environmental Risk and Extended Liability: The Case of Green Technologies', *Journal of Public Economics*, 2003, 87, p. 1025-1060.

Dixon 1992
Dixon, L.S., *Superfund and Transaction Costs: the Experiences of Insurers and Very Large Industrial Firms, Congressional Testimony*, CT-102, 39, 1992.

Dixon 1995
Dixon, L.S., 'The Transactions Costs Generated by Superfund's Liability Approach', in: Revesz, R. & Stewart, R. (eds.), *Analyzing Superfund: Economics, Science and Law*, Washington, D.C.: Resources for the Future, 1995, p. 171-185.

Dobbins 1994
Dobbins, J.C., 'The Pain and Suffering of Environmental Loss: Using Contingent Valuation to Estimate Nonuse Damages', *Duke Law Journal*, 1994, 43, p. 879-946.

Dougherty 1985
Dougherty, D.J., 'Impact of a Member's Insolvency or Bankruptcy on a Protection & Indemnity Club', *Tulane Law Review*, 1985, 59, p. 1466-1486.

Dubin & Rothwell 1990
Dubin, J. & Rothwell, G., 'Subsidy to Nuclear Power through Price-Anderson Liability Limit', *Contemporary Economic Policy*, 1990, 8, p. 73-79.

Duncan 2008
Duncan, E.J.C., 'The Problems and Gaps in the Nuclear Liability Conventions and an Analysis of How an Actual Claim Would Be Brought under the Current Existing Treaty Regime in the Event of a Nuclear Accident', *Denver Journal of International Law and Policy*, 2008, 35, p. 85-127.

Elliott 2005
Elliott, W.P., 'A Guide to Captive Insurance Companies (Part one)', *Journal of International Taxation*, 2005, 16, p. 22-31.

Emmerechts 2008
Emmerechts, S., 'Environmental Law and Nuclear Law: A Growing Symbiosis', *Nuclear Law Bulletin*, 2008, p. 91-110.

Epstein 1985
Epstein, R.A., 'Products Liability as and Insurance Market', *Journal of Legal Studies*, 1985, 14, p. 645-669.

Esty & Porter 2005
Esty, D.C. & Porter, M.E., 'National Environmental Performance: An Empirical Analysis of Policy Results and the Determinants', *Environment and Development Economics*, 2005, p. 391-434.

Farber 1986
Farber, D.A., 'Toxic Causation', *Minnesota Law Review*, 1986, 71, p. 1219-1262.

Faure 1995
Faure, M., 'Economic Models of Compensation for Damage Caused by Nuclear Accidents: Some Lessons for the Revision of the Paris and Vienna Conventions', *European Journal of Law and Economics*, 1995, p. 21-43.

Faure 2001
Faure, M., 'Tort Liability in France: An Introductory Economic Analysis', in: Deffains, B. & Kirat, T. (eds.), *Law and Economics in Civil Law Countries*, Amsterdam: Elsevier Science B.V., 2001, p. 169-181.

Faure 2002a
Faure, M., 'Environmental Damage Insurance in the Netherlands', *Environmental Liability*, 2002, p. 31-41.

Faure 2002b
Faure, M., 'Environmental Damage Insurance in Theory and Practice', in: Swanson, T. (ed.), *An Introduction to the Law and Economics of Environmental Policy: Issues in Institutional Design*, Oxford: Elsevier Science, 2002, p. 283-328.

Faure 2004
Faure, M., 'Alternative Compensation Mechanisms as a Remedy for Insurability of Liability', *The Geneva Papers on Risk and Insurance*, 2004, p. 455-489.

Faure 2007a
Faure, M., 'A Shift toward Alternative Compensation Mechanisms for Environmental Damage?', in: Faure, M. & Verheij, A. (eds.), *Shifts in Compensation for Environmental Damage*, Wien: Springer, 2007, p. 73-102.

Faure 2007b
Faure, M., 'Is Risk Differentiation on European Insurance Markets in Danger?', *Maastricht Journal of European and Comparative Law*, 2007, p. 83-100.

Faure 2010
Faure, M., 'Behavioural Accident Law and Economics', *The Journal of Applied Economy*, 2010, p. 11-68.

Faure & Bruggeman 2007
Faure, M. & Bruggeman, V., 'Causal Uncertainty and Proportional Liability', in: Tichy, L. (ed.), *Causation in Law*, Praha: E. Rozkotová, 2007, p. 105-121.

Faure & Bruggeman 2008
Faure, M. & Bruggeman, V., 'Catastrophic Risks and First-party Insurance', *Connecticut Insurance Law Journal*, 2008, 15, p. 1-52.

Faure & Fiore 2008a
Faure, M. & Fiore, K., 'The Civil Liability of European Nuclear Operators: Which Coverage for the New 2004 Protocols? Evidence from France', *International Environmental Agreements*, 2008, p. 227-248.

Faure & Fiore 2008b
Faure, M. & Fiore, K., 'The Coverage of the Nuclear Risk in Europe: Which Alternative?', *The Geneva Papers on Risk and Insurance-Issues and Practices*, 2008, p. 288-322.

Faure & Fiore 2009
Faure, M. & Fiore, K., 'An Economic Analysis of the Nuclear Liability Subsidy', *Pace Environmental Law Review*, 2009, 26, p. 419-450.

Faure, Goodwin & Weber 2010
Faure, M., Goodwin, M. & Weber, F., 'Bucking the Kuznets Curve: Designing Effective Environmental Regulation in Developing Countries', *Virginia Journal of International Law*, 2010, 51, p. 95-156.

Faure & Grimeaud 2003
Faure, M. & Grimeaud, D., 'Financial Assurance Issues of Environmental Liability', in: Faure, M. (ed.), *Deterrence, Insurability, and Compensation in Environment Liability: Future Development in the European Union*, Vienna: Springer, 2003, p. 7-256.

Faure & Hartlief 2003
Faure, M. & Hartlief, T., *Insurance and Expanding Systemic Risks*, Paris: OECD, 2003.

Faure & Hu 2011
Faure, M. & Hu, W., 'Towards a Reform of Environmental Liability in China: An Economic Analysis', *Asia Pacific Journal of Environmental Law*, 2011, 13, p. 225-247.

Faure & Liu 2012
Faure, M. & Liu, J., 'New Models for the Compensation of Natural Resources Damage', *Kentucky Journal of Equine, Agriculture & Natural Resources Law*, 2012, 4, p. 261-314.

Faure & Skogh 1992
Faure, M. & Skogh, G., 'Compensation for Damages Caused by Nuclear Accidents: A Convention as Insurance', *The Geneva Papers on Risk and Insurance*, 1992, p. 499-513.

Faure & Van den Bergh 2002
Faure, M. & Van den Bergh, R., 'Competition on the European Market for Liability Insurance and Efficient Accident Law', *Maastricht Journal of European and Comparative Law*, 2002, 9, p. 279-306.

Faure & Vanden Borre 2008
Faure, M. & Vanden Borre, T., 'Compensating Nuclear Damage: A Comparative Economic Analysis of the U.S. and International Liability Schemes', *William & Mary Environmental Law and Policy Review*, 2008, 33, p. 219-286.

Faure & Wang 2006a
Faure, M. & Wang, H., 'Economic Analysis of Compensation for Oil Pollution Damage', *Journal of Maritime Law and Commerce*, 2006, 37, p. 179-218.

Faure & Wang 2006b
Faure, M. & Wang, H., 'Financial Caps for Oil Pollution Damage: China and the International Conventions', in: Faure, M. & Hu, J. (eds.), *Prevention and Compensation of Marine Pollution Damage: Recent Developments in Europe, China and the U.S.*, Alphen aan den Rijn: Kluwer, 2006, p. 317-347.

Feess & Hege 2000
Feess, E. & Hege, U., 'Environmental Harm and Financial Responsibility', *Geneva Papers on Risk & Insurance*, 2000, 25, p. 220-234.

Fiore 2009
Fiore, K., 'No-fault Compensation Systems', in: Faure, M. (ed.), *Tort Law and Economics,* Cheltenham/Northampton: Edward Elgar, 2009, p. 406-432.

Fogleman 2005
Fogleman, V., *Environmental Liabilities and Insurance in England and the United States,* Livingston: Witherby & Co Ltd, 2005.

Fogleman 2009
Fogleman, V., 'The Environmental Damage Regulations: The New Regime', *Environmental Liability*, 2009, p. 147-176.

Foley & Nolan 2008
Foley, V. & Nolan, C., 'The Erika Judgment — Environmental Liability and Places of Refuge: A Sea Change in Civil and Criminal Responsibility that the Maritime Community Must Heed', *Tulane Maritime Law Journal*, 2008, 33, p. 41-78.

Force, Davies & Force 2011
Force, R., Davies, M. & Force, J.S., 'Deepwater Horizon: Removal costs, Civil Damages, Crimes, Civil Penalties, and State Remedies in Oil Spill Cases', *Tulane Law Review*, 2011, 85, p. 889-982.

Foster 1998
Foster, N.R., 'Marine Insurance: Direct Action Statues and Related Issues', *University of San Francisco Maritime Law Journal*, 1998, 11, p. 261-298.

Friedman 1982
Friedman, D., 'What is "Fair Compensation" for Death or Injury?', *International Review of Law and Economics*, 1982, p. 81-93.

Froggatt & Carroll 2010
Froggatt, A. & Carroll, S., 'Third Party Insurance: The Nuclear Sector's "Silent" Subsidy in Europe', in: Sokolski, H. (ed.), *Nuclear Power's Global Expansion: Weighing its Costs and Risks*, Carlisle: Strategic Studies Institute of the U.S. Army War College, 2010, p. 555-614.

Gergen 1994
Gergen, M.J., 'The Failed Promise of the "Polluter Pays" Principle: An Economic Analysis of Landowner Liability for Hazardous Waste', *New York University Law Review*, 1994, 69, p. 624-692.

Gilead 1997
Gilead, I., 'Tort Law and Internalization: The Gap between Private Loss and Social Cost', *International Review of Law and Economics*, 1997, p. 589-608.

Goelz 2009
Goelz, D., 'China's Environmental Problems: Is a Specialized Court the Solution?', *Pacific Rim Law & Policy Journal*, 2009, 18, p. 155-188.

Gordley 1997
Gordley, J., 'Tort Law in the Aristotelian Tradition', in: Owen D.G. (ed.), *Philosophical Foundations of Tort Law*, Oxford: Clarendon Press, 1997, p. 131-158.

Grad 1982
Grad, F.P., 'A Legislative History of the Comprehensive Environmental Response, Compensation and Liability Act of 1980', *Columbia Journal of Environmental Law*, 1982, 8, p. 1-36.

Greenstone & Gallagher 2008
Greenstone, M. & Gallagher, J., 'Does Hazardous Waste Matter? Evidence from the Housing Market and the Superfund Program', *The Quarterly Journal of Economics*, 2008, 123, p. 951-1003.

Gruber & Krueger 1991
Gruber, J. & Krueger, A.B., 'The Incidence of Mandated Employer-Provided Insurance: Lessons from Workers' Compensation Insurance', *Tax Policy and the Economy*, 1991, 5, p. 111-144.

Hamilton & Viscusi 1997
Hamilton, J. & Viscusi, K., 'The Benefit and Costs of Regulatory Reforms for Superfund', *Stanford Environmental Law Journal*, 1997, 16, p. 159-200.

Hamilton & Viscusi 1999
Hamilton, J. & Viscusi, K., *Calculating Risks: The Spatial and Political Dimensions of Hazardous Waste Policy*, Massachusetts: Institute of Technology, 1999.

Han & Wang 2010
Han, L. & Wang D., 'Discussion on Limitation of Liability and Compulsory Insurance of Compensation for Bunker Oil Pollution Damage from Ships in China', in: Faure, M. & Han, L. & Shan, H. (eds.), *Maritime Pollution Liability and Policy – China, Europe and the U.S.*, The Hague: Kluwer, 2010, p. 145-158.

Hanrilica 2009
Hanrilica, J., 'Harmonization of Nuclear Liability in the European Union: Challenges, Options and Limits', *Nuclear Law Bulletin*, 2009, p. 37-60.

Hanson & Logue 1990
Hanson, J.D. & Logue, K.D., 'The First-Party Insurance Externality: An Economic Justification for Enterprise Liability', *Cornell Law Review*, 1990, 76, p. 129-196.

Harrington 2000
Harrington, S.E., 'Rethinking Disaster Policy', *Regulation*, 2000, 23, p. 40-46.

Hathlia 2009
Hathlia, E., 'The 2004 Paris Convention — The Issue of Third Party Nuclear Liability Insurance and a Methodology for Calculating the Premium for Environmental Heads of Damage: A UK Perspective', in: Pelzer, N. (ed.), *European Nuclear Liability Law in a Process of Change*, Baden-Baden: Nomos, 2009, p. 75-85.

Hattan 2002
Hattan, E., 'The Environmental Liability Directive', *Environmental Liability*, 2002, p. 3-10.

Hawke & Hargreaves 2003
Hawke, N. & Hargreaves, P., 'Environmental Funds, Compensation and Liability', *Environmental Liability*, 2003, p. 39-47.

Hendrickx 2007
Hendrickx, R., 'Marine Oil Pollution: An Empirical Analysis', in: Faure, M. & Verheij, A. (eds.), *Shifts in Compensation for Environmental Damage*, Wien: Springer, 2007, p. 243-260.

Heyes & Liston-Heyes 1998
Heyes, A. & Liston-Heyes, C., 'Subsidy to Nuclear Power through Price-Anderson Liability Limit: Comment', *Contemporary Economic Policy*, 1998, p. 122-124.

Heyes & Liston-Heyes 2000
Heyes, A. & Liston-Heyes, C., 'Capping Environmental Liability: the Case of North American Nuclear Power', *The Geneva Papers on Risk and Insurance*, 2000, p. 196-202.

Hinteregger 2008
Hinteregger, M., 'Comparison', in: Hinteregger, M. (ed.), *Environmental Liability and Ecological Damage in European Law*, New York: Cambridge University Press, 2008 p. 579-636.

Hollaender & Kaminisky 2000
Hollaender, K. & Kaminisky, M.A., 'The Past, Present, and Future of Environmental Insurance including a Case Study of MTBE Litigation', *Environmental Forensics*, 2000, 1, p. 205-211.

Holstein-Childress 2002
Holstein-Childress, V., 'The Enforceability of Arbitral Clauses contained in Marine Insurance Contracts against no signatory direct action Claimants', *Tulane Maritime Law Journal*, 2002, 27, p. 205-266.

Hu & Bo 2006
Hu, J. & Bo, Y., 'Application of Law in Civil Liability for Oil Pollution Damage Caused by Coastal Vessels in China', in: Faure, M. & Hu, J. (eds.), *Prevention and Compensation of Marine Pollution Damage: Recent Developments in Europe, China and the U.S.*, Alphen aan den Rijn: Kluwer, 2006, p. 193-205.

Hutchison & Van't Veld 2005
Hutchison, M. & Van't Veld, K., 'Extended Liability for Environmental Accidents: What You See Is What You Get', *Journal of Environmental Economics and Management*, 2005, 49, p. 157-173.

Innes 2004
Innes, R., 'Enforcement Costs, Optimal Sanctions, and the Choice between Ex-post Liability and Ex-ante Regulation', *International Review of Law and Economics*, 2004, p. 29-48.

International Atomic Energy Agency 2006
International Atomic Energy Agency, *Environmental Consequence of the Chernobyl Accident and their Remediation: Twenty Years of Experience*, Report of the Chernobyl Forum Expert Group 'Environment', Vienna: IAEA, 2006.

Jans & Vedder 2012
Jans, J.H. & Vedder, H.H.B., *European Environmental Law after Lisbon*, 4th edn, Groningen: Europa Law Publishing, 2012.

Jendroska & Bar. 2007
Jendroska, J. & Bar, M., 'Transposition of Directive 2004/35/EC(ELD) in Poland', *Environmental Liability*, 2007, p. CS 43-CS 47.

Jones 2009
Jones, R., 'Nuclear Insurance: Where Does It Fit in the Green Generation?', *Journal of Reinsurance*, 2009, 16, p. 71-78.

Jose & Garza 2007
Jose, D. & Garza, M., 'The Complete Federal Preemption of Nuclear Safety Should Prevent Scientifically Irrational Jury Verdicts in Radiation Litigation', *Temple Journal of Science, Technology and Environmental Law*, 2007, 26, p. 1-26.

Judy & Probst 2009
Judy, M. & Probst, K., 'Superfund at 30', *Vermont Journal of Environmental Law*, 2009, 11, p. 191-248.

Kahan 1989
Kahan, M., 'Causation and the Incentives to Take Care under the Negligence Rule', *Journal of Legal Studies*, 1989, 18, p. 427-448.

Kahneman & Tversky 1979
Kahneman, D. & Tversky, A., 'Prospect Theory: An Analysis of Decision under Risk', *Econometrica: Journal of the Econometric Society*, 1979, p. 263-291.

Kambia-Chopin 2010
Kambia-Chopin, B., 'Environmental Risks, the Judgment-Proof Problem and Financial Responsibility', *European Journal of Law and Economics*, 2010, p. 77-87.

Kanner 2005
Kanner, A., 'The Public Trust Doctrine, Parens Patriae and the Attorney General as the Guardian of the State's Natural Resources', *Duke Environmental Law and Policy Forum*, 2005, 16, p. 57-115.

Kanner & Nagy 2005
Kanner, A. & Nagy, T, 'Measuring Loss of Use Damages in Natural Resource Damage Actions', *Columbia Journal of Environmental Law*, 2005, 30, p. 417-448.

Kanner & Ziegler 2006
Kanner, A. & Ziegler, M.E., 'Understanding and Protecting Natural Resources', *Duke Environmental Law and Policy Forum*, 2006, 17, p. 120-163.

Kaplow & Shavell 1996
Kaplow, L. & Shavell, S., 'Accuracy in the Assessment of Damages', *Journal of Law and Economics*, 1996, p. 191-210.

Katzman 1986
Katzman, M.T., 'Pollution Liability Insurance and the Internationalization of Environmental Risks', *Review of Policy Research*, 1986, p. 614-623.

Kaye 1982
Kaye, D., 'The Limits of the Preponderance of the Evidence Standard: Justifiably Naked Statistical Evidence and Multiple Causation', *American Bar Foundation Research Journal*, 1982, 7, p. 487-516.

Keenan & Rubin 1988
Keenan, D. & Rubin. P., 'Shadow Interest Groups and Safety Regulation', *International Review of Law and Economics*, 1988, p. 21-36.

Kehne 1986
Kehne, J., 'Encouraging Safety Through Insurance-Based Incentives: Financial Responsibility for Hazardous Waste', *Yale Law Journal*, 1986, 96, p. 403-427.

Kiern 2000
Kiern, L., 'Liability, Compensation, and Financial Responsibility under the Oil Pollution Act of 1990: A Review of the First Decade', *Tulane Maritime Law Journal*, 2000, 24, p. 481-590.

Kim 2002
Kim, I., 'Financial Responsibility Rules under the Oil Pollution Act of 1990', *Natural Resources Journal*, 2002, 42, p. 565-598.

Klass 2004
Klass, A.B., 'From Reservoirs to Remediation: the Impact of CERCLA on Common Law Strict Liability Environmental Claims', *Wake Forest Law Review*, 2004, 39, p. 904-970.

Kornhauser & Revesz 1989
Kornhauser, L.A. & Revesz, R.L., 'Sharing Damages among Multiple Tortfeasors', *Yale Law Journal*, 1989, 98, p. 831-884.

Kornhauser & Revesz 2009
Kornhauser, L.A. & Revesz, R.L., 'Joint Tortfeasors', in: Faure, M. (ed.), *Tort Law and Economics*, Cheltenham: Edward Elgar, 2009, p. 109-133.

Koziol & Zhu 2010
Koziol, H. & Zhu, Y., 'Background and Key Contents of the New Chinese Tort Liability Law', *Journal of European Tort Law*, 2010, p. 328-361.

Kraakman 1984
Kraakman, R.H., 'Corporate Liability Strategies and the Costs of Legal Controls', *Yale Law Journal*, 1984, 93, p. 857-900.

Krämer 2009
Krämer, L., 'The Environmental Complaint in the EU', *Journal for European Environmental & Planning Law*, 2009, p. 13-35.

Kronenberg 1995
Kronenberg, W., 'The Environmental Insurance Market in the U.S. and Western Europe: A U.S. Underwriter's Observations', *The Geneva Papers on Risk and Insurance*, 1995, p. 336-347.

Kunreuther 1987
Kunreuther, H., 'Problems and Issues of Environmental Liability Insurance', *The Geneva Papers on Risk and Insurance*, 1987, p. 180-197.

Kunreuther & Hogarth 1992
Kunreuther, H. & Hogarth, R., 'How Does Ambiguity Affect Insurance Decisions?', in: Dionne, G. (ed.), *Contributions to Insurance Economics*, Boston/Dordrecht/-London: Kluwer, 1992, p. 307-324.

Kunreuther, Hogarth & Meszaros 1993
Kunreuther, H., Hogarth, R. & Meszaros, J., 'Insurer Ambiguity and Market Failure', *Journal of Risk and Uncertainty*, 1993, p. 71-87.

Kuznets 1995
Kuznets, S., 'Economic Growth and Income Inequality', *American Economic Review*, 1995, p. 1-28.

Lahnstein 2003
Lahnstein., C., 'A Market-Based Analysis of the Financial Assurance Issues of Environmental Liability Taking Special Account of Germany, Austria, Italy and Spain', in: Faure, M. (ed.), *Deterrence, Insurability, and Compensation in Environment Liability: Future Development in the European Union*, Vienna: Springer, 2003, p. 303-330.

Lamm 2006
Lamm, V., 'The Protocol Amending the 1963 Vienna Convention', in: OECD Nuclear Energy Agency, International Atomic Energy Agency, *International Nuclear Law in the Post-Chernobyl Period*, Paris: OECD, 2006, p. 169-188.

Lan, Simonis & Dudek 2007
Lan, X., Simonis, U. & Dudek, D., 'Environmental Governance for China: Major Recommendations of a Task Force', *Environmental Politics*, 2007, 16, p. 669-676.

Larsson 1999
Larsson, M.-L., 'Legal Definitions of the Environment and of Environmental Damage', in: Wahlgren, P. (ed.), *Scandinavian Studies in Law, Volume 38, Legal Issues of the Late 1990s*, Stockholm: Stockholm Institute for Scandinavian Law, 1999 p. 155-176.

Lau 2009
Lau, C., 'Running Head: China NGOs, the Role of NGOs in China', *Quarterly Journal of Ideology*, 2009, 31(3&4), p. 1-18.

Lazarus 1986
Lazarus, R.J., 'Changing Conceptions of Property and Sovereignty in Natural Resources: Questioning the Public Trust Doctrine', *Iowa Law Review*, 1986, 71, p. 631-716.

Lee 2002
Lee, V.A., *The Natural Resource Damage Assessment Deskbook: A Legal and Technical Analysis*, Washington DC: Environmental Law Institute, 2002.

Lee 2008
Lee, C., '"Pollution First, Control Later" No More: Combating Environmental Degradation in China Through an Approach Based in Public Interest Litigation and Public Participation', *Pacific Rim Law & Policy Journal*, 2008, 17, p. 795-824.

Leiter 2001
Leiter, A.C., 'Environmental Insurance: Does it Defy the Rules?', *Harvard Environmental Law Review*, 2001, 25, p. 259-322.

Letourneau & Welmaker 2000
Letourneau, K.B. & Welmaker, W.T., 'The Oil Pollution Act of 1990: Federal Judicial Interpretation through the End of the Millennium', *University of San Francisco Maritime Law Journal*, 2000, 12, p. 147-226.

Levin & Pacala 2003
Levin, S.A. & Pacala, S.W., 'Ecosystem Dynamics', in: Maler, K.G. & Vicent, J.R. (eds.), *Handbook of Environmental Economics, Volume 1, Environmental Degradation and Institutional Responses*, Amsterdam: Elsevier Science B.V., 2003, p. 62-90.

Levmore 1990
Levmore, S., 'Probabilistic Recoveries, Restitution, and Recurring Wrongs', *Journal of Legal Studies*, 1990, 19, p. 691-726.

Lewis 2001
Lewis, B., 'It's been 4380 Days and Counting Since Exxon Valdez: Is it Time to Change the Oil Pollution Act of 1990?', *Tulane Environmental Law Journal*, 2001, 15, p. 97-128.

Li 2006
Li, Y., *A Study on Natural Resource Damage from Oil Pollution from Ships*, Shanghai Maritime University, Master Dissertation, 2006 (in Chinese).

Li 2011
Li, Y.W, 'Introduction: Challenges and Opportunities for NGOs in Different Parts of the World', in: Li, Y.W. (ed.), *NGOs in China and Europe: Comparisons and Contrasts*, Farnham: Ashgate, 2011, p. 1-20.

Li & Zhan 2011
Li, Z. & Zhan, S., 'Water Pollution Public Interest Litigation in Practice: Research Report on Relevant Cases in Guangzhou Maritime Court (1991-2009)', *Sun Yat-sen University Law Review*, 2011, 9, p. 240-251.

Libby 1952
Libby, G.E., 'Some Aspects of Protection and Indemnity Insurance', *Insurance Law Journal*, 1952, p. 684-709.

Lin 2006
Lin, A.C., 'The Unify Role of Harm in Environmental Law', *Wisconsin Law Review*, 2006, 3, p. 897-985.

Lindenbergh & Van Kippersluis 2009
Lindenbergh, S.D. & Van Kippersluis, P.P.M., 'Non-Pecuniary Losses', in: Faure, M. (ed.), *Tort Law and Economics*, Cheltenham/Northampton: Edward Elgar, 2009, p. 215-227.

Liu 2012
Liu, J.J., 'Environmental Justice with Chinese Characteristics: Recent Developments in Using Environmental Public Interest Litigation to Strengthen Access to Environmental Justice', *Florida A & M University Law Review*, 2012, 7, p. 229-271.

Liu & Faure 2012
Liu, J. & Faure, M., 'Compensating Nuclear Damage in China', *Washington University Global Studies Law Review*, 2012, 11, p. 781-816.

Liu, Faure & Wang 2013
Liu, J., Faure, M. & Wang, H., 'Compensating for Natural Resources Damage Caused by Vessel-Induced Marine Oil Pollution: Comparing the International Regime, the U.S. and China', *Journal of Environmental Law and Litigation*, 2013, forthcoming.

Lo 2006
Lo, C., 'Effective Regulations with Little Effect? The Antcedents of the Perceptions of Environmental Officials on Enforcement Effectiveness in China', *Environmental Management*, 2006, p. 388-410.

Ma & Du 2006
Ma, J. & Du, J., 'Discussion on the National Claim System for Oil Pollution Damage from Ships', in: Faure, M. & Hu, J. (eds.), *Prevention and Compensation of Marine Pollution Damage: Recent Developments in Europe, China and the U.S.*, Alphen aan den Rijn: Kluwer, 2006, p. 223-240.

Ma & Ortolano 2000
Ma, X. & Ortolano, L., *Environmental Regulation in China: Institutions, Enforcement, and Compliance*, Lanham: Rowman & Littlefield Publishers, 2000.

Maffei 1991
Maffei, M.C., 'The Compensation for Ecological Damage in the "Patmos" Case', in: Francioni, F. & Scavazzi, T. (eds.), *International Responsibility for Environmental Harm*, London: Graham & Trotman, 1991, p. 381-394.

Magnus 2001
Magnus, U., *Unification of Tort Law: Damages*, The Hague/London/Newyork: Kluwer Law International, 2001.

Magnus & Fedtke 2001
Magnus, U. & Fedtke, J., 'Country Reports: Germany', in: Rogers, W.V.H. (ed.), *Damages for Non-Pecuniary Loss in a Comparative Perspective*, Wien: Springer, 2001, p. 109-128.

Magnus, Rogers & Moréteau 2005
Magnus, U., Rogers, W. & Moréteau, O., 'Damages', in: The European Group on Tort Law (ed.), *Principles of European Tort Law: Text and Commentary*, Wien/New York: Springer, 2005, p. 149-181.

Markesinis & Unberath 2002
Markesinis, B.S. & Unberath, H., *The German Law of Torts: A Comparative Treaties*, Oxford/Portland: Hart Publishing, 2002.

Marticke 1992
Marticke, H., 'Liability for Ecological Damage — Report', *Environmental Policy and Law*, 1992, p. 28-31.

Mason 2003
Mason, M., 'Civil Liability for Oil Pollution Damage: Examining the Evolving Scope for Environmental Compensation in the International Regime', *Marine Policy*, 2003, p. 1-12.

McAllister, Van Rooij & Kagan 2010
McAllister, L.K., Van Rooij, B. & Kagan, R.A., 'Reorienting Regulation: Pollution Enforcement in Industrializing Countries', *Law & Policy*, 2010, p. 1-13.

McClure 1968
McClure, R., 'A Review of Nuclear Energy Insurance', *Proceedings of the Causalty Actuarial Society, Volume LV*, 1968, p. 255-294.

McMorrow 2004
McMorrow, A.P., 'CERCLA Liability Redefined: An Analysis of the Small Business Liability Relief and Brownfields Revitalization Act and its Impact on State Voluntary Cleanup Programs', *Georgia University Law Review*, 2004, 20, p. 1087-1128.

McMullin 2009
McMullin, J., 'Does Chinese Environmental Laws Work? A Study of Litigations as A Response to the Problem of Fishery Pollution in China', *UCLA Pacific Basin Law Journal*, 2009, 26, p. 142-183.

Mez & Piening 2002
Mez, L. & Piening, A., 'Phasing-out Nuclear Power Generation in Germany: Policies, Actors, Issues and Non-issues', *Energy & Environment*, 2002, p. 161-181.

Mol & Carter 2006
Mol, A. & Carter, N., 'China's Environmental Governance in Transition', *Environmental Politics*, 2006, 15, p. 149-170.

Monti 2001
Monti, A., 'Environmental Risk: A Comparative Law and Economics Approach to Liability and Insurance', *European Review of Private Law*, 2001, p. 51-79.

Moser & Yang 2011
Moser, A. & Yang, T., 'Environmental Tort Litigation in China', *Environmental Law Reporter*, 2011, 41, p. 10896-10901.

Ogus 2007
Ogus, A., 'Shifts in Governance for Compensation to Damages: A Framework for Analysis', in: Van Boom, W.H. & Faure, M. (eds.), *Shifts in Compensation between Private and Public Systems*, Wien: Springer, 2007, p. 31-41.

Palmer 2011
Palmer, V., 'The Great Spill in the Gulf... and a Sea of Pure Economic Loss: Reflections on the Boundaries of Civil Liability', *Penn State Law Review*, 2011, 116, p. 105-144.

Pánovics 2010
Pánovics, A., 'The Need for an EU Directive on Access to Justice in Environmental Matters', *Studia Iuridica Auctoritate Universitatis Pecs*, 2010, 147, p.135-161.

Papavizas & Kiern 2011
Papavizas, C.G. & Kiern, L.I., '2009-2010 Maritime Legislative Developments', *Journal of Maritime Law & Commerce*, 2011, 42, p. 291-453.

Patrick Thomas III 1992
Patrick Thomas III, M., 'Natural Resource Damages Under CERCLA: the Emerging Champion of Environmental Enforcement', *Pepperdine Law Review*, 1993, 20, p. 185-216.

Peabody 1994
Peabody, P.R., 'Taming CERCLA: A Proposal to Resolve the Trustee "Owner" Liability Quandary', *Administrative Law Journal of The American University*, 1994, 8, p. 405-460.

Peck 1999
Peck, J., 'Measuring Justice for Nature: Issues in Evaluating and Litigating Natural Resources Damages', *Journal of Land Use & Environmental Law*, 1999, 14, p. 275-306.

Peeters & Woerd 2006
Peeters, M. & Woerd, F.V.D., 'Financial Security Obligations to Prevent Orphan Damage: Some Preliminary Experiences from the Netherlands', *Environmental Liability*, 2006, p. 217-225.

Pelzer 2007
Pelzer, N., 'International Pooling of Operators' Funds: An Option to Increase the Amount of Financial Security to Cover Nuclear Liability', *Nuclear Law Bulletin*, 2007, p. 37-55.

Pelzer 2010
Pelzer, N., 'Deliberations on Compensation and Remediation of Nuclear Damage to the Environment', *Nuclear Law Bulletin*, 2010, p. 49-57.

Pendygraft, Plews, Clark, & Wright 1988
Pendygraft, G., Plews, G., Clark, J. & Wright, P., 'Who Pays for Environmental Damage: Recent Developments in CERCLA Liability and Insurance Litigation', *Indiana Law Review*, 1988, 21, p. 117-158.

Pitchford 1995
Pitchford, R., 'How Liable Should a Lender Be? The Case of Judgment-Proof Firms and Environmental Risk', *The American Economic Review*, 1995, 85, p. 1171-1186.

Poncelet 2012
Poncelet, C., 'Access to Justice in Environmental Matters – Does the European Union Comply with its Obligations?', *Journal of Environmental Law*, 2012, p. 287-309.

Priest 1987
Priest, G., 'The Current Insurance Crisis and Modern Tort Law', *Yale Law Journal*, 1987, 96, p. 1521-1590.

Radetzki & Radetzki 2000
Radetzki, M. & Radetzki, M., 'Private Arrangements to Cover Large-scale Liabilities Caused by Nuclear and Other Industrial Catastrophes', *The Geneva Papers on Risk and Insurance*, 2000, p. 180-195.

Randall 1983
Randall, A., 'The Problem of Market Failure', *Natural Resources Journal*, 1983, 23, p. 131-148.

Randle, Sands, McKillop & Levy 2009
Randle, R., Sands, L., McKillop, D. & Levy, J., 'Superfund and Natural Resource Damages Litigation 2009 Annual Report', *Environment, Energy and Resources Law: The Year in Review*, 2009, p. 132-147.

Ratliff 1999
Ratliff, J., 'Parens Patriae: An Overview', *Tulane Law Review*, 1999, 74, p. 1847-1858.

Rehbinder 2007
Rehbinder, E., 'Implementation of the Environmental Liability Directive in Germany', *Environmental Liability*, 2007, p. 199-203.

Reisch 2002
Reisch, S., 'The Brownfields Amendments: New Opportunities, New Challenger – Part I', *Colorado Lawyer*, 2002, 31, p. 99-103.

Reyners 2009
Reyners, P., 'Liability Problems Associated with the Current Patchwork Nuclear Liability Regime within the EU States', in: Pelzer, N. (ed.), *European Nuclear Liability Law in a Process of Change*, Baden-Baden: Nomos, 2009, p. 93-104.

Richardson 2001
Richardson, B.J., 'Mandating Environmental Liability Insurance', *Duke Environmental Law and Policy Forum*, 2001, 12, p. 293-330.

Rogers 2001
Rogers, W.V.H., 'Comparative Report of a Project Carried Out By the European Centre for Tort and Insurance Law', in: Rogers, W.V.H. (ed.), *Damages for Non-Pecuniary Loss in a Comparative Perspective*, Wien: Springer, 2001, p. 245-296.

Ronnerberg 1990
Ronnerberg, N.J., 'An Introduction to Protection & Indemnity Clubs and the Marine Insurance They Provide', *University of San Francisco Maritime Law Journal*, 1990, 3, p. 1-36.

Ru & Ortolano 2009
Ru, J. & Ortolano, L., 'Development of Citizen-Organized Environmental NGOs in China', *International Journal of Voluntary and Nonprofit Organization*, 2009, 20, p. 141-168.

Sandvik & Suikkari 1997
Sandvik, B. & Suikkari, S., 'Harm and Reparation in International Treaty Regimes: An Overview', in: Wetterstein, P. (ed.), *Harm to the Environment: The Right to Compensation and the Assessment of Damages*, Oxford: Clarendon Press, 1997, p. 57-72.

Schäfer 2000
Schäfer, H., 'Tort Law: General', in: Bouckaert, B. & De Geest, G. (eds.), *Encyclopedia of Law and Economics, Volume II, Civil Law and Economics*, Cheltenham: Edward Elgar, 2000, p. 569-596.

Schäfer & Müller-Langer 2009
Schäfer, H. & Müller-Langer, F., 'Strict Liability Versus Negligence', in: Faure, M. (ed.), *Tort Law and Economics*, Cheltenham/Northampton: Edward Elgar, 2009, p. 3-45.

Schall 2008
Schall, C., 'Public Interest Litigation Concerning Environmental Matters before Human Rights Courts: A Promising Future Concept?', *Journal of Environmental Law*, 2008, p. 417-453.

Schmid 2008
Schmid, K.C., 'The Depletion of the Superfund and Natural Resource Damages', *New York University Environmental Law Journal*, 2008, 16, p. 483-532.

Schwartz 2003
Schwartz, J., 'The Impact of State Capacity on Enforcement of Environmental Policies: The Case of China', *Journal of Environmental & Development*, 2003, p. 50-81.

Schwartz 2006
Schwartz, J., 'International Nuclear Third Party Liability Law: The Response to Chernobyl', in: OECD Nuclear Energy Agency, International Atomic Energy Agency, *International Nuclear Law in the Post-Chernobyl Period*, Paris: OECD, 2006, p. 37-72.

Schwartz 2009
Schwartz, J., 'Great Expectations: Where do We Stand with the International Nuclear Liability Conventions?', in: Pelzer, N. (ed.), *European Nuclear Liability Law in a Process of Change*, Baden-Baden: Nomos, 2009, p. 43-67.

Schwarze 2001
Schwarze, R., 'Environmental Liability and Accident Prevention: Preliminary Experiences in Germany', *European Environment*, 2001, p. 314-323.

Schwarze & Hoffmeister 2010
Schwarze, R. & Hoffmeister, O., 'The Winding Road to Industrial Safety: Evidence on the Effects of Environmental Liability on Accident Prevention in Germany', *The Geneva Papers on Risk and Insurance*, 2010, p. 416-434.

Seevers 1996
Seevers, J.S., 'NOAA's New Natural Resource Damage Assessment Scheme: It's Not about Collecting Money', *Washington and Lee Law Review*, 1996, 53, p. 1513-1570.

Shavell 1979a
Shavell, S., 'On Moral Hazard and Insurance', *The Quarterly Journal of Economics*, 1979, 93, p. 541-562.

Shavell 1979b
Shavell, S., 'Risk Sharing and Incentives in the Principal and Agent Relationship', *The Bell Journal of Economics*, 1979, 10, p. 55-73.

Shavell 1980
Shavell, S., 'Strict Liability versus Negligence', *The Journal of Legal Studies*, 1980, 9, p. 1-25.

Shavell 1982
Shavell, S., 'On Liability and Insurance', *The Bell Journal of Economics*, 1982, 13, p. 120-132.

Shavell 1984a
Shavell, S., 'Liability for Harm versus Regulation of Safety', *Journal of Legal Studies*, 1984, 13, p. 357-374.

Shavell 1984b
Shavell, S., 'A Model of the Optimal Use of Liability and Safety Regulation', *The Rand Journal of Economics*, 1984, 15, p. 271-280.

Shavell 1985
Shavell, S., 'Uncertainty over Causation and the Determination of Civil Liability', *Journal of Law and Economics*, 1985, 28, p. 587-610.

Shavell 1986
Shavell, S., 'The Judgment Proof Problem', *International Review of Law and Economics*, 1986, 6, p. 45-58.

Shavell 1987
Shavell, S., *Economic Analysis of Accident Law*, Cambridge: Harvard University Press, 1987.

Shavell 2000
Shavell, S., 'On the Social Function and the Regulation of Liability Insurance', *The Geneva Papers on Risk and Insurance*, 2000, p. 166-179.

Shogren, Herrige & Govindasamy 1993
Shogren, J., Herriges, J.A. & Govidasamy, R., 'Limits to Environmental Bonds', *Ecological Economics*, 1993, p. 109-133.

Sigman 2000
Sigman, H., 'The Pace of Progress at Superfund Sites: Policy Goals and Interest Group Influence', NBER Working Paper Series, Working Paper 7704.

Sigman 2010
Sigman, H., 'Environmental Liability and Redevelopment of Old Industrial Land', *Journal of Law and Economics*, 2010, 53, p. 289-306.

Skogh 1998
Skogh, G., 'Development Risks, Strict Liability and the Insurability of Industrial Hazards', *The Geneva Papers on Risk and Insurance*, 1998, p. 247-264.

Skogh 1999
Skogh, G., 'Risk-sharing Institutions for Unpredictable Losses', *Journal of Theoretical and Institutional Economics*, 1999, p. 505-515.

Skogh 2000
Skogh, G., 'Mandatory Insurance: Transaction Costs Analysis of Insurance', in: Bouckaert, B. & De Geest, G. (eds.), *Encyclopedia of Law and Economics, Volume II, Civil Law and Economics*, Cheltenham: Edward Elgar, 2000, p. 521-537.

Skogh 2008
Skogh, G., 'A European Nuclear Accident Pool', *The Geneva Papers on Risk and Insurance-Issues and Practice*, 2008, p. 274-287.

Slabbinck, Descamps & Bocken 2006
Slabbinck, R., Descamps, H. & Bocken, H., 'Implementation of the Environmental Damage Directive in Belgium (Flanders)', *Environmental Liability*, 2006, p. 3-12.

Slovic, Finucane, Peters & MacGregor 2004
Slovic, P., Finucane, M.L., Peters, E. & MacGregor, D.G., 'Risk as Analysis and Risk as Feelings: Some Thoughts about Affect, Reason, Risk and Rationality', *Risk Analysis*, 2004, p. 311-322.

Smets 1983
Smets, H., 'The Oil Spill Risk: Economic Assessment and Compensation Limit', *Journal of Maritime Law and Commerce*, 1983, 14, p. 23-44.

Smith 1993
Smith, J.T.II., 'Natural Resource Damages under CERCLA and OPA : Some Basic for Maritime operators', *Tulane Maritime Law Journal*, 1993, 18, p. 1-32.

Soanes & Stevenson 2003
Soanes, C. & Stevenson, A. (eds.), *Oxford Dictionary of English*, 2nd edn., Oxford: Oxford University Press, 2003.

Stafford 2003
Stafford, S.L., 'Assessing the Effectiveness of State Regulation and Enforcement of Hazardous Waste', *Journal of Regulatory Economics*, 2003, 23, p. 27-41.

Stein 1979
Stein, L.A., *Locus Standi*, Sydney: Law Book Company, 1979.

Stern 2010
Stern, R., 'On the Frontlines: Making Decisions in Chinese Civil Environmental Lawsuits', *Law & Policy*, 2010, 32, p. 79-103.

Stern 2011
Stern, R., 'From Dispute to Decision: Suing Polluters in China', *China Quarterly*, 2011, p. 294-312.

Summers 1983
Summers, J., 'The Case of the Disappearing Defendant: An Economic Analysis', *University of Pennsylvania Law Review*, 1983, 132, p. 145-185.

Sump 2011
Sump, D.H., 'The Oil Pollution Act of 1990: a Glance in the Rearview Mirror', *Tulane Law Review*, 2011, 85, p. 1101-1120.

Swanson 2001
Swanson, S.R., 'OPA 90 +10: The Oil Pollution Act of 1990 after Ten Years', *Journal of Maritime Law and Commerce*, 2001, 32, p. 135-174.

Taupitz 1993
Taupitz, J., 'The German Environmental Liability Law of 1990: Continuing Problems and the Impact of European Regulation', *Syracuse Journal of International Law and Commerce*, 1993, 19, p. 13-38.

Tetley 2006
Tetley, M., 'Revised Paris and Vienna Nuclear Liability Conventions – Challenges for Nuclear Insurers', *Nuclear Law Bulletin*, 2006, p. 27-30.

Tettenborn, Wilby & Bennett 2003
Tettenborn, A., Wilby, D. & Bennett. D., *The Law of Damages*, Belfast: LexisNexis, 2003.

Thomas 2007
Thomas, G.R., 'Some Novel Perspectives on Risk Classification', *The Geneva Papers on Risk and Insurance,* 2007, p. 105-132.

Tousley 1980
Tousley, D., 'Abolishing the "Extraordinary Nuclear Occurrence" Threshold of the Price-Anderson Act', *University of Michigan Journal of Law Reform*, 1980, 14, p. 609-630.

Trebilcock & Winter 1997
Trebilcock, M. & Winter, R.A., 'The Economics of Nuclear Accident Law', *International Review of Law and Economics*, 1997, 17, p. 215-243.

Turner & Smith 1983
Turner, T. & Smith, J., 'Environmental Damage Liability Insurance, a Primer', *The Business Lawyer*, 1983, 39, p. 333-354.

Tyran & Zweifeil 1993
Tyran, J. & Zweifel, P., 'Environmental Risk Internalization through Capital Markets (ERICAM): The Case of Nuclear Power', *International Review of Law and Economics*, 1993, p. 434-444.

Van Aaken 2005
Van Aaken, A. (2005), 'Making International Human Rights Protection more Effective: a Rational Choice Approach to the Effectiveness of Use Standi Provisions', *Max-Planck Institute for Research on Collective Goods*, Preprint No. 2005-16.

Van Boom 2008
Van Boom, W., 'Insurance Law and Economics: an Empirical Perspective', in: Faure, M. & Stephen, F. (eds.), *Essays in the Law and Economics of Regulation in Honour of Anthony Ogus*, Antwerp/Oxford/Portland: Intersentia, 2008, p. 253-276.

Vanden Borre 1999
Vanden Borre, T., 'Channelling of Liability: A Few Juridical and Economic Views on an Inadequate Legal Construction', in: Horbach, N. (ed.), *Contemporary Developments in Nuclear Energy Law: Harmonising Legislation in CEEC.NIS*, The Hague: Kluwer, 1999, p. 9-30.

Vanden Borre 2007
Vanden Borre, T., 'Shifts in Governance in Compensation for Nuclear Damage: 20 Years after Chernobyl', in: Faure, M. & Verheij, A. (eds.), *Shifts in Compensation for Environmental Damage*, Wien: Springer, 2007, p. 261-311.

Vanden Borre 2010
Vanden Borre, T., 'Nuclear Liability: an Anachronism in EU Energy Policy?, in Roggenkamp, M. & Hammer, U. (eds.), *European Energy Law Report VII*, Antwerp: Intersentia, 2010, p. 177-218.

Van den Broek 2009
Van den Broek, G.M., 'Environmental Liability and Nature Protection Areas: Will the EU Environmental Liability Directive Actually Lead to the Restoration of Damaged Natural Resources?', *Utrecht Law Review*, 2009, p. 117-131.

Van Rooij 2003
Van Rooij, B., 'Organization and Procedure in Environmental Law Enforcement: Sichuan in Comparative Perspective', *China Information*, 2003, 17, p. 36-64.

Van Rooij 2006
Van Rooij, B., 'Implementation of Chinese Environmental Law: Regular Enforcement and Political Campaigns', *Development and Change*, 2006, p. 57–74.

Van Rooij 2010
Van Rooij, B., 'The Peoples vs. Pollution: Understanding Citizen Action against Pollution in China', *Journal of Contemporary China*, 2010, p. 55-77.

Verheij 2007a
Verheij, A., 'Shifts in Governance: Oil Pollution', in: Faure, M. & Verheij, A. (eds.), *Shifts in Compensation for Environmental Damage*, Wien: Springer, 2007, p. 133-195.

Verheij 2007b
Verheij, A., 'Shifts in Governance: Soil Pollution', in: Faure, M. & Verheij, A. (eds.), *Shifts in Compensation for Environmental Damage*, Wien: Springer, 2007, p. 9-71.

Viscusi 1989
Viscusi, W.K. (1989), 'Toward a Diminished Role for Tort Liability: Social Insurance, Government Regulation, and Contemporary Risks to Health and Safety', *Yale Journal on Regulation*, 1989, 6, p. 65-108.

Viscusi & Hamilton 1999
Viscusi, K. & Hamilton, J., 'Are Risk Regulators Rational? Evidence from Hazardous Waste Cleanup Decisions', *The American Economic Review*, 1999, 89, p. 1010-1027.

Visscher 2009
Visscher, L.T., 'Tort Damages', in: Faure, M. (ed.), *Tort law and Economics*, Cheltenham/ Northampton: Edward Elgar, 2009, p.153-200.

Von Bar 2000
Von Bar, C., *Damage and Damages, Liability for and without Personal Misconduct, Causality, and Defences, The Common European Law of Torts, Volume 2*, Oxford: Clarendon Press, 2000.

Vorwerk 2002
Vorwerk, A., 'The 2002 Amendment to the German Atomic Energy Act Concerning the Phase-out of Nuclear Power', *Nuclear Law Bulletin*, 2002, p. 7-15.

Wagner 2007
Wagner, G., '(Un)insurability and the Choice between Market Insurance and Public Compensation Systems', in: Van Boom, W.H. & Faure, M. (eds.), *Shifts in Compensation between Private and Public Systems*, Wien: Springer, 2007, p. 87-112.

Wagner 2009
Wagner, G., 'Tort Law and Liability Insurance', in: Faure, M. (ed.), *Tort Law and Economics*, Cheltenham/Northampton: Edward Elgar, 2009, p. 377-405.

Wang 2006
Wang, A., 'The Role of Law in Environmental Protection in China: Recent Developments', *Vermont Journal of Environmental Law*, 2006, 8, p. 202-223.

Wang 2007
Wang, H., 'Shifts in Governance in the International Regime of Marine Oil Pollution Compensation: A Legal History Perspective', in: Faure, M. & Verheij, A. (eds.), *Shifts in Compensation for Environmental Damage*, Wien: Springer, 2007, p. 197-241.

Wang 2008
Wang, H., 'Transboundary Vessel-source Marine Pöllution--International Legal Framework and Its Application to China', in: Faure, M. & Song, Y. (eds.), *China and International Environmental Liability – Legal Remedies for Transboundary Pollution*, Cheltenham/Northampton: Edward Elgar, 2008, p. 64-106.

Wang 2011
Wang, H., *Civil Liability for Marine Oil Pollution Damage: A Comparative and Economic Study of the International, U.S. and Chinese Compensation Regime*, Alphen aan den Rijn: Kluwer, 2011.

Wang & Faure 2010
Wang, H. & Faure, M., 'Civil Liability and Compensation for Marine Pollution – Lessons to be Learned for Offshore Oil Spills', *Oil, Gas & Energy Law Intelligence*, 2010, p. 1-27.

Wang & Gao 2010
Wang, A. & Gao, J. 'Environmental Courts and the Development of Environmental Public Interest Litigation in China', *Journal of Court Innovation*, 2010, 3, p. 37-50.

Wang, Huang & Yan 2008
Wang, J., Huang, C. & Yan, H., 'Reflections from the transboundary pollution of Songhua River', in: Faure, M. & Song, Y. (eds.), *China and International Environmental Liability: Legal Remedies for Transboundary Pollution*, Cheltenham: Edward Elgar, 2008, p. 273-290.

Wang, Yu, Li & Li 2008
Wang, C., Yu, W., Li, D. & Li, J., 'Pondering over the Incident of Songhua River Pollution from the Perspective of Environmental Law', in: Faure, M. & Song, Y. (eds.), *China and International Environmental Liability: Legal Remedies for Transboundary Pollution*, Cheltenham: Edward Elgar, 2008, p. 291-314.

Warendorf, Thomas & Curry-Sumner 2009
Warendorf, H., Thomas, R. & Curry-Sumner, I., *The Civil Code of the Netherlands*, Alphen aan den Rijn: Kluwer Law International, 2009 (translation).

Weissenbacher 2005
Weissenbacher, M., 'Implementation of the Environmental Liability Directive in Austria', *Environmental Liability*, 2005, p. 199-203.

Wendel 2004
Wendel, L., 'Compensation in the Swedish Health Care Sector', in: Dute, J., Faure, M. & Koziol, H. (eds.), *No-fault Compensation in the Health Care Sector*, Wien: Springer, 2004, p. 367-391.

Wendland 2002
Wendland, W., *Effluent Release Performance Indicators – An Insurer's Perspective, Presentation at 12th Annual RETS/REMP Workshop*, 2002.

Wetterstein 1997
Wetterstein, P., 'A Proprietary or Possessory Interest: A *Conditio Sine Qua Non* for Claiming Damages for Environmental Impairment?', in: Wetterstein, P. (ed.), *Harm to the Environment: The Right to Compensation and the Assessment of Damages*, Oxford: Clarendon Press, 1997, p. 29-54.

White 1992
White, M.J., '"Legal Complexity and Lawyers" Benefit from Litigation', *International Review of Law and Economics*, 1992, p. 381-395.

White & Wittman 1983
White, M.J. & Wittman, D., 'A Comparison of Taxes, Regulation and Liability Rules under Imperfect Information', *The Journal of Legal Studies*, 1983, 12, p. 413-426.

Wilks 2002
Wilks, G., 'Evaluating of Risk From A Property Insurance Perspective', in: IAEA (ed.), *Topical Issues in Nuclear Safety: Proceedings of an International Conference,* Vienna IAEA, 2002, p. 327-331.

Wilkinson 1993
Wilkinson, D., 'Moving the Boundaries of Compensable Environmental Damage Caused by Marine Oil Spills: the Effect of Two New International Protocols', *Journal of Environmental Law*, 1993, p. 71-90.

Winter, Jans, Macrory & Krämer 2008
Winter, H., Jans, J.H., Macrory, R. & Krämer, L., 'Weighing up the Environmental Liability Directive', *Journal of Environmental Law*, 2008, p. 163-191.

Wissink & Van Boom 2001
Wissink, M.H. & Van Boom, W.H., 'Country Reports: Netherlands', in: Rogers, W.V.H. (ed.), *Damages for Non-Pecuniary Loss in a Comparative Perspective*, Wien: Springer, 2001, p. 155-172.

World Bank 1997
World Bank, *Clear Water, Blue Skies: China's Environmental in the New Century*, Washington D.C.: The World Bank, 1997.

Wood 1975
Wood, L.D., 'An Integrated International and Domestic Approach to Civil Liability for Vessel-Source Oil Pollution', *Journal of Maritime Law and Commerce*, 1975, 7, p. 1-68.

Wu 1996
Wu, C., *Pollution from the Carriage of Oil by Sea: Liability and Compensation*, London: Kluwer Law International, 1996.

Yang 2005
Yang, G., 'Environmental NGOs and Institutional Dynamics in China', *The China Quarterly*, 2005, p. 46-66.

Yin, Kunreuther & White 2011
Yin, H., Kunreuther, H. & White, M.W., 'Risk-Based Pricing and Risk-Reducing Effort: Does the Private Insurance Market Reduce Environmental Accidents?', *Journal of Law and Economics*, 2011, 54, p. 325-363.

Zhan & Tang 2011
Zhan, X. & Tang, S., 'Political Opportunities, Resource Constraints and Policy Advocacy of Environmental NGOs in China', *Public Administration*, 2011, p. 1-19.

Zhang 2010
Zhang, L., 'Compensation for the Domestic Oil Pollution in China's Coast: Which Law Shall Apply?', in: Faure, M., Han, L. & Shan, H. (eds.), *Maritime Pollution Liability and Policy – China, Europe and the U.S.*, The Hague: Kluwer, 2010, p. 359-369.

Zhang & Zhang 2012
Zhang, M. & Zhang, B., 'Specialized Environmental Courts in China: Status Quo, Challenges and Response', *Journal of Energy & Natural Resource Law*, 2012, 30, p. 351-378.

Zhao 2004
Zhao, Y., 'Environmental Dispute Resolution in China', *Journal of Environmental Law*, 2004, 16, p. 157-193.

Zhao & Jiang 2012
Zhao, P. & Jiang, P., 'The Analysis on the Acceptance of Cases in Environmental Courts in China', *Energy & Environment*, 2012, p. 16-17 (in Chinese).

Zhu 2007a
Zhu, L., *Compulsory Insurance and Compensation for Bunker oil Pollution Damage*, Berlin/Heidelberg: Springer, 2007.

Zhu 2007b
Zhu, X., *A Study of Socialized Indemnification for Ecological Damage: A Jurisprudential Perspective*, Beijing: China University of Political Science and Law Publisher, 2007 (in Chinese).

Zweifel & Tyran 1994
Zweifel, P. & Tyran, J., 'Environmental Impairment Liability as an Instrument of Environmental Policy', *Ecological Economics*, 1994, p. 43-56.

Zycher 1992
Zycher, B., Accounting for Costs and Cost Biases, *Cato's Letters*, 1992, 15(2), p. 2-6.

CURRICULUM VITAE

Jing Liu (born in Jingmen, China in 1985) studied Law in Wuhan, China from September 2003 till June 2007. She obtained a master's degree in environmental law in 2009 in Wuhan University. After that, she joined the Maastricht Institute for Transnational Legal Research (METRO) in Maastricht University to conduct her doctoral research.

IUS COMMUNE EUROPAEUM

A peer-reviewed book series in which the common foundations of the legal systems of the Member States of the European Union are the central focus.

The *Ius Commune Europaeum* series includes horizontal comparative legal studies as well as studies on the effect of treaties within the national legal systems. All the classic fields of law are covered. The books are published in various European languages under the auspices of METRO, the Institute for Transnational Legal Research at the Maastricht University.

Recently published:

Volume 102: *The Changing Role of the European Council in the Institutional Framework of the European Union*, F. EGGERMONT
Volume 103: *Linked Contracts*, I. SAMOY and M.B.M. LOOS (eds.)
Volume 104: *Constitutions Compared* (3rd ed.), A.W. HERINGA and PH. KIIVER
Volume 105: *Alternative Ways to Ius Commune*, A.L.M. KEIRSE and M.B.M. LOOS (eds.)
Volume 106: *Property Law Perspectives*, B. AKKERMANS and E. RAMAEKERS (eds.)
Volume 107: *Constitutional Conversations*, C. VAN DE HEYNING and M. DE VISSER (eds.)
Volume 108: *Enforcing Health and Safety Regulations. A Comparative Economic Approach*, L. TILINDYTE
Volume 109: *Administrative Law of the European Union, its Member States and the United States. A Comparative Analysis*, R. SEERDEN (ed.)
Volume 110: *Market Integration through 'Network Governance': The Role of European Agencies and Networks of Regulators*, M. ZINZANI
Volume 111: *Truth and Efficiency in Civil Litigation. Fundamental Aspects of Fact-finding and Evidence-taking in a Comparative Context*, C.H. VAN RHEE and A. UZELAC (eds.)
Volume 112: *Law and Security in Europe: Reconsidering the Security Constitution*, M. FICHERA and J. KREMER (eds.)
Volume 113: *Inside Police Custody: An Empirical Account of Suspects' Rights in Four Jurisdictions*, T. SPRONKEN, E. CAPE and J. HODGSON (eds.)
Volume 114: *Motive Matters! An Exploration of the Notion 'Deliberate Breach of Contract' and its Consequences for the Application of Remedies*, M. VAN KOGELENBERG
Volume 115: *Legal Education*, A.W. HERINGA
Volume 116: *Standing up for your Rights in Europe. A Comparative Study on Legal Standing (Locus Standi) before the EU and Member States' Courts*, M. ELIANTONIO, CH. BACKES, C.H. VAN RHEE, T. SPRONKEN and A. BERLEE
Volume 117: *European Union Property Law. From Fragments to a System*, E. RAMAEKERS